Lecture Notes in Mechanical Engineering

Lecture Notes in Mechanical Engineering (LNME) publishes the latest developments in Mechanical Engineering - quickly, informally and with high quality. Original research reported in proceedings and post-proceedings represents the core of LNME. Volumes published in LNME embrace all aspects, subfields and new challenges of mechanical engineering. Topics in the series include:

- Engineering Design
- Machinery and Machine Elements
- Mechanical Structures and Stress Analysis
- Automotive Engineering
- Engine Technology
- Aerospace Technology and Astronautics
- Nanotechnology and Microengineering
- Control, Robotics, Mechatronics
- MEMS
- Theoretical and Applied Mechanics
- Dynamical Systems, Control
- Fluid Mechanics
- Engineering Thermodynamics, Heat and Mass Transfer
- Manufacturing
- Precision Engineering, Instrumentation, Measurement
- Materials Engineering
- Tribology and Surface Technology

To submit a proposal or request further information, please contact the Springer Editor of your location:

China: Dr. Mengchu Huang at mengchu.huang@springer.com
India: Priya Vyas at priya.vyas@springer.com
Rest of Asia, Australia, New Zealand: Swati Meherishi at swati.meherishi@springer.com
All other countries: Dr. Leontina Di Cecco at Leontina.dicecco@springer.com

To submit a proposal for a monograph, please check our Springer Tracts in Mechanical Engineering at http://www.springer.com/series/11693 or contact Leontina.dicecco@springer.com

Indexed by SCOPUS. The books of the series are submitted for indexing to Web of Science.

More information about this series at http://www.springer.com/series/11236

Mohamed Kharrat · Mounir Baccar ·
Fakhreddine Dammak
Editors

Advances in Mechanical Engineering, Materials and Mechanics

Selected contributions from the 7th International Conference on Advances in Mechanical Engineering and Mechanics, ICAMEM 2019, December 16–18, 2019, Hammamet, Tunisia

Springer

Editors
Mohamed Kharrat
Laboratory of Electromechanical Systems
National School of Engineers of Sfax
Sfax, Tunisia

Mounir Baccar
Computational Fluid Dynamics
and Transfer Phenomena
National School of Engineers of Sfax
Sfax, Tunisia

Fakhreddine Dammak
Laboratory of Electromechanical Systems
National School of Engineers of Sfax
Sfax, Tunisia

ISSN 2195-4356 ISSN 2195-4364 (electronic)
Lecture Notes in Mechanical Engineering
ISBN 978-3-030-52070-0 ISBN 978-3-030-52071-7 (eBook)
https://doi.org/10.1007/978-3-030-52071-7

This Springer imprint is published by the registered company Springer Nature Switzerland AG
The registered company address is: Gewerbestrasse 11, 6330 Cham, Switzerland

Preface

The 7th edition of the "International Conference on Advances in Mechanical Engineering and Mechanics" (ICAMEM 2019) was held in Hammamet, Tunisia, from 16 to 18 December 2019. The conference was organized by the Laboratory of Electromechanical Systems (LASEM) at the National School of Engineers of Sfax (ENIS) and the Tunisian Scientific Society (TSS) in collaboration with a number of university institutions from Tunisia and outside Tunisia.

Since its first edition in 2002, the ICAMEM Conference has been held for several editions to bring together specialists, academicians, researchers and industrialists from international community working in research areas related to mechanical engineering and mechanics. ICAMEM 2019 has allowed the scientific community from several regions of the world to exchange research ideas and experiences. The conference was organized in seven specialized symposiums focused on the following themes:

- Applied Mechanics
- Industrial Applications and Technology Transfer
- Design and Manufacturing
- Thermal Sciences and Renewable Energy
- Systems and Dynamics
- Materials
- Fluid Mechanics

Eight keynote speakers, who are leaders in the field of mechanical engineering and mechanics at the international scale, have participated at ICAMEM 2019. They are namely:

- Pr. Abdessattar Abdelkefi, Mechanical & Aerospace Engineering, New Mexico State University, USA.
- Pr. Mohamed S. Ghidaoui, Department of Civil and Environmental Engineering, Hong Kong University of Science and Technology, Hong Kong.
- Pr. Antoine Chateauminois, Soft Matter Sciences & Engineering Laboratory (SIMM), ESPCI Paris, France.

- Pr. Bruno Brunone, Department of Civil and Environmental Engineering, Water Engineering Laboratory, University of Perugia, Italy.
- Pr. Walter Arnold, Department of Materials Science and Engineering, Saarland University, Germany.
- Pr. Mohamed Slim El Bahi, Laboratory of Microstructure Studies and Mechanics of Materials, University of Lorraine, France.
- Pr. Nizar Ben Salah, Laboratory of Mechanics, Materials and Processes, National Superior School of Engineers of Tunis, University of Tunis, Tunisia
- Pr. Amine Ammar, LAMPA, ENSAM Angers, France

More than 200 papers were accepted for ICAMEM 2019, and participants discussed during three days the latest advances in mechanical engineering and mechanics. Only 62 papers have been selected to be presented as chapters in this book. Each chapter was thoroughly reviewed by at least two reviewers from the considered topics. The organizers would like to thank all members of the Scientific Committee for their sacrifices. Our thanks go also to all the ICAMEM 2019 participants, authors and especially Springer for their support.

December 2019

Mohamed Kharrat
ICAMEM 2019 General Chair

Contents

Design and Manufacturing

Applied Mechanics

Dynamic and Post-buckling Analysis of Structures Like-Shell Using a Quadrilateral Shell Element with In-plane Drilling Rotational Degree of Freedom and a Conservative Implicit Time Integration Scheme

Djamel Boutagouga[1(✉)] and Said Mamouri[2]

[1] Civil Engineering Department, Laboratoire des Mines,
University Larbi Tébessi - Tébessa, BP 02, 12000 Tébessa, Algeria
d.boutagouga@yahoo.fr

[2] University de Tahri Mohamed - Bechar, Béchar, Algeria
said.mamouri@gmail.com

Abstract. In this work, post-buckling problems of structures like shell are investigated using an implicit conservative time integration dynamic scheme. We have proposed the use of a four nodded quadrilateral flat shell element, with the drilling rotation degree of freedom (D. Boutagouga., A new enhanced assumed strain quadrilateral membrane element with drilling degree of freedom and modified shape functions, Int. J. Numer. Meth. Eng. 2017; 110:573–600) and based upon the updated Lagrangian formulation. An implicit conservative scheme (K.J, Bathe., G, Noh., Insight into an implicit time integration scheme for structural dynamics. Comput. Struct. 2012; 98–99: 1–6) is chosen to obtain the solution to the nonlinear dynamic behaviour and geometric nonlinear post-buckling response.

Several numerical simulations are considered in the aim to investigate the ability of the considered time integration scheme to deal with unstable branches after the limit point in non-linear post-buckling response of shell structures with no prerequisite structural damping, though it is required when using conventional time integration schemes. The obtained results illustrate a very satisfying performance of the implicit conservative-dissipative direct time integration scheme combined with the quadrilateral shell element with drilling rotation.

Keywords: Nonlinear dynamic analysis · Post-buckling · Conservative dissipative Scheme · Quadrilateral shell element · Drilling rotation

1 Introduction

In geometrically nonlinear dynamic analysis, conservative and decaying time integration schemes with second-order accuracy are highly desired. A robust time integration algorithm must reveal the minimum numerical damping for low-frequency modes and maximum numerical damping for high-frequency parasitic modes.

M. Kharrat et al. (Eds.): ICAMEM 2019, LNME, pp. 3–10, 2020.
https://doi.org/10.1007/978-3-030-52071-7_1

Energy preservation appears to be the most important since it provides unconditional stability in the nonlinear regime if the total mechanical energy can be conserved. However, when dealing with nonlinear systems, especially complex structures like-shell, numerical dissipation becomes crucial. Indeed, high-frequency modes are purely numerical due to the implications of the discrete equations and they do not reflect the high frequencies of the physical problem.

In order to obtain such desired geometric nonlinear dynamic response, geometric nonlinearity of structures like-shell is considered in this work. The Updated Lagrangian Co-rotational Description of motion is adopted, in which, large translations and rotations are decomposed into large rigid and relatively small translations and rotations. For this purpose, starting from a quadrilateral membrane element with rotational *d.o.f* less sensitive to mesh distortion via the use of modified shape functions by Boutagouga (2017), a flat shell finite element is constructed after combining the membrane element with the quadrilateral discrete Kirchhoff plate element developed by Batoz and Ben-Tahar (1982). Then, based on the Co-rotational formulation for geometrically nonlinear static and dynamic analysis established by the authors in former works (Boutagouga et al. 2010; Boutagouga and Djeghaba 2014; Boutagouga and Djeghaba 2016; Mamouri et al. 2015), the nonlinear dynamic analysis is established based on a Conservative dissipative time integration algorithm presented by Bathe and Noh (2012).

2 The Nonlinear Dynamic Updated Lagrangian Formulation for the Flat Shell Element

The Co-rotational approach founded by (Wempner 1969; Belytschko and Hsieh 1973; Belytschko and Hsieh 1974), enables to simplify the ULD using a Co-rotational coordinate frame that translates and rotates with the finite element. The status of equilibrium of a deformable body at time $(t + \Delta t)$ is characterized by a minimum of energy expressed in standard form as:

$$\delta\pi = \delta W_{\text{int}} - \delta W_{ext} = 0 \tag{1}$$

In the Updated Lagrangian Description of motion, all kinematic and mechanical quantities are measured in reference to the last known configuration $C^{(t)}$. Considering the equilibrium of a deformable body at time $(t + \Delta t)$, the virtual displacements principle is written in $C^{(t)}$ as:

$$\int_{{}^{t}v} \left({}^{t}T_{ij} \cdot \delta_t \varepsilon_{ij}^{NL} + D_{ijkl} \cdot {}_t\varepsilon_{kl} \cdot \delta_t \varepsilon_{ij} \right) d\,{}^{t}v = {}^{t+\Delta t}W_{ext} - \int_{{}^{t}v} {}^{t}T_{ij} \cdot \delta_t \varepsilon_{ij} \cdot d\,{}^{t}v \tag{2}$$

In which:

$\Delta\varepsilon_{ij}$, $\Delta\varepsilon_{ij}$*: refer to the linear and nonlinear parts of the Green's strain increment, T_{ij} stands for Cauchy stress, and D_{ijkl} is the constitutive matrix.

The left term in (2) represents the tangent stiffness matrix $[K_T]$ that can be written as:

$$[K_T] = [K_0] + [K\sigma]$$

with $[K_0]$: small displacement matrix; $[K\sigma]$: geometric stiffness matrix.

In this work, in which de drilling rotation is involved in the shell element formulation, the geometric stiffness $[K\sigma]$ it is written as:

$$[K_\sigma] = \begin{bmatrix} K_\sigma^m & 0 \\ 0 & K_\sigma^b \end{bmatrix} \quad (3)$$

See Boutagouga and Djeghaba (2016) for more details.

The internal forces $\{F_{int}\}$ are such as:

$$^t f_{\text{int}} = \int_{^t v} {}^t T_{ij} \cdot {}^t B_{ij} \cdot d\,{}^t v \quad (4)$$

The incremental total external virtual work due to surface forces ${}_t^{t+\Delta t} t_k$ and body forces ${}_t^{t+\Delta t} f_k$ expression is:

$$^{t+\Delta t} W_{ext} = \int_{^t A} {}_t^{t+\Delta t} t_k \cdot \delta\, u_k \cdot d\,{}^t A + \int_{^t v} {}_t^{t+\Delta t} f_k \cdot \delta\, u_k \cdot d\,{}^t v \quad (5)$$

The term ${}_t^{t+\Delta t} f_k$ arises from dynamic analysis, in which the body force components are related to the mass inertia ρ and acceleration $\delta\, \ddot{u}_k$.

3 The Conservative Time Integration Scheme

Obviously, both implicit and explicit methods can be used to solve buckling and post-buckling problems. Explicit schemes are the most used because of their main advantage of being less expensive (Saigal et al. 1987; Hilburger et al. 1997; Kobayashi et al. 2015; White et al. 2015). However, these schemes present the disadvantage of being conditionally stable. In this investigation, we have considered the post-buckling response of thin shells by using an implicit conservative decaying dynamic scheme proposed by Bathe and Noh (2012). The fundamental concept of this composite time integration scheme can be summarized as follows:

1. The time step (Δt) is split into two equal sub-steps of span ($\Delta t/2$) each.
2. The trapezoidal rule is used over the first sub-step, which yields:

$$\begin{cases} {}^{t+\frac{\Delta t}{2}}\ddot{u} = \frac{16}{\Delta t^2}\left({}^{t+\frac{\Delta t}{2}}u - {}^t u\right) - \frac{8}{\Delta t}\,{}^t\dot{u} - {}^t\ddot{u} \\ {}^{t+\frac{\Delta t}{2}}\dot{u} = \frac{4}{\Delta t}\left({}^{t+\frac{\Delta t}{2}}u - {}^t u\right) - {}^t\dot{u} \end{cases} \quad (6)$$

$$M^{t+\frac{\Delta t}{2}}\ddot{u} + C^{t+\frac{\Delta t}{2}}\dot{u} + {}^{t+\frac{\Delta t}{2}}f = {}^{t+\frac{\Delta t}{2}}R \quad (7)$$

$$(b_1 M + b_3 C + K_T)^{t+\frac{\Delta t}{2}}u = {}^{t+\frac{\Delta t}{2}}R + M(b_1 u_t - b_2 \dot{u}_t + \ddot{u}_t) + C(b_3 u_t + \dot{u}_t) \quad (8)$$

with: $b_1 = \frac{16}{\Delta t^2}$; $b_2 = \frac{-8}{\Delta t}$; $b_3 = \frac{4}{\Delta t}$ and ${}^{t+\frac{\Delta t}{2}}R = {}^{t+\frac{\Delta t}{2}}F_{ext} - {}^{t+\frac{\Delta t}{2}}f_{\text{int}}$ is the residue. K, M and C are the structural stiffness, mass and damping matrices respectively.

3. With the resulting equations of the first sub-step in hand, the 3-point Euler backward method is used over the second sub-step, which gives:

$$\begin{cases} {}^{t+\Delta t}\dot{u} = \frac{1}{\Delta t}{}^{t}u - \frac{4}{\Delta t}{}^{t+\frac{\Delta t}{2}}u + \frac{3}{\Delta t}{}^{t+\Delta t}u \\ {}^{t+\Delta t}\ddot{u} = \frac{1}{\Delta t}{}^{t}\dot{u} - \frac{4}{\Delta t}{}^{t+\frac{\Delta t}{2}}\dot{u} + \frac{3}{\Delta t}{}^{t+\Delta t}\dot{u} \end{cases} \quad (9)$$

$$M^{t+\Delta t}\ddot{u} + C^{t+\Delta t}\dot{u} + {}^{t+\Delta t}f = {}^{t+\Delta t}R \quad (10)$$

$$(-3b_2 M - 3b_4 C + K_T)^{t+\Delta t}u = {}^{t+\Delta t}R + M(b_1{}^{t+\frac{\Delta t}{2}}u + b_2{}^{t}u + b_3{}^{t+\frac{\Delta t}{2}}\dot{u} + b_4{}^{t}\dot{u}) + C(b_3{}^{t+\frac{\Delta t}{2}}u + b_4{}^{t}u) \quad (11)$$

with: $b_1 = \frac{12}{\Delta t^2}$; $b_2 = \frac{-3}{\Delta t^2}$; $b_3 = \frac{4}{\Delta t}$; $b_4 = \frac{-1}{\Delta t}$ and ${}^{t+\Delta t}R = {}^{t+\Delta t}F_{ext} - {}^{t+\Delta t}f_{\text{int}}$.

4 Results

4.1 Non-linear Dynamic Analysis of a Clamped-Free Half-Cylinder Shell

We consider in this example the half-cylinder shell with clamped-free straight edges represented in Fig. 1. The Half-cylinder is defined by the following geometry: $R = 1.2$ m, $b = 2$ m, $h = 0.006$ m. The adopted mechanical properties are: Young's modulus $E = 73$ GPa, Poisson's ratio $\nu = 0.3$ and material density $\rho = 2700$ kg/m^3.

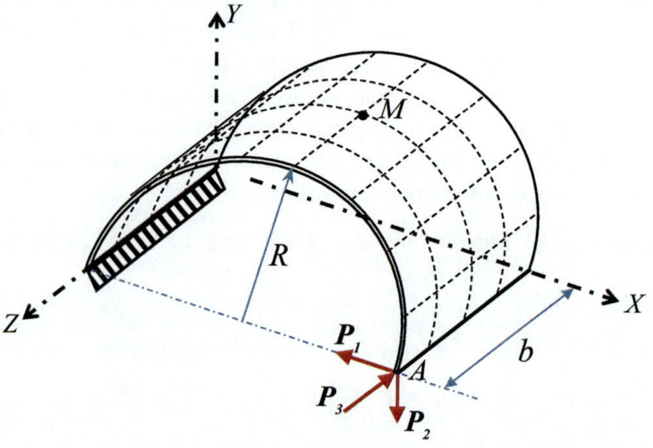

Fig. 1. Clamped-free half-cylinder: geometry and mesh

The Half-cylinder is subjected to the transient concentrated loading (P_1, P_2 and P_3), acting at point *A* presented in Fig. 1, with: $\{P_1 \quad P_2 \quad P_3\} = \{-1 \quad -1 \quad -1\} \cdot P_0(t)$.

$$P_0(t) = \begin{cases} 0.05(1 - \cos(\pi t)); & t \leq 2.0 \\ 0; & t > 2.0 \end{cases}$$

where *t* is in seconds and P_0 is in Newton. A mesh of 8 × 4 quadrilateral elements is used for modelling the half-cylinder. The example was performed with the Newmark's trapezoidal and Bathe conservative dissipative Schemes for a time period of *6* s, using a time step Δt = *0.05* s. The corresponding results are shown in Figs. 2 and 3.

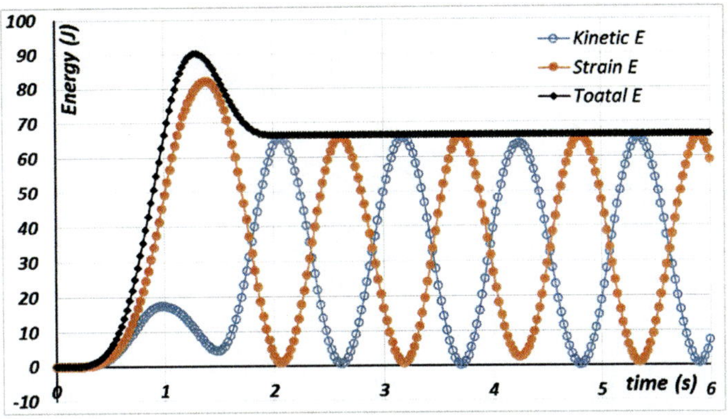

Fig. 2. Kinetic, strain and total energies variation with time

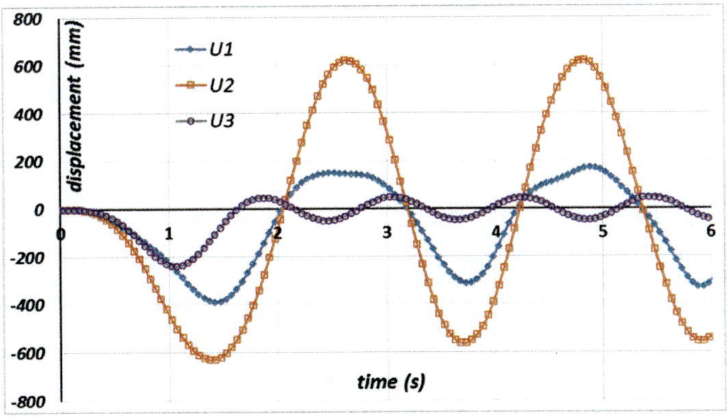

Fig. 3. Time history of the displacement components at point A

4.2 Post Buckling of a Spherical Shallow Shell

In this example, the efficiency of the considered algorithm combined with the co-rotational formulation in dissipating high frequencies is investigated, see Mamouri et al. (2015) for more details. It is used here to plot the post-buckling curve of the geometrically nonlinear static behaviour of a spherical shell with four articulated edges, subjected to the concentrated load *P*, as shown in Fig. 4. Due to the double symmetry of the problem, only one-quarter of the spherical shell is considered. The shell geometric and mechanical properties are: $R = 2540$ mm; $2a = 784.90$ mm; $h = 99.45$ mm; $E = 68950$ N/mm^2 and $\nu = 0.3$.

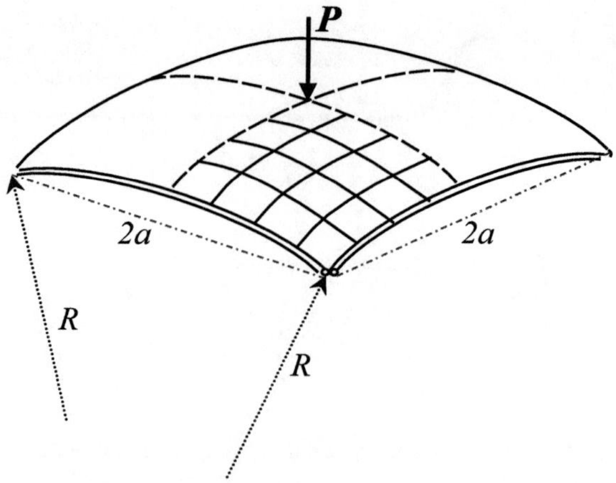

Fig. 4. Cylindrical shell: geometry and mesh

The static nonlinear response is plotted in the load-deflection curve represented in Fig. 5 using dynamic analysis algorithms.

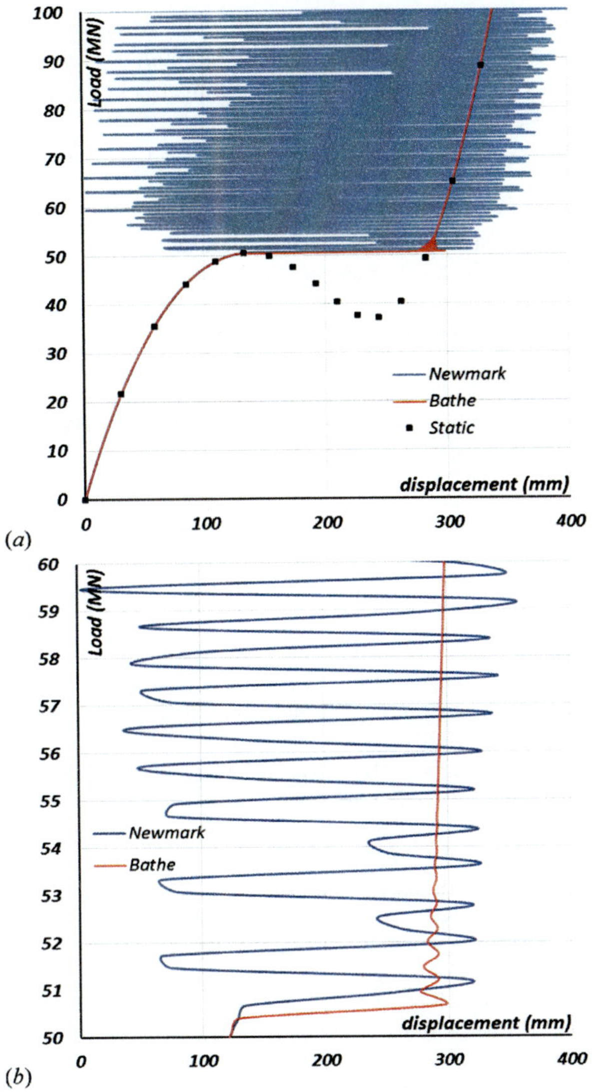

Fig. 5. Post buckling nonlinear response

5 Conclusion

This work aims to evaluate the capability of nonlinear dynamic analysis based upon the Co-rotational Lagrangian Formulation to obtain the solution to the geometric nonlinear post-buckling response of structures like-shell with conservative dissipative direct time integration algorithms. The studied Algorithm is tested for a quadrilateral flat shell finite element with drilling rotation. The scheme is found to be conservative, and efficient for damping out high-frequency modes and thus assuring the stability of the nonlinear response with no need for structural damping. The performances of the studied algorithm are illustrated through several numerical examples.

References

Boutagouga, D.: A new enhanced assumed strain quadrilateral membrane element with drilling degree of freedom and modified shape functions. Int. J. Num. Meth. Eng. **110**, 573–600 (2017)

Batoz, J.L., Ben-Tahar, M.: Evaluation of a new quadrilateral thin plate bending el-ement. Int. J. Num. Meth. Eng. **18**, 1655–1677 (1982)

Boutagouga, D., Gouasmia, A., Djeghaba, K.: Geometrically nonlinear analysis of thin shell by a quadrilateral finite element with in-plane rotational degrees of freedom. Eur. J. Comput. Mech. **19**(8), 707–724 (2010)

Boutagouga, D., Djeghaba, K.: Geometrically nonlinear dynamic analysis of thin shells by a four-node quadrilateral element with in-plane rotational degree of freedom. Eur. J. Comput. Mech. **23**(3–4), 161–177 (2014)

Boutagouga, D., Djeghaba, K.: Nonlinear dynamic co-rotational formulation for membrane elements with in-plane drilling rotational degree of freedom. Eng. Comput. **33**(3), 667–697 (2016)

Mamouri, S., Mourid, E., Ibrahimbegovic, A.: Study of geometric non-linear instability of 2D frame structures. Eur. J. Comput. Mech. **24**(6), 256–278 (2015)

Bathe, K.J., Noh, G.: Insight into an implicit time integration scheme for structural dynamics. Comput. Struct. **98–99**, 1–6 (2012)

Saigal, S., Yang, T.Y., Kapania, R.K.: Dynamic buckling of imperfection-sensitive shell structures. J. Aircr. **24**(10), 718–724 (1987)

Hilburger, M.W., Waas, A.M., Starnes Jr., J.H.: Modeling the dynamics response and establishing post-buckling/post snap-thru equilibrium of discrete structures via a transient analysis. J. Appl. Mech. **64**(3), 590–595 (1997)

Kobayashi, T., Mihara, Y., Fujii, F.: Path-tracing analysis for post-buckling process of elastic cylindrical shells under axial compression. Thin-Walled Struct. **61**, 180–187 (2015)

White, S.C., Weaver, P.M., Chauncey Wu, K.: Post-buckling analyses of variable stiffness composite cylinders in axial compression. Compos. Struct. **123**, 190–203 (2015)

Wempner, G.A.: Finite Elements, Finite rotations and small strains of flexible shells. Int. J. Solids Struct. **5**, 117–153 (1969)

Belytschko, T., Hsieh, B.J.: Nonlinear transient finite element analysis with covected co-ordinates. Int. J. Num. Meth. Eng. **7**, 255–271 (1973)

Belytschko, T., Hsieh, B.J.: Nonlinear transient analysis of shells and solids of revolution by convected elements. AIAA J. **12**(8), 1031–1035 (1974)

Influence of Material Gradient Index on Stress Distribution of Functionally Graded Dental Implants

Sameh Elleuch[1(✉)], Hanen Jrad[1], Mondher Wali[1,2], and Fakhreddine Dammak[1]

[1] Laboratory of Electromechanical Systems (LASEM), National Engineering School of Sfax, University of Sfax, Route de Soukra km 4, 3038 Sfax, Tunisia
{sameh.elleuch, fakhreddine.dammak}@enis.tn, hanen.j@gmail.com, mondherwali@yahoo.fr

[2] Department of Mechanical Engineering, College of Engineering, King Khalid University, Abha, Saudi Arabia

Abstract. Dental implant have important role in restoration of damaged or lost teeth, but some problem is found in their utilization. In fact, the chewing forces are applied directly in the bone system, which necessitates studying the relationship between the stress distributions and the characteristics of dental implant near bone system interface and dental implant. Recently functionally graded materials FGM play an important role on dental clinical implant application thanks to their abilities and advantages of their mechanical properties. The objective of this study is to focus on mechanical properties of FGM denal implant in order to investigate the effect of the material gradient index on the stress distribution around dental prosthesis. The FGM properties change gradually from titanium (Ti) to hydroxyapatite (HAP). The implant body is subjected to an axial load to simulate masticatory forces. The effect of FGM properties on reduction of stress distributions in the implant-prosthesis components are highlighted.

Keywords: Dental implant · Functionally graded material · Stress distributions · Finite element analysis

1 Introduction

Dental implant is a popular clinical surgery to solve problem of missing tooth. Dental implantation treatment has developed into one of the most successful prosthetic technologies. A critical progress made in this area was the development of biocompatible materials to enable an engineered device (implant) to integrate within its surrounding bony tissues. The challenge to face in prosthetics is to develop both biologically and mechanically compatible biomaterials for this purpose. Functionally graded materials (FGMs) are the advanced materials with varying properties in dimensions. These are made of two or more constituent phases with continuous and smoothly varying composition in preferred directions. In recent year, these materials are gaining attention have performed in many applications such as aeronautic industries, cutting tools, nuclear

M. Kharrat et al. (Eds.): ICAMEM 2019, LNME, pp. 11–17, 2020.
https://doi.org/10.1007/978-3-030-52071-7_2

engineering and dental implant. Many researchers have shown great interest in modelling FG materials, namely (Jrad et al. 2018a, b, 2019; Mallek et al. 2019a, b, c; Mellouli et al. 2019a, b, c and Bouhamed et al. 2019a, b) using different shells theories.

Few existing research has been reported to develop an optimized design of functionally graded material (FGM) dental implant for promoting a long-term success.

In the present study, finite element model is proposed to evaluate the mechanical performance of HAP/Ti FGM dental implant. The influence of gradient material index of FGM implant is examined. Stress distribution (maximum Von Mises stress) in dental implant and bone system which are general performance indicator are investigated. The present work can serve as guideline for the design of optimized FGM dental implant.

2 Modeling and Materials

2.1 Finite Element Analysis

The dental implant/supporting bone system presented in Fig. 1 comprises a Titanium/Hydroxyapatite implant, an abutment, and surrounding cortical bone and cancellous bone in the mandibular section. The mandible segment with an implant (16 mm in height and 4 mm in diameter) and a superstructure is modeled assuming an axisymmetric geometrical model. The cancellous bone is surrounded by a 2 mm thick cortical bone layer.

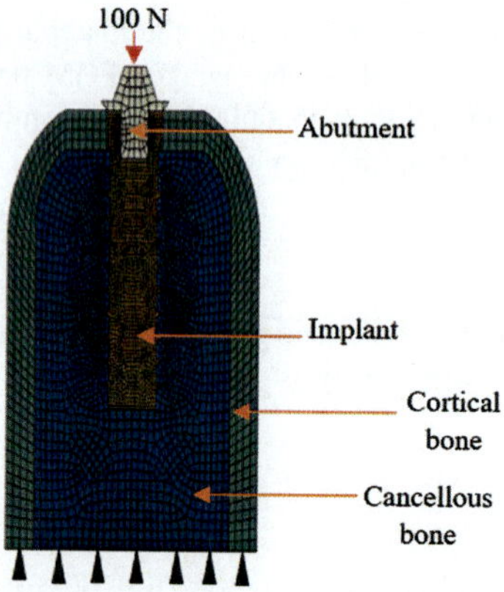

Fig. 1. Views of finite element model of FGM dental implant system

The material properties used in this study are taken from literature and are listed in Table 1. Implants and abutments are made of titanium alloy. For functional grading, hydroxyapatite is used. All materials are considered isotropic and homogenous. (Sazesh et al. 2017)

Table 1. Material properties.

Material	E(GPa)	υ	Ref
Abutment	110	0.35	Yang and Xiang 2007
Titanium (Ti)	110	0.35	Lin et al. 2006
HAP	40	0.27	Hedia and Mahmoud 2004
Cortical bone	14	0.30	Rho et al. 1993
Cancellous bone	3	0.30	Rho et al. 1993

ABAQUS software is used to analyze the model, the component are meshed with hexadehom elements. The number of elements was approximately 30,000, with approximately 26,000 nodes. The interface between the implant and the bones is assumed as perfectly bonded property. Loading of 100 N is applied (Fig. 1) to the center of abutment according to (Kaman and Celik 2013)

2.2 FGM Implant

In this study, dental implant is considered made of Ti/HAP FGM model. The implant properties change gradually from hydroxyapatite to titanium along longitudinal direction using a power law with parameter n as follows, (Sazesh et al. 2017)

$$V_c = \left(\frac{1}{2} + \frac{z}{h}\right)^n \tag{1}$$

$$V_c = 1 - V_m \tag{2}$$

Where V_c is volume fraction of HAP, V_m is volume fraction of titanium, n is material gradient index and h is height of dental implant. In this case, material properties can be obtained as:

$$E = V_c E_c + V_m E_m \tag{3}$$

$$\upsilon = V_c \upsilon_c + V_m \upsilon_m \tag{4}$$

E and υ denote respectively the Young modulus and Poisson's ratio. In addition, c and m symbolize respectively the ceramic (HAP) and metal (Ti) materials.

The material gradient index are considered as n = 0, 0.2, 0.5, 1, 2, 5 and 10.

3 Results and Discussions

The purpose of this section is to evaluate the stress distribution on FGM dental implant system. The maximum Von Mises stress with different material gradient index on the FGM dental implant and the bone system under static loading are presented in Fig. 2 and 3. It can be seen that the maximum Von Mises stress increase when the material gradient index increase.

The maximum stress is focused at the first thread of the implant (Fig. 2), which is a junction point with the bone.

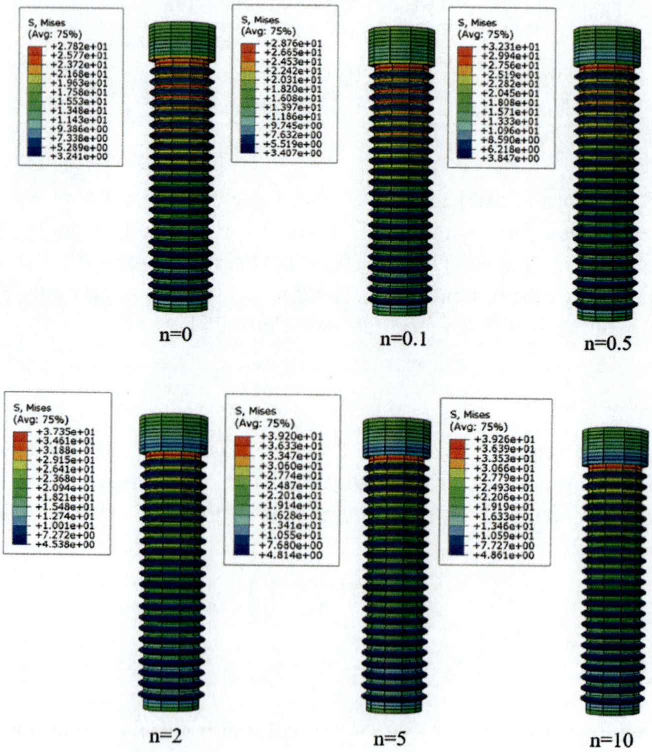

Fig. 2. Stress distribution on FGM dental implant for different material gradient index n

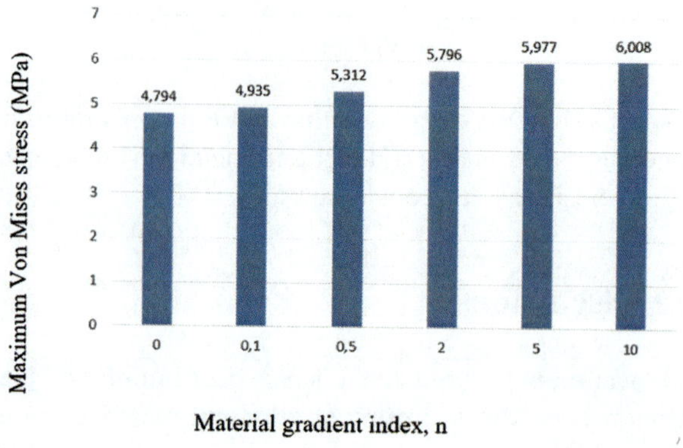

Fig. 3. Maximum Von Mises stress (MPa) on bone system for different material gradient index n

From Fig. 2 and 3, maximal values of stress in both FGM dental implant and bone system is respectively 39.26 MPa and 6 MPa obtained when the power index n = 10. Indeed, the maximum stress value on the bone system are smaller than FGM dental implant

Morever, it is observed that the material gradient index have significant effect on maximum Von Mises stress in static loading and will affect the stability of the implant system. The maximum displacements in FGM dental implant with different material gradient index are presented in Fig. 4. It is shown that by increasing material gradient index n, the displacement of FGM implant increases by reason of the decrease of elastic modulus. From Fig. 4 it appears that the value of displacement in implant is small under static loading and maximum value is 0.0067 mm obtained when material gradient index n = 10 that is to say when having an important amount of titanium. Although titanium and its alloys have been widely adopted as such materials due to their excellent biocompatibility. However, their mechanical prperties largely differ from those in host bony tissues. Therefore, the use of FGM Ti/HAP can reduce displacement, improve stress distribution and have better biocompatibility.

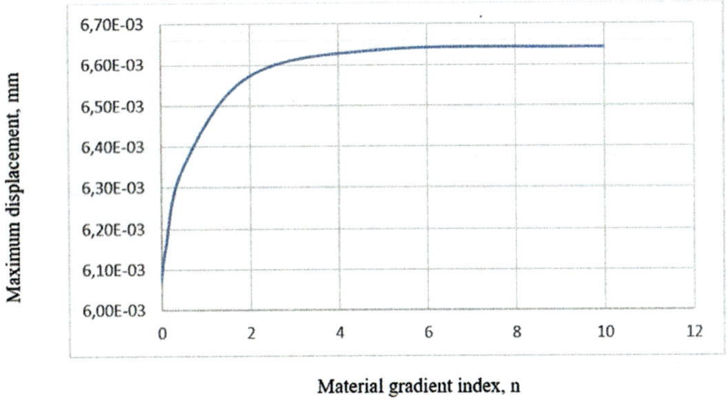

Fig. 4. Maximum displacement on FGM implant for different material gradient index n

4 Conclusion

This work aims to investigate numerically the effect of FGM to reduce stress concentration in the implant prosthesis components. Problem is solved using ABAQUS package program. Results are summarized as follows:

- Maximum Von Mises stress values increases with increases material gradient index n in both FGM dental implant and bone system.
- The maximum value of stress on the bone system are smaller than on dental implant.
- Maximum displacement on FGM dental implant increases with increases material gradient index.

Acknowledgments. This work is carried out thanks to the support and funding of Tunisian Ministry of Higher Education and Scientific Research through the Project to encourage young researchers under grant number 19PEJC10-08.

References

Bouhamed, A., Jrad, H., Jamel, M., Wali, M., Gamaoun, F., Dammak, F.: Homogenization of elasto-plastic functionally graded material based on representative volume element: application to incremental forming process. Int. J. Mech. Sci. **160**, 412–420 (2019a)

Bouhamed, A., Jrad, H., Said, L.B., Wali, M., Dammak, F.: A non-associated anisotropic plasticity model with mixed isotropic–kinematic hardening for finite element simulation of incremental sheet metal forming process. Int. J. Adv. Manuf. Technol. **100**(1–4), 929–940 (2019b)

Hedia, H.S., Mahmoud, N.A.: Design optimization of functionally graded dental implant. Bio-Med. Mater. Eng. **14**(2), 133–143 (2004)

Jrad, H., Mallek, H., Wali, M., Dammak, F.: Finite element formulation for active function-ally graded thin-walled structures. C.R. Mec. **346**(12), 1159–117 (2018a)

Jrad, H., Mars, J., Wali, M., Dammak, F.: An extended finite element method for modeling elastoplastic FGM plate-shell type structures. Struct. Eng. Mech. **68**, 299–312 (2018b)

Jrad, H., Mars, J., Wali, M., Dammak, F.: Geometrically nonlinear analysis of elastoplastic behavior of functionally graded shells. Eng. Comput. **35**(3), 833–847 (2019)

Kaman, M., Celik, N.: Effects of thread dimensions of functionally graded dental implants on stress distribution. World Acad. Sci. Eng. Technol. **78**, 2020–2026 (2013)

Lin, C.L., Wang, J.C., Kuo, Y.C.: Numerical simulation on the biomechanical interactions of tooth/implant-supported system under various occlusal forces with rigid/non-rigid connections. J. Biomech. **39**(3), 453–463 (2006)

Mallek, H., Jrad, H., Algahtani, A., Wali, M., Dammak, F.: Geometrically non-linear analysis of FG-CNTRC shell structures with surface-bonded piezoelectric layers. Comput. Meth. Appl. Mech. Eng. **347**, 679–699 (2019a)

Mallek, H., Jrad, H., Wali, M., Dammak, F.: Geometrically nonlinear finite element simulation of smart laminated shells using a modified first-order shear deformation theory. J. Intell. Mater. Syst. Struct. **30**(4), 517–535 (2019b)

Mallek, H., Jrad, H., Wali, M., Dammak, F.: Piezoelastic response of smart functionally graded structure with integrated piezoelectric layers using discrete double directors shell element. Compos. Struct. **210**, 354–366 (2019c)

Mellouli, H., Jrad, H., Wali, M., Dammak, F.: Geometrically nonlinear meshfree analysis of 3D-shell structures based on the double directors shell theory with finite rotations. Steel Compos. Struct. **31**(4), 397–408 (2019a)

Mellouli, H., Jrad, H., Wali, M., Dammak, F.: Meshfree implementation of the double director shell model for FGM shell structures analysis. Eng. Anal. Boundary Elem. **99**, 111–121 (2019b)

Mellouli, H., Jrad, H., Wali, M., Dammak, F.: Meshless implementation of arbitrary 3D-shell structures based on a modified first order shear deformation theory. Comput. Math. Appl. **77**, 34–49 (2019c)

Rho, J.Y., Ashman, R.B., Turner, C.H.: Young's modulus of trabecular and cortical bone material: ultrasonic and microtensile measurements. J. Biomech. **26**(2), 111–119 (1993)

Sazesh, S., Ghassemi, A., Ebrahimi, R., Khodaei, M.: Experimental and numerical analysis of Titanium/HA FGM for dental implantation. Int. J. Adv. Des. Manuf. Technol. **10**(1), 57–74 (2017)

Yang, J., Xiang, H.J.: A three-dimensional finite element study on the biomechanical behavior of an FGBM dental implant in surrounding bone. J. Biomech. **40**(11), 2377–2385 (2007)

Check for updates

Parameter Identification of a Viscohyperelastic Constitutive Model for Fiber Reinforced Thermoplastic Composites

M. Allouch[1(✉)], M. Kamoun[1], J. Mars[1], M. Wali[1,2], and F. Dammak[1]

[1] Laboratory of Electromechanical Systems (LASEM), National Engineering School of Sfax, University of Sfax, B.P W3038, Sfax, Tunisia
{marwa.allouch, fakhreddine.dammak}@enis.tn, kamoun_moez2@yahoo.fr, jamelmars@yahoo.fr, mondherwali@yahoo.fr

[2] Department of Mechanical Engineering, College of Engineering, King Khalid University, Abha, Saudi Arabia

Abstract. Short glass fiber reinforced thermoplastic materials exhibit visco-hyperelastic behavior which can be analyzed using different techniques such as instrumented indentation and dynamic mechanical analysis. In the present study, the mechanical response of polyamide (PA) as matrix reinforced by different ratio of glass fiber is studied experimentally and numerically. Force-displacement curves of composite materials are obtained through uniaxial tensile test. Moreover, as viscoelastic behavior takes place due to deviation from static deformation, stress relaxation test is investigated to give a good deal of information about the time dependence of the material's behavior. Then, a visco-hyperelastic constitutive model is proposed for the behavior of these orthotropic incompressible fiber-reinforced polymer composites. The experimental data are fit in order to obtain the unknown adjustable parameters in constitutive equations using available model in commercial FEM soft-ware, such as ABAQUS. A good agreement is shown between the experimental and numerical results for all compositions. Therefore, it can be noted that the adapted model can accurately predict the nonlinear behavior of molded composites.

Keywords: Reinforced polyamid 1 · Short glass fiber 2 · Vico-hyperelasticity 3 · FE simulation

1 Introduction

Historically, prediction of mechanical response for such thermoplastic materials still to be difficult for many industrial and medical purposes, predominately when the polymer is reinforced with fibers. In fact, the choice of a constitutive model represents one of the major difficulties to reproduce a complex behavior and it requires extensive experimental and theoretical works. Else, accurate identification of the material properties is not only important, but it becomes essential for specific applications in terms of preciseness and safety.

M. Kharrat et al. (Eds.): ICAMEM 2019, LNME, pp. 18–24, 2020.
https://doi.org/10.1007/978-3-030-52071-7_3

In the aircraft and automotive industry, the use of short glass fiber reinforced injection molded thermoplastic composites is constantly increasing due to a large number of advantages. In fact, these lightweight composite materials include ease of processing for complex geometries at high production rate, ability to reprocess, outstanding cost to performance ratio and high mechanical properties as are improved by the works of Andriyana et al. (2010); Mouhmid et al. (2006). Further-more, glass fibers are widely used to improve strength, elastic modulus and stability of neat thermoplastics as shown in the work by Güllü et al. (2006); Thomason (1999); Moriwaki (1996); Fu and Lauke (1996). However, reinforced thermoplastic materials usually show a complex behavior. For this purpose, optimum and accurate design of parts with thermoplastic materials are challenging and requires many test data at different conditions.

In previous works, some research assume that glass fiber-reinforced thermo-plastic composites are isotropic and hyperelastic. For example, the homogenization approaches are adapted to calculate the effective stored-energy function for reinforced materials with incompressible Neo-Hookean phases by deBotton et al. (2006). Furthermore, Agoras et al. (2009) develop a more general constitutive mod-el, which assume that the matrix and reinforcement phases are considered incompressible and isotropic solids. The model is derived mainly by means of the "second-order" homogenization theory.

Due to the viscoelastic nature of polymer, the purpose of this paper is to predict the response of thermoplastic material reinforced with different content of glass fiber (0, 10, 20, and 30% by weight) under static and quasi-static uniaxial load. Then, a well-known visco-hyperelastic model is proposed to simulate the behavior of molded composites referring to experimental results. For this purpose, a systematic identification procedure of optimal parameters for the associated model is implicated through a FEA code. Finally, a comparison between experimental and numerical results is effected to highlight the validity and the accuracy of the adapted model.

2 Specimens and Experimental Device

For this study, glass fiber-reinforced composites are elaborated primarily from two commercial polyamide: a virgin polyamide 66 and a reinforced polyamide with 30 wt% of short glass fiber. The mixture of these basic materials let to produce four compositions with 00, 10, 20 and 30 wt% of reinforcement. In fact, the dumb-bell test pieces are obtained by injection molding, see the works of Chebbi et al. (2016) and Mars et al. (2018).

In order to understand the mechanical behavior of the fiber-reinforced thermoplastic materials, experimental investigations are performed under different loading conditions. The tensile test is carried out on the injected specimens in order to characterize the mechanical responses of studied composites according to ISO 527-2 standard. Additionally, the relaxation test is performed in order to capture the time dependent properties by using the same instrument used in tensile test. Deformation rate is fixed at 1 mm/min for this experiment up to a strain of about 1% at an ambient temperature and humidity.

3 Constitutive Modeling

Hyperelastic behavior can be described by constitutive model, which are mathematical formulation between the stress and the strain. Generally, hyperelastic materials behavior are described in terms of a “strain energy potential. In fact, many different kinds of strain energy potentials are integrated in Abaqus code to describe incompressible isotropic hyperelastic material and trie to estimate as accurately as possible the materials ‘properties. In addition, FEA is able to evaluate the hyperelastic material behavior automatically by creating response curves using the optimal strain energy potential.

For isotropic and compressible reinforced thermoplastic, the form of the full polynomial strain energy potential can be given as

$$\psi = \sum_{i,j=1}^{n} Cij(I1-3)^{i}(I2-3)^{j} + \sum_{i,j=1}^{n} \frac{1}{Di}(Jel-3)^{2i} \tag{1}$$

where ψ = strain energy per unit of reference volume;
C_{ij} = temperature-dependent material parameters that control the shear behavior and can be determinated from uniaxial test;
D_i = material parameter that control bulk compressibility;
J_{el} = Elastic volume ratio;
n = Number of terms in strain energy function;

Since the material is incompressible, the D_i constants are set to zero. Then, polynomial model is adapted to simulate the elastic properties of four different fiber percentages. This model has a form

$$\psi = C_{10}(I1-3) + C_{01}(I2-3) + C_{11}(I1-3)(I2-3) + C_{20}(I1-3)^2 + C_{02}(I2-3)^2 \tag{2}$$

Moreover, the viscoelastic properties are identified from an experimental relaxation curve. This requires Abaqus to evaluate the Prony series, which represent a relaxation functions defined individually in terms of a series of exponentials.

$$K(\tau) = K_{\infty} + \sum_{i=1}^{n} K_{i}e^{-\frac{\tau}{\tau_i}t} \tag{3}$$

$$G(\tau) = G_{\infty} + \sum_{i=1}^{n} G_{i}e^{-\frac{\tau}{\tau_i}t} \tag{4}$$

where K_{∞} represent the long-term bulk and G_{∞} is the shear moduli.

4 Parameter Identification

Using abaqus code, the curve fitting process allows user to compare experimental test data with various hyperelastic material models. Moreover, the shear constants Cij, the compressibility constants Dij and pressure-volume ratio are optimized from experimental stress-strain data through a least-squares-fit procedure, which consist to minimize the relative error in stress.

Again, the same identification technique is applied for the viscoelastic behavior estimation of studied materials. In fact, the relaxation parameters can be defined by automatically creating response curves in conjunction with relaxation test data. After defining the instantaneous elastic response of the material, the Prony series terms are evaluated.

However, it is quite important to verify that a sufficient correlation exists between the behavior predicted by the material definition and the experimental data.

To find the suitable fit for the proposed model, the material parameters identification was achieved by means of least-square equation

$$E = \sum_{i=1}^{n} (1 - T_i^{th}/T_i^{\exp})^2 \tag{5}$$

where T_i^{th} is obtained from the model of the nominal stress expressions and $T_i^{\exp}$ represent a stress value from the experimental test.

5 Result and Discussion

The result of the fitting are presented for tensile test and f relaxation test respectively in Fig. 1 and Fig. 2. From experimental curves, it can be note that polyamide 66 reinforced with short glass fibers exhibit both a nonlinear hyperelastic behavior and a viscoelastic behavior under the quasi-static uniaxial test. In the following figure, the fitting process by Abaqus code shows a good agreement with the experimental results with relatively low number of parameter. The polynomial model is considered here to be most suitable choice for predicting the behavior of given composite because of its ability to much experimental data points at a small and large strain values. Else, the comparison of results show that Prony series expansion seems to be the best solution to present the experimental response.

Model parameters are identified using a least square method in conjunction with the experimental data via Abaqus tools. The hyperelastic parameter for polynomial model (N = 2) are represented in Table 1. Else, Prony series constants, which describe rate depending behavior of all compounds, are collected in Table 2.

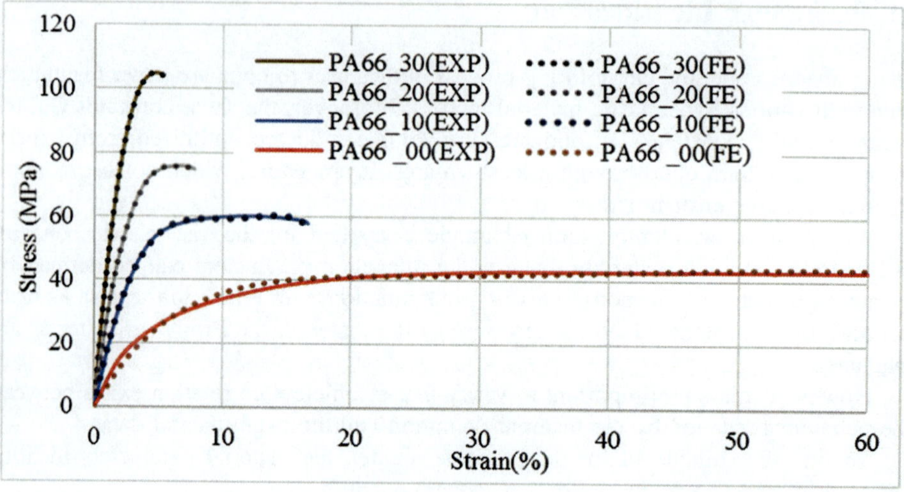

Fig. 1. Comparison between numerical model and experimental test data of tensile test for each compound

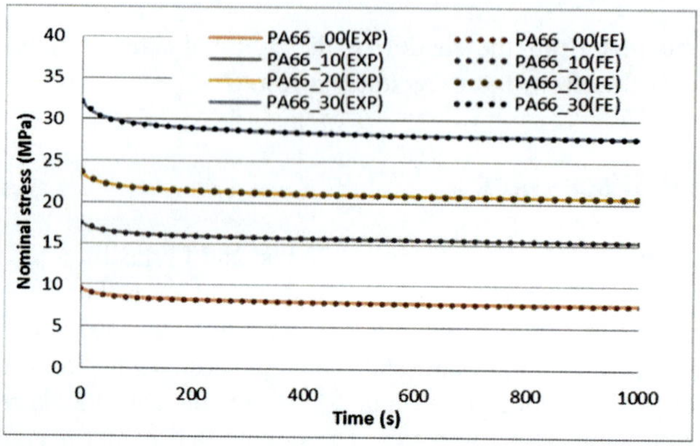

Fig. 2. Comparison between numerical model and experimental test data of stress relaxation test at 1% strain for each compound (MPa)

Table 1. Identification of material parameter from curve fitting process

Compound	C_{10}(MPa)	C_{01}(MPa)	C_{20}(MPa)	C_{11}(MPa)	C_{02}(MPa)
PA-00	–662.87	782.57	186.24	–674.36	882.79
PA-10	–2787.39	3121.47	–156582.74	351781.06	–196155.22
PA-20	–10188.099	10772.57	2172834.13	–4657157.93	25070801.89
PA-30	–16260.18	17190.72	67217661.38	–14120417.9	7434664.88

Table 2. Prony series constants for each compound

Compound	G_1(MPa)	G_2(MPa)	τ_1	τ_2	K
PA-00	9.60e-2	0.10	32.89	454.49	0
PA-10	7.54e-2	6.46e-2	39.26	766.55	0
PA-20	6.60e-2	6.4e-2	27.19	375.22	0
PA-30	7.16e-2	6.83e-2	27.73	387.32	0

6 Conclusion

The determination of visco-hyperelastic behavior of fiber reinforced thermoplastic materials requests some experimental tests under a simplified strain state. For this purpose, tensile and relaxation test are performed on four material composites reinforced with different content of fibers. Then, Abaqus code makes it easy to identify accurately unknown parameters for suitable model by using curve fittings from experimental data.

In order to prove the applicability of the involved model, macroindentation experiment will be simulated by finite element analysis in a future work. The comparison with experimental data and simulating results has to show reasonably agreement, which demonstrate that assigned model is capable of predicting accurately the time-dependent mechanical behavior of the glass fiber-reinforced thermoplastic materials.

Acknowledgments. This project is carried out under the MOBIDOC scheme, funded by the EU through the EMORI program and managed by the ANPR. The authors are grateful to S.T.E the LAB's members and to the Laboratory of Electromechanical Systems 'members for their support.

References

Andriyana, A., Billon, N., Silva, L.: Mechanical response of a short fiber-reinforced thermoplastic: experimental investigation and continuum mechanical modeling. Eur. J. Mech. A/Solids **29**(6), 1065–1077 (2010)

Mouhmid, B., Imad, A., Benseddiq, N., Benmedakhène, S., Maazouz, A.: A study of the mechanical behaviour of a glass fibre reinforced polyamide 66: experimental investigation. Polym. Test. **25**, 544–552 (2006)

Thomason, J.L.: The influence of fibre properties of the performance of glass-fibre-reinforced polyamide 6, 6. Compos. Sci. Technol. **59**(16), 2315–2328 (1999)

Moriwaki, T.: Mechanical property enhancement of glass fibre-reinforced polyamide composite made by direct injection moulding process. Compos. Part A Appl. Sci. Manuf. **27**(5), 379–384 (1996)

DeBotton, G., Hariton, I., Socolsky, E.A.: Neo-Hookean fiber-reinforced composites in finite elasticity. J. Mech. Phys. Solids **54**(3), 533–559 (2006)

Agoras, M., Lopez-Pamies, O., Castañeda, P.P.: A general hyperelastic model for incompressible fiber-reinforced elastomers. J. Mech. Phys. Solids **57**(2), 268–286 (2009)

Chebbi, E., Wali, M., Dammak, F.: An anisotropic hyperelastic constitutive model for short glass fiber-reinforced polyamide. Int. J. Eng. Sci. **106**, 262–272 (2016)

Mars, J., Chebbi, E., Wali, M., Dammak, F.: Numerical and experimental investigations of low velocity impact on glass fiber-reinforced polyamide. Compos. Part B Eng. **146**, 116–123 (2018)

Industrial Applications and Technology Transfer

The Importance of Ergonomic Risk and the Methods to Follow for a Problem of Assembly Line Balancing

Jihene Sdiri[1(✉)], Khaoula El Bedoui[2], and Kamel Mehdi[3]

[1] Production Mechanics and Energy Laboratory (LMPE), ENIT, BP 37 Belvedere, 1002 Tunis, Tunisia
jihene.sdiri@enit.utm.tn

[2] Research Laboratory in Computer Science Modeling and Processing of Information and Knowledge (LIMTIC), Higher Institute of Informatics, 2 Street Bayrouni, 2080 Ariana, Tunisia
khaoula.elbedoui@enicarthage.rnu.tn

[3] Production Mechanics and Energy Laboratory (LMPE), I.P.E.I. Manar, BP 244, 2092 Tunis, Tunisia
kamel.mehdi@ipeiem.utm.tn

Abstract. In this paper, we consider the problem of the assembly line balancing, which allows us to find the assignment of each task that respects fixed constraints while optimizing the repair of working time. However, some other factors can affect productivity and workers, such as repetitive work, handling heavy loads. In this last case, workers in the assembly are subject to ergonomic risks. Besides, it is very important to reduce these risks which is the aim of this research where we attempt to show the importance of ergonomic of an assembly line and to prove the influence of its risks on an assembly line balancing problem. In fact, poor division of labor between stations can lead to unnecessary idle time and unnecessary additional costs for each part produced. In addition, the ergonomic risks can be included in inactivity and this underlines the importance of reducing this risk and its effect on the line. Thus, we expose in this paper a mathematical model to perform the assembly line balancing problem with ergonomic risks which are to reduce.

Keywords: Assembly line balancing problem · Ergonomic risk · Optimization

1 Introduction

Technology imposes variation and massive production of products. Therefore, assembly lines with mixed products are solution. In this case, the products although similar, differ in the use of resources and components' consumption [1]. For these types of production systems, one of the problems to solve is the fact of correctly allocating the tasks to be performed on the various workstations [2]. The problem Assembly Line Balancing Problem (ALBP) considers the cycle time (c) and the number of stations (m). In the last decades, the researchers consider the space occupied at the post by the material to be provided for each task. So, a new kind of problem should be treated, in

M. Kharrat et al. (Eds.): ICAMEM 2019, LNME, pp. 27–31, 2020.
https://doi.org/10.1007/978-3-030-52071-7_4

which Time and Space constrained Assembly Line Balancing Problem (TSALBP) [7]. Other researches incorporate ergonomic parameters into the ALBP [8]. Considering other typologies of lines, other constraints or other objectives forms so-called general problems (GALBP) [4] are discussed such as restricted assignment of tasks or bulk assignment of certain tasks [5]. The general problems group all the problems that are not simple: thus, they are very varied.

A new family of problem TSALBP_erg [9] considers ergonomic risks. For example, musculoskeletal disorders (TMS) have repercussions on work relationships at a very high cost and lost productivity. This case is included in the ergonomic risks that have an impact on productivity. This gives us the importance of ergonomics in assembly lines (a proportional relationship between the importance of ergonomics and productivity). To assess ergonomic risks, authors of [1] have proposed a classification for the risk levels (Table 1).

Table 1. Classification of the level of risk by categories and actions to consider

Risk	Action
Level 1	No danger to the laborer
Level 2	Minor risk
Level 3	A degree of risk requiring medical intervention
Level 4	A high degree of risk that requires medical intervention to treat the disease

Ergonomic risks known in the industry are: Postural loads, Repetitive movements or manual handling. For the first one, the workers are exposed to lifting the weight, which leads to fatigue and stress of the worker. To analyze these risks, **RULA** (Rapid Upper Limb Assessment) and **REBA** (Rapid Entire Body Assessment) can solve the types of problem. For second risk, **OCRA** Check List (Occupational Repetitive Action) is the solution. The injuries caused by repetitive movements are one of this kind of risks. For the third risk, physical and difficult work leads to harm affects the workflow and for analyze it we use **NIOSH** equation (National Institute for Occupational Safety and Health) and the Tables from Snook and Ciriello.

This article will be divided as follow. Section 2 presents a literature review of the ALBP and the show the importance of ergonomic on the assembly lines. Section 3 presents the parameters and variables used in this problem. Finally, Sect. 4 presents the conclusion of the paper.

2 Literature Review

The simple assembly line balancing problem is a combinatorial optimization problem that has been formulated for the first time by [2]. This is to perform the tasks at stations while respecting various constraints to optimize a given criterion of efficiency. The problem with mono-objective (OP), we talk about Simple Assembly Line Balancing (SALBP) is proposed in the literature as the objectives to optimize (Table 2).

Table 2. Classification of the objectives to optimize in SALBP

	Cycle time	
Number of stations	Given	To minimize
Given	SALBP-F	SALBP-2
To minimize	SALBP-1	SALBP-E

In the last years, we talk to multi- objective problem (MOP) such as a minimization of variables and given parameters. In the literature we find the incorporation of the parameters like the cost of the stations and their attributions, the parameters of ergonomics in the lines of assembly, the consideration of the space or area of materials and tools, this problem so-called TSALBP are included in the family of problems of Time and Space [7].

3 Problem Formulations for Assembly Line Balancing Considering Ergonomic Risk

Parameters

J Set of elemental tasks (j $= 1, \ldots, |J|$)
K Set of workstations (k $= 1, \ldots, |K|$)
E Set of ergonomic risk factors ($e = 1, \ldots, |E|$)
t_j Processing time of an elemental task $j \in J$
C_K Maximum processing time given to the workstation $k \in K$
$X_{e,j}$ Ergonomic risk category associated to the task $j \in J$ regarding the risk factor $e \in E$
$E_{e,j}$ Ergonomic risk associated to the task $j \in J$ regarding the risk factor $e \in E$

$$R_{e,j} = X_{e,j} \cdot t_j$$

P_j Set of tasks that precedence the task j $\in J$

Variables

$X_{j,k}$ Binary variable equal to 1 if the task $j \in J$ is assigned to the workstation $k \in K$ and 0 otherwise.
$E^{max}_{e,k}$ Maximum ergonomic risk allowed at workstation $k \in K$ according to the ergonomic risk factor $e \in E$

With these parameters and variables, we present two mathematical formulas (without ergonomic parameters and with it) to give the importance of ergonomic on the assembly line.

$$\sum_{k=1}^{|K|} X_{j,k} = 1 \qquad (\mathrm{j} = 1, \ldots, |J|) \tag{1}$$

$$\sum_{k=1}^{|K|} k.(X_{i,k} - X_{j,k}) \leq 0 \quad \in i \in P_j, (\mathrm{j} = 1, \ldots, |J|) \tag{2}$$

$$\sum_{j=1}^{|J|} t_j.X_{j,k} \leq C_K \leq C_K \qquad (\mathrm{k} = 1, \ldots, |K|) \tag{3}$$

$$\sum_{j=1}^{|J|} E_{e,j}.X_{j,k} \leq E_{e,k}^{max} \qquad (\mathrm{k} = 1, \ldots, |K|) \text{ and } (e = 1, \ldots, |E|) \tag{4}$$

In this model the equality (1) forces the assignments of tasks to workstations in order. Constraint (2) imposes the precedence relations to carry out the work of line. Constraint (3) defines capacity that ensures the assignment of tasks j in these workstations k. Constraint (4) establishes the maximum limit of the ergonomic risk.

$$\min K + E \qquad \mathrm{k} \ = 1, \ldots, |K| \tag{5}$$

Depending on the given profile, we assume the importance of ergonomics on an assembly line. The reduction of the ergonomic risk reflects on the whole line, on the processing time, on the importance of the human being, on the cost… Moreover, if the number of stations is reduced, this means a reduction of the ergonomic risk. The latter is a set of movements made by the hands during the treatment on a station. Equation (5) models this problem, the objective of this profile is the minimization of both the number of stations and the ergonomic risk on an assembly line.

4 Conclusion

In this paper, we presented the problem of balancing assembly lines and we defined the ergonomics risk concept. The ergonomic risk increases, and the lost time is affecting the productivity. For this reason, this model defies this situation and for future research we propose a digital application that ensures the importance of avoiding an ergonomic risk while giving values to these risks, it helps industrialists to make decisions and limit the overall risk. By also giving costs to this parameter in order to study all possible cases.

References

1. Bautista, J., Batalla-Garcia, C., Alfaro-Pozo, R.: Models for assembly line balancing by temporal, spatial and ergonomic risk attributes. Eur. J. Oper. Res. **251**(2016), 814–829 (2016)

2. Solverson, M.E.: The assembly line balancing problem. J. Ind. Eng. **6**, 3 (1995)
3. Baybars, I.: A survey of exact algorithms for the simple assembly line balancing problem. Manage. Sci. **32**(8), 909–932 (1986)
4. Boysen, N., Fliedner, M., Scholl, A.: A classification for assembly line balancing problems. Eur. J. Oper. Res. **2007**(183), 674–693 (2007)
5. Battaïa, O., Dolgui, A.: Reduction approaches for a generalized line balancing problem. Comput. Oper. Res. **39**, 2337–2345 (2012)
6. Battaïa, O., Dolgui, A.: A taxonomy of line balancing problems and their solution approaches. Int. J. Prod. Econ. **142**, 259–277 (2013)
7. Chica, M., Cordón, O., Damas, S., Bautista, J.: Multiobjective constructive heuristics for the 1/3 variant of the time and space assembly line balancing problem: ACO and random greedy search. Inf. Sci. **180**, 3465–3487 (2010)
8. Otto, A., Scholl, A.: Incorporating ergonomic risks into assembly line balancing. Eur. J. Oper. Res. **212**, 277–286 (2011)
9. Bautista, J., Batalla, C., Alfaro, R.: Incorporating ergonomics factors into the TSALBP, comunicación. In: Proceedings of the (CD), International Conference on Advances in Production Management Systems (APMS 2012). Rodas (Grecia) (2012)

Check for updates

Modeling of the Cutting Effort

Taoufik Kamoun(✉) and Kamel Guemri

ISET de Sfax, Route de Mahdia, Sfax, Tunisia
taoufikkamoun@yahoo.fr, kamelguemri2001@yahoo.fr

Abstract. Cutting is a process for obtaining parts that are widely used in the press tool industry. The calculation of cutting force is generally done by the simplified formula which only describes the shear force of the sheet (Fc = P.e. Rc). In this formula, we only see the perimeter of the part, the thickness and shear strength of the sheet.

In practice, other factors can influence this effort, such as, the clearance between the punch and the die, the speed of the cutting operation, the thickness of the sheet, the friction between the cut pieces and the die, etc. This work consists in making an experimental study, according to a complete experimental plan, to study the effect of some parameters, such as, the play between the punch and the matrix, the speed of cutting and the thickness of the sheet, on the cutting effort. This study is completed by a statistical analysis of the results and a modeling of the Fc effort according to the parameters studied.

Keywords: Cutting force · Fc · Cutting · Clearance · Cutting speed

1 Introduction

Sheet metal cutting is a shaping process that is widely used in the industry. This process is confronted with an ever-greater miniaturization of the products manufactures and undergoes more and more competition.

Given these elements, mastering product quality, increasing productivity and reducing costs are essential for business development and survival. To cope with these constraints, manufacturers must analyze and optimize all aspects related to production operations.

In shaping by mechanical cutting, the cutting clearance (j), the cutting speed (Vc) and the thickness of the cutting sheet (e) are the main parameters to be controlled which influence the cutting force (Fc).

The main reason for this study lies in the desire to study the effect of these parameters on the cutting effort. The ultimate goal being, if possible, to propose a relation between these parameters and the effort and to propose a model, which expresses this effort according to the parameters.

M. Kharrat et al. (Eds.): ICAMEM 2019, LNME, pp. 32–37, 2020.
https://doi.org/10.1007/978-3-030-52071-7_5

2 Materials and Experiences

2.1 Strategy of the Study

The parameters studied are chosen in three levels for each parameter as shown in Table 1, according to a complete experimental design, which was followed during the experimental tests. The influences of the cutting parameters on the cutting effort were studied, based on the statistical method of the analysis of the variance "ANOVA".

Table 1. Matrix of levels

	Clearance: j (mm)	Thickness: e (mm)	Speed Vc (mm/min)
Level 1	0.15	1	20
Level 2	0.2	1.5	25
Level 3	0.25	2	30

2.2 Tests Conditions

As shown in Fig. 1, the tests were carried out with a cutting tool of a circular shape mounted on a traction machine, on parts cold rolled sheet (DC01). It is a sheet often used in obtaining parts by cutting and stamping.

Fig. 1. Cutting tool

After each test the WINDAP V1.6 software of the traction machine draws a curve, which contains some parameters such as the maximum force (Fmax) and the strain at Fmax (Fig. 2).

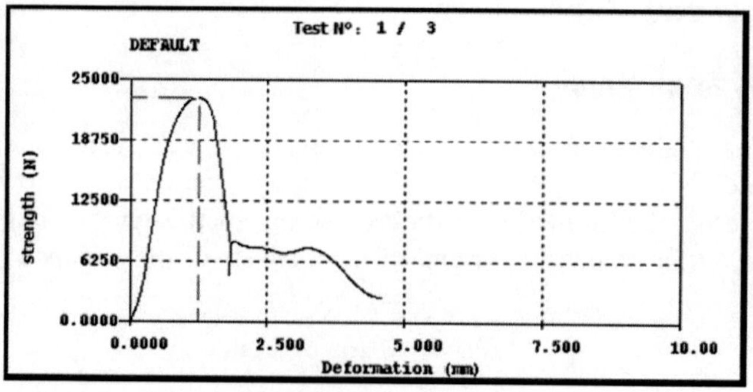

Fig. 2. Curves and values

3 Results and Discussion

3.1 Average Effects of Cutting Parameters

Table 2 presents the average of experimental Fc, in the 3 levels of each parameter, the calculated effect of each parameter, the delta, which represents the difference between the level max, the min level. and the rank that ranks the effects in descending order.

Table 2. Average effects on Fc.

	j (mm)	e (mm)	Vc (mm/min)
Level 1	17349	10554	17459
Level 2	17637	17784	17666
Level 3	17443	24091	17895
Delta	287	13537	436
Rank	3	1	2

The graphical representation of the average effects on Fc of Fig. 3 shows that the cutting force (Fc) is too influenced by the thickness of the cut sheet (e). The influence of the cutting speed (Vc) and that of the cutting clearance (j) are much smaller and less significant than the influence of (e).

3.2 Interactions of the Cutting Parameters

The graphs of the interactions between the parameters show segments of almost parallel lines. We can therefore conclude that the effects of the interactions between the parameters are small and negligible (Fig. 4).

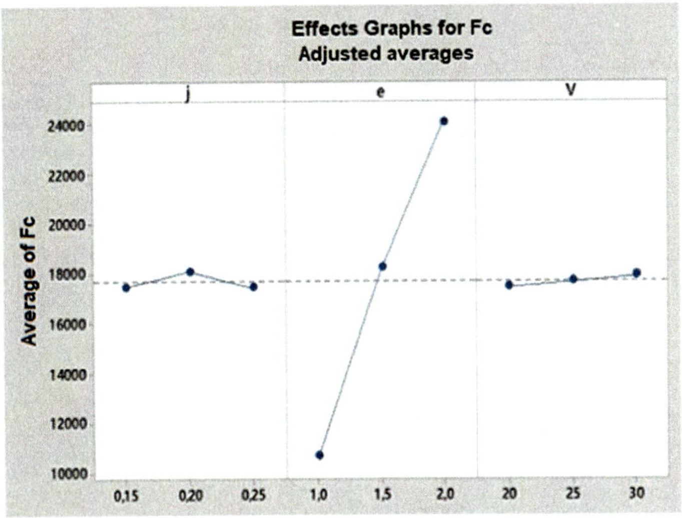

Fig. 3. Representation of the average effects on Fc

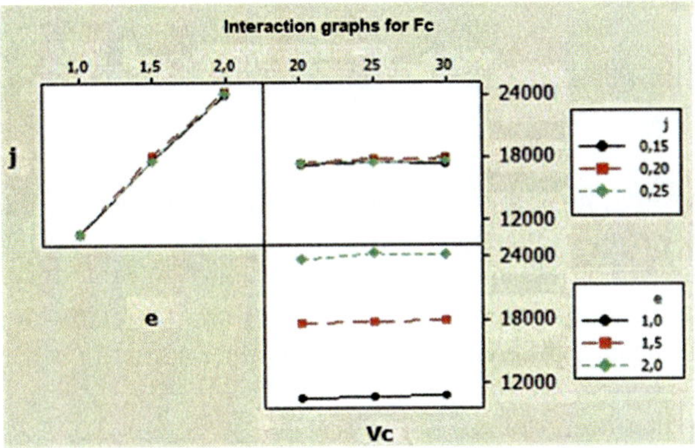

Fig. 4. Interaction graphs for Fc

3.3 Modeling of Fc According to the Parameters of Division

Fc can be modeled using a model that considers first-order interactions as follows:

$$Fc = b0 + b1 * j + b2\, e + b3 * vc + b12 * (j*e) + b13 * (j*vc) + b23 * (e*vc) + b123 * (j*e*vc) \quad (1)$$

The statistical analysis of this model leads to the results displayed on the following tables. Table 3 shows 3 stars, in front of the significance of the regression. Which clearly indicates the good quality of our initial choice.

Table 3. Fc variance analysis

Source of variation	Sum of squares	DDL	Middle square	Report	Significance
Regression	1.4289E09	7	2.041E08	1401.6401	<0.01 ***
Residues	3.4954E06	24	1.456E05		
Total	1.4324E09	31			

The statistical analysis of this model brings us to the Table of Analysis of Variance (ANOVA) Table 4 below: t. exp.

Table 4. Estimates and statistics of Fc coefficients.

Name	Coefficient nt	F. Infla	Standard deviation	t. exp.	Signif. %
b0	17220.3		67.5	255.25	<0.01 ***
b1	107.9	1.00	67.5	1.60	12.3
b2	6679.6	1.00	67.5	99.01	<0.01 ***
b3	77.9	1.00	67.5	1.15	26.0
b12	100.4	1.00	67.5	1.49	15.0
b13	–67.9	1.00	67.5	–1.01	32.4
b23	5.4	1.00	67.5	0.08	93.7
b123	–80.4	1.00	67.5	–1.19	24.5

The statistical analysis of this model, essentially, shows that the model used is well adjusted, since the sum of the squares due to the residues (3.4954E06) is too much smaller, than, the sum of the squares due to the regression (1.4289E09).

It can be seen that only the constant b0 and the term b2 (relative to the thickness of the sheet) are significantly different from zero. All other terms are weak and statistically insignificant.

Equation (1) then becomes:

$$\mathrm{Fc} = 17220.3 + 6679.6 * \mathrm{e} \tag{2}$$

Figure 5 shows the curves of Fc measured experimentally (in red), Fc calculated by the model (in blue) and the difference between the two (in green). It helps us to judge more accurately the quality of the choice made, and confirms that our fit, is of very good quality.

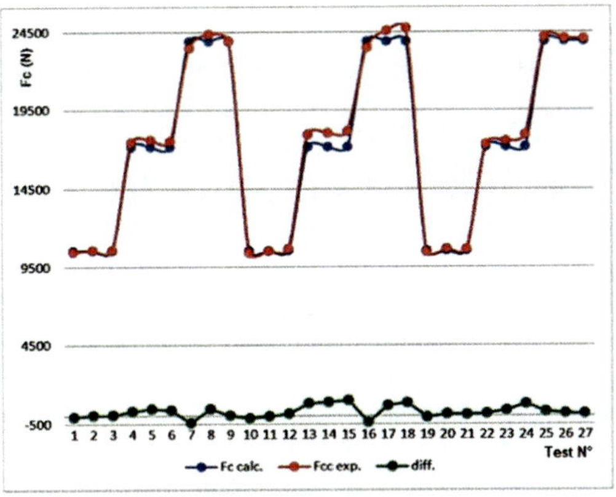

Fig. 5. Difference between Fcexp and Fccalc.

4 Conclusion

The interpretation of the experimental results allowed us to study the influence of the chosen parameters on the effort of cutting (Fc) and their interactions. It also permitted us to make a model of the Fc rate, taking into account the different interactions between these parameters.

Bibliography

1. Chabenat, A., Martin, R.: Sheet metal working, cutting - punching. Technical report, Engineering Techniques B 7561, Paris (1978)
2. Hambli, R.: Experimental, numerical and theoretical study of sheet metal cutting for the optimization of the process. Ph.D thesis, National School of Arts and Crafts of Angers, 5 October 1996
3. Ramde, S.: Experimental, numerical and theoretical study of sheet metal cutting for the optimization of the process. Ph.D thesis, National School of Arts and Crafts of Angers, 5 October 1996

Improving Performance of Production Lines: Integration of Maintenance and Quality Policies: A Literature Review

Z. Boumallessa[1(✉)], H. Chouikhi[2(✉)], M. Elleuch[3(✉)], and H. Bentaher[3(✉)]

[1] Laboratory of Electromechanical Systems (LASEM), National School of Engineers of Sfax, Sfax, Tunisia
zeineb.boumallessa@gmail.com
[2] Mechanical Engineering Department, College of Engineering, King Faisal University (KFU), Hofuf, Kingdom of Saudi Arabia
hchouikhi@kfu.edu.sa
[3] LASEM, Higher Institute of Industrial Management of Sfax, Sfax, Tunisia
mounir_tn_sf@yahoo.fr, bentaherh@yahoo.com

Abstract. Production scheduling, quality management and maintenance planning are the main functions of operational management in industrial plants. Although in reality they are interdependent, in practice these three fields are treated separately. This has generated as a fact independent models for each function that might yield suboptimal solutions. Hence, an appropriate understanding of the dependency between the mentioned parameters may improve the production systems efficiency and result in significant savings in costs. The objective of this study was to review the literature on models which deal with different integration issues in the production scheduling, quality management and maintenance planning, to point gaps which exist in current works to be able to deal with, in order to solve a simulation-based industrial system representing this integration.

Keywords: Integration · Quality · Maintenance · Production · Optimization

1 Introduction

Industrial companies must produce, ensure appropriate product quality, and sustain the maintenance of their equipment. This is why practical optimization solutions that integrate production planning, maintenance and quality are essential. These fields are usually treated independently, resulting, generally, in suboptimal solutions. A right treatment of this dependence may pave the way for more integrated models and ensure crucial savings in costs as well as efficiency for the whole production system. In this case, it is vital to offer a well-structured decision support tool, evaluating the behavior of the production process three axes at the same time, and providing answers to different questions.

The aim of this paper was to present a review on the models proposed by the literature addressing the different aspects of integration in the above-underlined fields.

M. Kharrat et al. (Eds.): ICAMEM 2019, LNME, pp. 38–45, 2020.
https://doi.org/10.1007/978-3-030-52071-7_6

Limitations of optimization methods that have been developed in this context were also involved to identify possible future research perspectives.

2 Analyses and Method

2.1 Evolution of the Integration Strategy of Production Policy, Maintenance and Quality

Many models have been proposed to model the production system behavior and the different maintenance actions that can be implemented on it. These models aim to find the optimal balance between preventive and corrective maintenance while respecting the planning and quality objectives set. It is then necessary to determine the time of the maintenance and the actions to be performed in order to optimize a given decision criterion. This criterion can be based either on cost as shown in the work by Castanier et al. (2005), system availability and maintenance of equipment or safety as shown in the study of Aarab et al. (2017).

The integration of production planning, quality and maintenance policies has been the subject of a great amount of scientific research for decades. Many researchers thought that an independent treatment of these models would provide a suboptimal solution. In this regard, interest in integrated models has increased and started to be the subject of several scientific research reviews as shown by Hadidi et al. (2012), Fakher et al. (2014), Bouslah et al. (2016), and Bahria et al. (2018).

In a literature review on this research topic, Hadidi et al. (2012) classify integrated business management models into two categories:

- **The concept of interdependence:** it concerns models seeking to optimize a single function among the production control, quality and maintenance, considering the other functions as imposed constraints not to optimize.

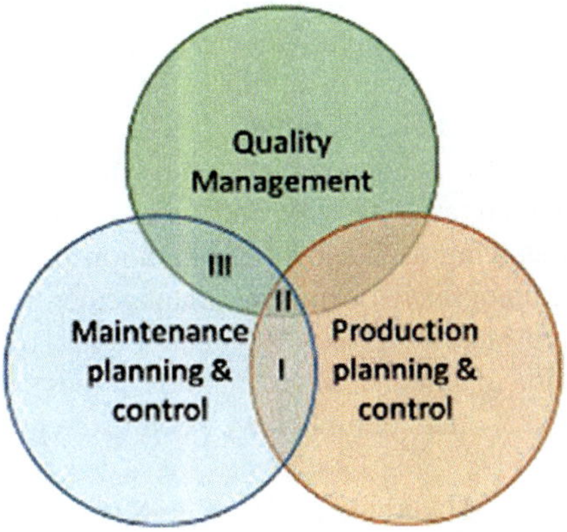

Fig. 1. Interaction of the main functions to plan and control production

– **The concept of integration (Fig. 1):** This is related to models where, at least, two functions among the production control, quality and maintenance are to be jointly optimized.

Based on the definition of integration proposed by Hadidi et al. (2012), we found out that the majority of integration models in the literature focus only on two functions at the same time.

Several integration models of production and maintenance policies have been proposed since the 1990s without considering the quality aspect. A detailed literature review of this work was proposed by Budai et al. (2008).

Further research on the integration of production and maintenance policies without taking quality management into account has also been published in such studies as those performed by (Géhan et al. 2014; Xia et al. 2012; Xia et al. 2015).

On the other hand, research on the integration of production and quality without considering the maintenance aspect was the subject of only two detailed reviews presented by Goyal et al. (1993) and Inman et al. (2013).

Thus, it conspicuous that the majority of models that have been developed for integrated operations management address only two functions in both production management, quality, and maintenance scheduling as shown by (Hadidi et al. 2012; Rivera-Gómez et al. 2013; Colledani et al. 2014).

However, only recently have scientific efforts been intensified to deal with the three mentioned axes simultaneously.

2.2 Integration of Production Control, Quality and Maintenance Policies; Developed Approaches

As already underlined, production, maintenance and quality are the basic elements of any production system. In practice, these three functions are generally managed separately by independent organizational units, although they are closely interrelated as shown by Ben-Daya and Rahim (2001).

Several research projects have been conducted to develop integrated operations management strategies that jointly optimize production control, quality and maintenance policies. Several literature reviews on this subject have been published, as shown in the work by Hadidi et al. (2012) and Inman et al. (2013) and Colledani et al. (2014) and Aarab et al. (2017). This tendency was motivated by the fact that integrated management policies improve productivity and product quality compared to the traditional approaches of managing each aspect of operations in isolation (Colledani et al. 2011). For example, the integrated operations management helps companies significantly to reduce costs and increase profits (Colledani and Tolio 2012).

Integrating production, maintenance and quality comes as a natural extension of two-way integration. Infact, some attempts have been made to model the integration of these three parameters.

The literature reviews of (Hadidi et al. 2012) and (Colledani et al. 2014), and the recently-published article by Bahria et al. (2018) offer fewer than a dozen articles that address the simultaneous integration of production control, quality and maintenance policies.

Table 1. Relevant publications

Publication	Proposed method	Corresponding industrial system	Publication's year
Rivera-Gómez et al. (2013)	A stochastic dynamic programming model	Single type of product and single machine	2013
Fakher et al. (2014)	A mathematical modeling and formulation of the profit maximization model which integrates lot sizing, sales planning, and preventive maintenance scheduling with certain aspects of the quality system in the same decision structure	Single machine that can produce several types of products	2014
Fakher et al. (2015)	A mathematical model and a genetic algorithm is used to solve it	A system of M machines assigned to production of P products	2015
Bouslah et al. (2016)	A stochastic mathematical model using a simulation optimization approach	Single production unit for single-product batches	2016
Bahria et al. (2018)	A mathematical model was developed to find the optimal strategy which minimizes the average total cost per time unit	Single machine producing one type of items	2018
Robert et al. (2018)	An anticipative maintenance strategy	Single production unit for single-product batches	2018
Khatab et al. (2018)	An integrated optimization model and an incremental heuristic method to solve it	A manufacturing system producing a single product at a constant rate to meet a constant and continuous demand rate	2018
Farahani et al. (2019)	An integrated optimization model using Markov chain	A job shop system in the form of a continuous time Markov chain	2019
Wang et al. (2019)	An integrated optimization model was proposed and a genetic algorithm was proposed to solve it	A capacitated lot-sizing production system subject to deterioration	2019

(*continued*)

Table 1. (*continued*)

Publication	Proposed method	Corresponding industrial system	Publication's year
Rivera-Gómez et al. (2019)	A simulation-based optimization approach is proposed, which combines the descriptive capabilities of simulation modeling with analytical models and statistical analysis	An unreliable production unit producing one part type	2019

Some models have recently discussed the joint integration of production and quality under different maintenance policies. The table below summarizes the most interesting ones (Table 1).

Despite the relevant results of the above-mentioned works, we should underline that these models are based on simplistic assumptions, even unrealistic, when modeling the degradation of a product quality and machine reliability. These models seem to have neglected the dynamic aspect of degradation, and the uncertain reality of industrial systems. However, several studies have emphasized the important impact of the dynamics of the operations of production on the deterioration intensity of quality and maintenance (Rivera-Gómez et al. 2018).

Works that relied on simulation as a method of resolution are no exception as they are also limited, as shown in the articles of Bouslah et al. (2016) and Rivera-Gómez et al. (2019). Furthermore, the systems that were subject of this research are simple and do not reflect reality.

3 Results and Discussion

3.1 Summary on Existing Models

Researchers contributed different activities and efforts to express integration between quality control, production and maintenance. Different methods were inferred and many theoretical solutions were involved. However, it seems that weak spots exist yet. In fact, the imperfections noted below prevent the applicability of the existing models in practice and yield as a result suboptimal solutions.

1) In current approaches, very specific framework conditions are assumed which cannot be generalized for industrial environments. Hence, the existing models are not valid.
2) Many models are based on assumptions which makes them limited and unable to prevail in realistic production scenarios. For this reason, researchers validate those models by numerical examples most of the time.

3) The combination of parallel and serial interlinked facilities is not considered neither did they take into account the labour side nor the workers intervention. The existing approaches refer generally to single machine with single product and do not take labor into account.
4) The existing planning algorithms calculate stochastic dates and use theoretical methods for maintenance. Different types of predictive maintenance are not considered in most of the presented approaches.
5) The used methods in most of the research are analytical. Still, the analytical solutions cannot be faithful to the real industrial system and can no longer produce adequate results because of the production processes complexity.

This leads us to propose other more efficient methods such as simulation.

3.2 Research Perspectives

The above presented review can be suggestive to several new research perspectives related to the field among these, the following directions for research can be considered:

- Integration of the production control, quality and maintenance of industrial systems that have production lines consisting of more than two machines, and manufacturing several types of products.
- Integration of other maintenance management policies and work on other quality-related issues.
- Extending optimization assumptions to take into consideration the dynamic reality of the industrial system, considering a variable and non-constant demand rate.
- Use simulation as a powerful optimization tool that takes into account the specificities of production and test different scheduling and management scenarios. The tested scenarios may lead to significant improvements in the performance of the production system.

4 Conclusion

This review focused on a better understanding of the interaction between production, maintenance and quality models, in order to be able to model a real industrial system that integrates these mentioned components. In general, the so far proposed integrated models consider very specific situations that cannot be generalized to other environments. Researchers suggested models that are mathematically appropriate; however, they are not associated with real-world examples. In addition, the different techniques which were proposed to integrate the main functions, have certain disadvantages and rely on simplistic assumptions.

As future work, we will focus on the study of the integration of production, quality and maintenance of a real industrial system. A simulation model can be a promising research perspective.

Acknowledgments. We would like to show our gratitude to Prof Abdelmajid Dammak from the National school of engineering of Sfax for efforts that he provided to revise the English language of the paper.

References

Aarab, S., El Barkany, A., El Khalfi, A.: The integration of maintenance plans and production scheduling taking account of outsourcing: a literature review. Int. J. Prod. Qual. Manage. **21** (1), 1–22 (2017)

Bahria, N., Chelbi, A., Dridi, I.H., Bouchriha, H.: Joint optimization of maintenance, inventory and quality control policies. In: TORS 2018 - Third International Conference of the Tunisian Operational Research Society, 7–9 April 2018, Sousse, Tunisia (2018)

Ben-Daya, M., Rahim, M.A.: Integrated Production, Quality & Maintenance Models: An Overview. Integrated Models in Production Planning, Inventory, Quality, and Maintenance, pp. 3–28. Springer, Boston (2001)

Bouslah, B., Gharbi, A., Pellerin, R.: Integrated production, sampling quality control and maintenance of deteriorating production systems with AOQL constraint. Omega **61**, 110–126 (2016)

Budai, G., Dekker, R., Nicolai, R.: Maintenance and Production: A Review of Planning Models. Complex System Maintenance Handbook. Springer, London (2008)

Castanier, B., Grall, A., Bérenguer, C.: A condition-based maintenance policy with non-periodic inspections for a two-series system. Reliab. Eng. Syst. Saf. **87**(1), 109–120 (2005)

Colledani, M., Tolio, T., Fischer, A., Iung, B., Lanza, G., Schmitt, R., Váncza, J.: Design and management of manufacturing systems for production quality. CIRP Ann. Manuf. Technol. **63**(2), 773–796 (2014)

Colledani, M., Tolio, T.: Integrated quality, production logistics and maintenance analysis of multi-stage asynchronous manufacturing systems with degrading machines. CIRP Ann. Manuf. Technol. **61**(1), 455–458 (2012)

Colledani, M., Tolio, T., Fischer, A., Iung, B., Lanza, G., Schmitt, R., Vancza, J.: Joint design of quality and production control in manufacturing systems. J. Manuf. Sci. Technol. **4**(3), 281–289 (2011)

Fakher, H.B., Nourelfath, M., Gendreau, M.: Profit maximization by integrating production planning, maintenance scheduling, quality aspects and sale decisions. In: International Conference on Industrial Engineering and Operations Management, Bali, Indonesia, 7–9 January 2014. Proceedings, pp. 1281–1292 (2014)

Fakher, H.B., Nourelfath, M., Gendreau, M.: Joint production-maintenance planning in an imperfect system with quality degradation. In: International Conference on Industrial Engineering and Systems Management (IESM) (2015)

Farahani, A., Tohidi, H., Shoja, A.: An integrated optimization of quality control chart parameters and preventive maintenance using Markov chain. Adv. Prod. Eng. Manage. **14**, 5–14 (2019)

Géhan, M., Castanier, B., Lemoine, D.: Integration of maintenance in the tactical production planning process under feasibility constraint. In: APMS (2014)

Goyal, K., Gunasekaran, A., Martikaien, T., Yli-Olli, P.: Integrating production and quality control policies: a survey. Eur. J. Oper. Res. **69**, 1–13 (1993)

Inman, R., Dennis, E., Huang, N., Li, J.: Survey of recent advances on the interface between production system design and quality. IIE Trans. **45**, 557–574 (2013)

Hadidi, L.A., Al-Turki, U.M., Rahim, A.: Integrated models in production planning and scheduling, maintenance and quality: a review. Int. J. Ind. Syst. Eng. **10**(1), 21–50 (2012)

Khatab, A., Diallo, C., Aghezzaf, E.H., Venkatadri, U.: Integrated production quality and condition-based maintenance optimisation for a stochastically deteriorating manufacturing system. Int. J. Prod. Res. **57**(8), 2480–2497 (2018)

Wang, L., Lu, Z., Han, X.: Joint optimisation of production, maintenance and quality for batch production system subject to varying operational conditions. Int. J. Prod. Res. **57**(10), 1–15 (2019)

Pandey, D., Kulkarni, M.S., Vrat, P.: Joint consideration of production scheduling, maintenance and quality policies: a review and conceptual framework. J. Adv. Oper. Manage. **2**, 1–24 (2010)

Rivera-Gómez, H., Gharbi, A., Kenné, J.P.: Joint production and major maintenance planning policy of a manufacturing system with deteriorating quality. Int. J. Prod. Econ. **146**(2), 575–587 (2013)

Rivera-Gómez, H., Montaño-Arango, O., Corona-Armenta, J.R., Garnica-González, J., Ortega-Reyes, A.O., Anaya-Fuentes, G.E.: JIT production strategy and maintenance for quality deteriorating systems. Appl. Sci. **9**(6), 1180 (2019)

Rivera-Gómez, H., Montaño-Arango, O., Corona-Armenta, J.R., Garnica-González, J., Hernández-Gress, E.S., Barragán-Vite, I.: Production and maintenance planning for a deteriorating system with operation-dependent defectives. Appl. Sci. **8**(2), 165 (2018)

Robert, G., Karner, M., Nemetha, T., Matyas, K., Sihn, W.: An approach for the integration of anticipative maintenance strategies within a production planning and control model. In: 11th CIRP Conference on Intelligent Computation in Manufacturing Engineering. Procedia CIRP, vol. 67, pp. 46–51 (2018)

Xia, T., Xia, L., Zhoua, X., Lee, J.: Dynamic maintenance decision-making for series– parallel manufacturing system based on MAM–MTW methodology. Eur. J. Oper. Res. **221**(1), 231–240 (2012)

Xia, T., Jin, X., Xi, L., Ni, J.: Production-driven opportunistic maintenance for batch production based on MAM–APB scheduling. Eur. J. Oper. Res. **240**(3), 781–790 (2015)

Design and Manufacturing

The Impact of Abrasive Water Jet Cutting on Tensile Behavior of Woven Fabric CFRP

Adel Abidi[1(✉)], Sahbi Ben Salem[1,2], Abderrezak Bezazi[3], and Haithem Boumediri[3]

[1] Lab-MAI, National School of Engineers of Tunis (ENIT), Tunis El Manar University, Tunis, Tunisia
adel.abidi8@gmail.com
[2] Preparatory Institute of Engineering, Mrezka Nabeul (IPEIN), Nabeul, Tunisia
[3] Laboratoire de Mécanique Appliquée des Nouveaux Matériaux (LMANM), Université 8 Mai 1945, Guelma, Algeria

Abstract. In the present work, the quasi-static tensile tests of woven carbon/epoxy (CFRP) laminates used for aeronautic structures were carried out using a tensile testing machine with a high performance. These tests were performed according to different specimen standards ASTM D5766 and ASTM D3039. This study aims to focus on the effect of the cutting process through the abrasive water jet (AWJ) on the tensile strength of carbon CFRP composite material. The laminate was tested with different orientation angles of plies that vary from 0 to 90° giving different composite stacking configurations. On another hand, the effect of stress concentration was verified with testing the specimens through open hole tensile (OHT). Test results show that the tensile behavior represents high strength and stiffness for different stacking configurations. This evaluation confirmed the quasi-isotropic behavior of stacking sequence configurations. However, the composite strength is affected significantly by stress concentration generated on the hole surface.

Keywords: CFRP · Abrasive water jet · Open hole tensile · Cutting · Carbon fabric

1 Introduction

Manufacturing composite parts needs after molding process machining operations such as cutting. Among the most used cutting processes, in the way to reduce the machining holes defects in composite parts, an advanced machining process like the Abrasive water jet (AWJ) which is widely used for cutting composite materials thanks to many advantages such as avoiding the thermal heat during the conventional cutting process. Saleem et al. (2013) examined the impact of various machining techniques on the surface quality of CFRP laminates. They noted that good hole surface quality improves the static tensile strength of laminates.

The AWJ technique parameters effect on the machined surface quality of CFRP composites were investigated (Azmir and Ahsan 2008; Ramulu et al. 2009). They observed that transverse feed rate, abrasive size and water pressure have the most impact.

M. Kharrat et al. (Eds.): ICAMEM 2019, LNME, pp. 49–56, 2020.
https://doi.org/10.1007/978-3-030-52071-7_7

Other studies (Hadded et al. 2014) interested in the cut surface quality using AWJ and conventional machining techniques and their impact on mechanical properties of carbon/epoxy laminates. But there are few studies in the literature which are interested to the fabric reinforcement CFRP composites machined by AWJ and their effects on the tensile strength. Kawai et al. (2006) examined the tensile behavior of a woven fabric of carbon/epoxy laminate. This investigation was performed with different ply orientations. They found that the stress-strain responses for orthogonal on and off-axis directions (0, 90°) were linear until damage failure. However, for off-axis tensile loading, a significant nonlinearity is observed, which is similar to the off-axis tensile behavior of unidirectional laminates (Kawai et al. 2001). Sadeghian et al. (2009) presented experimental studies about fiber orientation impact on mechanical characteristics of carbon/epoxy composites under tensile loading. It was demonstrated that the tensile behavior depended basically on the fiber orientations. The behavior in on-axis orientation was shown to be perfectly linear-elastic with brittle rupture, but in off-axis orientation, it was marked to be fully nonlinear with high ductility. Zhou et al. (2017) studied the failures of open-hole specimens with different stacking sequences under tensile loading. It is observed that stress–strain graphs of open hole tensile tests lead to determine matrix damage initiation because of the variation of orthotropic ratio. Toubal et al. (2005) examined the stress concentration factor caused by a circular hole in CFRP test coupons on mechanical strength. They checked that the stress concentration located near to the hole had the maximum level and it decreased as the distance from hole edge increased. This observation confirms that the elevation of tensile strength for open holes plate.

The aim of this experimentally work is to investigate the influence of AWJ machining process on the tensile static strength of woven CFRP composite plates. On the other side, the measuring of strength will be evaluated as a function of stacking orientation plies. Then this kind of tests will be investigated for two different geometries of specimens depending on the existence of holes.

2 Experimental Work

2.1 Material and Specimens

The material used in this study is carbon fabric reinforced epoxy. Eight layers of the material are used with fiber volume fraction was about 65%. The laminate was prepared through hand layup and vacuum bag process. The initial curing of the laminate was carried out at one bar ambient pressure. After curing, the laminate had a nominal thickness of 4.8 mm. The layers are orientated in two directions alternatively $[\pm 45, 0/90, \pm 45, 0/90]_{2S}$, as shown in Fig. 1. Four kinds of coupon specimens with different fiber orientations (0, 30, 45, and 90°) were cut from $500 \times 500\ mm^2$ laminate panels. The dimensions of the specimens are based on the testing standards ASTM D3039 (2004) as shown in (Fig. 2A).

The geometry for the open hole tensile (OHT) specimens is presented in Fig. 2B. The test open hole specimens had a ratio of hole size 8 mm with respect to specimen width (25 mm) equal to (0.32). Based on ASTM D5766 (2002) the test procedure and specimen geometries were prepared Fig. 3. The laminate stacking sequences were quasi-isotropic; four plies oriented at (±45°) and four orientated at (0/90°).

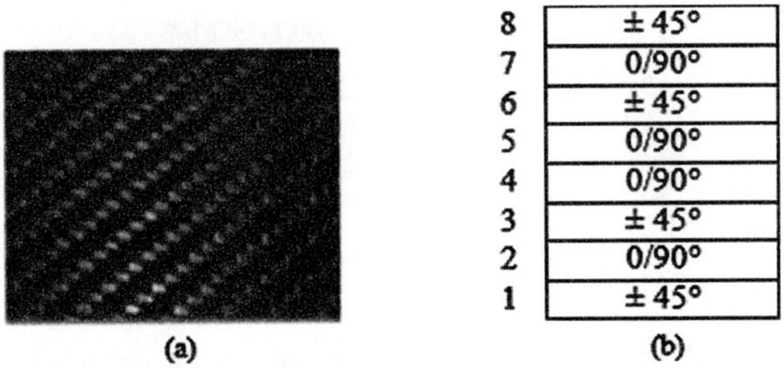

Fig. 1. (a) 2 × 2 Twill weave CFRP (600 g/m^2), (b) CFRP laminate layup

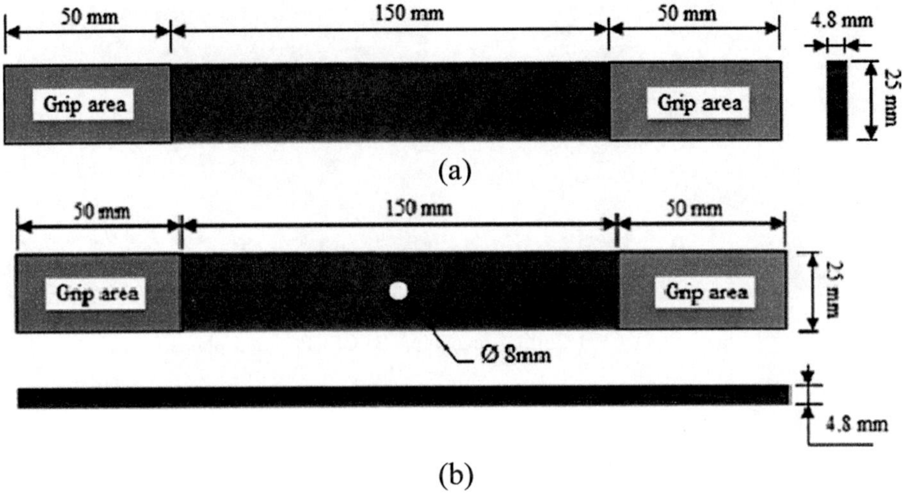

Fig. 2. Typical geometry of coupon specimen: (a) ASTM D3039 and (b) ASTM D5766

2.2 Experimental Set-up

The cutting of experimental specimens was conducted on a numerically controlled three-axis machine (MECANUMERIC model) abrasive water jet machine (Fig. 3A). It is equipped with a system with an ultra-high-pressure pump capable of providing maximum water pressure of 350 MPa and maximum abrasive flow 1 kg/min and maximum traverse speed 30.000 mm/min. Cutting was performed on CFRP plate with thicknesses 4 mm at different orientations (Fig. 3B).

Preliminary tests were conducted in order to use the best AWJ cutting conditions of traverse speed, stand of-distance, abrasive mass flow. These parameters are required to obtain a good hole surface roughness quality. Various combinations of these cutting parameters were tested using the mentioned machine above. The best cutting parameters selected to cut the specimens for the tensile test are shown in next Table 1.

Table 1. Abrasive water jet parameters

Parameters	Code	Level
Traverse speed	V (mm/min)	2000
Abrasive mass flow rate	Q (g/min)	400
Stand off-distance	S (mm)	2

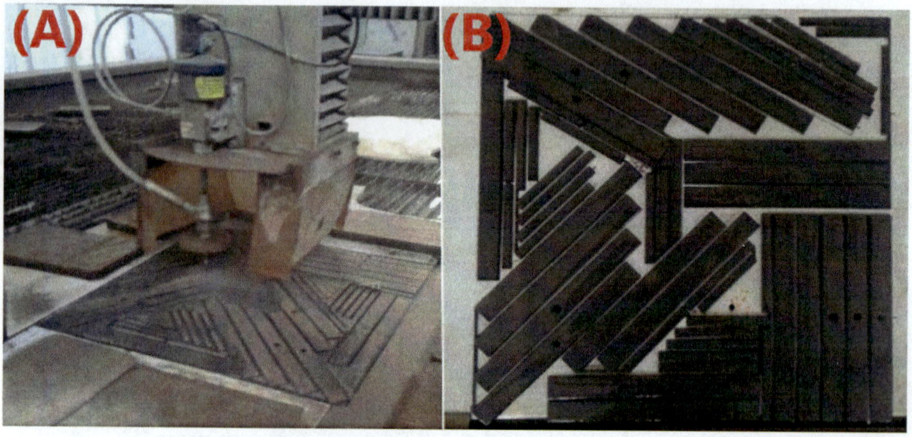

Fig. 3. (A) AWJ Cutting machine, (B) CFRP cutted plates.

2.3 Testing Procedure

The tensile trials were realized at ambient temperature using a tensile machine with high performance. The load was applied using an hydraulic universal testing machine (Walter and Bai) equipped with a load cell of 100 kN was used to determine the applied load at a constant speed rate of 2 mm/min to each sample. The tensile data were obtained and analyzed from the machine using DION 7 Pro software. For every testing configuration, many samples were performed at least three results were taken in consideration from specimens of same stacking configuration, in which have the approximately the same curves. Yield strength, apparent modulus of elasticity, stress and strain rates was determined. Three coupons of each orientation laminate stacking sequence were tested to failure to determine the average tensile strength. The deformation of tested specimens to failure was measured using two axial extensometers, one for the gauge length and the other for the transversal range as shown in Fig. 4. To measure the longitudinal and transverse displacement in the most solicited zone around the hole, the extensometers still attached to laminate in two sides of specimens until the final failure. Each extensometer had a gauge equal to 25 mm and was placed on the specimen so that the hole was in the middle of the gauge length. The use of hydraulically grips with specially faces for gripping the composite materials ensured that all laminates failed in the medium length of tested specimen and in the gauge length.

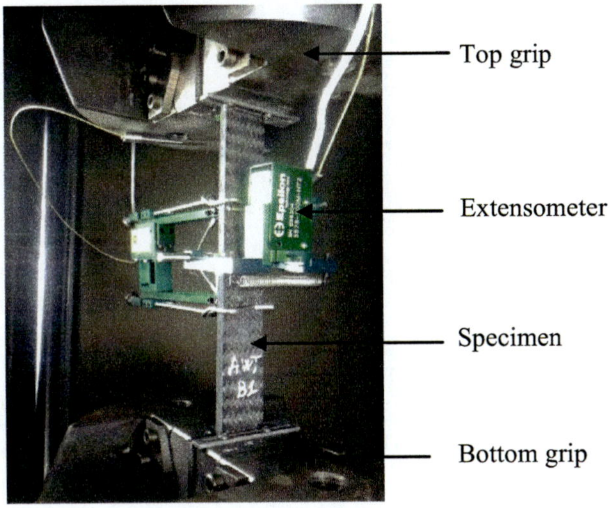

Fig. 4. Tensile specimen test set-up

3 Experimental Results and Discussion

The tensile stress-strain responses of the woven carbon fabric composite specimens for different orientations cut at room temperature are illustrated in Fig. 5. The stress-strain results for all directions are linear and smooth up to failure with the same stiffness level. These results confirm the quasi-isotropic material regardless of the cutting directions for coupons. Therefore, the woven fabric strength is equilibrated between warp and weft direction which is rarely found in previous works thanks to the symmetric plies orientation for both fiber angles ±45° and 0/90°.

For off-axis tensile loading (θ = 30, 45°), the stress-strain responses are linear and give the same relationship as the on-axis tensile which is be different to many results (Kawai and Taniguchi 2006) which found a significant nonlinearity behavior accompanied by a reduction in the strength. Consequently, the laminates stacking configurations formed by the alternative orientation plies for (θ = ±45°) and (θ = 0/90°) permit to avoid this non-linearity and improve the composite material strength. For the ultimate strain the good quality of cutting specimens by the AWJ process such as the hole circularity with a low level of surface roughness help to obtain the same stiffness level comparing to other cutting processes like drilling.

For the ultimate strain values until failure, the orientation (θ = 30°) had the least level less than 0.09 and reduced ultimate stress up to 280 MPa while the other orientations had the similar of ultimate strain level about 1.1% with maximum stress equal to around 350 MPa.

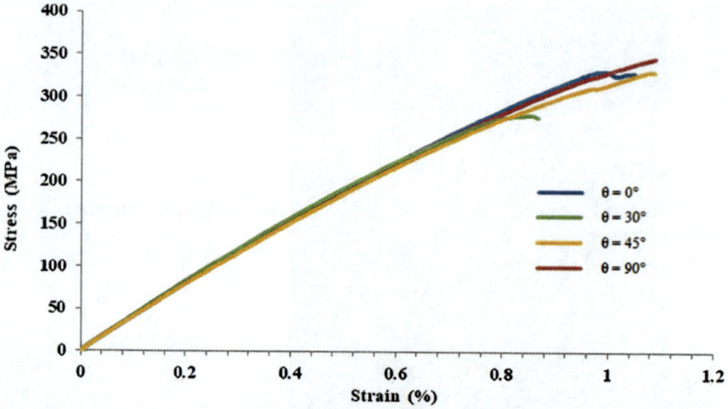

Fig. 5. Open hole tensile (OHT) test for different cutting directions with AWJ

For stress-strain tensile curves without a hole (Fig. 6), the ultimate stress is similar to OHT behavior which for orientations (θ = 0, 45, 90°) which attain a maximum value about 350 MPa. But in this case, the failure was not brittle such as the OHT. The responses had non-linearity behavior started at 250 MPA with 1.2% of strain. This non-linearity is explained by the predominant supporting of the off-axis tensile behavior which leads to torsion solicitation (Fig. 7).

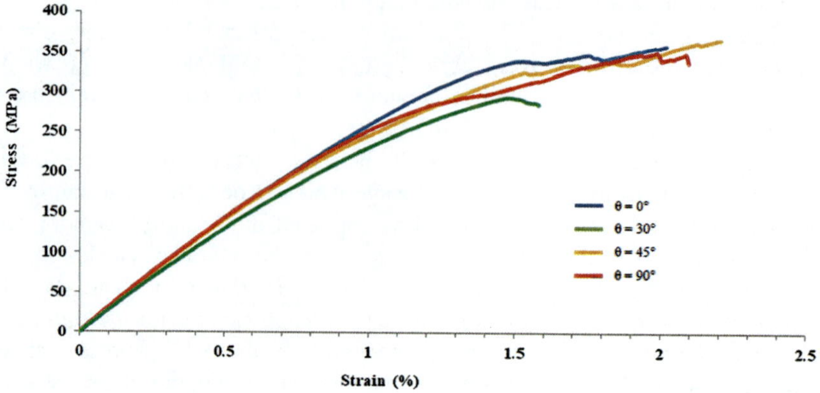

Fig. 6. Tensile test for AWJ cutting (without hole).

Fig. 7. Specimen deformation after tensile loading for the orientation (θ = 90°).

The experimental stress-strain responses (Fig. 8) demonstrate for CFRP without hole (unnotched) and OHT specimens for each of orientation angles. It is clear to conclude from these results that both specimen geometries had the same tensile strength value unless the coupons oriented at 30° had lower ultimate tensile stress up to 300 MPa. Furthermore, the OHT specimens have significantly higher stiffness than unnotched specimens which can be explained by the existence of high residual stress concentration at the hole surface according to many authors (Toubal et al. 2005) confirm these phenomena. However, the unnotched CFRP specimens have a greater ultimate strain which helps to absorb more strain energy.

In the case of coupons geometries without holes, the residual stress is eliminated and the deformation is distributed on all loaded section which larger than the OHT tests.

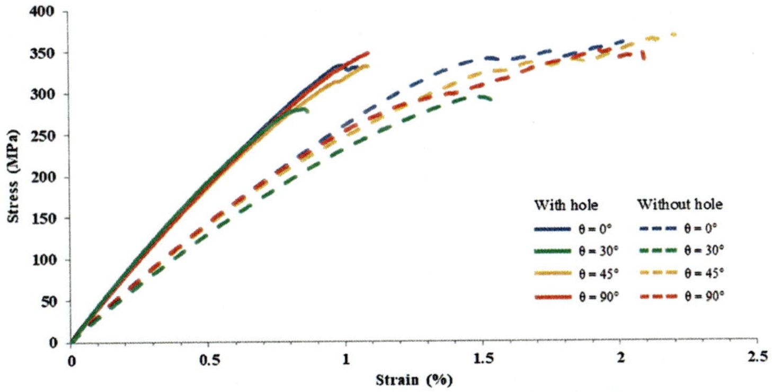

Fig. 8. Hole effect on CFRP tensile behavior.

4 Conclusion

This work presents experimental results of drilled holes composite specimens with abrasive water jet processes with quasi-isotropic stacking sequence made from 2/2 twill woven fabric carbon/epoxy composite with an open hole and solicited under tensile loading. Based on this experimental analysis, the following conclusions were concluded:

- The composite CFRP stacking has a quasi-isotropic behavior which confirmed the same tensile strength and stiffness for all cutting orientations. For the strain variation of OHT with respect to different orientations is about 1.1% for all orientation unless the specimens orientation of $\theta = 30°$ which has less deformation up to 0.85%.
- The tensile strength of specimens without holes is lower than the OHT strength. Consequently, the composite material became less rigid which confirms the influence of residual stress generated by hole. While the loading test, the failure regression was propagated along the off-axis besides the small torsion solicitation of coupons around the tensile loading direction. Then, the decreasing of the tensile strength of samples without hole lead to the elevation of ultimate strain values for all orientations up to 2%.

Further work is in a way to be developed in order to study the performance of structural assemblies made from composite plates with drilled holes with different processes and loaded under different kinds of solicitation such as bending.

References

Saleem, M., Toubal, L., Zitoune, R., Bougherara, H.: Investigating the effect of machining processes on the mechanical behavior of composite plates with circular holes. Compos. A **55**, 169–177 (2013)

Azmir, M.A., Ahsan, A.K.: Investigation on glass/epoxy composite surfaces machined by abrasive water jet machining. J. Mater. Process. Technol. **198**, 122–128 (2008)

Ramulu, M., Hwang, I., Isvilanonda, V.: Quality issues associated with abrasive water jet cutting and drilling of advanced composites. In: American Water Jet Technology Association Conference and Expo, Houston, US, 18–20 August 2009

Haddad, M., Zitoune, R., Bougherara, H., Eyma, F., Castanie, B.: Study of trimming damages of CFRP structures in function of the machining processes and their impact on the mechanical behaviour. Compos. B **57**, 136–143 (2014)

Kawai, M., Taniguchi, T.: Off-axis fatigue behavior of plain weave carbon/epoxy fabric laminates at room and high temperatures and its mechanical modeling, Compos. Part A Appl. Sci. Manuf. **37**, 243–256 (2006)

Kawai, M., Yajima, S., Hachinohe, A., Takano, Y.: Off-axis fatigue behavior of unidirectional carbon fiber-reinforced composites at room and high temperatures. J. Compos. Mater. **35**(76), 545–876 (2001)

Sadeghian, P., Rahai, A.R., Ehsani, M.R.: Effect of fiber orientation on nonlinear behavior of CFRP composites. J. Reinf. Plast. Compos. **28**(18), 2261–2272 (2009)

Zhou, S., Sun, Y., Chen, B., Tay, T.E.: Material orthotropy effects on progressive damage analysis of open-hole composite laminates under tension. J. Reinf. Plast. Compos. **36**, 1473–1486 (2017)

Toubal, L., Karama, M., Lorrain, B.: Stress concentration in a circular hole in composite plate. Compos. Struct. **68**, 31–36 (2005)

ASTM D5766. Standard test method for open hole tensile strength of polymer matrix composite laminates. Annual book of ASTM standards, 15 March 2002

ASTM D3039. Standard test method for open hole tensile strength of polymer matrix composite laminates. Annual book of ASTM standards (2004)

Modeling of Heat Transfer and Transport Phenomena During Laser Welding Of Aluminum/Magnesium Alloys

S. Ben Halim[1(✉)], S. Bannour[1], K. Abderrazek[1], W. Kriaa[1], and M. Autric[2]

[1] Laboratoire de Thermique et Thermodynamique des Procédés Industriels, Ecole Nationale d'Ingénieurs, Monastir, Tunisia
sabrinebenhalim8@gmail.com

[2] Aix-Marseille Univ, CNRS, IUSTI, Marseille, France
michel.autric@univ-amu.fr

Abstract. The weld geometry is an important factor which determines the quality of the weld. This work proposes a transitional and three-dimensional numerical model performed by the hydrodynamic software Ansys-Fluent. We are interested in the heterogeneous assembly of Al alloy and Mg alloy by laser beam welding. Aluminum and Magnesium plates are joined in butt configuration. The objective of this work is to simulate the temperature field, velocity field, fluid flow during the interaction between the laser beam and the material and to predict the dimensional characteristics of the laser welded joint. This makes it possible to better understand different physical phenomena (e.g. heat transfer, mass transport) generated in the molten pool. The results show that the simulated weld bead profile is in good agreement with those obtained by experiment.

Keywords: Fluid flow · Ansys · Thermal transfer · Diffusion · Dissimilar metals

1 Introduction

Laser welding is an assembly technique widely used in various sectors such as the aerospace, automotive and petroleum industries. Indeed, this technique has many advantages such as high speed welding, deep penetration, small heat affected zone and good mechanical performance of the weld. Despite the large amount of research on laser welding of similar materials (Cao et al. 2006), studies on heterogeneous assembly are still limited (Phanikumar et al. 2004 and Babalová et al. 2014). However, welding of dissimilar materials is now a very interesting field of research. Among the work done on laser welding of dissimilar materials is the parametric study of laser weldability of low carbon steel and stainless steel performed by Esfahani et al. (2015) who have tested different specific point energy (10 J, 17 J and 50 J). In this study, authors found that the high energy laser makes the convective motions more important in the fusion zone. This greatly affects the dilution of the weld bead and the homogeneity of a mixture of dissimilar materials. Indeed, specific energies of 17 J

M. Kharrat et al. (Eds.): ICAMEM 2019, LNME, pp. 57–62, 2021.
https://doi.org/10.1007/978-3-030-52071-7_8

and 50 J result in a higher temperature gradient and consequently, the speed of the fluid flow becomes faster, which makes it possible to obtain an intense mixture of chemical elements of alloys. Moharana et al. (2016) studied experimentally the mechanical and metallurgical properties of stainless steel dissimilar welded AISI 304 and pure copper. A homogeneous weld is obtained by CO_2 laser welding with a tensile strength of 109 MPa. A microstructure analysis revealed the macro-segregation of phases rich in Cu, Fe and Cr in different molten zones and the formation of the heat-affected zone of thickness of 10 μm and which becomes narrower from the side of stainless steel that on the copper side.

For numerical resolution, we used the software 'Ansys' which is based on the finite-volume method. Several phenomena such as convection, heat transfer and diffusive flux are involved in this study. The present numerical study aims to reproduce the development of melted zone between dissimilar materials during continous laser welding mode.

2 Experiences

In our experimental study (Bannour et al. 2014), base materials used are (A5754) Aluminum alloy and (AZ31) Magnesium alloy whose chemical composition and physical properties are indicated in Tables 1 and 2, respectively. The experiments were carried out with a continuous ND: YAG laser with a wavelength of 1030 nm. The radius of the laser beam is 0.45 mm. To protect the quality of the weld, an Argon shielding gas with a flow rate of 20 l/min was used. The welding experiments were carried out with an incident laser power of 2500 W and a welding speed of 5 m/min.

Table 1. Chemical composition of base materials.

Element	Mg	Mn	Si	Fe	Cu	Cr	Zn	Ti	Al	Ca
Aluminum	2.6–3.6	<0.5	<0.4	<0.4	0.14	<0.3	<0.2	<0.15	Bal.	–
Magnesium	Bal.	0.17	≤0.1	≤0.005	≤0.05	–	0.6–1.4	–	2.5–3.5	≤0.04

Table 2. Properties of materials used for numerical simulation.

Physical Property	Aluminum	Magnesium	Symbol/unity
Density of solid phase	2880	1810	ρ [kg/m^3]
Thermal conductivity of solid phase	202	146	k [W/(m·K)]
Specific heat of solid phase	728	1020	c_p [J/(Kg·K)]
Melting heat	397 × 10^3	370 × 10^3	10^3 L_m [J/Kg]
Solidus temperature	880	740	$T_s[k]$
Liquidus temperature	922	870	T_l $[k]$

3 Description of Numerical Procedure

The sheets of Magnesium and Aluminum alloys were joined with butt configuration. The dimensions of these two pieces are $20 \times 10 \times 2$ mm^3. The schematic model of laser welding is presented in Fig. 1. The mesh adopted is the same for each piece. The mesh contains 260 000 elements. In this model, the following assumptions have been adopted:

- The surface of the weld pool is flat.
- The workpiece moves in the y direction with a constant welding speed.
- The fluid flow in the molten pool is an incompressible and laminar flow.
- A Gaussian heat flux is applied on the upper surface of the sheets.

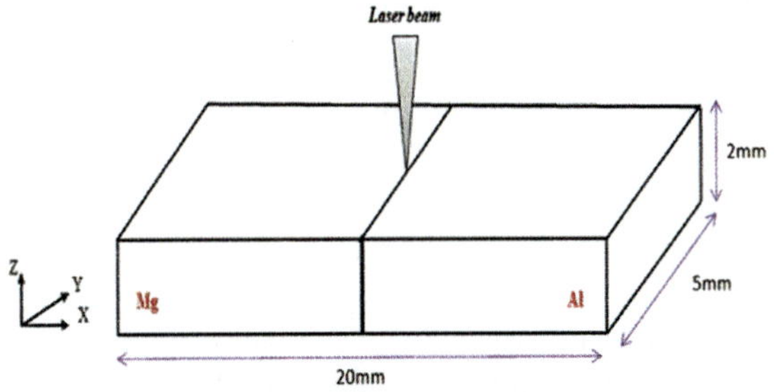

Fig. 1 Schematic representation of laser spot butt welding.

With a fixed grid system, the equations of continuity (1), momentum (2), energy (3) and species (4) can be written as follows in a Cartesian system (Xue et al. (2013)):

Equation of continuity

$$\frac{\partial(\rho)}{\partial t} + \nabla.(\rho\vec{v}) = 0 \tag{1}$$

Where $\vec{v}$ is the molten pool velocity in respective directions.

Equation of momentum

$$\frac{\partial(\rho\vec{v})}{\partial t} + \nabla.(\rho\vec{v}\vec{v}) = -\nabla p + \nabla.(\mu\nabla\vec{v}) + \rho\vec{g} + \vec{S} \tag{2}$$

Where p is the pressure, μ is the viscosity and $\vec{S}$ is the source term of momentum equation.

Equation of Energy

$$\frac{\partial(\rho H)}{\partial t} + \nabla.(\rho\vec{v}H) = \nabla.(k\nabla T) + S_E \tag{3}$$

Where H is the enthalpy of material, and S_E is the energy source term.

Equation of species

$$\frac{\partial(\rho C)}{\partial t} + \nabla.(\rho\vec{v}C) = \nabla.(\rho D \nabla C) \tag{4}$$

Where C is the concentration and D is the solutal diffusivity.

On all surfaces, heat loss due to convection and radiation was considered and given by the following expression:

$$k\frac{\partial T}{\partial \vec{n}} = q_{rad} + q_{conv} = \varepsilon\sigma(T^4 - T_0^4) - h_{conv}(T - T_0) \tag{5}$$

Where T_0 is the ambient temperature, h_{conv} is the convection coefficient, ε is the emissivity of the surface and σ is the Stefan-Boltzmann constant.

4 Results

Figure 2. shows the temperature contours at different weld zones. Note that the maximum temperature reaches 2490 K and is located at the center of laser beam. The view (x–y) shows the asymmetry of heat distribution around the laser source during dissimilar laser welding of Al and Mg alloys. The cause of this asymmetry is the difference in physical properties of Aluminum alloy and Magnesium alloy, especially the specific heat and the viscosity.

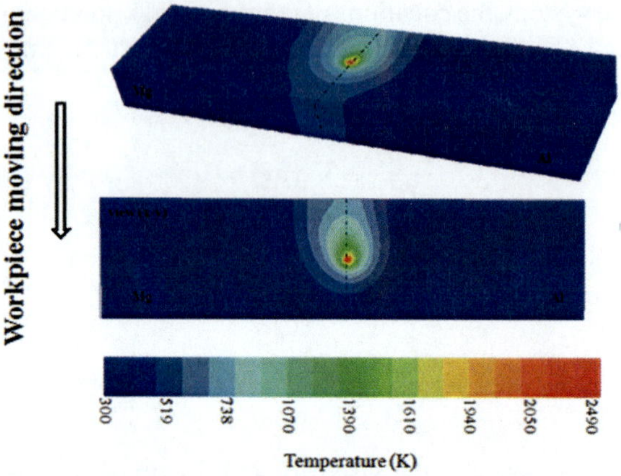

Fig. 2 Temperature contours for a laser power of 2500W and a welding speed of 5m/min

Figure 3. illustrates the top view (x–y) and the two sections (x–z) and (y–z) of the velocity field distribution during the laser welding process. It is noted that the flow velocity is higher at the surface of the weld. This increases the convective movements in the mixing zone.

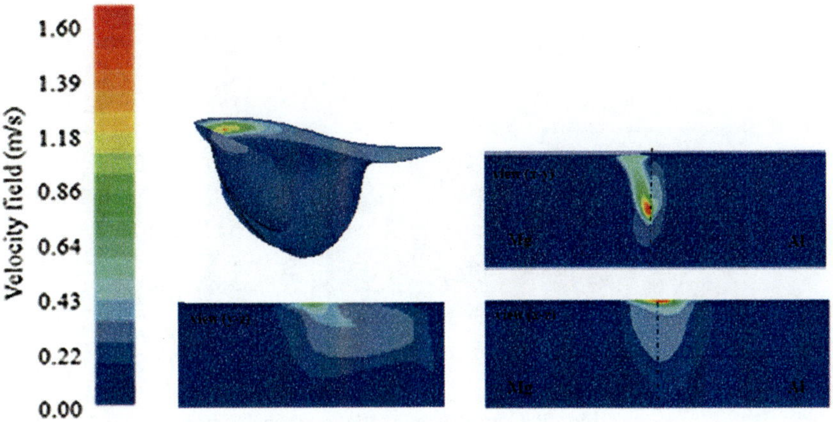

Fig. 3 Distribution of the velocity field in different planes.

Figure 4 shows that the fluid flow transports Magnesium towards Aluminum side via mass diffusion. In fact, melting at Mg side is more rapid than melting at Aluminum side. Liquid metal has a tendency to flow from Magnesium side to Aluminum side which can be explained by difference in surface tension of two liquid metals.

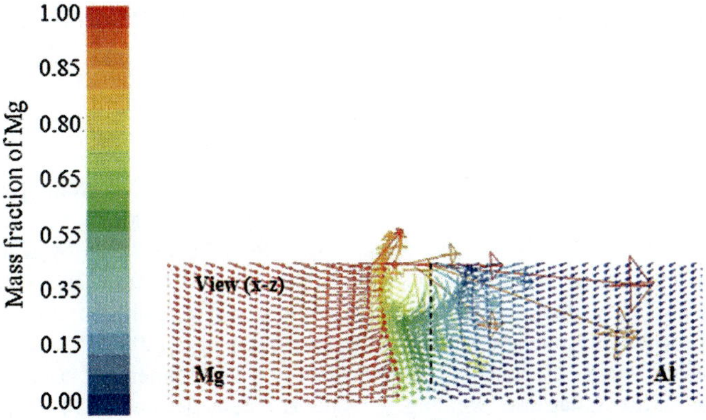

Fig. 4 Distribution of velocity vectors in the xz plane.

5 Comparison Model/Experience

To validate our model, a weld bead cross section was created to determine the dimensions of the weld formed as shown in Fig. 5. This figure shows a good qualitative agreement between simulated profile weld bead and that obtained experimentally for the same welding conditions. We found that the depth of the weld bead is obtained over the entire thickness of the piece.

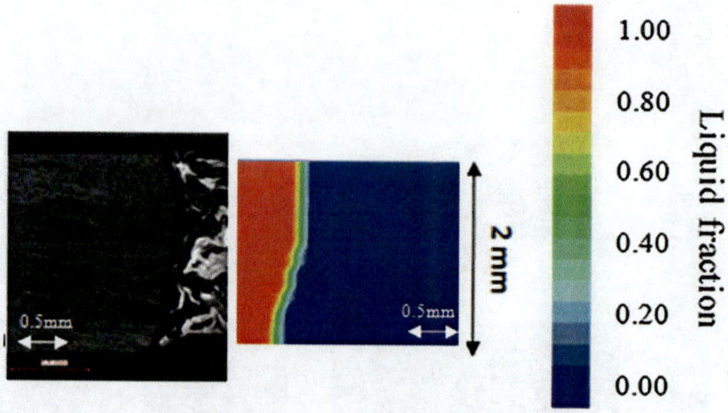

Fig. 5 Comparison Model/experience

6 Conclusion

The weldability of dissimilar Aluminum- Magnesium alloys by laser beam was numerically studied and validated with experimental results. The developed model showed that the welding of these materials causes the heat distribution asymmetry, the generation of convective movements in the molten pool during laser beam-material interaction and the mass transport of molten materials. The emergence of these physical phenomena is mainly due to the difference in physical properties of metals.

References

Cao, X., Jahazi, M., Immarigeon, J.P., Wallace, W.: A review of laser welding techniques for magnesium alloys. J. Mater. Process. Technol. **171**, 188–204 (2006)

Babalová, E., Behúlová, M.: Numerical simulation of temperature fields by welding of Ti-Al alloys applying volumetric heat source. Adv. Mater. Res. **887**(888), 1280–1283 (2014)

Phanikumar, G., Chattopadhyay, K., Dutta, P.: Computational modeling of laser welding of Cu-Ni dissimilar couple. Metall. Mater. Trans. B **35**(2), 339–350 (2004)

Nekouie Esfahani, M.R., Coupland, J., Marimuthu, S.: Numerical simulation of alloy composition in dissimilar laser welding. J. Mater. Process. Technol. **224**, 135–142 (2015)

Bannour, S., Autric, M., Masse, J.E., Matteï, S., Mhiri, H.: Experimental investigation on mechanical and microstructural properties of AISI 304 to Cu joints by CO2 laser. Eng. Sci. Technol. Int. J. **19**, 684–690 (2016)

Bannour, S., Autric, M., Masse, J.E., Matteï, S. Mhiri, H.: Microstructure and properties of welds between 5754 Al alloys and AZ31 Mg alloys using a Yb:YAG laser. In: 20th International Symposium on High Power Laser Systems& Applications, Chengdu, Sichuan Province, China (2014)

Xue, Z., Hu, S., Zuo, D., Cai, W., Lee, D., Elijah Jr., K.A.: Molten pool characterization of laser lap welded copper and aluminum. J. Phys. D Appl. Phys. **46**, 495501 (2013)

Optimal Planning of Multi-pass Turning Operations

Toufik Ameur(✉)

Mechanical Engineering Department, Faculty of Applied Sciences,
University Kasdi Merbah, Ouargla, Algeria
ameurt@hotmail.com

Abstract. This study presents a decision-making methodology for multi-pass turning operations that integrates tool life and cutting conditions optimization. In this approach, the initial machining plan considers all possible roughing passes and a finishing pass. At the end of the computation process, the existence of a roughing pass in the final program of the operation is evaluated using a combinatorial variable associated to this pass. On the other hand, the tool replacement is considered in the optimal selection of the cutting conditions so that this tool can achieve an exact number of operations. However, the developed optimization model which minimizes production cost is able to predict the number of required passes, the optimal cutting conditions for each pass, and the number of parts per tool. Finally, an example is presented to discuss the influence of the integration of combinatorial variables on the planning of operations. For several different total depths of cut, the obtained results show the capacity to produce a complete machining program in which even the tool replacement moment is clearly determined.

Keywords: Multi-pass turning operations · Optimization · Planning

1 Introduction

The objective of the planning of machining operations is to provide the necessary information for the achievement of the cutting program that includes all the technical details. This preparatory phase guarantees, normally, the security of the machining process and the obtaining of the best techno-economic results.

In turning, the cutting operations aim at removing a quantity of material measured by the total depth of cut. These operations are often completed in several passes because the depth of cut, for each pass, is limited by the technical characteristics of the cutting tool. Indeed, the implementation plan must determine the number of passes required and the cutting conditions related to each one.

At the beginning, the researchers presented a simple reflex based on the resolution of this problem in two stages (Agapiou 1992a; Shin and Joo 1992; Guptra et al. 1994; Al-Ahmari 2001; Djenane et al. 2016). The first is dedicated to selecting optimal cutting conditions for all possible depths of cut. These results are used in the second step to determine the optimal number of passes.

M. Kharrat et al. (Eds.): ICAMEM 2019, LNME, pp. 63–69, 2020.
https://doi.org/10.1007/978-3-030-52071-7_9

Thereafter, they progress to solve the problem in a single step. In this sense, a large number of researches assume that the achievement of a turning operation is completed through a number of identical roughing passes and a finishing pass (Lee and Tarng 2000; Belloufi et al. 2012; Aryanfar and Solimanpur 2012; Yildiz 2013). This approach has made it possible to reduce considerably the size of the problem by reducing the number of variables. A second part of researches propose to solve the problem in its original form but, it proceeds to the reduction of the size of the problem either by considering the small total depths of cut (Belloufi et al. 2012; Aryanfar and Solimanpur 2012; Jabri et al. 2017) which results automatically in a reduced number of roughing passes or by the direct limitation of this number (Liang et al. 2001; Mgwatu 2013).

On the other hand, we find that the economic costs associated to the cutting tool are often distributed over the passes according to their consumption. This assumes that the tool is used to its wear limit and is replaced regardless of its location. Knowing that it is impossible to replace a tool when engaged in the piece, the proposed hypothesis is less realistic (Liang et al. 2001). Therefore, it is normal to look for a planning method that determines exactly the tool replacement time. This method will certainly lead to a more realistic machining parameters prediction.

The present study proposes an optimal planning approach for multi-pass turning operations minimizing the production cost based not only on determining the number of passes required and optimal cutting conditions, but also on determining the number of parts per tool.

2 Problem Formulation

The proposed approach assumes a cutting process with all possible roughing passes and only one finishing pass.

Each roughing pass is presented by a number p_i having two possible values 0 or 1 and three variables representing the cutting conditions; Cutting speeds V_r (m/min), feed rate f_r (mm/rev), and depth of cut d_r (mm) related to this pass (V_{ri}, f_{ri}, d_{ri}). If p_i equal to 1 the pass exists (Deb 2001). The finishing pass is presented only by the three variables representing the cutting conditions; Cutting speeds V_s (m/min), feed rate f_s (mm/rev), and depth of cut d_s (mm).

The objective function representing the unit production cost can be given by (Ameur and Assas 2012).

$$Cm = C_O\left[t_h + \left(tm_s\left(1+\frac{t_{cs}}{T_s}\right)+t_r\right) + \sum\nolimits_{i=1}^{n} p_i\left(tm_{ri}\left(1+\frac{t_{cs}}{T_{ri}}\right)+t_r\right)\right] + C_t\left[\frac{tm_s}{T_s} + \sum\nolimits_{i=1}^{n} p_i\frac{tm_{ri}}{T_{ri}}\right] \tag{1}$$

Where C_O, C_t are the machining cost ($/min) and the tool edge cost ($), respectively. t_h, t_r, t_{cs} are the handling time (min/piece), the rapid traversing time (min), and the tool replacement time (min/edge), respectively. n is the number of roughing passes.

The cutting time for roughing passes and finishing pass is given by (Agapiou 1992b).

$$\begin{cases} tm_{ri} = \frac{\pi D_i L}{1000 V_{ri} f_{ri}}; D_i = D - 2\sum_{k=1}^{i-1} d_{rk} \\ tm_s = \frac{\pi D_s L}{1000 V_s f_s}; D_s = D - 2\sum_{i=1}^{n} d_{ri} \end{cases}$$

L and D are the length and the initial diameter of workpiece (mm), respectively. D_i, D_s are the diameters of workpiece in the ith roughing and the finishing pass (mm), respectively.

The tool life in the roughing passes and the finishing pass is given by (Agapiou 1992b).

$$\begin{cases} T_{ri} = \frac{C^{1/\alpha}}{V_{ri}^{1/\alpha} f_{ri}^{\beta/\alpha} d_{ri}^{\gamma/\alpha}} \\ T_s = \frac{C^{1/\alpha}}{V_s^{1/\alpha} f_s^{\beta/\alpha} d_s^{\gamma/\alpha}} \end{cases}$$

C is constant and α, β, γ are exponents.

From previous experience on the machine-tool-workpiece system, physical limitations on cutting conditions should be identified (Shin and Joo 1992):

For the variables, there are feasible ranges of values one can choose for the problem. This type of constraints is a simple lower and upper boundary.

$$\begin{cases} 0 \le p_i \le 1, V_{min} \le V_{ri} \le V_{max}, f_{min} \le f_{ri} \le f_{max}, d_{rmin} \le d_{ri} \le d_{rmax} \\ V_{min} \le V_s \le V_{max}, f_{min} \le f_s \le f_{max}, d_{smin} \le d_s \le d_{smax} \end{cases} \quad (2)$$

Cutting force in roughing and finishing passes is limited by the resistance of the holding device, the limit deflection of workpiece, and the limit of chatter.

$$\begin{cases} F_{ri} = k_1 f_{ri}^{\mu} d_{ri}^{\nu} \le F_{max} \\ F_s = k_1 f_s^{\mu} d_s^{\nu} \le F_{max} \end{cases} \quad (3)$$

F_{max} is the allowable cutting force (N). k_1 is constant and μ, ν are exponents.

Cutting power in roughing and finishing passes is limited by available power of machine-tool P_{max} (kW). Power consumption can be obtained by dividing the product of cutting force and the cutting speed by the efficiency of the machine tool.

$$\begin{cases} P_{ri} = \frac{F_{ri} V_{ri}}{60000\eta} = \frac{k_1 f_{ri}^{\mu} d_{ri}^{\nu} V_{ri}}{60000\eta} \le P_{max} \\ P_s = \frac{F_s V_s}{60000\eta} = \frac{k_1 f_s^{\mu} d_s^{\nu} V_s}{60000\eta} \le P_{max} \end{cases} \quad (4)$$

Surface roughness is one of the most important factors of machining quality. It is generally affected by various parameters, but the most dominant effects are those of feed and nose radius r_e (mm). For roughing passes and finishing pass, this surface quality is limited.

$$\begin{cases} R_{ri} = \frac{32.1 f_{ri}^2}{r_e} \leq R_{rmax} \\ R_s = \frac{32.1 f_s^2}{r_e} \leq R_{smax} \end{cases} \tag{5}$$

R_{rmax} and R_{smax} are the surface roughness requirements in roughing and finishing pass (µm), respectively.

The total depth of cut d_t (mm) should be equal to the sum of all roughing and finishing depths.

$$d_t = d_s + \sum_{i=1}^{n} p_i d_{ri} \tag{6}$$

For organizational reasons, the tool can be replaced only at the end of an operation. Indeed, the number of parts per tool, given by the inverse of total used tool life, must be exact. Therefore, this equality must be satisfied.

$$\sin\left(\frac{\pi}{\frac{tm_s}{T_s} + \sum_{i=1}^{n} p_i \frac{tm_{ri}}{T_{ri}}}\right) = 0 \tag{7}$$

The maximum number of possible roughing passes can be expressed by:

$$n = \frac{(d_t - d_{smin})}{d_{rmin}} \tag{8}$$

3 Application Example

The used multi-pass turning example is given by Shin and Joo (1992). Numerical data related to the technological parameters of the operation are shown in Table 1.

Table 1. Numerical data for machining example.

Parameter	Value	Parameter	Value	Parameter	Value
L, D (mm)	303, 50	V_{min}, V_{max} (m/min)	50, 500	R_{smax}, R_{rmax} (µm)	2.5, 25
r_e (mm)	1.2	f_{min}, f_{max} (mm/rev)	0.1, 0.9	F_{max} (N)	1960
C	227	d_{rmin}, d_{rmax} (mm)	1, 4	η, P_{max} (kW)	0.85, 5
α, β, γ	0.2, 0.35, 0.15	d_{smin}, d_{smax} (mm)	0.5, 2	t_h (min/piece)	0.75
k_1	1058	C_0 (\$/min)	0.5	t_{cs} (min/edge)	1.5
μ, ν	0.75, 0.95	C_t (\$/edge)	2.5	t_r (min)	0.5

Here we are faced with a nonlinear problem having a set of equality and inequality constraints that can be highly non-linear. Also, the presence of large number of variables creates additional difficulties. In this case, the choice of a stochastic method becomes more than a necessity. Indeed, the selection of the particle swarm optimization (PSO) method is made for two reasons. The first is the direct use of the objective function instead of additional information such as gradients. The second is the exclusive requirement of simple mathematical operators (Ameur and Assas 2012).

Similar to the other evolutionary computation methods, the PSO is based on the research by group where individuals are called particles and the group of individuals is called swarm.

In this algorithm, the ith particle (individual) of the swarm is characterized by two parameters:

The current position in the hyperspace of research: $X_i^k = \left(x_1^k, x_2^k, \ldots\ldots, x_m^k\right)$

The current velocity: $V_i^k = \left(v_1^k, v_2^k, \ldots\ldots, v_m^k\right)$

The PSO search process is based on the assumption of social sharing of information where individuals pursue their research using shared information about the best solution found by the group (Ameur et al. 2007). In this way, the new position of the particle is determined according to three trends; propensity to follow his own way, his tendency to return to his best position (*Pbest$_i$*), and the tendency to move towards the best position found by the whole group (*Gbest*).

The computation of the new velocity for each particle is based on its previous velocity, the particle's location at which the best fitness has been achieved (*Pbest$_i$*) so far, and the best particle among the group (*Gbest*) at which the best evaluation has been achieved so far (Ameur et al. 2007).

$$V_i^{k+1} = w^k V_i^k + c_1 r_1 \left(Pbest_i - X_i^k\right) + c_2 r_2 \left(Gbest^k - X_i^k\right) \tag{9}$$

With $w^k = w_{max} - \left(\frac{w_{max} - w_{min}}{Nite}\right) k$

Where c_1, c_2 are the acceleration constants, r_1, r_2 are random constants, and w_{max}, w_{min} are the maximum and minimum inertia weight, respectively.

The update of each particle's position in the solution hyperspace is based on its previous position and its new velocity.

$$X_i^{k+1} = X_i^k + V_i^{k+1} \tag{10}$$

The application of PSO's algorithm includes the following steps (Ameur et al. 2007):

1. Initialize an array of m particles with random positions and velocities. This constitutes the initial swarm.
2. Evaluate each particle of the swarm by calculate the objective function in the optimization variables.
3. Update the best personal position $Pbest_i$ for each particle of the swarm by comparing the current evaluation with the best position found by the same particle ($Pbest_i$). If the current value is less than the value of $Pbest_i$, the particle has achieved a new best personal position. Then, the value of $Pbest_i$ will be equal to the current value and the position of $Pbest_i$ will be equal to the current position.

4. Update the best global position Gbest for the swarm by comparing the evaluation of the current particle with the best solution found by the swarm (Gbest). If the current value is less than the value of Gbest, the swarm has achieved a new best solution. Then, the value of Gbest will be equal to the current value and the position of Gbest will be equal to the current position of this particle.
5. Establish the evolution by computing the velocity and position for each particle by using Eqs. 9 and 10.
6. If the maximum iterations are not satisfied go to Step 2.

The swarm size and the maximum number of iterations were selected as 100 and 500, respectively. For the evolution, the following parameters were established: $c_1 = c_2 = 1$, $r_1 = r_2 = 0.2$, $w_{min} = 0.4$, $w_{max} = 0.9$.

The computation results for different total depths of cut (from 6 to 12 mm) are presented in Table 2. These results show that for each depth of cut, the tool replacement time can be exactly determined. This time is presented by the number of parts per tool. Also, we find that each depth of cut requires a different number of roughing passes. Then we note that it is possible to have for different total depths of cut the same number of roughing passes or parts per tool but totally different production costs. This is verified by the decomposition of the production cost in machine cost and tool cost.

Table 2. Simulation results.

d_t (mm)	Roughing passes				Finishing pass			Cu ($)	Parts/tool
	n	d_r (mm)	V_r (m/min)	f_r (mm/rev)	d_s (mm)	V_s (m/min)	f_s (mm/rev)		
6	2	2.37	133.17	0.5402	1.43	205.57	0.1919	2.1857	11
		2.18	118.31	0.6110					
7	2	3.03	147.19	0.3627	1.13	183.24	0.2755	2.2202	11
		2.85	127.16	0.4687					
8	3	3.08	131.23	0.4636	0.56	235.27	0.1538	2.6907	09
		2.57	138.79	0.5574					
		1.77	123.45	0.8311					
9	3	3.09	133.46	0.3555	1.00	226.10	0.2442	2.9144	07
		2.93	142.88	0.4444					
		1.98	152.53	0.4186					
10	3	3.25	140.50	0.4878	1.57	207.62	0.1947	3.0266	05
		2.86	163.93	0.4659					
		2.30	190.15	0.2770					
11	4	2.70	144.15	0.3968	1.35	187.78	0.2233	3.5408	05
		2.44	144.14	0.5195					
		2.34	167.42	0.5841					
		2.15	165.98	0.2839					
12	4	3.17	162.00	0.3464	1.13	264.51	0.1259	3.6390	05
		2.82	170.24	0.3721					
		2.75	114.80	0.4375					
		2.12	143.41	0.5970					

4 Conclusion

In this work, optimal selection of cutting conditions for multi-pass turning operations problem is presented. The existence of a set of complex constraints in the developed model and the excessive number of optimization variables presented additional difficulties to the nonlinear nature of the problem. The use of particle swarm optimization algorithm has effectively overcome these difficulties. The discussion of the obtained results confirmed the effectiveness of the presented approach.

Indeed, adapting this approach to other production planning problems can be an important research topic.

References

Agapiou, J.S.: An optimization of multi-stage machining system, part I: mathematical solution part 2: the algorithm and application. Trans. ASME J. Eng. Ind. **114**, 524–538 (1992a)

Shin, Y.C., Joo, Y.S.: Optimization of machining conditions with practical constraints. Int. J. Prod. Res. **30**(12), 2907–2919 (1992)

Guptra, R., Batra, J.L., Lal, G.K.: Profit rate maximization in multi-pass turning with constraints: a geometric programming approach. Int. J. Prod. Res. **32**(7), 1557–1569 (1994)

Al-Ahmari, A.M.A.: Mathematical model for determining machining parameters in multi-pass turning operations with constraints. Int. J. Prod. Res. **39**(15), 3367–3376 (2001)

Djenane, M., Djari, D., Benbouta, R., Assas, M.: Multi-pass optimization of cutting conditions by using the genetic algorithms. Res. J. Appl. Sci. Eng. Technol. **13**(3), 223–231 (2016)

Lee, B.Y., Tarng, Y.S.: Cutting-parameter Selection for maximizing production rate or minimizing production cost in multistage turning operations. J. Mater. Process. Technol. **105**, 61–66 (2000)

Belloufi, A., Assas, M., Rezgui, I.: Optimization of cutting conditions in multi-pass turning using hybrid genetic algorithm-sequential quadratic programming. J. App. Mech. Eng. **1**, 1–10 (2012)

Aryanfar, A., Solimanpur, M.: Optimization of multi-pass turning operations using genetic algorithms. In: Proceedings of the 2012 International Conference on Industrial Engineering and Operations Management, 3–6 July 2012, Istanbul, Turkey (2012)

Yildiz, A.R.: Optimization of multi-pass turning operations using hybrid teaching learning-based approach. Int. J. Adv. Manuf. Technol. **66**, 1319–1326 (2013)

Jabri, A., El Barkany, A., El Khalfi, A.: Multipass turning operation process optimization using hybrid genetic simulated annealing algorithm. Model. Simul. Eng. **2017**, 1–10 (2017)

Liang, M., Mgwatu, M., Zuo, M.: Integration of cutting parameter selection and tool adjustment decisions for multi-pass turning. Int. J. Adv. Manuf. Technol. **17**, 861–891 (2001)

Mgwatu, M.I.: Integrated approach for optimizing machining parameters, tool wear and surface quality in multi-pass turning operations. Adv. Prod. Eng. Manage. **4**, 209–218 (2013)

Ameur, T., Assas, M.: Modified PSO algorithm for multi-objective optimization of the cutting parameters. Prod. Eng. Res. Dev. **6**(6), 569–576 (2012)

Agapiou, J.S.: The optimization of machining operation based on a combined criterion, part I: the use of combined objectives in single-pass operations. Trans. ASME J. Eng. Ind. **114**, 500–507 (1992b)

Deb, K.: Applications of Multi-objective Evolutionary Algorithms in Multi-Objective Optimization Using Evolutionary Algorithms. Wiley, England (2001)

Ameur, T., Assas, M., Mazouz, H.: Optimisation des paramètres de coupe en utilisant l'algorithme à essaim de particules. Matériaux Tech. **95**, 371–380 (2007)

Ductile Damage Prediction of DD13 Sheet Material in Single Point Incremental Forming Process

A. Bouhamed[1(✉)], L. Ben Said[1,3], H. Jrad[1], M. Wali[1,2], and F. Dammak[1]

[1] Laboratory of Electromechanical Systems (LASEM), National Engineering School of Sfax, University of Sfax, Route de Soukra km 4, 3038 Sfax, Tunisia
{abir.bouhamed, fakhreddine.dammak}@enis.tn, hanen.j@gmail.com

[2] Department of Mechanical Engineering, College of Engineering, King Khalid University, Abha, Saudi Arabia
mondherwali@yahoo.fr

[3] Mechanical Engineering Department, College of Engineering, Hail University, Hail, Saudi Arabia
bensaidrmq@gmail.com

Abstract. The purpose of this study is devoted to an "advanced" coupled ductile damage model that describes the elastoplastic behavior of DD13 (Hot rolled steel) sheet material deformed using for the first time a single point incremental forming (SPIF) process. This numerical approach is carried out using a non-associated plasticity model coupled with continuum ductile damage. The constitutive elastoplastic equations and the calculation algorithms are implemented on Abaqus/Explicit via the user interface (VUMAT) subroutine. The numerical simulation is scrutinized two phenomena in this study: the damage behavior and the springback of DD13 sheet material when a conical shape is manufactured using the SPIF process to assess where the damage can accumulate. The simulation results included the damage area of the DD13 sheets, the Hill'48 stress distribution as long as a quadratic function of Hill'48 is used in this formulation, and a comparative figure to explain the final shape of the manufactured part after springback.

Keywords: DD13 sheet material · Single point incremental forming · Non-associated plasticity · Ductile damage · Springback

1 Introduction

Hot rolled steels (DD13) have been widely used in automotive industries because of their capability to attain very deep products. These steels are considered as a soft steels that characterized by their specific ductility owing to their ferritic composition and their excellent formability for application involving severe drawing.

Many researchers have determined the behavior of DD13 sheet material in Deep drawing processes, as reported on Özdilli et al. (2018) and Ghorbel et al. (2019), but

M. Kharrat et al. (Eds.): ICAMEM 2019, LNME, pp. 70–76, 2020.
https://doi.org/10.1007/978-3-030-52071-7_10

this present investigation will be performed as a first attempt the behavior of this material in single point incremental forming (SPIF) process. As is well known, the SPIF operation is a manufacturing technology characterized by their malleability, their ability to be defended against attack, and their flexibility to deformed thin sheet strategically. This process is characterized by the absence of specific and expensive tooling and their enormous formability compared with conventional processes. Many researchers interested in SPIF process have been worked to ameliorate both the accomplishment and the quality of manufactured products, as published by the authors (Gatea et al. 2016; Guo et al. 2018; Ben Said et al. 2016; Bouhamed et al. 2019a).

Modeling and numerical simulation of ductile damage mechanisms have an increasing attention by several researchers to predict ductile fracture and to study the effect of anisotropy in failure of aluminum sheet metal during the SPIF process, as shown in Ben Said et al. (2017).

The present work focus to investigate the ductile damage behavior and the deformation mechanism of a DD13 sheet material during SPIF process using a non-associated plasticity model that interested in the separation between the yield function and the plastic potential function with quadratic yield criteria of Hill'48 coupled with ductile damage. This formulation is implemented in user defined material (VUMAT) subroutine to predict the accumulation of the damage throughout the deformed part and to show the springback phenomena of this material.

2 Plasticity of Non-associated Model with Ductile Damage

A little description of non-associated anisotropic model coupled with continuum ductile damage is presented in this section. Elasto-plastic constitutive equations are illustrated in Table 1, to give an idea about the behavior of DD13 sheet material during the SPIF process and detailed in Ghorbel et al. (2019) and Bouhamed et al. (2019b; 2019c). The plastic flow f and the plastic potential flow F are presented as separated function for the non-associated plasticity model as below.

In Table 1, σ_Y is the yield strength, $\boldsymbol{\sigma}$ is the Cauchy stress tensor, d is the damage variable, $\boldsymbol{\alpha}_k$ and $\boldsymbol{X}_k$ are the kinematic hardening parameters. ε^p is the plastic strain tensor, ε^e is the elastic strain tensor, r and R are the isotropic hardening parameters. $\boldsymbol{D}$ is the general elastic matrix, a_k, b_k, β_1, β_1, Q_1 and Q_2 are material parameters, $\boldsymbol{P}$ and $\boldsymbol{Q}$ present the anisotropic operators defined the plastic flow and the plastic potential flow respectively.

$$\boldsymbol{P} = \boldsymbol{P}(F,G,H,N,M,L) = \begin{bmatrix} H+G & -H & -G & 0 & 0 & 0 \\ & H+F & -F & 0 & 0 & 0 \\ & & F+G & 0 & 0 & 0 \\ & & & 2N & 0 & 0 \\ & Sym & & & 2M & 0 \\ & & & & & 2L \end{bmatrix} \tag{8}$$

$$\boldsymbol{Q} = \boldsymbol{Q}\left(F',G',H',N',M',L'\right) \tag{9}$$

Table 1. Damage constitutive equations.

a/Plastic flow	$f=\frac{1}{(1-d)^{0.5}}\left[\varphi_f(\xi)-R\right]-\sigma_Y$ $\varphi_f(\xi)=\lVert\xi\rVert_P=\sqrt{\xi^t\,\boldsymbol{P}\,\xi}\;;\xi=\boldsymbol{\sigma}-\boldsymbol{X}$ $R(r)=Q_1\left(1-e^{\beta_1 r}\right)+Q_2\left(1-e^{\beta_2 r}\right)$	(1)
b/Plastic potential flow	$F=\frac{1}{(1-d)^{0.5}}\left[\varphi_F(\xi)-R\right]-\sigma_Y+F_d(Y,d)$ $+\frac{1}{2(1-d)}\left[\sum_{k=1}^{N}\frac{\beta_k}{Q_k}R_k^2+\sum_{k=1}^{M}\frac{b_k}{a_k}\boldsymbol{X}_k:\boldsymbol{X}_k\right]$ $\varphi_f(\xi)=\lVert\xi\rVert_P=\sqrt{\xi^t\,\boldsymbol{Q}\,\xi}$	(2)
c/Plastic strain tensor	$\dot{\boldsymbol{\varepsilon}}^p=\dot{\gamma}\frac{\partial F}{\partial\boldsymbol{\sigma}}=\frac{\dot{\gamma}}{(1-d)^{0.5}}\boldsymbol{n};\;\;\boldsymbol{n}=\frac{1}{\varphi_F}\boldsymbol{Q}\xi$	(3)
d/Damage equivalent stress	$Y=\frac{1}{2}\boldsymbol{\varepsilon}^e:\boldsymbol{D}:\boldsymbol{\varepsilon}^e+\frac{1}{2}\left(\sum_{k=1}^{N}Q_k\,r_k^2+\sum_{k=1}^{M}a_k\,\boldsymbol{\alpha}_k:\boldsymbol{\alpha}_k\right)$	(4)
e/Damage evolution	$\dot{d}=\dot{\gamma}\frac{\partial F}{\partial Y}=\dot{\gamma}\overline{Y};\;\;\overline{Y}=\frac{\partial F_d}{\partial Y}$	(5)
f/Kinematic hardening	$\boldsymbol{X}=\sum_{k=1}^{M}\boldsymbol{X}_k;\dot{\boldsymbol{X}}_k=\frac{a_k\dot{\gamma}}{(1-d)^{\delta-1}}\boldsymbol{n}-\dot{\gamma}\left(b_k+\frac{\overline{Y}}{1-d}\right)\boldsymbol{X}_k$	(6)
g/Isotropic hardening	$R=R(r)$	(7)

The components of the P and Q tensors are represented the anisotropy of DD13 sheet material in different directions (0°, 45°, 90°). The values of these parameters are given in Table 2 and the details of how these terms were obtained are illustrated in the work of Gorbel et al. (2019).

3 Finite Element Modeling

In this part, the current model is investigated by a numerical example to show their capability to predict the damage behavior of DD13 sheet metal in both surfaces: yield surface and plastic potential surface. The properties of DD13 sheet material are presented in Table 2. Besides, a squared sheet of 200 × 200 mm^2 and 2.5 mm thickness of DD13 sheet metal is used. The blank is assumed anisotropic and meshed by S3R shell

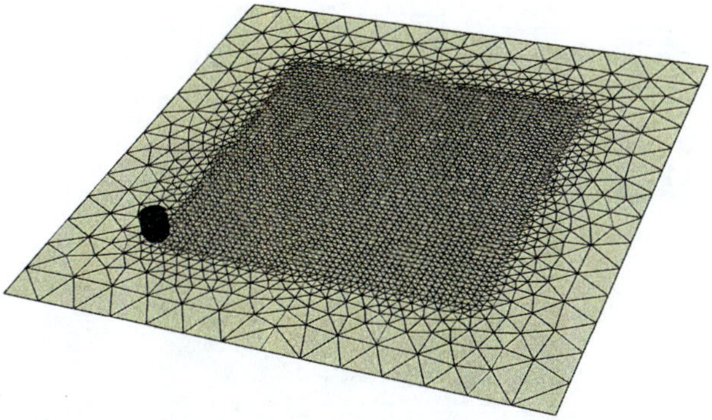

Fig. 1. Finite element mesh of SPIF

Table 2. Properties of DD13 sheet metal (Ghorbel et al. 2019)

Young modulus (GPa)	220				
Poisson's ratio	0.3				
Anisotropic coefficients	Yield function coefficients	***F***	***G***	***H***	***N***
		0.43	0.47	0.53	1.49
	Potential function coefficients	***F'***	***G'***	***H'***	***N'***
		0.64	0.47	0.53	1.77
Double exponential law (MPa)	$\sigma_Y + R(r) = 278 + 437\left(1 - e^{1.35r}\right) + 50\left(1 - e^{30r}\right)$				
Damage parameters		β	s	***S***	Y_0
		5	1.5	200	15

element, as presented in Fig. 1, with five Simpson integration points through the thickness. The hemispherical forming tool is designed as discretized rigid body, following a forming strategic path during the simulation. Moreover, in order to justify the contact condition, surface-to-surface contact is assumed between the sheet material and the punch.

4 Numerical Results

Numerical results have been considered in this section to show the ductile damage occurrence and the springback of DD13 sheet metal after SPIF process. Figure 2 depicts the damaged area of the sheet metal. The damage area is located approximately in the punch path. The highest effort of the localized contact force can explain this damage phenomenon. DD13 sheet metal shows a higher elongation to failure, therefore, it has a very good ductility. Thus, this material has the ability to be subjected in a certain amount of plastic strain without fracture. The numerical modeling of ductile damage is indeed crucial to predict damage area before fracture.

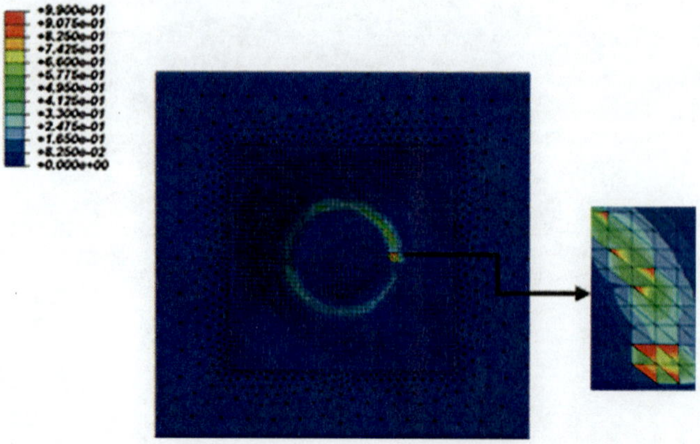

Fig. 2. Damaged area (SDV4)

According to the distribution of Hill'48 stresses (SDV2), Fig. 3 illustrates that the maximum Hill'48 distribution is located mainly along the cone tool path. The concentration of stresses at this zone can explain the maximum value of damage observed previously (SDV4).

Fig. 3. Distribution of Hill'48 stresses (SDV2)

Prediction of springback in SPIF process is important to show the desired final geometrical quality of parts after forming when the forming tools are removed. Figure 4 presents the deformed shape obtained after springback. The results show that the proposed model significantly improves the prediction of springback. It is obviously observed that the amount of springback decrease when the penetration depth increase. The supposed model has the ability to predict the general shape of sidewall more accurately.

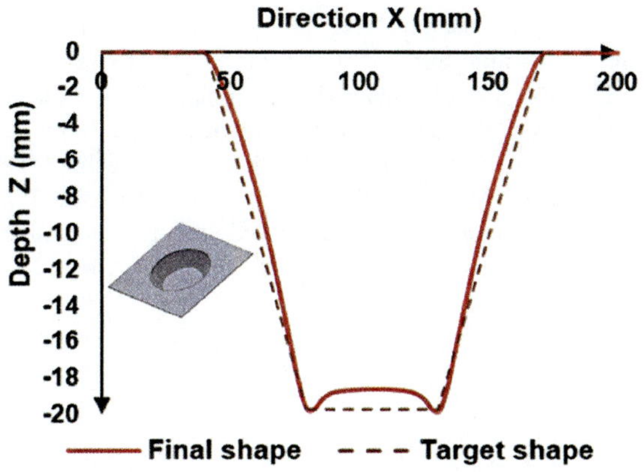

Fig. 4. Deformed shape after Springback

5 Conclusion

This research work makes the opportunity to determine the behavior of DD13 sheet metal during the SPIF process. The accuracy of the proposed model is demonstrated in the prediction of the damage behavior, the anisotropy, and the springback in sheet metal forming simulation. In fact, the current non-associated model coupled with continuum ductile damage illustrates an excellent option to enhance and optimize the SPIF process. This study offers an advanced insight to improve the quality of manufactured products obtained by DD13 sheet metal, which characterized by their better ductility.

References

Özdilli, Ö., Erdin, M.E.: Comparison of common deep drawing steel sheets in terms of blank holder force and friction conditions. Int. J. Autom. Sci. Technol. **2**(3), 36–41 (2018)

Ghorbel, O., Koubaa, S., Mars, J., Wali, M., Dammak, F.: Non associated-anisotropic plasticity model fully coupled with isotropic ductile damage for sheet metal forming applications. Int. J. Solids Struct. **166**, 96–111 (2019)

Gatea, S., Ou, H., McCartney, G.: Review on the influence of process parameters in incremental sheet forming. Int. J. Adv. Manuf. Technol. **87**, 479–499 (2016)

Guo, X., Gu, Y., Wang, H., Jin, K., Tao, J.: The Bauschinger effect and mechanical properties of AA5754 aluminum alloy in incremental forming process. Int. J. Adv. Manuf. Technol. **94**, 1387–1396 (2017)

Ben Said, L., Mars, J., Wali, M., Dammak, F.: Effects of the tool path strategies on incremental sheet metal forming process. Mech. Ind. **17**(4), 411 (2016)

Bouhamed, A., Jrad, H., Ben Said, L., Wali, M., Dammak, F.: Finite element simulation of single point incremental forming process of aluminum sheet based on non-associated flow rule. In: Advances in Mechanical Engineering and Mechanics (2019a). https://doi.org/10.1007/978-3-030-19781-0_15

Ben Said, L., Mars, J., Wali, M., Dammak, F.: Numerical prediction of the ductile damage in single point incremental forming process. Int. J. Mech. Sci. **131**, 546–558 (2017)
Bouhamed, A., Jrad, H., Ben Said, L., Wali, M., Dammak, F.: A non-associated anisotropic plasticity model with mixed isotropic–kinematic hardening for finite element simulation of incremental sheet metal forming process. Int. J. Adv. Manuf. Technol. **100**, 929–940 (2019a)
Bouhamed, A., Jrad, H., Jamel, M., Wali, M., Gamaoun, F., Dammak, F.: Homogenization of elasto-plastic functionally graded material based on representative volume element: application to incremental forming process. Int. J. Mech. Sci. **160**, 412–420 (2019b)

Effect of Shot Peening Parameters on Parabolic Leaf Spring's Residual Stress

Sari Guimaji[1,2,3](✉), Akram Khalifa[1,2,3], and Raouf Fathallah[1,2]

[1] Unité de Génie de Production Mécanique et Matériaux (UGPM2), Ecole Nationale d'Ingénieurs de Sfax (ENIS), Université de Sfax, Sfax, Tunisia
[2] Ecole Nationale d'ingénieur de Sousse, Université de Sousse, BP 264 Sousse Erriadh, Sousse, Tunisia
[3] Caveo Automotive Tunisia, Z.I. Borj Cedria BP 912, Rue Nilson Mandella -1164 - Hammam Chott, Ben Arous, Tunisia
{S.guimaji,A.khalifa}@caveoautomotive.com

Abstract. Shot peening applied on parabolic leaf spring is a mechanical surface treatment intended to improve the fatigue strength by generating residual compressive stress. However, a systematic investigation of the effect of shot peening parameters on residual stresses is still required.

The aim of the present work is to improve the knowledge of the role of shot peening in manufacturing leaf springs vehicles, through analyzing effects of different shots peening parameters by X-ray diffraction analysis. Among this investigation the parameters of the shot penning machine; turbine speed, amperage presented by the Almen intensity and the deflection of spring has been varied to optimal performance regarding residual stress. X-ray diffraction analysis shows that; on the one hand; the increase of shots speed and subsequently the Almen Intensity improves the residual stress at depth but the surface stress decreases. On the other hand, spring deflection improves the residual stress profile.

Keywords: X-ray diffraction · Residual stress · Shot peening · Almen intensity · Deflection

1 Introduction

The suspension system of a vehicle refers to the group of mechanical components that connect the wheels to the structure of vehicles. Consequently, this system is designed to absorb vibration caused by irregularities in the road and to improve vehicle ride and handling along, with driver safety and comfort [2].

There are mainly three basic suspension systems components, such as, springs, bumpers and stabilized bars. In fact, springs, supports the vehicle weight and absorb the energy generated by track irregularities while driving [1–6].

Actually, springs exists in several forms such as coil springs, leaf springs and, air springs. Leaf springs which are multi-leaves or single-leaf are the most commonly used for commercial vehicle (buses or trucks). There are two types of leaf springs, which are namely conventional and parabolic spring. Parabolic springs are characterized by a

M. Kharrat et al. (Eds.): ICAMEM 2019, LNME, pp. 77–83, 2020.
https://doi.org/10.1007/978-3-030-52071-7_11

parabolic profile while conventional springs have the same thickness along its length and it is found generally assembled as packages; however, parabolic springs can be produced as mono-leaf springs. In fact, spring industry usually uses the 51CrV4 steel for leaf springs manufacturing because it combines spring characteristics with excellent resistance to vibration and shock [5].

The aim of the present work is to examine the effects of the variation of leave's deflection and the shot peening parameters, particularly the impact of the Almen intensity of the shots, on the introduction of residual stress distributions in the parabolic single-leaf springs made of 51CrV4 steel. In the present investigation, X-ray diffraction and SEM examinations have been used to evaluate the effects of the Almen intensity and the leaf deflection on the residual stress fields produced on the 51CrV4 leaves of the same batch.

2 Experimental Procedure

The parabolic single-leaf springs used in this study are made of Chromium-Vanadium steel 51CrV4 and they are produced in the same batch in order to get the same mechanical characteristics. The work material chemical composition is reported in Table 1.

Table 1. Chemical composition of the 51CrV4 steel (wt%)

Element	C	Si	Mn	P	S	Cr	V
% of the element	0.510	0.245	0.8	0.0009	0.004	1.1	0.134

The experimental samples consist of 6 leaf-springs divided into two groups of three leaves where the first one is used to evaluate the effect of the leaf deflection whereas the second group is employed to assess the effect of the Almen intensity on the residual stress distributions. Before shot peening treatment, all the workpieces were firstly formed by taper rolling which is carried out at a temperature of 1000 °C in order to improve the stress distributions and to reduce the weight of the spring. Then, the samples were subjected to heat treatment, which consists of austenitization at 1000 °C and oil quenching (T = 80 °C) followed by tempering at around 450 °C. The resulting Brinell hardness is about 480 HB.

In the current study, the shot peening experiments were conducted on a pre-stressing shot peening machine with two-turning wheels. The specimens were shot-peened with cast steel shot S330 under a strained configuration. Under the effect of centrifugal force, shots were driven and accelerated towards the samples that were passed into the tunnel as shown in Fig. 1.

During processing, the shot media level was kept constant to guarantee that all experimental conditions are the same. The particle size analysis is reported in the Table 2.

Fig. 1. Shot peening under strained configuration

Table 2. The S330 shot media levels

Size	%
1.18	3%
1	69%
0.85	91%
0.71	98%
0.6	
0.5	100%
0.42	
0.35	

In this work, the effect of the Almen intensity on the residual stress distributions in the processed 51CrV4 samples was studied. As one of the essential ways to ensure process reproducibility, the Almen intensity is the measure of the shot's energy, which is basically related to the compression stress. In fact, the Almen intensity can be increased by increasing the projection speed or by increasing the exposure time [3]. The rotation speed of the shot peening machine wheels is kept constant during the process. Therefore, only two parameters were varied: the amperage of the wheels and the speed of the conveyer. In fact, the amperage controls the shot flow (the amount of shots per unit of time) and the speed conveyer controls the exposure time of leaf to shots. Consequently, the different Almen intensities are listed in Table 3.

Table 3. Different Almen intensity

Tests	Almen intensity	Almen arc (mm)
1	F60A	0.6 mm
2	F74A	0.74 mm
3	F84A	0.84 mm

The near-surface residual stress profiles of the shot peened 51CrV4 steel samples were analyzed by X-ray diffraction technique using the $\sin^2 \Psi$ method by means of a stress X machine, under the conditions listed in Table 4. The in-depth residual stress measurements were established using the electrochemical etching technique.

Table 4. System setting of stress machine

Parameters	Value
Anticathode	Cr
λ	2.289Å
Filter	Vanadium
2Θ	[144°, 164°]
Detector	LPSD
X-ray generator	3000 W
Working conditions	25 Kv–4 mA
Collimator diameter	2 mm
Ψ	[−40, 40]

3 Results and Discussion

Firstly; to study the influence of the Almen Arc we varied two parameters; the amperage of the turbines and the speed of the conveyer. The combination of these two parameters gives us the following Almen intensities; F80A, F73A and F70A (Table 5).

Table 5. Almen intensities

Tests	Almen intensity
1	F60A
2	F74A
3	F84A

An X-ray diffraction analysis is relized on the specimens and here are the results:

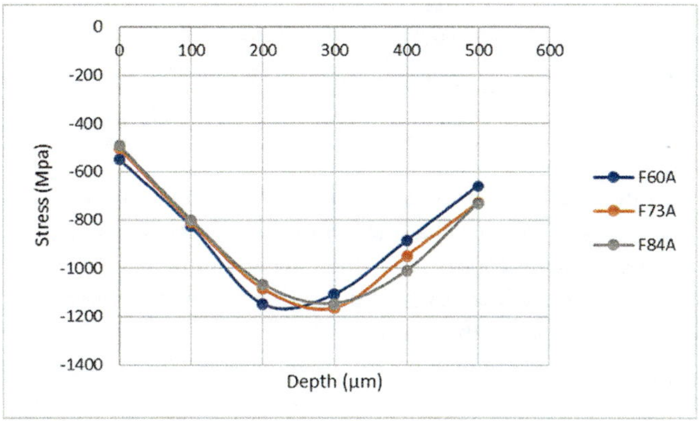

Fig. 2. Influence of Almen intensity on the residual stress profile

The three curves (Fig. 2) shows the evolution of compressive stresses as a function of Almen intensity; the profile of the stresses moves in depth but we note that the surface stress is getting degraded ones we exceed certain limits. This phenomenon can be explained by the over-shot peening which can cause the effect of overlaps (Fig. 3).

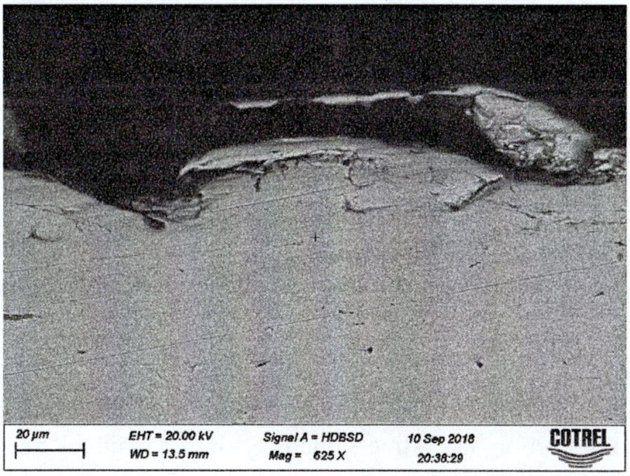

Fig. 3. Overlaps detected on the shot peened surface

Secondly, Concerning the deflection parameter, we measured the residual stress for 3 deflections presented in Table 6.

Table 6. Deflections

Deflection	COTE A (mm)	COTE B (mm)
1	190	215
2	250	250
3	275	275

The distribution of residual stress along depth with different prestresses states were illustrated in Fig. 4. It can be seen that the compressive residual stress (CRS) increased to the maximum point then dropped with increasing depth. The CRS and the position of the maximum value both increased with the deflection. But the excessive prestress can induce the micro-cracks in the surface [4].

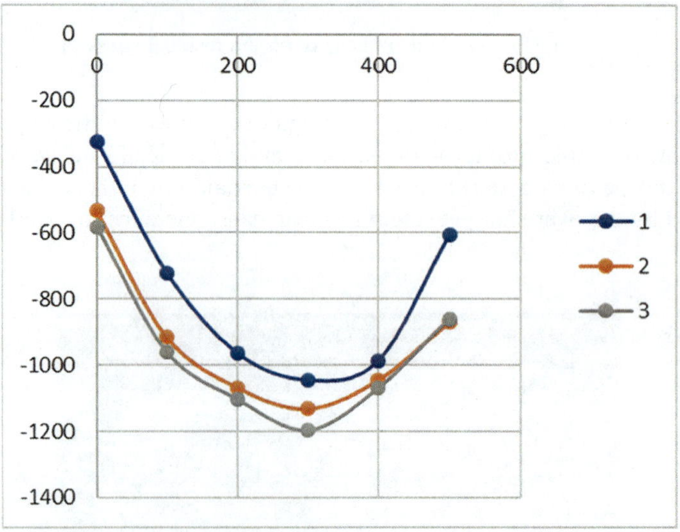

Fig. 4. Influence of deflections on the residual stress profile

4 Conclusions

The effects of the deflection and Almen Intensity on the residual stress of the 51CrV4 steel were investigated. The results revealed that CRS was improved when we increase the deflection of the spring at the moment of shot peening. The value of CRS reached the maximum when the prestress was 1600 Mpa (Deflection 3). Concerning Almen Intensity, the stress profile moves in depth and the maximum stress increases.

Acknowledgments. This work has been done with the COTREL-CAVEO working team. My thanks for my boss Mr. Imed Ben Rzouga; he trusted me and he encouraged me a lot. And for Mr. Raouf Fathallah and Akrem Khalifa for their supervision.

References

1. Crolla, D.A. (ed.): Automotive Engineering: Powertrain, Chassis System and Vehicle Body. Butterworth-Heineman, Oxford (2009)
2. Robert Tata, P.E.: Automotive Suspension System (2012). B.S.M.E
3. Metal Improvement Company, Inc., 'shot peening'- Eighth Edition, Compagnie spécialisée dans la protection des composants contre les pannes courantes et la réduction des coûts de maintenance pour l'industrie
4. Kobayashi, M., Matsui, T., Murakami, Y.: Mechanism of creation of compressive residual stress by shot peening. Int. J. Fatigue **20**, 351–357 (1998)
5. Scuracchio, B.G., de Lima, N.B., Schön, C.G.: Role of residual stresses induced by double peening on fatigue durability of automotive leaf springs. Mater. Des. **47**, 672–676 (2013)
6. Morello, L., Rossini, L.R., Pia, G., Tonoli, A.: The automotive body, vol. 1. Springer, Dordrecht (2011)

Influence of Sliding Speed and Normal Loads on the Wear Resistance of Hardox 500 Steel Ground Surfaces

Kamel Bensaid(✉) and Nabil Ben Fredj

Laboratoire de Mécanique, Matériaux et Procédés LR99ES05, Ecole Nationale Supérieure d'Ingénieurs de Tunis, Université de Tunis, 5 Avenue Taha Hussein, Montfleury, 1008 Tunis, Tunisia
kamelbensaid75@yahoo.com, benfredjnabil@gmail.com

Abstract. This work focuses on the characterization of the tribological behavior of Hardox 500 steel in the ground state with different types of lubrication (dry and soluble oil), by using the technique of single cycle scratch testing. A CSM scratch testing machine equipped with a 50 μm radius Rockwell indenter was used for the experiments. The scratch length was set to 5 mm and different scratching loads (5 N, 10 N and 15 N) and speeds (10 mm/min, 50 mm/min and 100 mm/min) were considered. Results show that the friction coefficient and worn volume increase with the increase of the scratching load and speed and that samples ground using soluble oil have a martensitic microstructure characterized by very fine martensite laths that offers better wear resistance than sample ground under dry conditions. Thus, it could be conducted that the wear resistance of the Hardox 500 ground surfaces depends not only on the sliding speed and scratching normal load, but on the material microstructure also.

Keywords: Hardox 500 · Friction coefficient · Wear volume · Wear coefficient

1 Introduction

The need for very high wear-resistant steels in industry is increasing to improve the life of mechanical components. Their use is currently very extensive due to their excellent mechanical properties and wear resistance. However, there is still a great lack of research work on Hardox 500 material, particularly the effects of machining parameters of processes such as drilling, milling, turning or grinding on the machined surfaces integrity are not elucidated. Indeed, the available data related to machining of the Hardox 500 material are published by this material maker. These data are not optimized and effects of the recommended machined parameters on the wear resistance of the machined surfaces are not investigated. Some investigations focused on studying the effects of thermo-chemical treatments such as nitriding on wear and corrosion resistance of the Hardox 500 (Ilyas et al. 2018).

In this work, we investigate the effects of the cooling conditions on the sliding wear resistance of ground surfaces. Influences of the sliding speed and normal load are studies.

M. Kharrat et al. (Eds.): ICAMEM 2019, LNME, pp. 84–90, 2020.
https://doi.org/10.1007/978-3-030-52071-7_12

2 Experimental Set-up

Wear tests were carried out on a CSM scratch test machine equipped with a Rockwell indenter 50 μm radius. Scratch length is 5 mm corresponding to a single scratch and tested samples are made from Hardox 500 that were ground under dry and soluble oil conditions.

The chemical composition of this steel grade is reported in Table 1.

Table 1. Chemical composition of Hardox 500 material.

Constituent	C	Si	Mn	P	S	Cr	Ni	Mo	B
Composition (wt%)	0.30	0.7	1.6	0.02	0.01	1.5	1.5	0.6	0.005

Table 2 shows the mechanical properties of Hardox 500.

Table 2. Mechanical properties of Hardox 500 material.

Hardness [Hv]	530
Yield strength: Re [MPa]	1336
Tensile strength: Rp [MPa]	1623
Elongation: A [%]	10

Scratch wear test parameters are summarized in Table 3.

Table 3. Scratch wear test parameters.

Samples	Sliding wear [mm/min]	Loads [N]
Dry ground part	10, 50, 100	5, 10, 15
Specimens ground with soluble oil	10, 50, 100	5, 10, 15

3 Results

In this section, we present results of tribological tests of Hardox 500 steel in the ground state under dry and oil soluble.

The coefficient of friction is defined as the ratio between tangential and normal applied forces. It is expressed by Eq. (1)

$$\mu = \frac{F_t}{F_n} \tag{1}$$

With Ft: tangential force, Fn: normal force

The expression of the wear coefficient is given by Eq. (2)

$$K = \frac{V.H_v}{F_n.L} \tag{2}$$

V: wear volume (mm^3),
Hv: hardness Vickers,
L: scratch length (mm).

Figure 1 shows the effects of the scratch normal load and the sliding speed on the friction coefficient. This figure shows that the coefficient of friction increases with the increase in the normal load and the sliding speed. This increase is explained by the increase in the contact surface between the indenter and the work-piece. It can be seen that the ground sample using conventional lubrication (soluble oil) has the lower friction coefficient than the dry ground sample. This can be explained by the increase in surface hardness in the case of ground parts with soluble oil (Riadh et al. 2014).

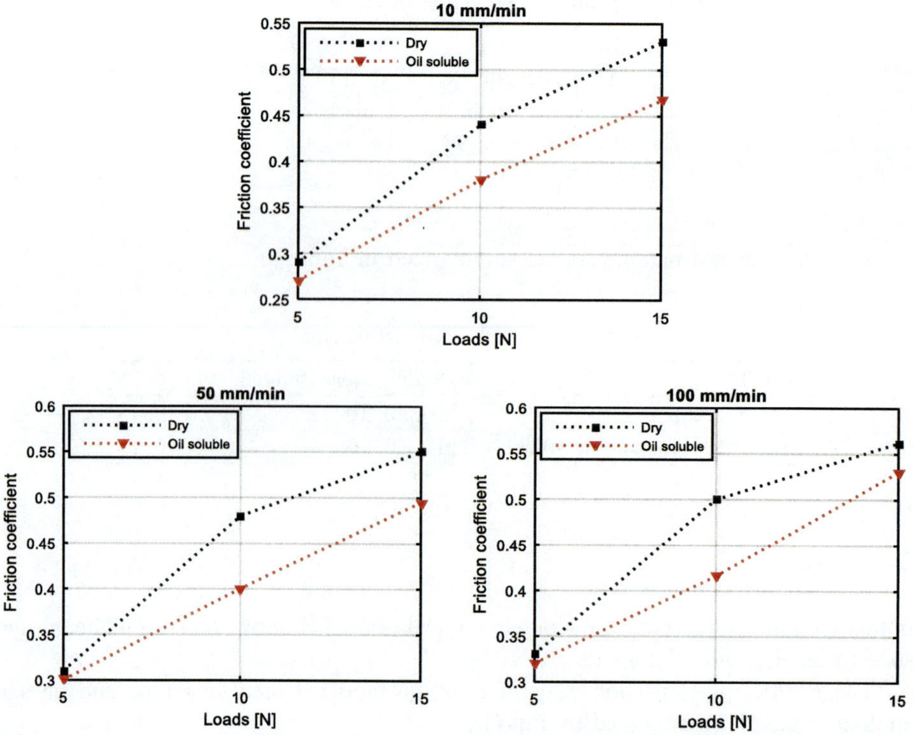

Fig. 1. Variation of friction coefficient as function of the scratching test normal load and sliding speed

Figure 2 shows a comparison of the results of wear volume evaluated for different loads (5, 10, 15 N) and sliding speeds (10, 50, 100 mm/min) and with a scratch length equal to 5 mm. Normal load and sliding speed have significant effect on the wear volume. Indeed, It can be seen that the wear volume increases with the increase in the normal load and the sliding speed. The wear volume is lower in the case of the ground sample with soluble oil. This is explained by the higher value of the surface hardness compared to the hardness of samples ground under dry condition.

Moreover, a microstructure with very fine martensite laths that forms grains of the ground surface under soluble oil was noticed. The ground sample with soluble oil offers better wear resistance.

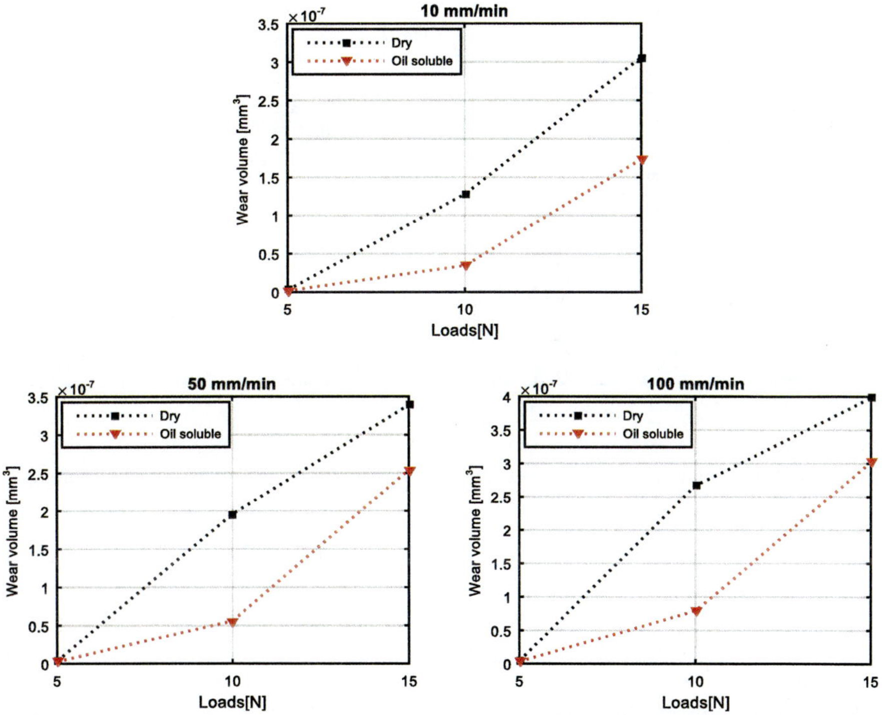

Fig. 2. Variation of wear volume as function of normal scratching load and sliding speed

The evolution of the wear coefficient as a function of normal loads and sliding speeds for the two ground samples with different types of lubrication (dry and soluble oil) is shown in Fig. 3.

It can be seen that the wear coefficient is lower for the two normal forces applied (10 N and 15 N) in the case of the ground sample with soluble oil. However, in the case of 5 N normal scratching loads and for the three levels of the scratching speed used in this investigation we observe a wear coefficient that is almost identical.

This means that the material that has a very fine microstructure and high hardness offers better resistance to wear mainly under high loading condition (Vega-Moron et al. 2018).

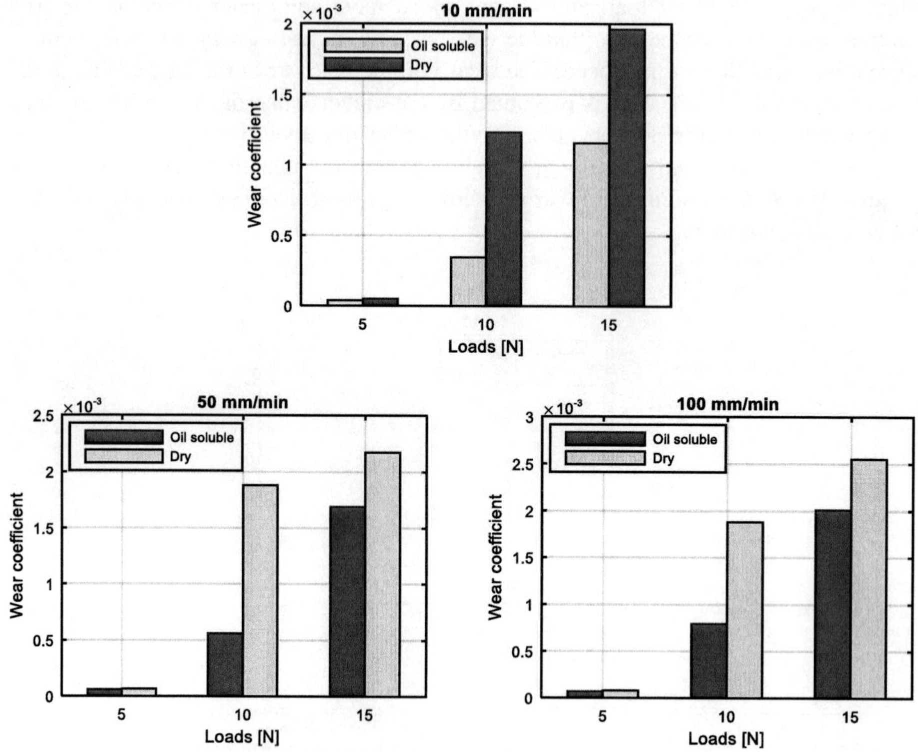

Fig. 3. Variation of wear coefficient as function of normal scratching load and sliding speed

The optical microscopy images of Fig. 4 show results of the wear tracks of the two ground samples (dry and soluble oil), and for a load equal to 5 N and sliding speed equal to 10 mm/min, 10 N at sliding speed of 50 mm/min and 15 N at a sliding speed of 100 mm/min. We notice that the width of the wear track increases with the increase of normal load.

Under all conditions the width of the lowest wear track is obtained in the case of ground specimen with soluble oil in comparison with ground sample under dry condition. This indicates that the former has better wear resistance.

The two samples used for the tribological tests did not undergo any phase transformation with respect to the reference state (as received state), but we notice a change in the microstructure such as the appearance of finer martensite laths which build small grains compared to the ground state under dry lubrication.

In conclusion, the use of conventional lubrication with soluble oil during grinding of Hardox 500 offers better wear resistance.

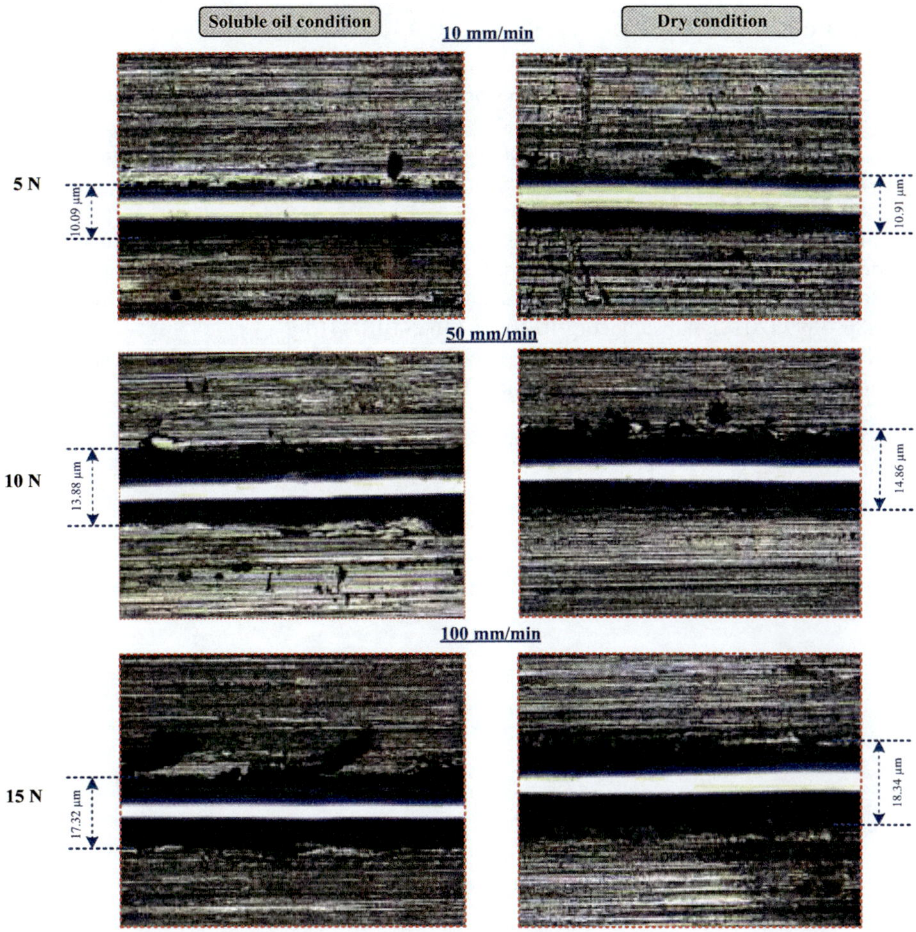

Fig. 4. Optical micrographs of wear track for different normal scratching loads and sliding speeds.

4 Conclusion

Work presented in this paper has been devoted to the scratch testing in a single cycle of two different samples made from Hardox 500 and ground under different cooling conditions (dry and soluble oil). Scratching tests were conducted with normal load of 5, 10 and 15 N and sliding speeds of 10, 50 and 100 mm/min.

Results show that:

- Friction coefficient and worn volume increase with increasing of the normal force and sliding speed.
- The ground sample with soluble oil has a microstructure that presents very fine martensite lamellas. These lamellas constitute very small grains that are more wear resistant.
- Poor wear resistance is achieved in the case of dry ground parts.

References

Hacisalihoğlu, İ., Yildiz, F., Çelik, A.: Tribocorrosion behavior of plasma nitrided Hardox steels in NaCl solution. Tribol. Int. **120**, 434–445 (2018)

Autay, R., Kchaou, M., Dammak, F.: Friction and wear behaviour of induction hardened ISO 42CrMo4 low-alloy steel under reciprocating sliding conditions. Proc. Inst. Mech. Eng. Part J J. Eng. Tribol. **229**(2), 115–125 (2014)

Vega-Morón, R.C., Rodríguez-Castro, G.A., Jiménez-Tinoco, L.F., Meneses-Amador, A., Méndez-Méndez, J.V., Escobar-Hernández, J., Reséndiz-Calderón, C.D., Nava-Sánchez, J.L.: Multipass Scratch Behavior of Borided and Nitrided H13 Steel. J. Mater. Eng. Perform. **27**(8), 3886–3899 (2018). https://doi.org/10.1007/s11665-018-3410-y

Autay, R., Kchaou, M., Dammak, F.: Experimental investigation of the tribological behaviour of carbon and low-alloy steels sliding against HSS. Mech. Ind. **16**(1), 109 (2015)

Development of STEP-NC Compliant Manufacturing for Machining Strategies of Aeronautical Components

Romdhane Ben Khalifa(✉) and Noureddine Ben Yahia

Mechanical, Production and Energy Laboratory (LMPE), National School of Engineering of Tunis (ENSIT), University of Tunis, Tunis, Tunisia
romdhane.benkhalifa@hotmail.com,
nourdine.benyahia@yahoo.com

Abstract. Aeronautics is certainly one of the most demanding industries of the world on a technical level. With materials that are increasingly difficult to machine, drastic specifications and ever tighter deadlines, the manufacture of aeronautical parts must exceed many limits while the production rate is constantly increasing. This paper aims to provide complete solutions for the machining of typical parts. Indeed, the evolution of CNC technology, combined with advances in computer systems, provided the basis for re-evaluating the way computer-assisted systems, can be used to enable universal manufacturing. The G code no longer meets the current expectations of programming more and more sophisticated machines. In reality, the STEP-NC standard (STEP-Numerical Control) provides new solutions and integration of manufacturing within the complete digital chain. This paper proposes a new numerical channel CAD/CAM based on a STEP-NC conceptual high-level standard for the machining of aeronautical parts, which aims to develop intelligent and powerful tools for data sharing and integration in the numerical chain of CAD-CAD/CAM systems.

Keywords: STEP-NC · CAD/CAM · CAPP · Aeronautical parts · CNC

1 Introduction

The aeronautics industry has always been a pioneer in terms of innovation, to meet the needs of requirements and the need to develop technical performance and efficiency aircraft safety. It also faces increased economic competition in the field, it is trying more than ever to reduce production costs by focusing on new materials, such as composites, new processes such as (HSM) and the adoption of new strategies (Peng et al. 2011) such as the application of DFM techniques (Johan et al. 2013) and the use of virtual prototyping (Nee et al. 2013).

Indeed, the machining of aeronautical parts is often characterized by severe tolerances and difficult-to-machine metals typical of engine components, as well as the complex contours typical of the two shapes of the turbine blade and aero structural components constituting an aircraft shape. Because the machined contours are so typical, five-axis machining is a routine in this sector, including long-distance five-axis

M. Kharrat et al. (Eds.): ICAMEM 2019, LNME, pp. 91–98, 2021.
https://doi.org/10.1007/978-3-030-52071-7_13

machine tools that can accommodate the aircraft's largest structural components. Aeronautical design methods, which did not take too much into account the manufacture of the part, have the effect of increasing machining problems and complicating the task of the engineer, who is mainly confronted with complex shapes, very difficult to produce or requiring the use of special and costly means.

The progress of machine technologies in terms of axis multiplication and dynamics is accompanied by the improvement of the control part through error compensation, tool positioning in five axis machining, etc. However, all these developments make programming more complex and machine tools less adaptable. Despite several research projects to improve NC structures, users are always looking for a common language within the CAD-CAM-CN digital chain, avoiding the loss of information between each unit and the possibility of sending information back to the CAD (Julio Garrido et al. 2010).

However, the G code is concerned with programming a path while respecting the movements of the machine axes rather than focusing on the machining needs while respecting the part (Sanjeev et al. 2007). In particular, we can note the main shortcomings that pose problems during programming and more generally in the integration of machining into the CAD-CAM-CN chain. Indeed, the G-code language is insufficient for manufacturing automation (Janusz et al. 2017). Nowadays, a new data standard known as STEP-NC, which makes it possible to easily define all the data of a product throughout its life cycle so that it can be accessed by any CAD/CAM system. In reality, STEP-NC offers new possibilities, not only in terms of machining, but also in terms of the restructuring it entails within the CAD/CAM chain. Also, STEP-NC makes the significant decrease in time in CAD (Xu et al. 2006) and in data preparation with the existence of a single file, the decrease in the number of files sent from CAD to CAM and the decrease in machining times for small and medium series of parts by intelligent optimization within the numerical control. It adjusts the machine tool freely and allows you to keep only one file that takes into account all the changes made (Christophe et al. 2016).

This paper describes the proposal for a new digital channel based on a CAD/CAM standard of high STEP-NC conceptual level for the machining process of aeronautical parts, which aims to develop intelligent and powerful sharing tools, as well as the integration of data in the digital chain of CAD-CAM/CAM/CN systems. The first part of this paper is dedicated to the identification of machining entities for parts, as well as, the second part is a presentation of the development of the STEP-NC module of machining strategies for aeronautical parts by VBA programming under CATIA.

2 Modeling of Machining Strategies

2.1 Aeronautical Parts

All aircraft structures are designed to withstand high stresses aerodynamics, mechanical strength and weight. These are three conflicting constraints because the improvement in the aerodynamics of the shape leads to a decrease in stiffness and the increase in stiffness leads to an increase in weight. The shape and total rigidity of the structure gives parts in different shapes and sizes (Phillips et al. 2011). Indeed, the shapes are very much influenced by their location in the structure, they can be very simple or extremely complex from a manufacturing point of view and which can have several sizes. The complexity due to geometry, as shown in the Fig. 3. The complexity of the form comes from the fact that during design, the aerodynamic shape and rigidity are very strong. In addition, there are assembly and maintenance constraints which require the creation of passages either by adopting the forms or by creating openings. Therefore, the geometry of the parts contains flat, left, and right sides of revolution or any other kind of revolution (Fig. 1).

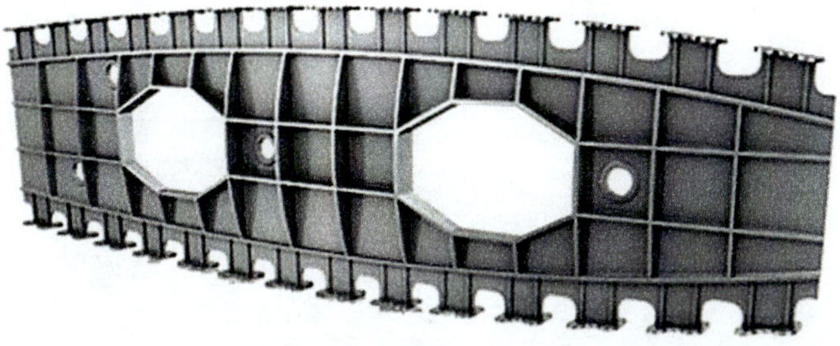

Fig. 1. Example of aeronautical part (Wing rib)

2.2 Machining Features

A machining feature is a combination of a basic geometric shape with a set which consists of a machine tool, a tool and a machining cycle. The latter will to determine the type of operation (milling, drilling, etc.). Obtaining a classification of entities requires it to take into account the three problems (machining strategies, range treatments and treatments geometric and feature identification technique). The Fig. 2 gives the general disaggregation of the generated entity classes.

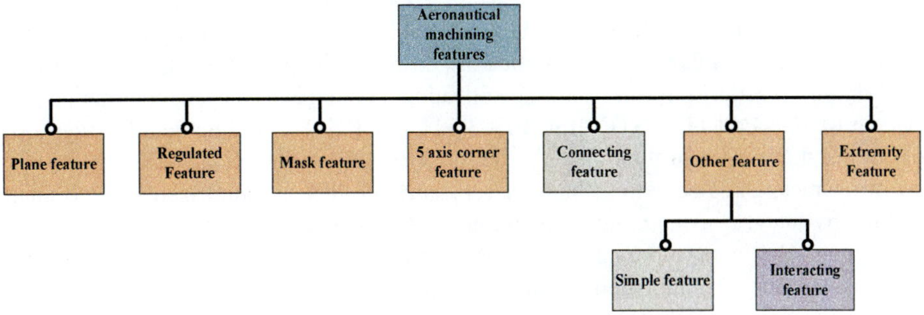

Fig. 2. Classification of machining features

The machining strategy consists of choosing the machining mode to generate each of the surfaces of the part to be machined, choosing the appropriate cutting tool, grouping the surfaces that can be machined during the same operation and ordering the resulting operations. This is achieved by minimizing machining time, respecting the quality criteria associated with the part and minimizing the necessary poses.

The machining strategy consists of choosing the machining mode to generate each of the surfaces of the part to be machined, choosing the appropriate cutting tool, grouping the surfaces that can be machined during the same operation and ordering the resulting operations. This is achieved by minimizing machining time, respecting the quality criteria associated with the part and minimizing the necessary poses (Zirmi et al. 2006).

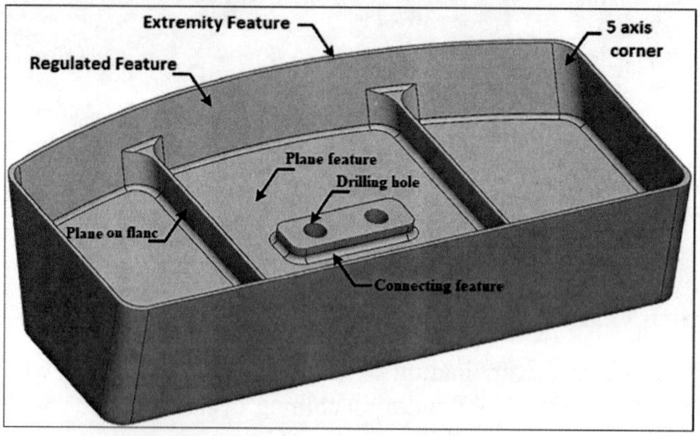

Fig. 3. Aeronautical machining features

2.3 Modeling by Express-G

The EXPRESS-G language was initially defined to define data models in the technical fields, with the aim of exchanging data that reliably and unambiguously represents this information and all this within the framework of the STEP project (Haddad et al. 2015).

For this reason, a formal data specification language has been determined and standardized as Part 11 of STEP[ISO10303-11]. An EXPRESS-G model identifies a set of entities that represent the objects to be modeled. These entities are determined by a set of characteristics called attributes. Every attribute has an authorized value domain defined by integrity constraints. The entities are arranged in an orderly manner. After assembling the data type, attributes and constraints necessary to define model-compliant properties, the model can be processed automatically and also compiled using programming. The appropriate representation of the EXPRESS-G is difficult to read (Suh et al. 2002) A graphical model for EXPRESS-G, has been determined to provide a synthetic view of the data models and facilitate their design in the initial phases of analysis of the problems to be modelled. The Fig. 4 shows the graphical modeling associated with an EXPRESS-G model (Christophe et al. 2016).

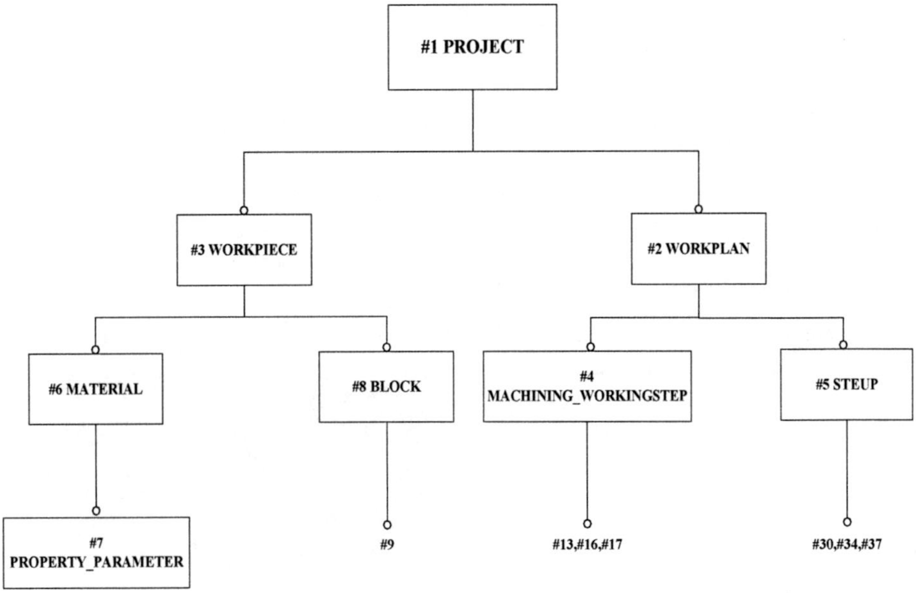

Fig. 4. Express-G model of STEP-NC

3 Generation of STEP-NC for Aeronautical Workpiece

The application proposed in this work, to validate the computer system developed for generating the STEP-NC, consists of an aeronautical part called "Fish Head" containing five machining entities (Fig. 5). This is an example of a five-axis test section provided by Airbus and manufactured by high-speed and conventional machining.

After completing the simulation phase of the program with CATIA machining, the generation of the APT file can be achieved as shown in the Fig. 5. Indeed, the APT file format corresponds to a "Metalanguage" of numerical control programming. It includes all instructions for all machine operations and is readable by everyone (its structure is clear). Its disadvantage is that it is not usable as it is because the CNC machine tool does not recognize this language. From this file, however, it is easy to recreate a STEP-NC file, directly understandable and usable by the machine, just create a small program that connects the APT file, eliminates unnecessary information, transcribes all machine instructions and copies them into another file. The STEP-NC file includes a geometry for the starting part, final part and tools, as well as tools for roughing, finishing, tool parameters, tolerances and reference points. The Fig. 5 shows the process of developing the STEP-NC program for machining strategies for aeronautical parts, the simulation and development of the program was carried out by the VBA under Catia, as well as validation by the STEP-Machines simulation software.

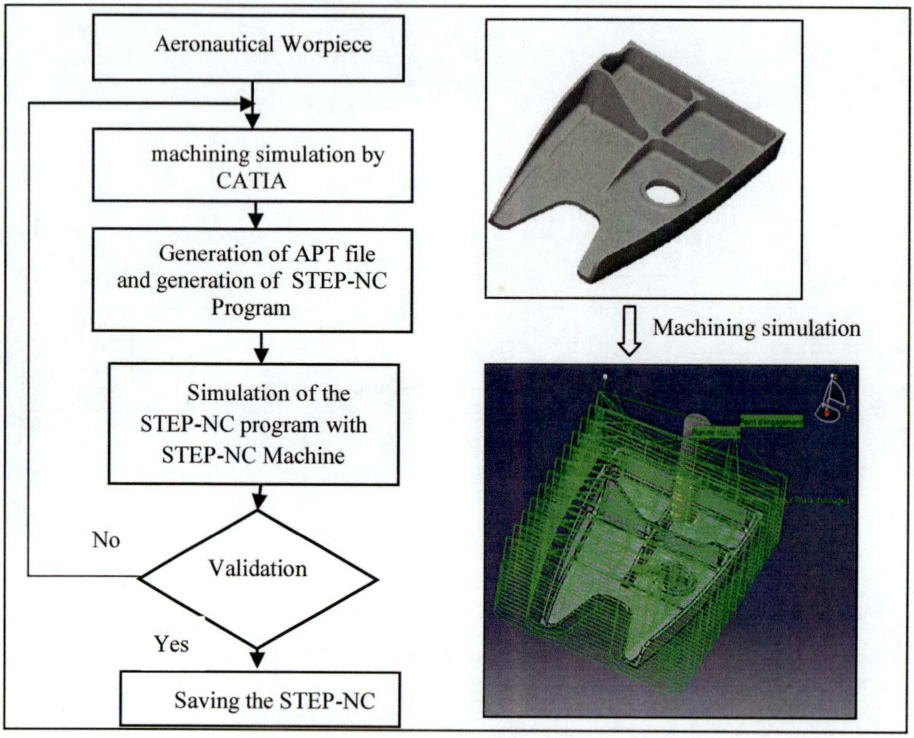

Fig. 5. Development of STEP-NC for aeronautical parts

4 Conclusion

The main achievement of this paper was to develop a data interface for process chain, by integrated strategy of machining features for aeronautical parts, between CAD and the manufacturing of the developed product on a computerized numerically controlled machine tool (CNC). Indeed, after studying the programming of the STEP-NC numerical control system, as well as the modelling of the STEP-NC program with the EXPRESS-G programming language, we know all the data and information necessary to prepare this program, such as machining strategies, workpiece dimensions and the choice of tools and machine tools, as well as the programming steps. We have also developed a computer system created to improve communication between CAD/CAM systems and DCNs by developing the STEP-NC program, by serving common databases, this prototype developed to generate the STEP-NC program from the part manufacturing data. The model is developed under CATIA V5 using its Visual Basic (VBA) programming environment. As prospects we can be considered the complete and automated integration of additive manufacturing processes. This would make it possible to work on a totally multi-process environment and would pave the way for the development and validation of inter-process simulation modules within the extended NC. We also propose the addition of new processes within the STEP-NC standards (incremental forming, material addition processes,…, etc.).

Appendix

APT file and STEP-NC program file

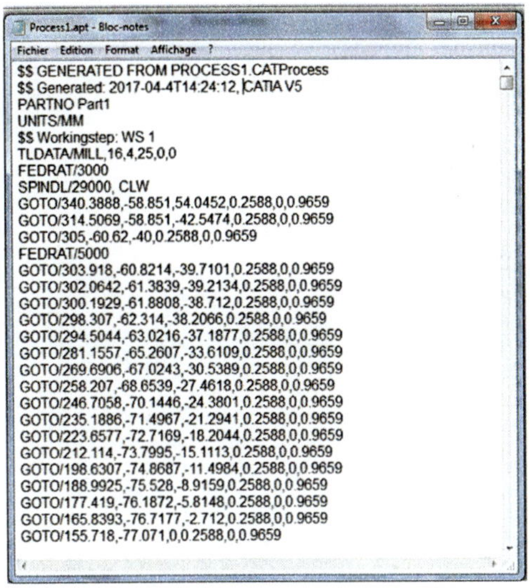

```
$$ GENERATED FROM PROCESS1.CATProcess
$$ Generated: 2017-04-4T14:24:12, CATIA V5
PARTNO Part1
UNITS/MM
$$ Workingstep: WS 1
TLDATA/MILL,16,4,25,0,0
FEDRAT/3000
SPINDL/29000, CLW
GOTO/340.3888,-58.851,54.0452,0.2588,0,0.9659
GOTO/314.5069,-58.851,-42.5474,0.2588,0,0.9659
GOTO/305,-60.62,-40,0.2588,0,0.9659
FEDRAT/5000
GOTO/303.918,-60.8214,-39.7101,0.2588,0,0.9659
GOTO/302.0642,-61.3839,-39.2134,0.2588,0,0.9659
GOTO/300.1929,-61.8808,-38.712,0.2588,0,0.9659
GOTO/298.307,-62.314,-38.2066,0.2588,0,0.9659
GOTO/294.5044,-63.0216,-37.1877,0.2588,0,0.9659
GOTO/281.1557,-65.2607,-33.6109,0.2588,0,0.9659
GOTO/269.6906,-67.0243,-30.5389,0.2588,0,0.9659
GOTO/258.207,-68.6539,-27.4618,0.2588,0,0.9659
GOTO/246.7058,-70.1446,-24.3801,0.2588,0,0.9659
GOTO/235.1886,-71.4967,-21.2941,0.2588,0,0.9659
GOTO/223.6577,-72.7169,-18.2044,0.2588,0,0.9659
GOTO/212.114,-73.7995,-15.1113,0.2588,0,0.9659
GOTO/198.6307,-74.8687,-11.4984,0.2588,0,0.9659
GOTO/188.9925,-75.528,-8.9159,0.2588,0,0.9659
GOTO/177.419,-76.1872,-5.8148,0.2588,0,0.9659
GOTO/165.8393,-76.7177,-2.712,0.2588,0,0.9659
GOTO/155.718,-77.071,0,0.2588,0,0.9659
```

```
ISO-10303-21;
HEADER;
/* Generated by macro VBA_CATIA_V5
 */
FILE_NAME(
/* name */ 'tete_de_poisson',
/* time_stamp */ '2017-04-18T20:39:07+01:00',
/* author */ (''),
/* organization */ (''),
/* preprocessor_version */ 'Macro_VBA_CATIAV5',
/* originating_system */ 'Various',
/* authorisation */ '');

FILE_SCHEMA (('INTEGRATED_CNC_SCHEMA'));
ENDSEC;

DATA;

/*************************************************
 * Application object: PROJECT (#10)
 * ITS_ID: #10, #11, #12, ['Tete_poisson']
 * ITS_WORKPIECE [*]: #10, #13, #20
 * MAIN_WORKPLAN: #10, #14, #15, #18
 */
#10=PRODUCT_DEFINITION('','',#11,#16);
#11=PRODUCT_DEFINITION_FORMATION('','',#12);
#12=MACHINING_PROJECT('Tete_poisson','',$,(#17));
#13=MACHINING_PROJECT_WORKPIECE_RELATIONSHIP('','','',#10,#20);
#14=PROCESS_PRODUCT_ASSOCIATION('','',#10,#15);
#15=PRODUCT_DEFINITION_PROCESS('machining','',#18,'');
#16=PRODUCT_DEFINITION_CONTEXT('CNC Machining',$,'manufacturing');
#17=PRODUCT_CONTEXT('CNC Machining',$,'manufacturing');

/*************************************************
 * Application object: WORKPLAN (#18)
 * ITS_ID: #18, ['fishhead_bordeaux_v2_finishing_20080206l']
 * ITS_ELEMENTS [1]: #18, #19, #23, ['machining_workplan']
```

References

Nee, A.Y.C., Ong, S.K.: Virtual and augmented reality applications in manufacturing. IFAC Proc. Volumes **46**(9), 15–26 (2013)

Christophe, D., Le Julien, D., Benoît, E.: Closed-loop manufacturing, a STEP-NC process for data feedback: a case study. Proc. CIRP **41**, 852–857 (2016)

Haddad, E., Khalifa, R.B., Yahia, N.B., Zghal, A.: Intelligent generation of a STEP-NC program for machining prismatic workpiece. In: Chouchane, M., Fakhfakh, T., Daly, H., Aifaoui, N., Chaari, F., (eds) Design and Modeling of Mechanical Systems - II. Lecture Notes in Mechanical Engineering, pp. 103–114. Springer, Cham (2015)

Julio, G.C., Ricardo, M.M.: Modelling and implementing circular sawblade stone cutting processes in STEP-NC. Robot. Comput. Integr. Manuf. **26**(6), 602–609 (2010)

Johan, V., Julia, M., Rikard, S., Kristina, W.: An approach for producibility and DFM-methodology in aerospace engine component development. Proc. CIRP **11**, 151–156 (2013)

Janusz, P., Sergiusz, S.: Extension of STEP-NC data structure to represent manufacturing process structure in CAPP system. Proc. Manuf. **11**, 1692–1699 (2017)

Peng, L., Tianliang, H., Chengrui, Z.: STEP-NC compliant intelligent process planning module: architecture and knowledge base. Proc. Eng. **15**, 834–839 (2011)

Phillips, M.: Agile manufacturing in the aerospace industry: an industrial viewpoint. Int. J. Agile Manage. Syst. **1**(1), 17–22 (1999)

Suh, S.-H., Cheon, S.-U.: A framework for an intelligent CNC and data model. Int. J. Adv. Manuf. Technol. **19**(10), 727–735 (2002)

Sanjeev, K., Aydin, N., Stephen, T.N., Richard, D.A., Manoj, K.T.: Process control in CNC manufacturing for discrete components: a STEP-NC compliant framework. Robot. Comput. Integr. Manuf. **23**(6), 667–676 (2007)

Xu, X.W., Newman, S.T.: Making CNC machine tools more open, interoperable and intelligent —a review of the technologies. Comput. Ind. **7**(2), 141–152 (2006)

Zirmi, O.: Analyse de fabricabilité en conception de gammes d'usinage pour l'aéronautique. Thèse de doctorat. Institut National Polytechnique de Grenoble INPG (2006)

Numerical Study for the Cutting of Unidirectional CFRP

A. Hassouna[1(✉)], S. Mzali[1], F. Zemzemi[2], and S. Mezlini[1]

[1] Ecole Nationale d'Ingénieurs de Monastir, Laboratoire de Génie Mécanique (LGM), Université de Monastir, Rue Ibn Eljazzar, 5019 Monastir, Tunisie
Hassounaamira0@gmail.com, Salah.mezlini@gmail.com, Mzalislah@hotmail.com

[2] Ecole Nationale d'Ingénieurs de sousse, Laboratoire de Mécanique de Sousse (LMS), Université de Sousse, BP 264 Cité Erriadh, 4023 Sousse, Tunisie
Fzemzemi@gmail.com

Abstract. Carbon Fiber Reinforced Polymer (CFRP) composites are heavily used for advanced structural applications owing to their special mechanical properties like high strength, corrosion resistance and slight weight, etc. The intrinsic characteristics of these materials generate several defects during machining such as fiber pull-out, matrix crack and burr. The understanding of the machining operation of these materials are essential to eliminate the workpiece damage and to achieve a good surface quality. In this work, the effect of the fiber orientation angle and the cutting speed on the chip formation mechanisms has been analyzed. For this reason, a finite element model with equivalent homogeneous material (EHM) was developed using Abaqus/Explicit code. A hashin criterion was adopted to predict the damage of CFRP. The results revealed that the numerical and experimental findings are in good agreement. The fiber orientation angle and the cutting speed have a notable effect on the chip formation mechanisms.

Keywords: CFRP · EHM · Finite element model · Cutting speed · Fiber orientation angle

1 Introduction

Composite materials exhibit high properties compared to conventional metals, which increase their applications in many fields. The cutting of composites is difficult and leads to various defects (Zitoune et al. 2005). Therefore, various experimental and numerical researches are oriented toward studying the machinability of composites. Koplev et al. (1983) examined experimentally the influence of fiber orientation angle on material removal process when cutting the unidirectional Carbon Fiber Reinforced Polymer (UD-CFRP). The results showed that a series of ruptures generates the formation of chips. Li et al. (2016) tested the effect of the cutting depth and fiber orientation on chip morphology. They found that the cutting depth and the fiber orientation angle strongly influence the chip morphology of CFRP. For a small cutting depth (DOC $\leq$ 0.1 mm), chips are powder, whereas become blocky for a high cutting depth.

M. Kharrat et al. (Eds.): ICAMEM 2019, LNME, pp. 99–104, 2020.
https://doi.org/10.1007/978-3-030-52071-7_14

As experimental studies on composite machining are expensive and time consuming, numerical studies were developed using different scales: macromechanical scale (Santiuste et al. 2010; Usui et al. 2014; Zhang and Strenkowski 2018) and micromechanical scale (Nayak et al. 2005).

Micromechanical modeling is more intricate and costly than macromechanical one, since it requires specific details regarding matrix, fibers and fiber-matrix interface. Thus, many studies are concentrated on macromechanical modeling to study the orthogonal cutting of composites. Soldani et al. (2011) investigated the influence of tool edge radius on cutting forces, chip morphology and subsurface damage during the orthogonal cutting of glass fiber reinforced-polymer composites (GFRP). The results revealed that the cutting forces and the subsurface damage rise with the tool edge radius, but this parameter shows an insignificant dependence on the chip morphology. Another study was conducted by Zhang and Strenkowski (2018) to study the effect of the relief angle, rake angle and cutter edge radius on the workpiece damage. The maximum stress and the Puck's fracture criteria were used for modeling the onset of damage of UD-CFRP, while a new damage propagation algorithm taken into account the spring back effect was considered. The results revealed that the boost of the relief angle generates the decrease of the subsurface damage. However, the rake angle as well as the tool edge radius do not considerably affect the machining induced damage.

Despite works on optimization of cutting parameters and tool geometry, numerical studies on the characterization of chip morphology with varying the cutting speed are rarely discussed. This paper is concentrated on determining the sensitivity of the chip morphology to the fiber orientation angle and cutting speed when orthogonal cutting of UD-CFRP. To reach this aim, a 2D finite element model, which is based on the homogenization technique was developed and validated with the experimental finding of the literature.

2 Macromechanical Model

2.1 Description of the FE Model

To model the cutting of composite, a 2D equivalent homogeneous model was established using ABAQUS/Explicit software. Figure 1 displays the model geometry, boundary conditions and mesh. The cutting tool is assumed to be a rigid part. A cutting speed (Vc) of 550 mm/s was applied in the reference point of the tool and a cutting depth of 0.04 mm was fixed. The bottom surface of the workpiece was clamped, and both left and right sides of the workpiece were restrained from translation in the machining direction.

The interaction between the workpiece and the tool is carried out with a surface-to-node contact with a penalty function. A Coulomb friction law with a friction coefficient of 0.5 was applied (Santiuste et al. 2010).

Quadrilateral elements, with plane stress, reduced integration, hourglass control (CPS4R) and element deletion were adopted. The workpiece mesh was refined in the contact zone (mesh size of about 4 μm), whereas a coarse meshes were adopted far enough, but does not exceed 40 μm.

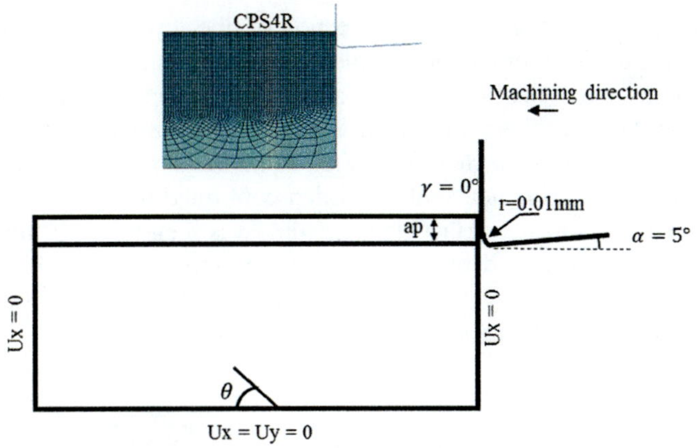

Fig. 1. Numerical model: geometry, boundary conditions and meshing.

2.2 Materiel Behavior

Unidirectional carbon/epoxy composite was used as a workpiece. It is supposed to be an equivalent homogenous materiel with orthotropic elastic behavior and brittle fracture damage. For simulating the chip formation of composites, Lasri et al. (2009) have shown that the Hashin failure criterion gives cutting forces closer to the experimental results than the maximum stress and the Hoffman failure criteria. Thus, Hashin's failure criterion (Jia et al. 2016) was used to predict the shown damage initiation of CFRP.

The properties of the simulated UD-CFRP are displayed in Table 1 (Soden et al. 1998).

Table 1. The parameters of UD-CFRP

Parameters	E1	E2	G12	υ_{12}	X_T	X_C	Y_T	Y_C	S
Value (GPa)	126	11	6.6	0.28	1.95	1.48	0.048	0.2	0.079

3 Results and Discussion

3.1 Influence of Fiber Orientation Angle

Figure 2 reveals the chip morphologies versus the fiber orientation angles (45°, 90° and 135°). It is obvious that the fiber orientation angle highly affect the material removal process. For a 45° fiber orientation angle, the chip was constituted through two combination modes of fracture. A primary one occurred in the matrix at the tool tip and then a secondary one propagated in the direction of fiber by separation of fiber/matrix interface. For a 90° fiber orientation angle, the compression of material under the advance of tool generated the bending of fibers, then their shearing perpendicular to the cutting direction. For a 135° fiber orientation angle, the material was bent and the chip was separated in mode II. It is clear from Fig. 2 that the simulated chip morphologies agree well with the experimental finding, which confirmed the FEM model. It is also noticed that the maximum subsurface damage (dm) was seen for a 90° fiber orientation angle. Such result is explained by the fact that the chip is formed only if the fibers perpendicular to the cutting direction are completely cut-off, which leads to fiber/matrix interfacial failure, and deep cracks beneath the cutting plane.

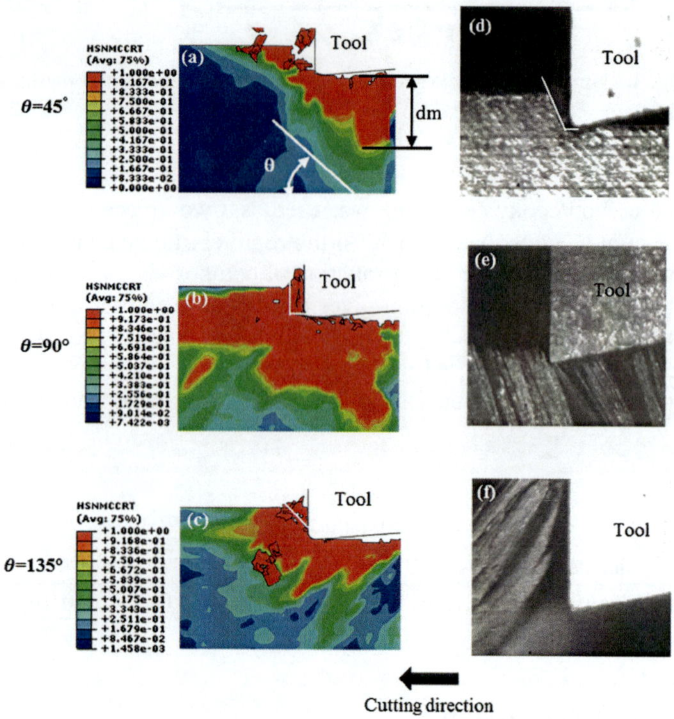

Fig. 2. (a, b, c) Simulated and (d, e, f) experimental (Iliescu et al. 2010) chip morphologies for different fiber orientation angles.

3.2 Influence of Cutting Speed

The chip morphologies for a 45° fiber orientation angle and for different cutting speeds are depicted in Fig. 3. It is observed that the cutting speed highly influence the chip morphologies. For a small cutting speed, a long and continuous chip is formed during the cutting of CFRP. This is explained by the fact that after cut off the fibers by the tool, the shear deformation in the matrix (Mode II fracture) is difficult to occur and so no chip separation in the matrix is happening. For a high cutting speed, the removed material was completely fractured and a fragmented chip was appearing. As mentioned in the experimental study of (Zimmermann et al. 2016) when machining of CFRP, that high cutting speeds (1000 mm/s and 1666 mm/s) generate small chips or even dust chips. However, the small cutting speed (60 mm/s) gives continues chips.

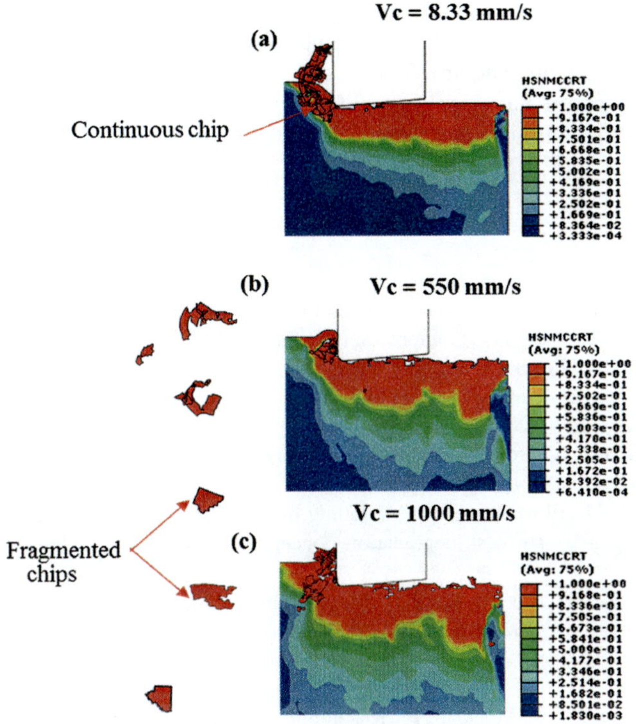

Fig. 3. Chip morphologies for different cutting speeds: a) 8.33 mm/s b) 550 mm/s c) 1000 mm/s.

It is also observed that the subsurface damage (dm) increased when the cutting speed was varied from 8.33 to 1000 mm/s. Such trend is due to the large pressure at the interface tool/chip, i.e. the large impact on the carbon fibers.

4 Conclusion

Finite element analysis has been achieved to explore the influence of fiber orientation angle and cutting speed on material removal mechanisms during the machining of CFRP. The Main conclusion that can be drawn from this paper:

- Numerical and experimental chip morphologies are in good agreement whatever the fiber orientation angle.

- The chip morphologies and the subsurface damage are greatly dependent on the fiber orientation angle. For a fiber orientation angle of 45°, chips are generated by crushing and shearing. However, for 90 and 135° fiber orientation angles, chips are formed by bending and shearing. The maximum subsurface damage was noticed for a fiber orientation angle of 90°.
- The chip morphologies and the subsurface damage are sensitive to the cutting speed. For low values, the chip is long and continuous. Nevertheless, for high cutting speeds, chips become fragmented.
- The cutting speed has a considerable influence on the subsurface damage. The increase of the cutting speed generates the increase of this machinability output.

References

Iliescu, D., Gehin, D., Iordanoff, I., Girot, F., Gutiérrez, M.E.: A discrete element method for the simulation of CFRP cutting. Compos. Sci. Technol. **70**(1), 73–80 (2010)

Jia, Z., Youliang, S., Niu, B., Zhang, B., Wang, F.: The interaction between the cutting force and induced sub-surface damage in machining of carbon fiber-reinforced plastics. J. Reinf. Plast. Compos. **35**(9), 712–726 (2016)

Koplev, A., Lystrup, A., Vorm, T.: The cutting process, chips, and cutting forces in machining CFRP. Composites **14**(4), 371–376 (1983)

Lasri, L., Nouari, M., El Mansori, M.: Modelling of chip separation in machining unidirectional FRP composites by stiffness degradation concept. Compos. Sci. Technol. **69**(5), 684–692 (2009)

Li, H., Qin, X., He, G., Jin, Y., Sun, D., Price, M.: Investigation of chip formation and fracture toughness in orthogonal cutting of UD-CFRP investigation of chip formation and fracture toughness in orthogonal cutting of UD-CFRP. Int. J. Adv. Manuf. Technol. **82**(5–8), 1079–1088 (2016)

Nayak, D., Bhatnagar, N., Mahajan, P.: Machining studies of Ud-Frp composites part 2: finite element analysis. Mach. Sci. Technol. **9**(4), 503–528 (2005)

Santiuste, C., Soldani, X., Miguélez, M.H.: Machining FEM model of long fiber composites for aeronautical components. Compos. Struct. **92**(3), 691–698 (2010)

Soden, P.D., Hinton, M.J., Kaddour, A.S.: Lamina properties, lay-up configurations and loading conditions for a range of fibre reinforced composite laminates. Fail. Criteria Fibre Reinf. Polym. Compos. **58**, 1011–1022 (1998)

Soldani, X., Santiuste, C., Muñoz-Sánchez, A., Miguélez, M.H.: Influence of tool geometry and numerical parameters when modeling orthogonal cutting of LFRP composites. Compos. A **42** (9), 1205–1216 (2011)

Usui, S., Wadell, J., Marusich, T.: Finite element modeling of carbon fiber composite orthogonal cutting and drilling. Procedia CIRP **14**, 211–216 (2014)

Zhang, S., Strenkowski, J.S.: An eulerian orthogonal cutting model for unidirectional fiber-reinforced polymers. J. Manufact. Sci. Eng. **140**(2) (2018). 021017-1-021017-8

Zimmermann, M., Heberger, L., Schneider, F., Effgen, C.: Investigation of chip formation and workpiece load when machining. Procedia Manufact. **6**, 124–131 (2016)

Zitoune, R., Collombet, F., Lachaud, F., Piquet, R., Pasquet, P.: Experiment-calculation comparison of the cutting conditions representative of the long fiber composite drilling phase. Compos. Sci. Technol. **65**(3–4), 455–466 (2005)

2-Axis Tool Strategy Applied on NC Lathe Machine to Manufacture Revolved Parts by Means of SPIF Process

L. Ben Said[1,3](✉), A. Bouhamed[3], M. Wali[2,3], and F. Dammak[3]

[1] Mechanical Engineering Department, College of Engineering, Hail University, Hail, Saudi Arabia
bensaid_rmq@yahoo.fr
[2] Department of Mechanical Engineering, College of Engineering, King Khalid University, Abha, Saudi Arabia
mondherwali@yahoo.fr
[3] Electro-Mechanical System's Laboratory (LASEM), National Engineering School of Sfax, University of Sfax, B.P 1173, 3038 Sfax, Tunisia
abirbouhamed@gmail.com, fakhreddinedammak@enis.rnu.tn

Abstract. ISF process is a promising flexible process that attracts several researches during the last decade. Applied on a CNC machine thin sheet are shaped through incremental plastic deformation trigged by a tool path programmed with a G-Code. This paper presents an experimental investigation on Single Point Incremental Forming applied on NC turning machine to manufacture revolved parts made of AA1050-H14 material. The idea is to use a 2-axis tool path strategy through the program of a paraxial roughing cycle used in machining to bore cylindrical parts. In fact, using this program, sheet metal can be shaped by incremental plastic deformation to the desired part. The effect of penetration increment, the shape of the final part on formability, appearance of cracks and the part finishing quality is analyzed. Results carried out show the efficiency of this tool strategy to manufacture parts usually manufactured through a 3-axis milling strategy in SPIF process. Certainly, SPIF applied in NC lathe gain the advantage of a rapid procedure and a low cost.

Keywords: SPIF process · NC lathe · 2-axis · Formability

1 Introduction

The principle of incremental processes is to deform locally a sheet metal to gradually obtain the desired shape. Indeed, in SPIF process one of these incremental processes, hemispherical tool follows a path generated by a CNC program implemented to a 3-axis milling machine. Without a die or any support, it is usually very practical to apply SPIF operations on this type of CNC machine. The advantage is the high flexibility, because the same tool can be used for a large number of family of pieces. The material investment is indeed reduced compared to stamping since the used tools are usually simple, recoverable and inexpensive. However, this type of process requires good knowledge of the material behavior used since the springback is much more important

M. Kharrat et al. (Eds.): ICAMEM 2019, LNME, pp. 105–111, 2020.
https://doi.org/10.1007/978-3-030-52071-7_15

compared to TPIF process that use a die or support under the sheet metal. ISF process attracted a lot of interest during the few last decades. Considering the last years, many contributions enhanced the experience in this field. Ndip-Agbor et al. (2019) provided a method for predicting the rigid body motion (RBM) during the multi-pass SPIF process at the aim to investigate the 3D contact patch between the tool and the sheet. The resulting MSPIF part demonstrated an increased formability and the absence of stepped features, thus confirming the validity of the prediction methodology. Choi and Lee (2019) verified the feasibility of a mathematical model proposed to predict the thickness distribution and formability during hybrid forming method, which combines incremental forming with stretch forming. Through their model, the authors found that hybrid forming inducts more uniform thickness distribution than pure incremental forming. Preforming in hybrid incremental process has significant influence on minimum thickness and formability (Jagtap and Kumar 2019). Double-side incremental forming (Otsu et al. 2018), can improve formability through the action of a second hemispherical tool under the blank sheet. An artificial neural network (ANN) based prediction model was developed in Mulay et al. (2019) to evaluate average surface roughness and the effect of a set of parameters on formability in SPIF process of aluminum sheets.

A good knowledge of the material behavior of sheet manufactured by SPIF process is crucial to control the different parameters of this process. Bouhamed et al. (2019) applied homogenization of elasto-plastic functionally graded material based on representative volume element to simulate SPIF process of FGM composite material. A fully coupled elasto-plastic-damage model based on Hill' 48 yield criteria with non-linear mixed isotropic/kinematic hardening has been proposed in Ben Said et al. (2017). At the aim to simulate SPIF process of AA1050-H14 and to analyze with more accuracy the effect of a set of parameters on formability. Ilyas et al. (2019) and Guzman et al. (2018) used the Gurson-Tvergaard-Needleman (GTN) model to predict failure in SPIF process. These two models predict well the parameters related to the incremental forming compared to simple models considering aspects of elasto-plasticity with isotropic or kinematic hardening like what it is proposed in Ben Said et al. (2015, 2016, 2017) and Wei et al. (2019).

In ISF filed, we can consider that all the experimental and numerical studies have been based on the use of 3-axis milling strategies to perform incremental operations. The idea in this paper is to investigate, through several experimental results, the efficiency of a 2-axis turning strategy to manufacture with SPIF process some revolved parts.

2 Experimental Setup

Two shapes presented along this section, were made by SPIF process applied to the NC lathe machine (2-axis NC REALMECA Num 1060). Figure 1 deploys the experimental setup used to apply ISF operation on this type of CNC machines. The conventional single pass strategy was used to manufacture these parts. Through a set of results, we try to analyze the effect of a set of parameters on the formability, the appearance of crack and the quality of the manufactured part.

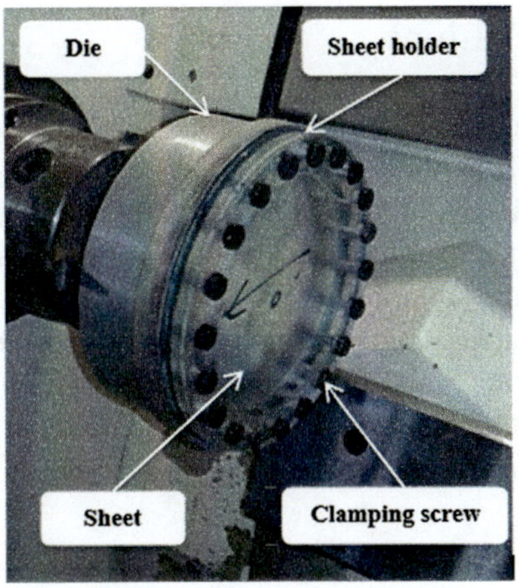

Fig. 1. Experimental setup.

Figure 2 draws the dimensional specifications as well as the toolpath for two cases of revolved parts: the hyperbolic cone and the truncated cone. In fact, the toolpath is automatically generated by the emulator of the 2-axis NC turning machine, through the manual programming of a paraxial roughing cycle. The same cycle used for boring operation in machining.

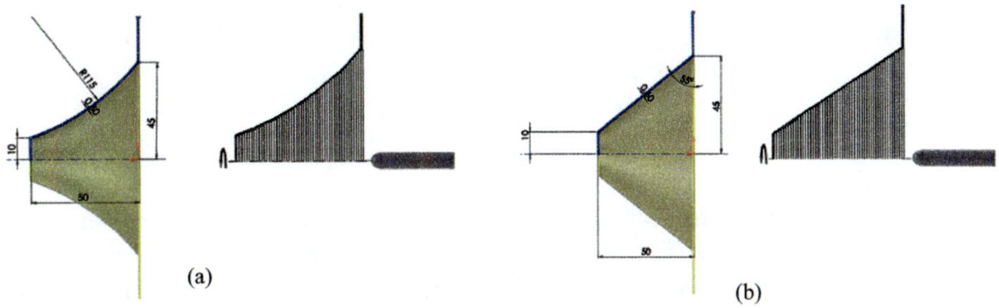

Fig. 2. Final shape and toolpath; (a) hyperbolic cone, (b) truncated cone.

The studied specimens are obtained from disc blanks having an initial diameter of 150 mm cut from an aluminum sheet AA 1050-H14 0.6 mm thick. Before the shaping operation, circles grids mark the blank sheet. This grid marking will be the tool to measure the major and minor strains in order to subsequently trace the forming limit diagram "FLD". A hemispherical tool of 10 mm diameter make a lubricated contact with the disc blanks during the paraxial roughing cycle.

3 Results and Discussions

Figures 3, 4 and 5 analyze the effect of increasing the tool penetration increment on the quality of the part as well as the ability of the operation to avoid the appearance of cracks. For the case of the hyperbolic cone, three penetration increments were tested 0.02, 0.1 and 0.2 mm. In fact, for this part the effect of increased penetration during the roughing cycle worsens the thinning of the sheet (which is proved by Fig. 3) which causes the appearance of cracks accompanied by a poor surface quality; an orange peel effect is observed at the outer wall of the hyperbolic cone. With a fairly high increment, we can see a total break in the bottom of the manufactured part (see Fig. 4.c, the case of a penetration value 0.2 mm).

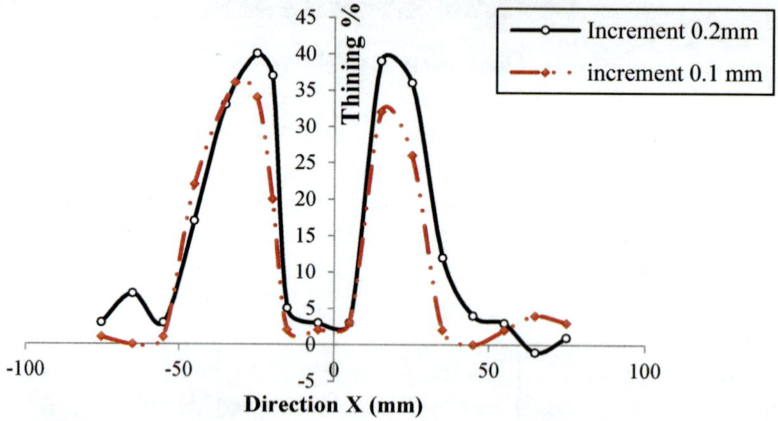

Fig. 3. Effect of increasing the penetration value on the thinning of the sheet

For the second part (truncated cone), increasing the penetration increment only reduces the surface quality (Fig. 5). In fact, we can see more traces of toolpaths that form wrinkles on the entire wall of the manufactured truncated cone (more remarkable for the highest penetration value, case of Fig. 5.c). In addition, a polishing effect has been observed in the inner wall of the manufactured parts (see Fig. 5.a) caused by a buffing phenomenon during the movement of the tool in lubricated contact with the inner wall of the workpiece.

Treating the formability of the manufactured blank sheets, the forming limit diagram FLD plotted in Fig. 6 shows that the final geometry obtained by SPIF-T has a direct effect on the main deformation levels reached. In fact, it can be denoted from this figure that the two conic and hyperbolic geometries produce deformation states close to the plane deformation, located in the right quadrant of the FLD diagram because the level of the major deformation exceeds remarkably the minor deformation. Although for the case of the hyperbolic cone, the major deformation exceeds that reached for the truncated cone.

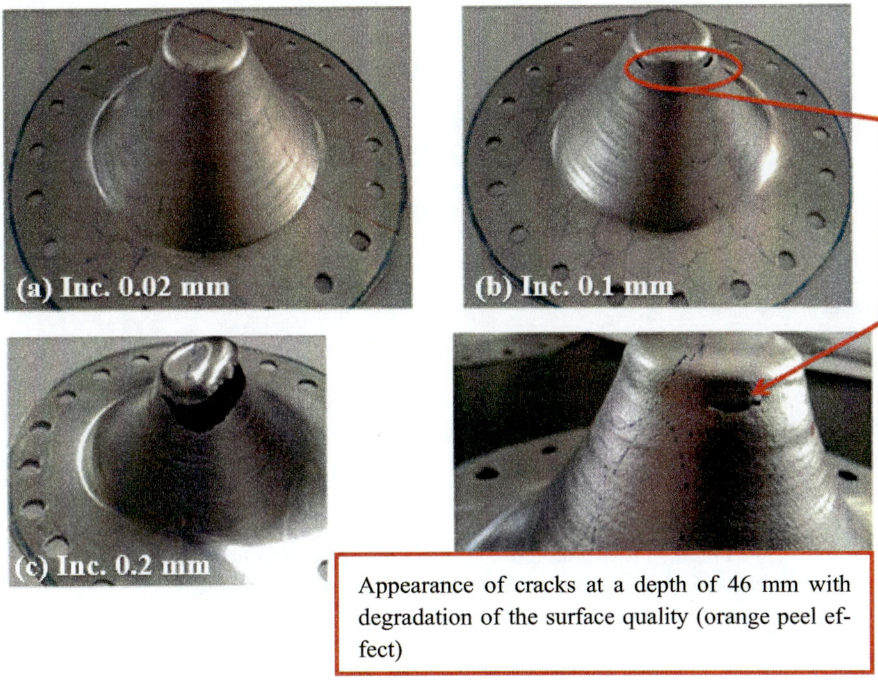

Fig. 4. Hyperbolic cone manufactured with different penetration increment values

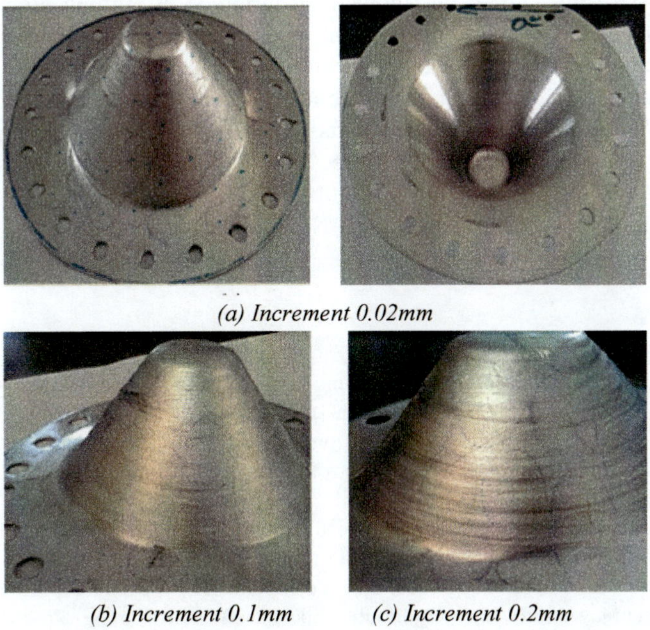

Fig. 5. Truncated cone manufactured with different penetration increment values

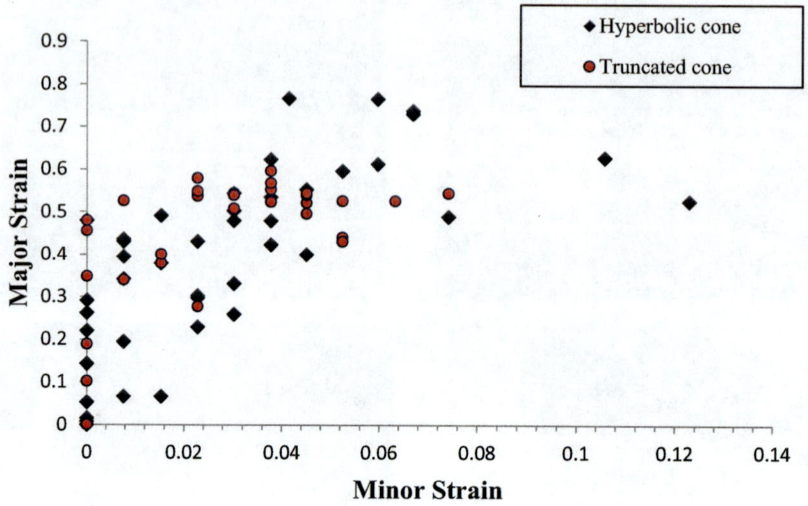

Fig. 6. FLD forming limit diagram for manufactured parts

4 Conclusion

The present paper presents an experimental investigation on the use of a 2-axis tool path strategy applied on NC lathe. That replace 3-axis strategy usually applied on NC milling machines to manufacture sheet metal by incremental forming processes such as SPIF process. Results carried out shows that during SPIF-Turning process the penetration increment has a great effect on the part finishing quality and the gravity of thinning. Changing the shape of the wall of a conic part affects the formability of the manufactured sheet. However, it can be confirmed that NC lathe can be considered in ISF process to manufacture revolved parts which is similar to spinning process but without any support or special tools. Moreover, complex parts can be manufactured in SPIF-T using turning centers with four and five axes, certainly with a low cost and a productive time compared to the use of a NC milling machine.

References

Ndip-Agbor, E., Cheng, P., Moser, N., Ehmann, K., Cao, J.: Prediction of rigid body motion in multi-pass single point incremental forming. J. Mater. Process. Technol. **269**, 117–127 (2019)

Choi, H., Lee, C.: A mathematical model to predict thickness distribution and formability of incremental forming combined with stretch forming. Robot. Comput. Integr. Manuf. **55, Part B**, 164–172 (2019)

Jagtap, R., Kumar, S.: An experimental investigation on thinning and formability in hybrid incremental sheet forming process. Procedia Manuf. **30**, 71–76 (2019)

Otsu, M., Nagai, S., Miura, T., Okada, M., Yoshimura, H., Matsumoto, R., Muranaka, T.: Forming accuracy improvement by double-side incremental forming. Procedia Manuf. **15**, 1177–1183 (2018)

Mulay, A., Ben, B.S., Ismail, S., Kocanda, A.: Prediction of average surface roughness and formability in single point incremental forming using artificial neural network. Archiv. Civ. Mech. Eng. **19**(4), 1135–1149 (2019)
Bouhamed, A., Jrad, H., Mars, J., Wali, M., Dammak, F.: Homogenization of elasto-plastic functionally graded material based on representative volume element: application to incremental forming process. Int. J. Mech. Sci. (2019). https://doi.org/10.1016/j.ijmecsci.2019.07.005
Ben Said, L., Mars, J., Wali, M., Dammak, F.: Numerical prediction of the ductile damage in single point incremental forming process. Int. J. Mech. Sci. **131–32**, 546–558 (2017)
Ilyas, M., Hussain, G., Espinosa, C.: Failure and strain gradient analyses in incremental forming using GTN model. Int. J. Lightweight Mater. Manuf. **2**(2), 177–185 (2019)
Guzmán, C.F., Yuan, S., Duchêne, L., Flores, E.S., Habraken, A.M.: Damage prediction in single point incremental forming using an extended Gurson model. Int. J. Solids Struct. **151**, 45–56 (2018)
Ben Said, L., Mars, J., Wali, M., Dammak, F.: Effects of the tool path strategies on incremental sheet metal forming process. Mech. Ind. **17**, 411 (2016)
Ben Said, L., Belhassen, L., Mars, J., Wali, M.: On the use of NC milling and turning machines in SPIF process of asymmetric parts: numerical investigation. In: International Conference Design and Modeling of Mechanical Systems—III (CMSM 2017), pp. 269–279 (2017)
Ben Said, L., Wali, M., Dammak, F.: Numerical simulation of incremental sheet metal forming Process. In: Conference on Multiphysics Modelling and Simulation for Systems Design (MMSSD 2014), pp. 489–496 (2015)
Wei, H., Zhou, L., Heidarshenas, B., Ashraf, I.K., Han, C.: Investigation on the influence of springback on precision of symmetric-cone-like parts in sheet metal incremental forming process. Int. J. Lightweight Mater. Manuf. **2**(2), 140–145 (2019)

Check for updates

Evaluating Assembly Sequences with the Assembly Tools Operation Space Criteria

Akram Bedeoui(✉), Riadh Benhadj, Moez Trigui, and Nizar Aifaoui

Mechanical Engineering Laboratory, National Engineering School of Monastir (LGM_ENIM), University of Monastir, 5 Av. Ibn Eljazzar, 5019 Monastir, Tunisia
bedeouiakram@hotmail.fr

Abstract. The generation of assembly sequences (AS) is a topical subject aimed to help designers for the determination and the simulation of the different stages of a mechanical product assembly. So, it is considered as an important step in the process of designing a new mechanical product. In addition, it allows minimizing the design cost and reducing the time of maintenance. For that, researches are developed several methods to optimize the determination of assembly sequences.

In this paper, an approach based on the evaluation of AS, generated by our old work, is proposed. The generated AS are feasible and respect the stability between components during the assembly operation. The criterion of the Assembly Tools Operation Space (ATOS) is used to deduce assembly sequences with high quality. Also, those assembly sequences will be more conformed to industrial practices. For that reason, the algorithm will detect fastener, calculate the movement length (ML), of the last one, to be demounted and select the appropriate tool (AT) to make the necessary tests.

An illustrative example is adopted to highlight the importance of this proposed approach.

Keywords: Assembly sequences · CAD model · Assembly tools operation space · Assembly tools

1 Introduction

Over the last three decades, the requirements of users for mechanical products have become more and more severe, resulting in increasingly sophisticated and complex products. In order to remain competitive, companies have furthering reduce production costs while improving quality and productivity right from the design step. This leads searchers to develop new approaches.

They are used several criteria, such as, the feasibility of assembly (Iacob et al. 2013; Su 2007), reducing the cost of assembly by Pintzos et al. (2016), the assembly directions by Hyoung and Gemmill (2001), the identification of subassemblies (Belhadj et al. 2016; Laperriere and Lavoie 1997) and the criterion of stability of AS/DS (Smith et al. 2001; Bahubalendruni and Biswal 2016; Kumar et al. 2016; Bedeoui et al. 2019).

M. Kharrat et al. (Eds.): ICAMEM 2019, LNME, pp. 112–117, 2020.
https://doi.org/10.1007/978-3-030-52071-7_16

Nevertheless, the assembly tools operation space is a criterion, which is not well explored in the generation of assembly sequences (Liu et al. 2013).

In this paper, an approach based on ATOS is developed. All steps of this approach will be explained through an illustrative example.

2 Developed Approach

This developed approach has three main steps, presented in the following Fig. 1:

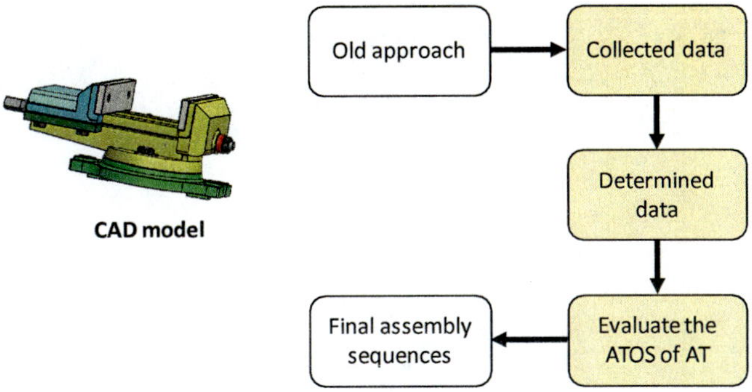

Fig. 1. Different steps of the developed approach

- Collected data
- Determined data
- Evaluate the ATOS of AT

The "swivel vice" is the illustrative example that will be used to explain this approach. It is composed by 22 components, presented in Fig. 2.

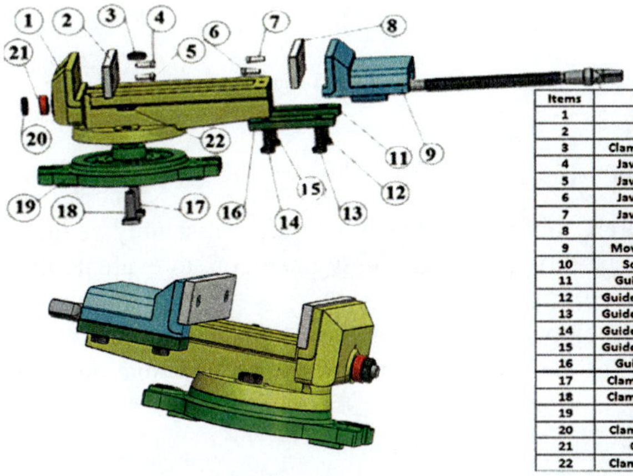

Items	Name
1	Body
2	Jaw 1
3	Clamping nut 2
4	Jaw screw 1
5	Jaw screw 2
6	Jaw screw 3
7	Jaw screw 4
8	Jaw 2
9	Moveable jaw
10	Screw rod
11	Guide strip 2
12	Guide strip bolt 4
13	Guide strip bolt 2
14	Guide strip bolt 1
15	Guide strip bolt 3
16	Guide strip 3
17	Clamping bolt 1
18	Clamping bolt 2
19	Base
20	Clamping nut 3
21	Collar 2
22	Clamping nut 1

Fig. 2. Illustrative example "swivel vice"

The Fig. 3 presents the flowchart of this developed approach.

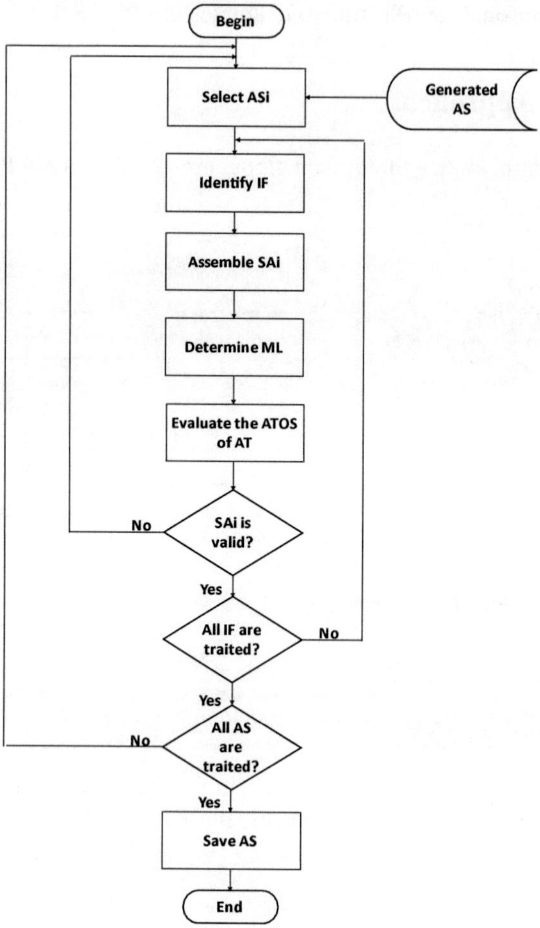

Fig. 3. The flowchart of the developed approach

2.1 Collected Data

From our past research, this approach will use these assembly sequences, which are feasible and respect stability during assembly operation, to evaluate them one by one. Those AS are presented in the Fig. 4.

In addition, the developed algorithm will explore the information of the constraints between components of the illustrative example. That information is presented in the Fig. 5.

Seq 1	(1)	(19)	(18)	(17)	(3)	(22)	(2)	(4)	(5)	(9)	(8)	(6)	(7)	(11)	(12)	(15)	(16)	(14)	(13)	(10)	(21)	(20)
Seq 2	(1)	(3)	(19)	(18)	(17)	(22)	(2)	(4)	(5)	(9)	(8)	(6)	(7)	(11)	(12)	(15)	(16)	(14)	(13)	(10)	(21)	(20)
Seq 3	(1)	(19)	(3)	(17)	(18)	(22)	(2)	(4)	(5)	(9)	(8)	(6)	(7)	(11)	(12)	(15)	(16)	(14)	(13)	(10)	(21)	(20)
Seq 4	(1)	(3)	(22)	(19)	(18)	(17)	(2)	(4)	(5)	(9)	(8)	(6)	(7)	(11)	(12)	(15)	(16)	(14)	(13)	(10)	(21)	(20)
Seq 5	(1)	(19)	(18)	(22)	(17)	(3)	(2)	(4)	(5)	(9)	(8)	(6)	(7)	(11)	(12)	(15)	(16)	(14)	(13)	(10)	(21)	(20)
Seq 6	(1)	(3)	(19)	(22)	(18)	(17)	(2)	(4)	(5)	(9)	(8)	(6)	(7)	(11)	(12)	(15)	(16)	(14)	(13)	(10)	(21)	(20)
Seq 7	(1)	(19)	(3)	(22)	(18)	(17)	(2)	(4)	(5)	(9)	(8)	(6)	(7)	(11)	(12)	(15)	(16)	(14)	(13)	(10)	(21)	(20)
Seq 8	(1)	(3)	(19)	(18)	(22)	(17)	(2)	(4)	(5)	(9)	(8)	(6)	(7)	(11)	(12)	(15)	(16)	(14)	(13)	(10)	(21)	(20)
Seq 9	(1)	(19)	(3)	(17)	(22)	(18)	(2)	(4)	(5)	(9)	(8)	(6)	(7)	(11)	(12)	(15)	(16)	(14)	(13)	(10)	(21)	(20)
Seq 10	(1)	(19)	(18)	(22)	(3)	(17)	(2)	(4)	(5)	(9)	(8)	(6)	(7)	(11)	(12)	(15)	(16)	(14)	(13)	(10)	(21)	(20)
Seq 11	(1)	(19)	(18)	(17)	(3)	(22)	(2)	(4)	(5)	(9)	(11)	(12)	(15)	(8)	(6)	(7)	(16)	(14)	(13)	(10)	(21)	(20)
Seq 12	(1)	(3)	(19)	(18)	(17)	(22)	(2)	(4)	(5)	(9)	(11)	(12)	(15)	(8)	(6)	(7)	(16)	(14)	(13)	(10)	(21)	(20)
Seq 13	(1)	(19)	(3)	(17)	(18)	(22)	(2)	(4)	(5)	(9)	(11)	(12)	(15)	(8)	(6)	(7)	(16)	(14)	(13)	(10)	(21)	(20)
Seq 14	(1)	(3)	(22)	(19)	(18)	(17)	(2)	(4)	(5)	(9)	(11)	(12)	(15)	(8)	(6)	(7)	(16)	(14)	(13)	(10)	(21)	(20)
Seq 15	(1)	(19)	(18)	(22)	(17)	(3)	(2)	(4)	(5)	(9)	(11)	(12)	(15)	(8)	(6)	(7)	(16)	(14)	(13)	(10)	(21)	(20)
Seq 16	(1)	(3)	(19)	(22)	(18)	(17)	(2)	(4)	(5)	(9)	(11)	(12)	(15)	(8)	(6)	(7)	(16)	(14)	(13)	(10)	(21)	(20)
Seq 17	(1)	(19)	(3)	(22)	(18)	(17)	(2)	(4)	(5)	(9)	(11)	(12)	(15)	(8)	(6)	(7)	(16)	(14)	(13)	(10)	(21)	(20)
Seq 18	(1)	(3)	(19)	(18)	(22)	(17)	(2)	(4)	(5)	(9)	(11)	(12)	(15)	(8)	(6)	(7)	(16)	(14)	(13)	(10)	(21)	(20)
Seq 19	(1)	(19)	(3)	(17)	(22)	(18)	(2)	(4)	(5)	(9)	(11)	(12)	(15)	(8)	(6)	(7)	(16)	(14)	(13)	(10)	(21)	(20)
Seq 20	(1)	(19)	(18)	(22)	(3)	(17)	(2)	(4)	(5)	(9)	(11)	(12)	(15)	(8)	(6)	(7)	(16)	(14)	(13)	(10)	(21)	(20)
Seq 21	(1)	(19)	(18)	(17)	(3)	(22)	(2)	(4)	(5)	(9)	(11)	(12)	(15)	(16)	(14)	(13)	(8)	(6)	(7)	(10)	(21)	(20)

Fig. 4. Assembly sequences generated by old work

Constraint	i	j	k	r1	r2	Component 1	Component 2
COINCIDENT	0	1	0	0	0	19	1
COINCIDENT	0	-1	0	0	0	19	17
CONCENTRIC	0	1	0	47.5	0	19	17
CONCENTRIC	0	1	0	47.5	0	19	18
COINCIDENT	0	-1	0	0	0	19	18
CONCENTRIC	0	-1	0	8	0	1	17
CONCENTRIC	0	-1	0	8	0	1	18
CONCENTRIC	0	-1	0	8	0	18	22
COINCIDENT	0	1	0	0	0	1	22
COINCIDENT	0	1	0	0	0	1	3
CONCENTRIC	0	-1	0	8	0	17	3
COINCIDENT	0	1	0	0	0	1	9
PARALLEL	-1	0	0	0	0	1	9
CONCENTRIC	-1	0	0	12.5	0	1	9
COINCIDENT	0	-1	0	0	0	9	11
COINCIDENT	0	-1	0	0	0	9	16
COINCIDENT	-1	0	0	0	0	1	2
PARALLEL	0	1	0	0	0	1	2
CONCENTRIC	-1	0	0	7	0	2	4
COINCIDENT	-1	0	0	0	0	2	4
CONCENTRIC	-1	0	0	5	0	2	5
COINCIDENT	-1	0	0	0	0	2	5

Fig. 5. Constraints data between components

2.2 Determined Data

For each AS, the algorithm starts by the identification of fastener (IF). Then, it assembles a subassembly (SAi), starting with the initial component until the IF.

In this moment, the algorithm moves the IF, to calculate the movement length that will do the IF to be disassembled.

After that, depend of the IF's type, the algorithm will choose the appropriate tool.

2.3 Evaluate the ATOS of AT

After the extraction and the determination of necessary data for this approach, the algorithm performs the required tests with reference to type of IF.

For the bolt, the algorithm makes two interference tests. The first one is the rotation test. For this one, the algorithm assembles the AT with the IF, according to the direction determined by the constraint's vector between the IF and its component to assemble. Then, the AT rotates 60° in the assembly direction. This rotation is done after the determination of the optimum position to start this action. This position is determined, using interference test, during the rotation of AT in the disassembly direction.

In addition, the algorithm translates the AT in the disassembly direction, with the ML.

During the rotation and translation, if the AT interferes with a component, the domain of the ATOS in not valid.

For the screw, the algorithm makes only the translation interference test. If there is not an interference during the movement of the AT, the domain of the ATOS is valid.

For the proper functioning of the AT, the algorithm determines the component (Ci) which will be assembled by the IF, as well as all fasteners, which will be used to assemble the Ci. After that, the developed algorithm checks, in the ASi, if those fasteners are all assembled after this Ci.

Otherwise, the algorithm will select another ASi. This step reduces the change of AT, which allows to reduce the assembly operation time.

All those steps are repeated until all identified fasteners and all AS are treated.

3 Results

Following the procedures of determination of the assembly sequences, the old approach obtained at the beginning 21 AS which respect the stability during the assembly operation. Subsequently, by adding the criterion of ATOS, there remain at the end only two sequences, which respect all the constraints imposed in the two developed approaches. Those AS are presented in Fig. 6.

AS 1
(1) → (19) →(18) → (22) → (17) → (3) → (2) → (4) → (5) → (9) → (8) → (6) → (7) → (11) → (12) → (15) → (16) → (14) → (13) → (10) → (21) → (20)
AS 2
(1) → (19) → (18) → (22) → (17) → (3) → (2) → (4)→ (5) → (9) → (11) → (12) → (15) → (16) → (14) → (13) → (8) →(6) → (7) → (10) → (21) → (20)

Fig. 6. AS respecting the criterion of ATOS

4 Conclusion

In this paper, an approach based on the assembly tools operation space is proposed. This criterion is used to ameliorate the quality and the efficiency of the assembly sequences, which are generated by our old work. To achieve that, the developed algorithm examines all these AS. For each one, it identifies fastener, the movement length and the appropriate tool. Then, interference tests are made, according to fastener's type, to deduce in the end the validated assembly sequences.

The execution time of this algorithm is high, because of interference test, which will be increased according to the number of fasteners, is the AS. This problem will be our future search.

References

Iacob, R., Popescu, D., Carutasu, N.: Development of assembly/disassembly process simulation platform. J. Proc. Manuf. Syst. **8**, 15–24 (2013)

Su, Q.: Computer aided geometric feasible assembly sequence p and optimizing. Int. J. Adv. Manuf. Technol. **33**, 48–57 (2007)

Pintzos, G., Triantafyllou, C., Papakostas, N,. Mourtzis, D., Chryssolouris, G.: Assembly precedence diagram generation through assembly tiers determination. Int. J. Comput. Integr. Manuf. **29**(10), 1045–1057 (2016)

Hyoung, R.L., Gemmill, D.D.: Improved methods of assembly sequence determination for automatic assembly systems. Euro. J. Oper. Res. **131**(3), 611–621 (2001)

Belhadj, I., Trigui, M., Benamara, A.: Subassembly generation algorithm from a CAD model. Int. J. Adv. Manuf. Technol. **87**, 2829–2840 (2016)

Laperriere, L., Lavoie, A.: Evaluating subassembly stability using stability directed subgraphs. Int. J. Comput. Appl. Technol. **10**(5/6), 348–360 (1997)

Smith, S.S.F., Smith, G.C., Liao, X.: Automatic stable assembly sequence generation and evaluation. J. Manuf. Syst. **20**(4), 225–235 (2001)

Bahubalendruni, M.V., Biswal, B.B.: Computer aid for stability testing between parts towards automatic assembly sequence generation. J. Comput. Technol. Appl. **7**(1), 22–26 (2016)

Kumar, M., Bahubalendruni, M.V., Biswal, B.B., Nayak, R.: Identification of stable configurations between constituent parts of an assembly. Appl. Mech. Mater. **852**, 595–601 (2016)

Bedeoui, A., Hadj, R.B., Hammadi, M., Trigui, M., Aifaoui, N.: Assembly sequence plan generation of heavy machines based on the stability criterion. Int. J. Adv. Manuf. Technol. **102**, 2745–2755 (2019)

Liu, X., Liu, Y., Xu, B.: A converse method-based approach for assembly sequence planning with assembly tool. Int. J. Adv. Manuf. Technol. **69**, 1359–1371 (2013)

Evaluating the Capability Index of a Process Integrating Sampling Plan and the Measurement System Number of Distinct Categories NDC

Adel Brik[1,2](✉), Mohamed Goddi[1], and Nabil Ben Fredj[1]

[1] Laboratory of Mechanics, Materials and Processes (LR99ES05), ENSIT, University of Tunis, 5 Av Taha Hussein Montfleury, 1008 Tunis, Tunisia
adel_brik@yahoo.fr, mohamed.goddi@gmail.com, benfredjnabil@gmail.com

[2] Esprit School of Engineering, Pôle Technologique El Ghazela, B.P. 160 Z.I. Chotrana II, 2083 Ariana, Tunisia

Abstract. To assess the capability of a process, the collected data are influenced by the errors of the used measurement system. This study aims to integrate these errors in determining the acceptance threshold value of the capability index C_p that is referred as in this work. An expression of that depends on the Number of Distinct Categories (*NDC*), type I and type II errors, the index of the capability process corresponding to both rejectable and acceptable quality levels (C_{pRQL}) and (C_{pAQL}), respectively, was developed. These parameters effects and their existing interactions on the variability in the defective parts per million *ppm* (Δppm) were examined using a response surface experimental design with a total of 81 runs. Analysis of the results of this experimental design showed that it is possible to propose a technique for overcoming the inaccuracy of the used measurement system by targeting higher C_{pAQL} values as the same value of can be obtained by several combinations of C_{pAQL} and *NDC*.

Keywords: C_p index of process capability · Errors of measurement system · *NDC* · Of threshold capability index value

1 Introduction

Indices of Process Capability Indices (IPC) are used for monitoring the variability of manufacturing processes to prevent the production of non-conforming parts. One of the mostly used PCI is C_p that compares the process variability with the required specification tolerance (Kane 1986). In spite of their popularity, IPCs are modeled under some assumptions that are difficult to respect in practice. In this sense, Kane (1986) calculated the sampling acceptance threshold value C_0 useful to adjudicate the process capability. This value is expressed considering risks α and β, the size of the sample n while neglecting the errors of the measurement system. Moreover, Pearn and Wu (2007) developed an acceptance sampling plan centered on the cumulative distribution function of the estimated capability index without considering the measurement system real state. Nevertheless, recent investigations established that the estimation accuracy of

M. Kharrat et al. (Eds.): ICAMEM 2019, LNME, pp. 118–123, 2020.
https://doi.org/10.1007/978-3-030-52071-7_17

PCIs is widely dependent on the sampling conditions and the measurement system state (de-Felipe and Benedito (2017); Ben Amara et al. (2017); Mahshid et al. (2018); Brik et al. (2019)). Indeed, Ben Amara et al. (2017) offered a technique that establishes the true index of the process capability (C_{pTrue}) based on the observed capability index (C_{pObs}) and a metric characterizing the used measurement. Brik et al. (2019) proposed a method to adjust the sampling plan depending on the real measurement system state evaluated by the *GR&R* metric (Gage Repeatability and Reproducibility). However, this metric is not the only one used for evaluating the measurement systems performances. Indeed, the AIAG (2010) (Automotive Industry Action Group) supports using three other metrics that are the Number of Distinct Categories (*NDC*), the Discrimination Ration (*DR*) and the Precision to Tolerance Ration (*PTR*). As these metrics have different formulations and threshold values it becomes important, from industrial point of view, to give expression of the acceptance threshold value of C_p (C^*_{p0}) depending on each metric. In the present work, *NDC* is the metric considered for the evaluation of the measurement system error and an expression of C^*_{p0} depending on type I and II errors, the rejectable and the acceptable quality level (C_{pRQL} and C_{pAQL} respectively) is developed.

2 Consideration of Sampling Involving Errors of the Measurement System

This paper is based on the following assumptions: process with an only one normally distributed quality characteristic is considered; the capability index is evaluated using variable sampling plan; the measurement system state is assessed by *NDC*.

Kane (1986) showed that the capability index C_p can be estimated by variable sampling plan developed using the following hypothesis testing:

$$\begin{cases} H_0 : \hat{C}_p \leq C_{p0},\ a\,not\,capable \\ H_1 : \hat{C}_p > C_{p0},\ a\,capable\,process \end{cases}$$

This test of hypothesis is subjected to the occurrence of type I and II errors expressed by $\alpha = P(reject\,H_0/C_p = C_{pRQL})$ and $\beta = P(accept\,H_0/C_p = C_{pAQL})$. Considering that $(n-1)\left(\frac{\hat{\sigma}}{\sigma}\right)^2$ follows a chi-square distribution with *(n-1)* degree of freedom, the acceptance critical value C_{p0} is equal to:

$$C_{p0} = C_{pRQL}\sqrt{\frac{(n-1)}{\chi^2_{\alpha,(n-1)}}} \quad (1)$$

The variable sampling plan is established using couples of values (C_{pRQL},) and (C_{pAQL}, β) and is characterized by n (the sample size) and the distance in units of the standard deviation between the acceptance limit and the specification limit *(k)*.

2.1 Measurement System Errors Integration

Number of distinct categories (*NDC*) is expressed by the *AIAG* as the ratio between the variability of the process σ_p and the measurement system variability σ_m.

$$NDC = \sqrt{2}\frac{\sigma_p}{\sigma_m} \tag{2}$$

As

$$\sigma_t^2 = \sigma_p^2 + \sigma_m^2 \tag{3}$$

It can be established that:

$$C_{pObs} = \frac{Tolerance}{6\sigma_t} = \frac{Tolerance}{6\sqrt{\sigma_p^2 + \sigma_m^2}} \tag{4}$$

$$C_{pAct} = \frac{Tolerance}{6\sigma_p} \tag{5}$$

Hence,

$$\frac{C_{pAct}}{C_{pObs}} = \frac{\sigma_t}{\sigma_p} = \sqrt{\frac{\sigma_p^2 + \sigma_m^2}{\sigma_t^2 - \sigma_m^2}} \tag{6}$$

Using Eqs. (2) and (6) the following expression can be given:

$$C_{pAct} = C_{pObs}\sqrt{1 + \frac{2}{NDC^2}} \tag{7}$$

2.2 The Threshold Capability Index Calculation Considering *NDC*

As the actual index of capability C_{pAQL} must be greater than C_{p0} ($C_{pAQL} > C_{p0}$) to conclude about the process capability and using Eqs. (7) and (1) leads to:

$$C_{pObs} > C_{pRQL}\sqrt{\frac{(n-1) \times NDC^2}{\chi^2_{\alpha,(n-1)} \times (NDC^2 + 2)}} \tag{8}$$

Therefore, a new critical threshold value C_{p0}^* can be established and the observed index of capability must exceeds this value to assess the process capability while integrating the state of the measurement system:

$$C_{p0}^{*} = C_{pRQL}\sqrt{\frac{(n-1) \times NDC^2}{\chi^2_{\alpha,(n-1)} \times (NDC^2 + 2)}} \tag{9}$$

Equation (9) shows that C_{p0}^{*} depends on C_{pRQL}, α, NDC and n (the sample size) of the sampling plan that is determined using couples (C_{pRQL}, α) and (C_{pAQL}, β). For these reason, the effects of α, β, C_{pRQL}, C_{pAQL} and NDC on C_{p0}^{*} will be discussed in the following.

3 Results and Discussion

3.1 Effect of Occurred Measurement System Errors on *ppm* Defective Items

As previously indicated, C_{p0} and C_{p0}^{*} are the acceptance threshold values of the capability index that must be used when the measurement system is assumed to be perfect ($NDC \geq 5$) or having errors ($NDC < 2$), respectively. These two threshold values can be associated with the non-conforming parts per million ppm_{Cp0} and $ppm_{C_{p0}^{*}}$ corresponding to C_{p0} and C_{p0}^{*}, respectively. The relationship between *ppm* and the capability index C_p can be found in Montgomery (2009). Therefore, it is possible to quantify the number of under evaluated defective *ppm* considering a perfect measurement system by using the parameter $\Delta ppm = ppm_{C_{p0}^{*}} - ppm_{Cp0}$. It is obvious that neglecting the measurement system errors impacts dramatically the process capability evaluation and the *ppm*. This may lead to false decision-making with the associated extra-costs. As the variation of *ppm* depends on the different parameters C_{pRQL}, C_{pAQL}, α, β and NDC, it's important to investigate the dependence of Δppm on these parameters and their eventual interactions by using a design of experiments (*DOE*).

3.2 *DOE* Establishment

The *DOE* technique is used because it allows to identify the most significant parameters and interactions affecting Δppm. Response surface method (*RSM*) was selected in this study because it's recognized as design for dealing with continuous treatments when describing the responses and finding optima are goals (Myers et al. 2016). The response surface design selected in this work is a quadratic model strongly supported by the *RSM* with great optimization performances (Dean et al. 2017). Therefore, 2^4 factorial design with an added center point (level 0) was used. Thus a design with a total of 3^4 runs and 15 and 66 freedom degrees (*df*) linked to the quadratic model and to the error, respectively was developed.

3.2.1 Results

Table 1 shows that all factors have significant effects on Δppm and that *NDC* interacts significantly with the other factors (P-value < 0.05). In the followings, a particular attention will be given to the interaction between C_{pAQL} and *NDC* to explain how the value of C_{pAQL} can be adjusted to overcome the lack of the measurement system accuracy.

Table 1. Variance Analysis for Δppm.

Source	Sum of Squares	df	Mean Square	F-Ratio	P-Value
α	18518	1	18518	8.69	0.0044
β	24180.6	1	24180.6	11.34	0.0013
C_{pAQL}	378181	1	378181	177.42	0.0000
NDC	867137	1	867137	406.8	0.0000
α NDC	21015.9	1	21015.9	9.86	0.0025
β NDC	27501.1	1	27501.1	12.9	0.0006
C_{pAQL} NDC	430848	1	430848	202.12	0.0000
NDC NDC	233636	1	233636	109.6	0.0000
Total error	140687	66	2131.62		
Total (corr.)	2.15167E6	80			

3.2.2 Measurement System Errors Effects on C^*_{p0}

In the followings, we make use of the results of the experimental design to propose a method for overcoming the lack of measurement system precision. Hence, measurement systems with low values of *NDC* can be used to evaluate accurately the capability of a process. For this purpose, contour lines of C^*_{p0} are plotted depending on *NDC* and C_{pAQL} (Fig. 1).

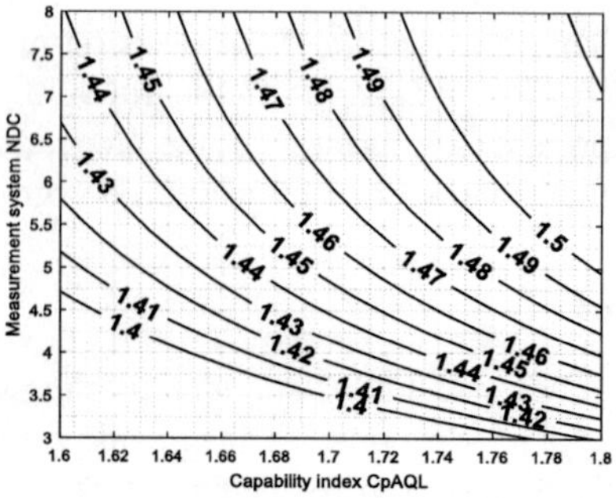

Fig. 1. Contour plots of C^*_{p0} depending on *NDC* and C_{pAQL} (C_{pRQL}= 1.33, α = 0.055, β = 0.04)

These contours show that the same value of C^*_{p0} can be obtained by different combinations of *NDC* and C_{pAQL}. Moreover, the lack of the measurement system accuracy can be compensated by high values of C_{pAQL}. For example, for $C^*_{p0} = 1.43$ when an accurate measurement system is being used (i.e. $NDC = 6.7$) selecting a value of C_{pAQL}= 1.6 is enough to judge accurately the process capability. Meanwhile, when an inaccurate measurement system is used (i.e. $NDC = 3.5$) higher value of C_{pAQL} must be targeted (C_{pAQL}= 1.77) to achieve similar conclusion about the process capability given by the accurate system ($NDC = 6.7$). This finding is very useful in practice as it is possible to continue evaluating the process capability with inaccurate measurement systems characterized by low values of *NDC*. It is important to notice that in comparison with the previous work of Brik et al. (2019), the required adjustments of C_{pAQL} to consider the errors of the measurement system assessed using *GR&R* and *NDC* are totally different suggesting that these two metrics have to be investigated separately.

4 Conclusion

In this paper, a new threshold acceptance capability index C^*_{p0} is developed considering the measurement system error characterized by *NDC*, type I and II sampling errors, C_{pAQL} and C_{pRQL}. To study the interaction effects of these parameters on the process capability, a 3^4 factorial *DOE* was developed. Results showed that *NDC* and C_{pAQL} interact strongly when evaluating C^*_{p0}. Based on this finding, contour of C^*_{p0} depending on *NDC* and C_{pAQL} were plotted. Therefore, a method for overcoming the inaccuracy of the measurement system by adjusting the value of C_{pAQL} was proposed. These results are very useful as they offer the possibility to practitioners to continue using unreliable measurement system to establish reliable decisions regarding the process capability.

References

AIAG Work Group: Measurement Systems Analysis: Reference Manual, 4th edn. Automotive Industry Action Group, Detroit (2010)

Ben, A.S., Dhahri, J., Ben, F.N.: Process true capability evaluation with the consideration of measurement system variability and expected quality loss. Qual. Reliab. Eng. Int. **33**(5), 937–944 (2017)

Brik, A., Goddi, M., Dhahri, J., Fredj, N.B.: Assessing process capability index using sampling plan in the presence of measurement system errors. Int. J. Adv. Manuf. Technol. **102**, 3301–3313 (2019)

Dean, A., Voss, D., Draguljić, D.: Design and Analysis of Experiments, 6th edn. Springer, Cham (2017)

de-Felipe, D., Benedito, E.: A review of univariate and multivariate process capability indices. Int. J. Adv. Manuf. Technol. **92**, 1687–1705 (2017)

Kane, V.E.: Process capability indices. J. Qual. Technol. **18**(1), 41–52 (1986)

Mahshid, R., Mansourvar, Z., Hansen, H.N.: Tolerance analysis in manufacturing using process capability ratio with measurement uncertainty. Precis. Eng. **52**, 201–210 (2018)

Montgomery, D.C.: Introduction to Statistical Quality Control: A Modern Introduction, 6th edn. Wiley, New York (2009)

Myers, R.H., Montgomery, D.C., Anderson-Cook, C.M.: Response Surface Methodology, Product and Process Optimization Using Experimental Design, 4th edn. Wiley, New York (2016)

Pearn, W.L., Wu, C.-W.: An effective decision making method for product acceptance. Omega **35**(1), 12–21 (2007)

Simulation of Low Velocity Impact of Epoxy-Flax Fibers Composite Using Solid Shell Finite Element

A. Chaker[1,2](✉), S. Koubaa[1], J. Mars[1], F. Gehring[2], F. Dammak[1], and A. Vivet[2]

[1] Electro-Mechanical System's Laboratory (LASEM), National Engineering School of Sfax, University of Sfax, B.P 1173, 3038 Sfax, Tunisia
aidacheker811@gmail.com, koubaasana@yahoo.fr, jamelmars@yahoo.fr, fakhreddine.dammak@enis.tn

[2] Normandie Univ, ENSICAEN, UNICAEN, CEA, CNRS, CIMAP, 14000 Caen, France
{florian.gehring,alexandre.vivet}@unicaen.fr

Abstract. The objective of the present paper is to predict numerically the effect of low velocity impact on the composite epoxy reinforced with quasi-unidirectional flax fabrics. The composite circular plate is meshed with a solid shell finite element (SS). Only one element in the thickness direction is adopted. This SS element is implemented into the user element (UEL) interface of ABAQUS in order to overcome numerical locking problems that may occurred when using the ABAQUS solid and shell elements. The epoxy/flax composite is characterized by a elasto-visco-plastic behavior. The Hill's anisotropic criterion of plasticity with isotropic hardening and the Johnson-Cook viscoplastic model are applied. The simulations are accomplished via ABAQUS/Standard. The obtained results present the efficiency of the developed model using one solid shell element in the thickness direction from a computational point of view.

Keywords: Low velocity impact · Solid-shell · Bio-composite · Flax

1 Introduction

In the last few decades, the development of an efficient and free-locking finite element becomes the main challenge in mechanical engineering. Indeed, the conventional solid and shell elements present in some cases numerical locking problems. To avoid these problems, an enhanced solid shell element was proposed by Klinkel et al. (2006) and Vu-Quoc and Tan (2003). The 3D mixed formulation of this element is developed by combining two enhancing methods (ANS and EAS methods) For the case of orthotropic materials, this element is widely applied for the case of static analysis (Leonetti and Nguyen-Xuan 2019; Wang et al. 2017). However, to the knowledge of authors, the solid shell element was

M. Kharrat et al. (Eds.): ICAMEM 2019, LNME, pp. 124–130, 2021.
https://doi.org/10.1007/978-3-030-52071-7_18

rarely applied for the impact problems. Zhi and Tay (2019) applied this element to analyze the matrix cracking and delamination of a multilayered composite under low velocity impact solicitation. This composite is characterized by an elastic orthotropic behavior coupling with a 3D anisotropic damage model. Referring to previous studies, the low velocity impact problem was widely investigated adopting conventional elements for the case of composite reinforced with flax fibers. Abida et al. (2018a) investigated numerically the impact loading on epoxy reinforced with quasi-unidirectional flax fabric. This composite is characterized by an orthotropic visco-elastoplastic behavior. The composite circular plate was meshed with ten C3D8 elements in the thickness direction which is expensive in computational times (CPU). To solve this problem, Chaker et al. (2020) implemented a solid shell element (SS) into the ABAQUS UEL interface. Only one element through the thickness direction is adopted to mesh the impact circular plates. The finite element model was validated in the case of aluminum circular plate. It was noticed that this finite element model applying SS element allows to save more than 10 CPU times compared to conventional solid element. Nonetheless, this model was limited to isotropic materials.

In this current paper the application field of the solid shell element is extended to orthotropic visco-elastoplastic composite epoxy reinforced with quasi-unidirectional flax fabric.

2 Finite Element Formulation

In this work, an hexahedral eight nodes solid shell finite element (SS) with only three degrees of freedom is derived. The formulation of this element, projected by Klinkel et al. (2006) and Vu-Quoc and Tan (2003), is procured by the mixture of two enhancing methods, Enhanced Assumed Strain (EAS) and Assumed Natural Strain (ANS) methods. The Green-Lagrange strain tensor by $\mathbf{E}$ is expressed as function of the compatible strain tensor $\mathbf{E}^c$ and the incompatible strain tensor $\tilde{\mathbf{E}}$ as:

$$\mathbf{E} = \mathbf{E}^c + \tilde{\mathbf{E}} \tag{1}$$

The Hu-Whashizu variational principle takes the following expression as function of the displacement field $\mathbf{u}$ and the incompatible strain tensor $\tilde{\mathbf{E}}$:

$$\Pi(\mathbf{u}, \tilde{\mathbf{E}}) = \int_V \Big(\psi(\mathbf{E}^c + \tilde{\mathbf{E}})\Big) dV - \int_V \mathbf{F}_V.\mathbf{u} dV - \int_{\partial V_f} \mathbf{F}_S.\mathbf{u} dA \tag{2}$$

ψ is the Helmotz strain energy function, $\mathbf{F}_V$ and $\mathbf{F}_S$ are respectively the body and surface forces.

The ANS method is developed by Dvorkin and Bathe (1984) for the purpose of solving transverse shear locking problems. This method consists on enhancing the shear terms (E^c_{13}, E^c_{23}) and the transverse term (E^c_{33}) of the compatible strain tensor as:

$$\begin{bmatrix} 2E^c_{13} \\ 2E^c_{23} \end{bmatrix} = \begin{bmatrix} (1-\eta)E^B_{13} + (1+\eta)E^D_{13} \\ (1-\xi)E^A_{23} + (1+\xi)E^C_{23} \end{bmatrix} \tag{3}$$

$$E_{33}^{c} = \sum_{L=1}^{4} \frac{1}{4} \left(1 + \xi_L \xi\right) \left(1 + \eta_L \eta\right) E_{33}^{L} \tag{4}$$

where A, B, C, D and A_i (i = 1,2,3,4) are the transverse shear and transverse strain interpolation points.

The EAS method was developed by Simo and Rifai (1990) in order to avoid volumetric locking problems. This method consists on define the enhancing incompatible strain tensor $\tilde{\mathbf{E}}$ as:

$$\tilde{\mathbf{E}} = \frac{\det \mathbf{J}_0}{\det \mathbf{J}} \left(\mathbf{T}_S^0\right)^{-T} \tilde{\bar{\mathbf{E}}} \tag{5}$$

where $\mathbf{J}$ is the Jacobian matrix, $\mathbf{T}_S$ is a matrix transformation from the local covariant base to the global cartesian base and $\tilde{\bar{\mathbf{E}}}$ is the incompatible strain tensor in the local base. The details of this formulation are depicted in Chaker et al. (2020) and Hajlaoui et al. (2019)

3 Visco-Elastoplastic Constitutive Equations

In this study, the composite epoxy reinforced with flax fibers is considered with orthotropic visco-elastoplastic behavior (Abida et al. 2018a; 2018b). The Green-Lagrange strain tensor is decomposed into elastic and plastic strain tensors as:

$$\mathbf{E} = \mathbf{E}^e + \mathbf{E}^p \tag{6}$$

This two tensors are decoupled. This assumption is valid also for the case of the updated Lagrangian formulation. The second Piola-Kirchhoff and the Green-Lagrange strain tensorstress tensor can be written respectively as σ and ε. The elastic strain tensor is related to the stress tensor as:

$$\sigma = \mathbf{D} : \varepsilon^e \tag{7}$$

Where $\mathbf{D}$ is the elastic part of the stiffness matrix. It's expressed as the inverse of the compliance matrix $\mathbf{S}$ where:

$$\mathbf{S} = \begin{bmatrix} \frac{1}{E_{11}} & -\frac{\nu_{12}}{E_{11}} & -\frac{\nu_{13}}{E_{11}} & 0 & 0 & 0 \\ & \frac{1}{E_{22}} & \frac{-\nu_{23}}{E_{22}} & 0 & 0 & 0 \\ & & \frac{1}{E_{33}} & 0 & 0 & 0 \\ & Sym & & \frac{1}{G_{12}} & 0 & 0 \\ & & & & \frac{1}{G_{13}} & 0 \\ & & & & & \frac{1}{G_{23}} \end{bmatrix} \tag{8}$$

The plastic surface is defined as:

$$f = \varphi\left(\sigma\right) - \sigma_p < 0 \tag{9}$$

$\varphi(\sigma)$ is the equivalent stress and σ_p is the yield stress. The equivalent stress takes the following expression, applying the Hill criterion:

$$\varphi(\sigma) = \|\sigma\|_{\mathbf{H}} = \sqrt{\sigma^T \mathbf{H}\, \sigma} \tag{10}$$

$$\mathbf{H} = \begin{bmatrix} H+G & -H & -G & 0 & 0 & 0 \\ & H+F & -F & 0 & 0 & 0 \\ & & F+G & 0 & 0 & 0 \\ & & & 2N & 0 & 0 \\ & Sym & & & 2M & 0 \\ & & & & & 2L \end{bmatrix} \tag{11}$$

Where F, G, H, L, M are the Hill material parameters. The yield stress is expressed for the case of isotropic hardening and when applying the visco-plastic Johnson-Cook model as following:

$$\sigma_p = [\sigma_Y + Q(1 - \exp(-\beta\varepsilon_p)) + k\varepsilon_p] \left[1 - CLog\left(\frac{\dot{\varepsilon}}{\dot{\varepsilon}_0}\right)\right] \tag{12}$$

k is defined as the asymptotic hardening modulus, Q and β are the cause of nonlinearity of the tensile curve when the plasticity achieves. C and $\dot{\varepsilon}_0$ refers respectively to the sensitivity coefficient of the strain rate and the reference strain rate. Finally, the Euler implicit algorithm and the Newton-Raphson method were applied to determine the plastic part of the tangent modulus. The details of the present formulation are presented in Abida et al. (2018b); Koubaa et al. (2017). The composite material orthotropic elastic properties and the Hill'48 material constants and the Johnson-Cook model parameters are illustrated respectively in the Table 1 and the Table 2.

Table 1. Flax/epoxy Mechanical properties (Cherif et al. (2016))

$E_{11}(GPa)$	$E_{22}(GPa)$	$G_{12} = G_{13}(GPa)$	G_{23}	$\nu_{12} = \nu_{13}$	ν_{23}
29.2 ± 1.0	4.7 ± 0.3	1.86 ± 0.17	1.66 ± 0.15	0.37 ± 0.01	0.4 ± 0.06

Table 2. Hill's material constants and hardening parameters (Abida et al. 2018b)

F	G	H	N	M	L	$\sigma_Y(MPa)$	$Q(MPa)$	β	$k(MPa)$	C
4.38	0.50	0.50	3.20	3.20	3.20	33	6.015	1210	20684	0.152

4 Application to Low Velocity Impact

The main objective of this part is to utilize a SS element to model the low velocity impact problems on the orthotropic materials as the epoxy/flax composite with visco-elastoplastic behavior. The impact model was validated by

Chaker et al. 2019; 2020 for the case of the isotropic materials using aluminum circular plates. The impact model is made of a clamped flax/epoxy circular plate of 60 mm radius and 1 mm thickness and a cylindrical projectile with hemispherical nosed. The plate is meshed with 4888 SS elements and only one element in the thickness is adopted. The projectile of 6.35 mm and 28.6 g mass is assumed rigid. It's subjected to an initial velocity $V_0 = 5\,\text{m/s}$. Owning to the geometry symmetry, the simulations are operated with the one quarter of this geometry (presented in Fig. 1). The simulations were carried on via ABAQUS/Standard and the temporel responses of the impact force were determined. The obtained results with (SS) element are compared to the case of conventional elements (10 C3D8 elements in the thickness direction). It can be noticed that our model is well validated with the numerical one when adopting 10 hexahedral elements through the thickness direction in the case of the composite epoxy reinforced with quasi-unidirectional flax fabric characterized by an orthotropic visco-elastoplastic behavior. Referring to the Table 3, this solid shell element provides accurate results and allows to save six times of CPU time (see Table 3) (Fig. 2).

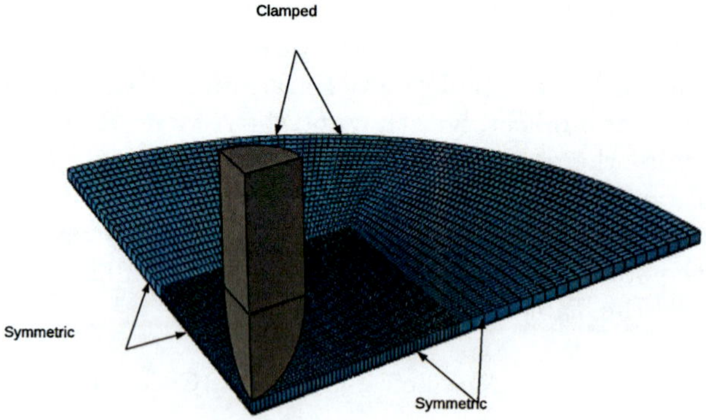

Fig. 1. Description of the low velocity impact model

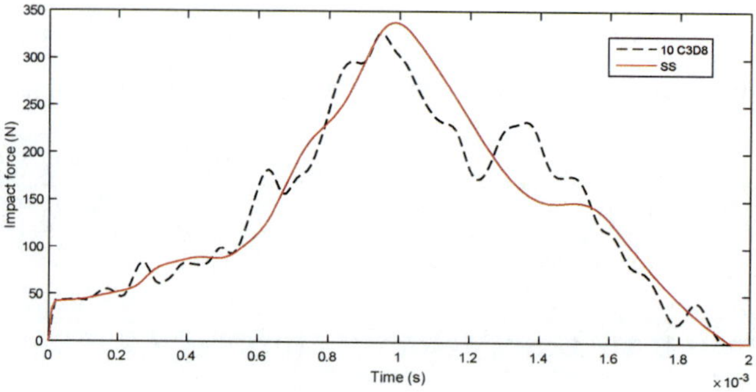

Fig. 2. Temporel response of the impact force

Table 3. CPU time when adopting solid shell and solid elements

Element	48880 C3D8	4888 SS
CPU time (s)	6082	1099
Saving time		6

5 Conclusions

The purpose of this work is to employ the solid shell element to predict the response of low velocity on orthotropic composite materials as the epoxy reinforced with quasi-unidirectional flax fabric. The circular plates are meshed with one solid shell finite element in the thickness. This composite is characterized by an orthotropic visco-elastoplastic mechanical behavior. The present contribution consists on validating our finite element model in the case of orthotropic materials. The impact obtained results are well correlated with the numerical ones in which ten solid element are applied in the thickness direction. Further experimental tests will be performed in future works.

References

Abida, M., Mars, J., Gehring, F., Vivet, A., Dammak, F.: Anisotropic visco-elastoplastic modeling of quasi-unidirectional flax fiber reinforced epoxy behavior: an investigation on low-velocity impact response. J. Renew. Mater. **6**, 464–476 (2018a)

Abida, M., Mars, J., Gehring, F., Vivet, A., Dammak, F.: Anisotropic elastic-viscoplastic modelling of a quasi-unidirectional flax fibre-reinforced epoxy subjected to low-velocity impact. In: Haddar, M., Chaari, F., Benamara, A., Chouchane, M., Karra, C., Aifaoui, N. (eds.) Design and Modeling of Mechanical Systems–III, pp. 171–178. Springer, Cham (2018b)

Chaker, A., Koubaa, S., Mars, J., Vivet, A., Dammak, F.: An ABAQUS implementation of a solid-shell element: application to low velocity impact. In: Aifaoui, N., et al. (eds.) Design and Modeling of Mechanical Systems - IV, pp. 770–777. Springer, Cham (2019). https://doi.org/10.1007/978-3-030-27146-6_84

Chaker, A., Koubaa, S., Mars, J., Vivet, A., Dammak, F.: An efficient ABAQUS solid shell element implementation for low velocity impact analysis of FGM plates. Eng. Comput. (2020). https://doi.org/10.1007/s00366-020-00954-8

Cherif, Z.E., Poilâne, C., Vivet, A., Ben Doudou, B., Chen, J.: About optimal architecture of plant fibre textile composite for mechanical and sorption properties. Compos. Struct. **140**, 240–251 (2016)

Dvorkin, E.N., Bathe, K.: A continuum mechanics based four node shell element for general non linear analysis. Eng. Comput. **1**, 77–88 (1984)

Hajlaoui, A., Chebbi, E., Dammak, F.: Buckling analysis of carbon nanotube reinforced FG shells using an efficient solid-shell element based on a modified FSDT. Thin-Walled Struct. **144**, 106254 (2019)

Klinkel, S., Gruttmann, F., Wagner, W.: A robust non-linear solid shell element based on a mixed variational formulation. Comput. Methods Appl. Mech. Eng. **195**, 179–201 (2006)

Koubaa, S., Mars, J., Wali, M., Dammak, F.: Numerical study of anisotropic behavior of Aluminum alloy subjected to dynamic perforation. Int. J. Impact Eng. **101**, 105–114 (2017)

Leonetti, L., Nguyen-Xuan, H.: A mixed edge-based smoothed solid-shell finite element method (MES-FEM) for laminated shell structures. Compos. Struct. **208**, 168–179 (2019)

Simo, J.C., Rifai, M.S.: A class of mixed assumed strain methods and the method of incompatible modes. Int. J. Numer. Methods Eng. **29**, 1595–1638 (1990)

Vu-Quoc, L., Tan, X.G.: Optimal solid shells for non-linear analyses of multilayer composites. I. statics. Comput. Methods Appl. Mech. Eng. **192**, 975–1016 (2003)

Wang, P., Chalal, H., Abed-Meraim, F.: Quadratic prismatic and hexahedral solid-shell elements for geometric nonlinear analysis of laminated composite structures. Compos. Struct. **172**, 282–296 (2017)

Zhi, J., Tay, T.-E.: Explicit modeling of matrix cracking and delamination in laminated composites with discontinuous solid-shell elements. Comput. Methods Appl. Mech. Eng. **351**, 60–84 (2019)

Finite Element Residual Stress Computation for Combined Grinding/Burnishing Applied to 100Cr6 Steel

Yasmine Charfeddine[1,2](✉), Sawsen Youssef[1], Salem Sghaier[1], and Hédi Hamdi[2]

[1] University of Monastir, ENIM, LGM, Avenue Ibn El Jazzar, 5000 Monastir, Tunisia
yasmine.charfeddine29@gmail.com, sawsen.youssef@gmail.com

[2] University of Lyon, ENISE, LTDS, 58 Rue Jean Parot, 42023 Saint-Etienne, France
hedi.hamdi@enise.fr

Abstract. Combining grinding with ball burnishing: "REGAL" is a newly developed technique consisting in grinding and burnishing performing simultaneously by placing the burnishing tool on the grinding machine. This combined manufacturing process ensures obtaining a better surface quality by taking advantage of the workpiece being thermally affected by the grinding process, reducing surface damages obtained after grinding like tensile residual stress results and enhancing roughness results by the burnishing process. REGAL ensures also reducing manufacturing energy and time by combining grinding and burnishing acting simultaneously instead of each one acting separately. Further, the surface integrity would be enhanced by REGAL process in terms of hardness, surface roughness, and compressive residual stress obtained at a more important depth compared to burnishing process, which extends workpiece lifetime, decreases fatigue and improves wear resistance. In this paper, the residual stress results are studied thanks to 3D finite element simulations of the grinding, ball burnishing and combined grinding/ball-burnishing processes. The finite element method helps to compare the residual stress results obtained after combining two different processes one thermomechanical (grinding) and the other purely mechanical (burnishing) both with different manufacturing parameters to regular grinding and burnishing process.

Keywords: Combined processes · Grinding · Ball-burnishing process · 3D FEM simulation · Residual stress · 100cr6 (AISI 52100)

1 Introduction

Grinding is a thermomechanical cutting process used to get good surface finish with shape and dimension precision. This process ensures also obtaining a smooth surface but introduces tensile residual stresses (Hamdi et al. 2004) responsible for fatigue (Majzoobi et al. 2016) and low wear life resistance of the workpiece. To remedy to this,

M. Kharrat et al. (Eds.): ICAMEM 2019, LNME, pp. 131–137, 2020.
https://doi.org/10.1007/978-3-030-52071-7_19

it is necessary to add burnishing process. The burnishing ball or roller pushes in the valleys the microscopic pics left by the grinding wheel to get a better surface roughness (Prakash et al. 2014). In the other hand, imposing a high pressure to the surface and subsurface by the burnishing tool ensures getting a plastic deformation (Grochala et al. 2017), increases workpiece hardness (Malleswara Rao et al. 2011) and introduces compressive residual stress in surface and subsurface (Sanchez et al. 2017).

"REGAL" is the french abbreviation for grinding and burnishing. "REGAL" process is a promising solution since it can offer a gain of time and energy by launching these two processes at the same time. REGAL offers as well a better surface quality by profiting from the thermal energy issued from the grinding process. The aim of this work is to study residual stress induced by REGAL process, and to compare them to residual stresses resulting from both the grinding and burnishing process separately. The residual stress results of grinding, burnishing and combined grinding/ball-burnishing process are obtained thanks to finite element simulations conducted with ABAQUS software.

2 Simulation of the Combined Grinding/Ball-Burnishing

2.1 Numerical Model

The REGAL model consists on combining the grinding with the burnishing model considering the real distance existing between the burnishing tool and the grinding wheel. The grinding process is modeled by a moving triangular heat flux (Malkin and Guo 2007) and the burnishing process is represented with a burnishing force applied to the center of a rigid ball. The rigid ball and the heat flux are placed adjacently and moving with the same workpiece speed (Fig. 1).

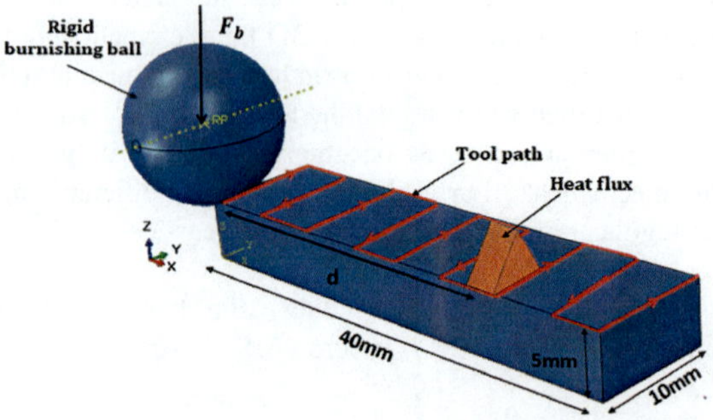

Fig. 1. 3D model of combined grinding/ball-burnishing process

2.2 Finite Element Simulation

The 3D finite element simulation of the combined Grinding-ball burnishing is proposed thanks to the software ABAQUS™. The triangular heat flux and the ball burnishing move along the specimen with the same speed, which is the workpiece speed according to a pre-defined path.

The boundary conditions for the workpiece are an initial temperature of 20 °C, a convective heat change coefficient of 4.10^4 $Jm^{-2}s^{-1}C^{-1}$ and a friction coefficient between the ball and the workpiece equal to 0.3. The workpiece mesh is composed of 39600 linear hexahedral elements with 43512 nodes (Fig. 2).

Fig. 2. Meshed workpiece

The grinding simulation parameters are cited below in Table 1.

Table 1. Grinding simulation parameters

Grinding simulation parameters	
D_s [mm]	250
b [mm]	10
V_s [m/s]	37.5
a_p[μm]	50
L_c[mm]	3.6
V_w[mm/s]	250

The workpiece is considered deformable and is fixed at its bottom. The burnishing ball is considered rigid free to rotate along the workpiece surface and a burnishing force is applied along Z direction at its center. This burnishing force is calculated from the pressure applied to the burnishing tool Eq. 1.

$$F_b = \frac{\Pi D_b^2 P}{4} \tag{1}$$

Where: F is the burnishing force, D_b is burnishing ball diameter and P is the pressure applied to the burnishing tool.

The burnishing simulation parameters are cited in Table 2.

Table 2. Burnishing simulation parameters

Burnishing simulation parameters	
D_b [mm]	13
P [MPa]	30
F_b [N]	3982
V_w[mm/s]	250
over step [mm]	0.2

The workpiece material 100Cr6 (AISI 52100) steel follows a Johnson-Cook's elasto-plastic behavior following the Eq. 2 below:

$$\sigma = (A + B\varepsilon^n)\left(1 + C\,ln\frac{\dot{\varepsilon}}{\dot{\varepsilon}_0}\right)(1 - \left(\frac{T - T_{room}}{T_{melt} - T_{room}}\right)^m) \quad (2)$$

Where:

A: Initial yield stress (MPa), B: Hardening modulus (MPa), n: Work hardening exponent, C: Dependent strain rate coefficient, m: Thermal softening coefficient, ε: plastic strain, $\dot{\varepsilon}$: Plastic strain rate, $\dot{\varepsilon}_0$: Reference plastic strain rate (s^{-1}) (Table 3).

Table 3. Johnson-cook's parameters for the 100Cr6 steel

A[MPa]	B[MPa]	n	C	m
688.17	150.82	0.3362	0.003	2.7786

3 Residual Stress Results

Grinding permits having a good quality of surface with high dimension accuracy but undesirable tensile residual stresses known to be bad for the workpiece wear resistance may appear. After finite element simulation of the grinding process, tensile residual stresses are obtained on the workpiece surface estimated at 670 MPa on the grinding direction S22 and 480 MPa for the residual stress perpendicular to the grinding direction S11 (Fig. 3).

Burnishing is a mechanical surface and subsurface treatment allowing to obtain a harder and smoother workpiece surface then the one got after cutting processes, presenting generally compressive residual stresses results on the surface. The simulation of burnishing process shows that compressive residual stresses equal to −100 MPa are obtained on the surface in the direction that is perpendicular to the burnishing one (tangential residual stresses S11) but tensile residual stress are obtained on the surface of the workpiece in the burnishing direction (Axial residual stresses S22) (Fig. 4).

After REGAL, the maximum compressive residual stress results equal to −580 MPa are obtained for the axial direction and according to z axis reached a depth of 1.25 mm. The maximum compressive residual stress results in the tangential direction are equal to −240 MPa at a depth of 0.15 mm (Fig. 5). At the surface and reaching a depth of 0.15 mm, the axial residual stress results S22 are tensile followed by compressive residual stress results (Fig. 5).

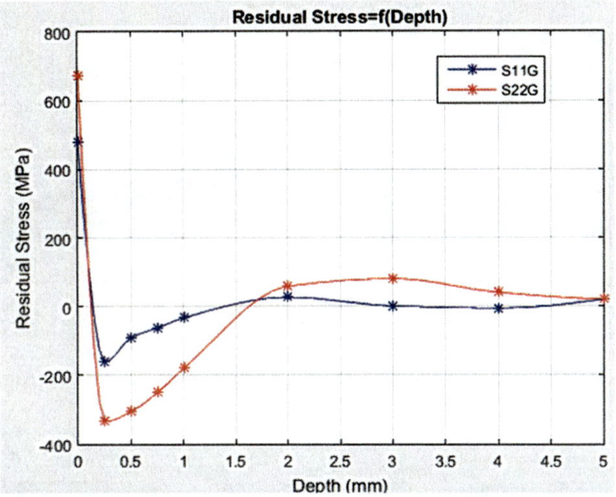

Fig. 3. Axial residual stress S22 and tangential S11 residual stress results versus workpiece depth after grinding

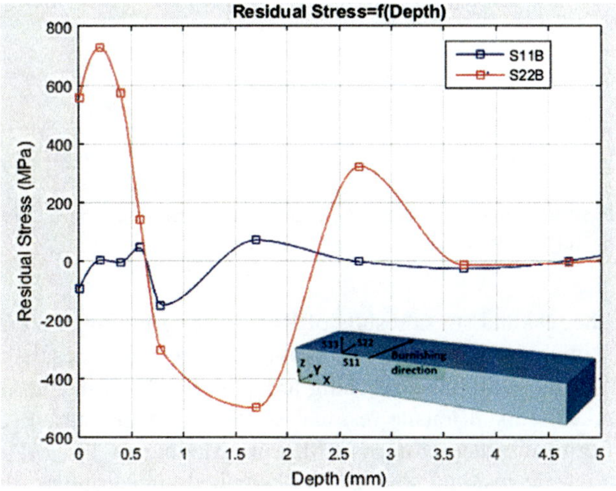

Fig. 4. Axial residual stress S22 and tangential S11 residual stress results versus workpiece depth after burnishing

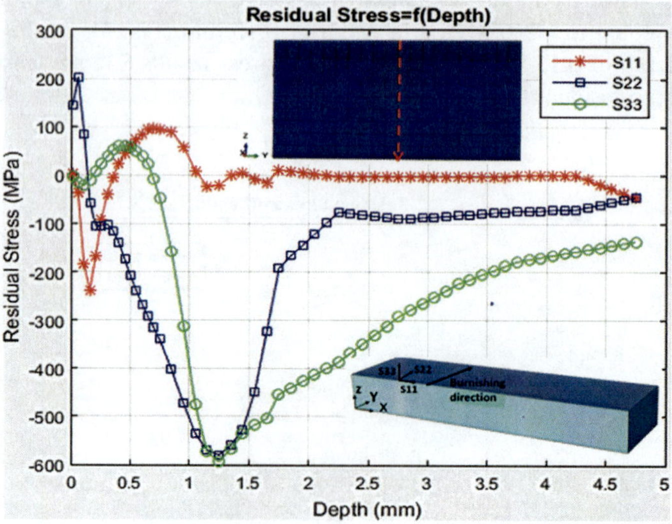

Fig. 5. REGAL Residual stress results (Axial S22, tangential S11 and according to Z axis S33)

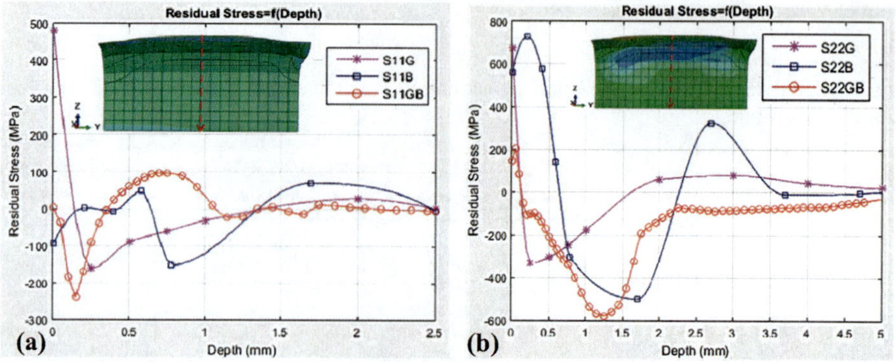

Fig. 6. Residual stress versus workpiece depth after grinding, burnishing and REGAL (a) tangential S11 and (b) Axial S22

Comparing the residual stress results of REGAL to grinding and burnishing acting separately, it is obvious that the compressive residual stress results affects a more important depth of the workpiece reaching a depth of 2.25 mm (Fig. 6(b)).

After REGAL, the axial tensile residual stress S22 at the surface decreased from 555.3 MPa in burnishing and from 670 MPa in grinding to 144 MPa after REGAL (Fig. 6(b)). The tensile residual stresses S11 obtained after grinding in the tangential direction estimated at 480 MPa became purely compressive after REGAL process (Fig. 6(a)).

The absolute value of the compressive residual stress and the affected depth beneath the surface are clearly more important after REGAL, which would increase the lifetime of the workpiece.

4 Conclusions

The tensile residual stresses obtained after grinding are reduced in surface and became compressive in subsurface after the combination of grinding and ball-burnishing process. The value of the compressive residual stress results and its induced affected depth are more important after REGAL if compared to grinding and burnishing process acting separately.

In this study, only a burnishing pressure of 30 MPa was considered which conducts to study the effect of the burnishing pressure and the step over on the residual stress results obtained after REGAL.

In the subsequent work, the focus will be on obtaining purely compressive residual stresses at the surface and subsurface after REGAL by optimizing REGAL parameters.

Acknowledgments. The authors are grateful to the Program "Investissements d'Avenir"-labelled by the French Government and operated by the National Research Agency (ANR) – for financial support to the LabEx MANUTECH-SISE of Université de Lyon (France).

References

Grochała, D., Berczyński, S., Grządziel, Z.: Modeling of burnishing thermally toughened X42CrMo4 steel with a ceramic ZrO2ball. Arch. Civ. Mech. Eng. **17**(4), 1011–1018 (2017)

Hamdi, H., Zahouani, H., Bergheau, J.M.: Residual stresses computation in a grinding process. J. Mater. Process. Technol. **147**(3), 277–285 (2004)

Majzoobi, G.H., Zare Jouneghani, F., Khademi, E.: Experimental and numerical studies on the effect of deep rolling on bending fretting fatigue resistance of Al7075. Int. J. Adv. Manuf. Technol. **82**(9), 2137–2148 (2016)

Malkin, S., Guo, C.: Thermal analysis of grinding. CIRP Ann. Manuf. Technol. **56**(2), 760–782 (2007)

Malleswara Rao, J.N., Chenna Kesava Reddy, A., Rama Rao, P.V.: The effect of roller burnishing on surface hardness and surface roughness on mild steel specimens. Int. J. Appl. Eng. Res. **1**(3), 777–785 (2011)

Prakash, S., Satpal, D., Mishra, R.K.: Effects of roller burnishing process parameters on surface roughness of A356/5% SiC composite using response surface methodology. Adv. Manuf. Technol. **2**, 303 (2014)

Sanchez, L.E.A., Giaretta, F., Nogueira, L.G., Ingraci Neto, R.R.: Effect of hot burnishing aided by infrared radiation on the modification of surface and subsurface of AISI 1045 Steel. Procedia CIRP **58**, 463–468 (2017)

Inspection on a Three-Dimensional Measuring Machine for a Virtual Model for Additive Manufacturing

Hacene Ameddah(✉), Rabia Selloum, and Mourad Brioua

Laboratory of Innovation in Construction, Eco-Design, and Seismic Engineering (LICEGS), University of Batna 2, 05000 Batna, Algeria
hacamed@gmail.com, selloumrabia@gmail.com, brio_ray@hotmail.com

Abstract. Today, and to quickly meet the high demands of variability, supply chain efficiency and energy optimization, business markets are looking for modern manufacturing technologies and as a solution, industry 4.0 is using the benefits of integrating modern manufacturing technologies and information systems to promote production capabilities. In this context, intelligent industry represents a new generation of automatic production systems based on the concepts of intelligent industry, intelligent manufacturing, control and intelligent inspection, such as inspection on coordinate measuring machines (CMMs). This technology allows many machines to be integrated into a plant and controlled online using the MBD (Model Based Design) quality system. The problem of conformity of parts with complex geometry is becoming more and more important. The objective of this work is to present a 3D inspection technique on a virtual model (MBD: Model Based Design), using a coordinate measuring machine equipped with a "POWER INSPECT" measurement and inspection software. The interest of this technique is to show the impact of the dimensional inspection and geometric tolerance process of the CAD model for the CAI (Computer aided Inspection) approach on the fidelity of the finished product for additive manufacturing (AM) including intelligent industry.

Keywords: 3D inspection · Tolerancing · Reverse engineering · Smart industry · Computer aided design and inspection

1 Introduction

With increasingly difficult global competition, increasing product quality and lowering production costs have become a constant concern for manufacturers. Modern products are increasingly characterized by complex shapes, small volumes, great diversity, tight tolerances and very high quality. The production system must therefore be able to react in terms of reliability, productivity and flexibility.

Obtaining a quality mechanical product requires that the components be dimensionally and geometrically inspected to ensure that they correspond to the design characteristics and functionalities. The inspection results, on the other hand, make it possible to control the manufacturing process and thus adjust different parameters to

M. Kharrat et al. (Eds.): ICAMEM 2019, LNME, pp. 138–143, 2020.
https://doi.org/10.1007/978-3-030-52071-7_20

obtain the desired results (Frechette 2011). The inspection of a manufactured part consists in verifying the accuracy of the part against a given set of specifications (tolerances), it is necessary that the method and technique used to carry out this inspection itself be very precise. However, the problem arises when facing any surface of complex shape or geometry, as the design and manufacture of surfaces of complex geometry shapes has become a common practice in industry and occupies the vast majority of companies' inspection and metrology services.

These techniques have undergone a rapid evolution in recent years thanks to the introduction of the coordinate measuring machine. The objective of this work is to present the interest of three-dimensional measurement and the control of dimensions and geometric tolerances in the industrial field, by applying the 3D inspection technique on a virtual model under the POWER INSPECT measurement and inspection software.

In this context, intelligent industry uses the advantages of the integration of modern manufacturing technologies and information systems to promote production capacity and increasing competition on the world market is the main motivation to build a Model Based Enterprise (MBE).

2 Acquisition Systems for 3D Inspection

3D inspection is a process that verifies the accuracy of a manufactured part against a given set of specifications (Tong et al. 2004). Thus, the 3D inspection allows the control of metrological characteristics. The objective is to produce a specification verification report "Fig. 1" and/or a numerical representation of the part.

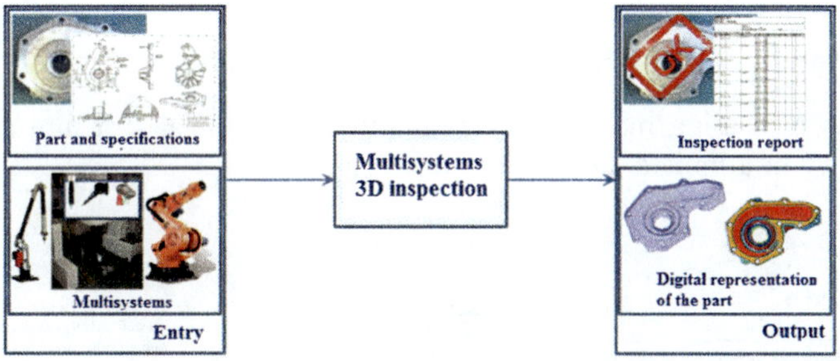

Fig. 1. Inspection 3D of objects (Frechette 2011)

This objective leads to the identification of three major issues related to 3D digitization "Fig. 1" the choice of digitization systems, the definition of a coherent digitization strategy and data processing. We can distinguish two categories of sensors: contact sensors, optical non-contact sensors. We present here the major technologies of active optical acquisition systems. The most common optical acquisition systems in the inspection field are triangulation sensors. To overcome this problem, a number of studies have focused on CAIP (Computer-Aided Inspection Planning). The main purpose of this work is to automatically integrate the control results into the manufacturing loop to modify the manufacturing parameters (Frechette 2011). Most of the work is done in the context of on Machine Measurement (Kojung et al. 2007); or in the context of on-line inspection in the broad sense (Kojung et al. 2007), the purpose of which is to integrate inspection data into the manufacturing process through a return chain in the process "Fig. 2".

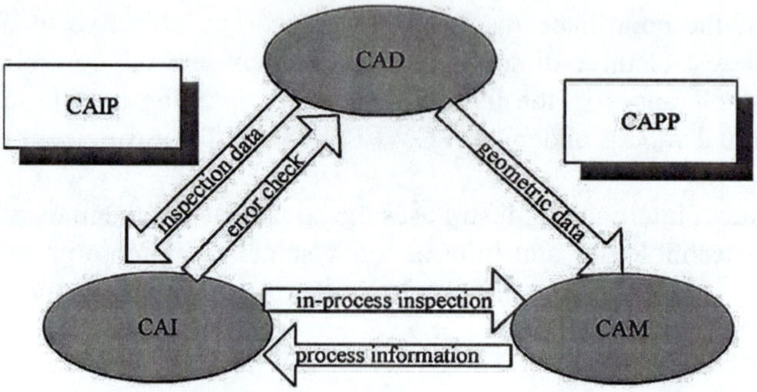

Fig. 2. Manufacturing loop including inspection phase (Kojung et al. 2007)

3 Coordinate Measuring Machine (MMT)

Coordinate measuring machines (CMMs) are important units in dimensional and geometrical inspection systems for mechanical parts. They make it possible to check the conformity of complex shaped parts in 3D, quickly and with great precision.

3.1 Description of a Coordinate Measuring Machine

A coordinate measuring machine is essentially a mechanical structure that allows a probe to move in a limited amount of work "Fig. 3".

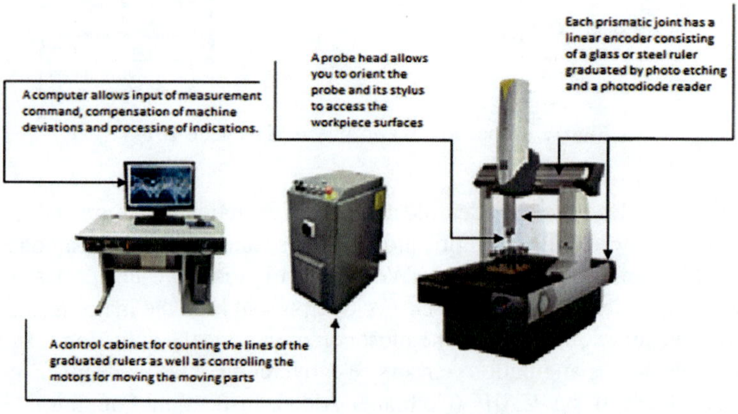

Fig. 3. Complete system of a coordinate measuring machine

4 Results for Measurement and Inspection of the Model Part

In this part, we will start a measurement operation on a certain type of part (virtual model) "Fig. 4". The measurement operation is carried out directly on the software using a CAD model "Fig. 5".

After completing the probing operation, the software displays all the probed points "Fig. 5" and compares the measurement results with the CAD data and uses colors to indicate the deviation from the CAD model.

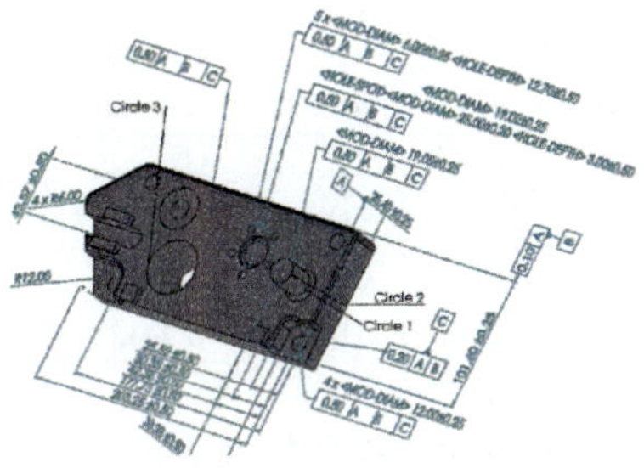

Fig. 4. Virtual part definition with the dimensional and geometric specifications

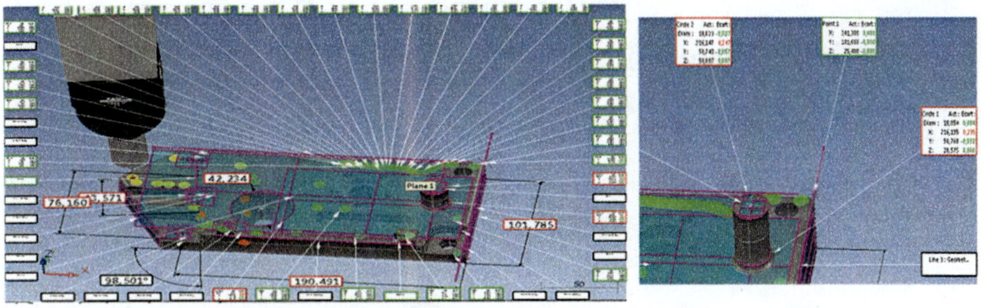

Fig. 5. The coordinates of all measured points.

4.1 Measurement and Inspection Report

The report displays details of the elements and groups in the inspection sequence. The display and the information it contains depend on our measurement report, and the settings and measurement states of each element "Table 1" and "Fig. 6".

"Table 1" Shows the measurement results of group 2 (circles 1, circle 2 and circles 3):

– For circles 1 and 2, the measured value of point x is displayed in red, so this value is outside the tolerance zone with a deviation of 0.235 and an error of 0.135 for circle 1 and a deviation of 0.247 and an error of 0.147 for circle 2, but the values of the results of point y, z and the diameter value of each circle are displayed in green, so the results are acceptable (the values are within the tolerance zone)

Table 1. Alignment results.

Geometric Group 2							
Circle 1 (Datum – Geometric PLP Alignment 1)							
		Hi- Tol	Lo- Tol	Nominal	Measured	Deviation	Error
Centre	X	0.100	−0.100	215.900	216.135	0.235	0.135
	Y	0.100	−0.100	50.800	50.768	−0.032	–
	Z	0.100	−0.100	28.575	28.575	0.000	–
Diameter		0.100	−0.100	19.050	19.054	0.004	–
Number of probed points: 6							
Circle 2 (Datum – Geometric PLP Alignment 1)							
		Hi- Tol	Lo- Tol	Nominal	Measured	Deviation	Error
Centre	X	0.100	−0.100	215.900	216.147	0.247	0.147
	Y	0.100	−0.100	50.800	50.743	−0.057	–
	Z	0.100	−0.100	50.800	50.807	0.007	–
Diameter		0.100	−0.100	19.050	19.023	−0.027	–
Number of probed points: 6							
Circle 3 (Datum – Geometric PLP Alignment 1)							
		Hi- Tol	Lo- Tol	Nominal	Measured	Deviation	Error
Centre	X	0.200	−0.200	101.600	101.905	0.305	0.105
	Y	0.200	−0.200	34.290	34.451	0.161	–
	Z	0.200	−0.200	25.400	25.400	0.000	–
Diameter		0.200	−0.200	42.029	42.234	0.205	0.005
Number of probed points: 97							

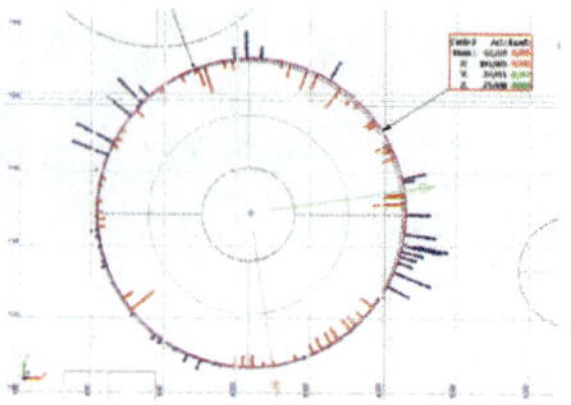

Fig. 6. Detail of measure for circle 3.

– For circle 3 the values of point x and the diameter of the circles are outside the tolerance zone with a deviation of 0.305 and an error of 0.105 for point x and a deviation of 0.205 and an error of 0.005 for the diameter, so these values are acceptable according to the desired accuracy.

To demonstrate the effectiveness of the method we have added an angular type geometric tolerance measure "Fig. 7".

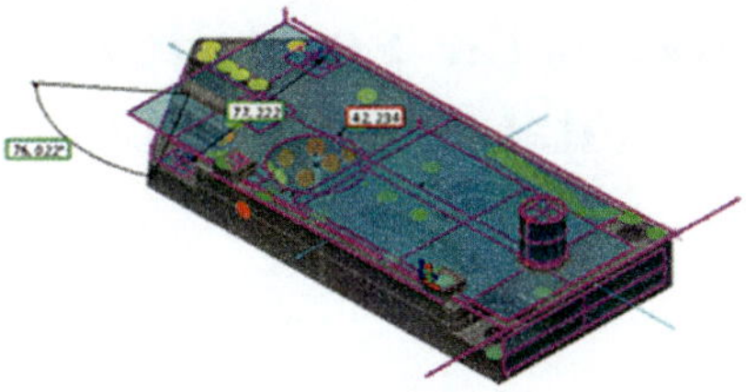

Fig. 7. Angle between two lines 1 (references SCP (Referee CAD)::Plan (XOY), line2:geometric PLP alignment 2, line 5)

5 Conclusion

The role of CMM software has become increasingly important in recent years. Today, they are often linked to CAD databases for interactive control. The most interesting new feature is, without a doubt, the CAD/MMT connection. The software displays in real time, on screen, a CAD part model, and the points controlled by the machine are displayed live on the model in question. It is also possible to orientate the controlled areas from the screen. Usefulness the immediate visualization of the geometric defects of the part in relation to its model. With regard to the method used and the measurement and inspection results obtained, it can be said that the method used for surface inspection is a general method for the inspection of mechanical parts with left surfaces and/or elementary surfaces. The virtual model is a model that allows direct control of advanced functional conditions. It generalizes and formalizes the different concepts of standardized geometric tolerancing.

References

Frechette, S.: Model based enterprise for manufacturing. In: Proceedings of the 44th CIRP International Conference on Manufacturing Systems (2011)

Tong, K., Lehtihet, E.A., Joshi, S.: Software compensation of rapid prototyping machine. J. Precis. Eng. **28**, 280–292 (2004)

Kojung, T.J., Park, W., Kim, H.S., Kim, S.H.: On-machine measurement using a noncontact sensor based on a cad model. Int. J. Adv. Manuf. Technol. **32**, 739–746 (2007). https://doi.org/10.1007/s00170-005-03834

Simulation of the Local Heating Effect on Incremental Sheet Forming Process

M. Sbayti(✉), R. Bahloul, and H. Belhadjsalah

Laboratory of Mechanical Engineering (LGM),
National Engineering School of Monastir (ENIM), University of Monastir,
Av. Ibn El Jazzar, 5000 Monastir, Tunisia
manelasbayti@gmail.com, riadhbahloul4@gmail.com,
hedi.belhadjsalahl@gmail.com

Abstract. Incremental sheet forming is a process which is primarily used for prototyping, small series production and unique parts. The combination of this technology with a heating system can improve the formability of existing alloys known by poor room-temperature formability. In order to further study the performance of the numerical tools, the present work deals with finite element modeling of heat-assisted incremental sheet forming process using ABAQUS/EXPLICIT software. Thus, the numerical simulation for this thermomechanical problem was developed to state and clarify the temperature distribution during sheet metal forming and its major influence on the enhancement of process formability and the avoidance of the part failure. In this case, the heat loss due to radiation and natural convection boundary conditions were neglected. This FE model is utilized to control the damage parameter evolution at different temperature values. Also, the two process technologies (incremental sheet forming combined with global/local heating) have been evaluated. Consequently, the results showed that the Ti6Al4V titanium alloy exhibits good hot incremental forming ability.

Keywords: Metal forming · Titanium alloy · CAD/CAM · Simulation · Stiffness degradation

1 Introduction

Several innovative manufacturing techniques have been developed to produce highly customized products with a reasonable manufacturing cost. The Incremental Sheet Forming (ISF) process, as one of these new technologies, has aroused the interest of the academic and industrial community during the last decades due to its greater flexibility, increased formability and materials and energy consumption (Ingarao et al. 2012; Racz et al. 2018). During ISF, a punch progressively forms the sheet material, which is fixed at its boundaries, following a path previously created by a CAD/CAM software. This emerging forming process is suitable for small batch production and rapid prototype development. However, many studies point out the low accuracy as a major obstacle for its industrial implementation (Sbayti et al. 2019). Researches carried out for further optimization. Some authors used optimization techniques based on

M. Kharrat et al. (Eds.): ICAMEM 2019, LNME, pp. 144–151, 2020.
https://doi.org/10.1007/978-3-030-52071-7_21

numerical and experimental studies to improve the accuracy (Attanasio et al. 2008; Sbayti et al. 2019). Others records to couple the ISF process with heating devises to enhance the formability of hard-to-form materials. In this case, two heat-assisted ISF variants, including the localized heating and the global heating were studied. Hence, this research work aims at numerically evaluating the feasibility of hot ISF. So, thermomechanical finite element modeling has been presented in order to control the temperature evolution during the process and its impact on the improvement of the final product quality.

2 Process Simulation

In this work, FE simulations were carried out in ABAQUS 6.14- Explicit software package taking into account temperature dependence. The materials used for the hemispherical tool and the sheet were the SM30 tungsten carbide and the Ti6Al4V alloy, respectively. The thermo-physical materials parameters and the parameters for the Johnson-Cook (J-C) constitutive model used for the numerical simulation have been presented in detail in the last work (Sbayti et al. 2016).

The process parameters applied are 0.2 mm vertical step size, 8 mm tool diameter and the standard material sheet used is 1 mm thick Ti6Al4V. The initial shape of the titanium alloy sheet has been considered square for different temperature value with a size of 90 mm × 90 mm. For each simulation, the entire clamped sheet has been maintained at room temperature. Compared to the previous work (Sbayti et al. 2016), we consider in this study that the sheet is locally heated, thus, the punch initially subjected to a heating temperature. Figure 1 describes the boundary conditions (clamped sheet, punch temperature, tool path).

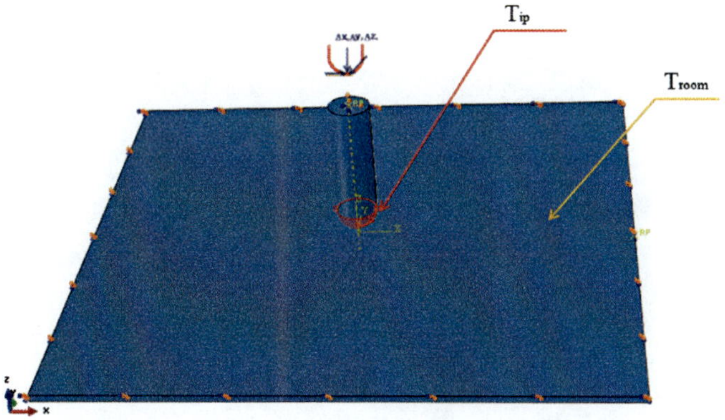

Fig. 1. FE model and boundary conditions

The toolpath generated by CATIA and integrated into ABAQUS was presented in Fig. 2. In the FE model, the sheet blank is meshed with eight-node, coupled temperature-displacement, reduced integration, hourglass control (C3D8RT) elements. On the other hand, for the forming tool, 4-node linear elements (C3D4T) are chosen.

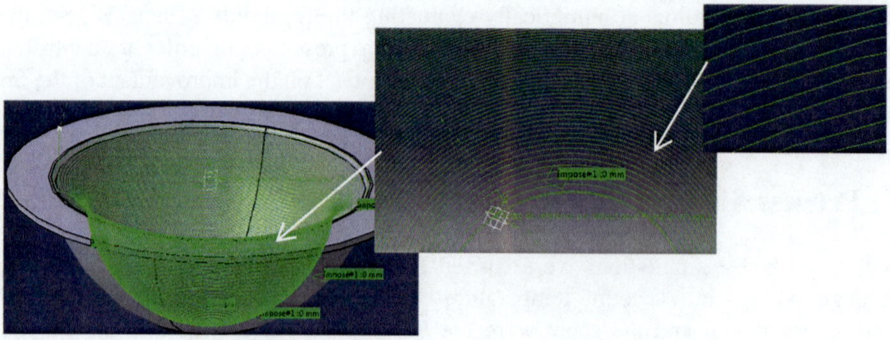

Fig. 2. Schematic description of the simulated tool path

In addition, two FEM techniques: element deletion and mass scaling are deployed to enable the convergence of the FEM solution in hot incremental forming modeling. When used appropriately, mass scaling can often improve the computational efficiency while retaining the necessary degree of accuracy required for a particular problem class. For that, many trials were made on mass scaling adjustment to obtain the suitable values for the hot SPIF in the case of our study, in which the ratio of kinematic energy to total energy can be controlled within a limited value (Fig. 3).

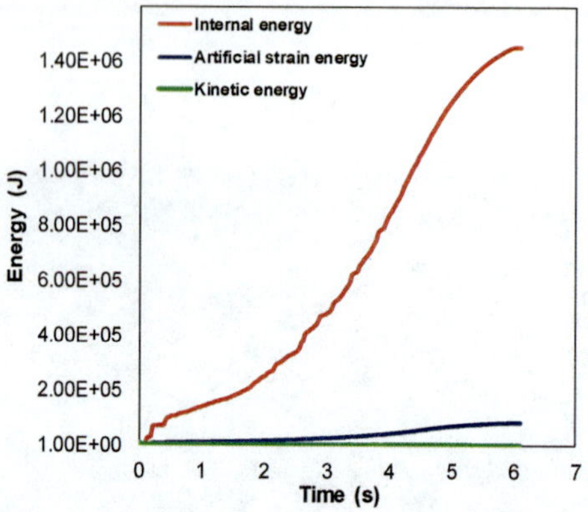

Fig. 3. Comparison between internal, artificial and kinetic energy

3 Results and Discussions

3.1 Temperature Evolution During ISF Process

As mentioned in (Fan et al. 2010; Liu et al. 2015) the formability of Ti6Al4V is improved in the range of 400 °C to 600 °C. So, in this case the initial heating temperature was fixed to T_p = 500 °C. Figure 4 shows the representative temperature contour during the hot SPIF process. The thermally affected zone corresponds to the contact area punch/sheet. The evolution of this thermal parameter was controlled in the element number 6987 localized in the center of the inclined wall areas and it was depicted in Fig. 4.

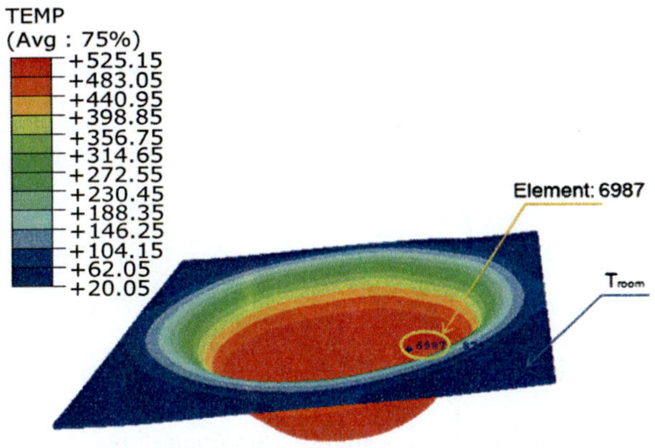

Fig. 4. Representative temperature contour plot during hot SPIF process

It can be noted from Fig. 5 that the temperature at this considered element is still near room temperature value for the first contours. From t = 1 s to t = 5 s the temperature increases rapidly due to the accumulated heat during several and successive punch/sheet contact. At the last second of step time, the temperature seems to be stable at the target temperature (T_p = 500 °C). So, the punch temperature is completely transferred to the sheet.

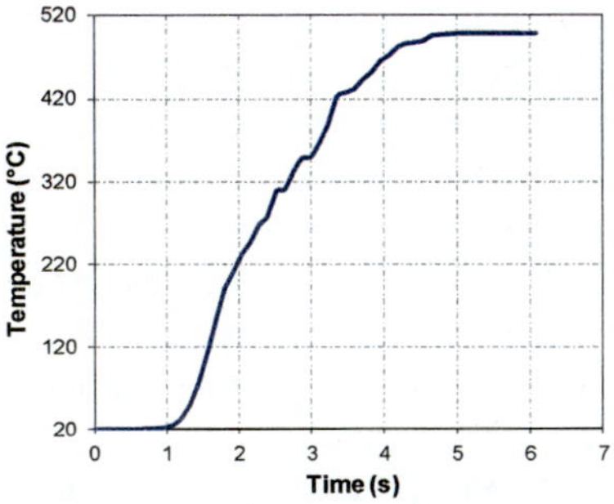

Fig. 5. Evolution of temperature during SPIF process

Figure 6 shows the evolution of the temperature in the sheet along the radial path at T_p = 500 °C. As it can be concluded from this figure, the calculated final temperature at the end of the process simulations increases due to the heat conduction from the punch and to the heat generation due to friction phenomena during the process. Besides, the punch position's reliance on the distribution of the final temperature against the sheet surface, and after several incremental deformations passes, proves to be reasonable.

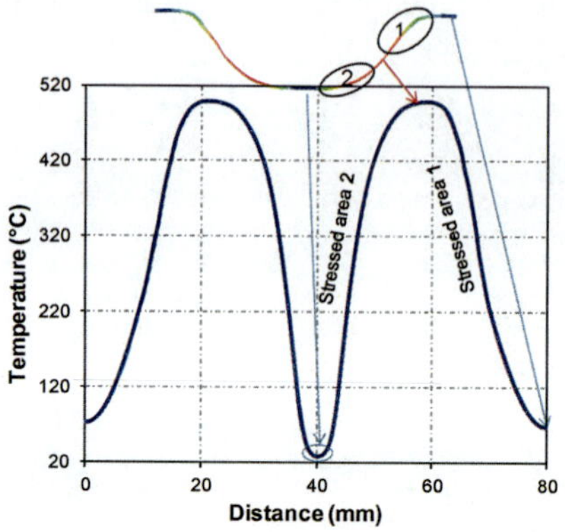

Fig. 6. Temperature distribution versus radial distance formed at T_p = 500 °C

In fact, the temperature evolution is slightly important in the stressed areas (zones 1 and 2). In the central region of the bottom of the part there is no heat and T = T_{room} = 20 °C. It can also be noted that the temperature at the extremities can reach a value of T = 345 °K, which is different from the room temperature. Whereas, in the wall region which corresponds to the position of the material point on the sheet, great evolution of the temperature can be observed.

3.2 Damage Field Distribution

At the recent time, the deformation and fracture mechanism of SPIF has not well understood. Thus, revelation of the failure mechanism causing the fracture in SPIF is very important in order to control processing parameters for obtaining the maximum achievable depth. Figure 7 shows the distribution of the Stiffness Degradation parameter SDEG during ISF process at four temperature values. As reported by these results, the critical damage appears at T_p = 20 °C since the SDEG value is very close to 1 in the punch-sheet contact zone. As expected, the damage parameter increased which confirm the previous results (Fan et al. 2010; Göttmann et al. 2011) where the forming of parts from Ti6Al4V titanium alloy doesn't occur at room temperature. It is further observed from the SDEG contours that a localized crack develops at the tool/sheet contact area essentially at T_p = 20 °C and T_p = 200 °C. This gives us an insight that the suitable forming temperature of processing can improve the formability in incremental forming.

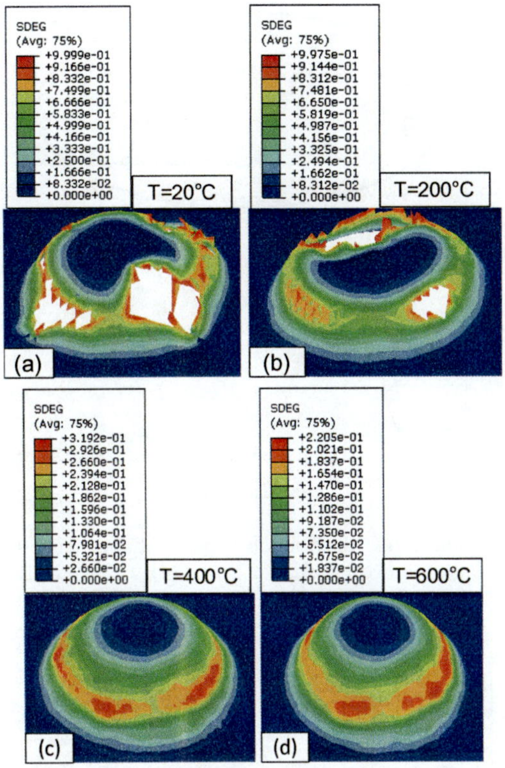

Fig. 7. Distribution of the Stiffness Degradation variable SDEG for different temperatures

The evolution of the Stiffness Degradation SDEG along the radial path at T_p = 500 °C was plotted in Fig. 8. The evolution curve started with SDEG = 0 until t = 0.5 s. It can be noted that the central region of the bottom of the part and the extremities which aren't in contact with the punch still at SDEG = 0 due to no presence of deformation. While, in the tool trajectory zone, this damage parameter is different from a zero value and it evolves in agreement and in a proportional way with the sheet local plastic deformation (see Fig. 8). The damage doesn't occur during process forming (SDEG << 1) at T_p = 500 °C.

3.3 Incremental Forming with Global and Local Heating

Incremental sheet forming processes supported by heating devices were developed to allow economic production of titanium components. Basically, several heating methods have been proposed in ISF reaching local or global heating. In this context, the modeling of these two technologies was studied in this section. Figure 9 presented the iso-values of the temperature distribution obtained by FEM calculation for SPIF with: a) local heating at T_p = 500 °C and b) global heating at T_i = 500 °C. Concerning the first strategy, the moving tool is initially heated and the blank is locally warmed by heat transfer from the tool surface (Fig. 9.a). So, the local heating aims at

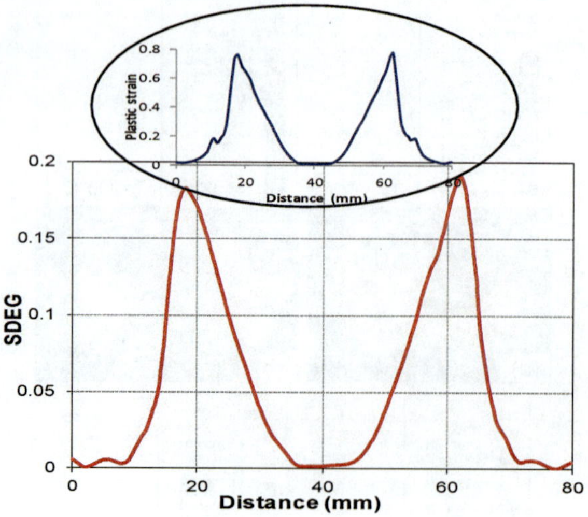

Fig. 8. Stiffness degradation SDEG distribution versus radial distance formed at T_p = 500 °C: Correlation with plastic strain

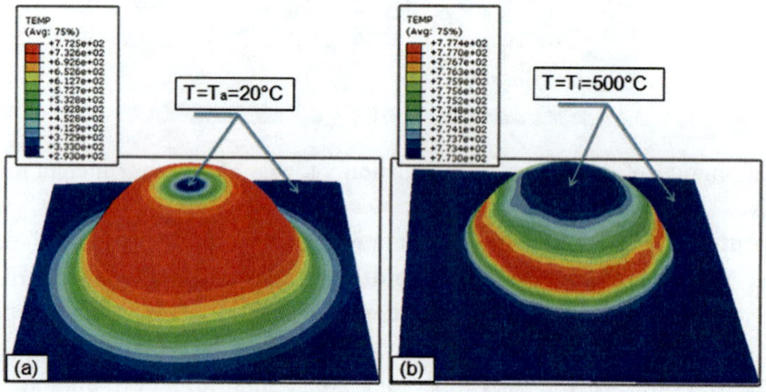

Fig. 9. 3D iso-values of the temperature distribution obtained by FEM calculation for SPIF with: a) local heating at T_p = 500 °C, b) global heating at T_i = 500 °C

increasing the temperature in the actual forming zone of the sheet during the forming process. Then, the ductility of the blank around the tool-contact point becomes higher and the synergistic effect of the incremental forming and the local heating enhances the forming limit of the blank remarkably. In the global heating approach (Fig. 9.b), the whole sheet metal blank is globally and uniformly heated up. Both approaches provide an increase of the flow stress and the material formability in the actual forming zone. However, the ISF technology combined with local heating has the advantage of reducing energy consumption, since only the deformation zone needs to be heated. Additionally, as described by Duflou et al. (2007), it is theoretically possible to reach high formability in the actual forming zone and higher flow resistance in its vicinity, which reduces the local springback.

4 Conclusion

The results herein obtained demonstrate that the heat-assisted ISF seemed to be an effective procedure in avoiding the parts failure. As shown in previous sections, as the temperature increases, the scalar damage parameter decreases and the localized cracks disappear. Especially, the Ti6Al4 V formability is improved at $T_p = 500\ ^{\circ}C$. In addition, both process variants (ISF coupled with local or global heating) proved their efficiency for titanium alloy part forming when appropriate heating temperature was chosen.

References

Attanasio, A., Ceretti, E., Giardini, C., Mazzoni, L.: Asymmetric two points incremental forming: improving surface quality and geometric accuracy by tool path optimization. J. Mater. Process. Technol. 7, 59–67 (2008). https://doi.org/10.1016/j.jmatprotec.2007.05.053

Duflou, J.R., Callebaut, B., Verbert, J., De Baerdemaeker, H.: Laser assisted incremental forming: formability and accuracy improvement. CIRP Ann. Manuf. Technol. **56**, 273–276 (2007). https://doi.org/10.1016/j.cirp.2007.05.063

Fan, G., Sun, F., Meng, X., Gao, L., Tong, G.: Electric hot incremental forming of Ti-6Al-4V titanium sheet. Int. J. Adv. Manuf. Technol. **49**, 941–947 (2010). https://doi.org/10.1007/s00170-009-2472-2

Göttmann, A., Diettrich, J., Bergweiler, G., Bambach, M., Hirt, G., Loosen, P., Poprawe, R.: Laser-assisted asymmetric incremental sheet forming of titanium sheet metal parts. Prod. Eng. **5**, 263–271 (2011). https://doi.org/10.1007/s11740-011-0299-9

Ingarao, G., Ambrogio, G., Gagliardi, F., Di Lorenzo, R.: A sustainability point of view on sheet metal forming operations: material wasting and energy consumption in incremental forming and stamping processes. J. Clean. Prod. **29–30**, 255–268 (2012). https://doi.org/10.1016/j.jclepro.2012.01.012

Liu, R., Lu, B., Xu, D., Chen, J., Chen, F., Ou, H., Long, H.: Development of novel tools for electricity-assisted incremental sheet forming of titanium alloy. Int. J. Adv. Manuf. Technol. (2015). https://doi.org/10.1007/s00170-015-8011-4

Racz, S.G., Breaz, R.E., Tera, M., Gîrjob, C., Biris, C., Chicea, A.L., Bologa, O.: Incremental Forming of Titanium Ti6Al4V Alloy for Cranioplasty Plates—Decision-Making Process and Technological Approaches (2018). https://doi.org/10.3390/met8080626

Sbayti, M., Bahloul, R., Belhadjsalah, H.: Efficiency of optimization algorithms on the adjustment of process parameters for geometric accuracy enhancement of denture plate in single point incremental sheet forming Finite element. Neural Comput. Appl. (2019). https://doi.org/10.1007/s00521-019-04354-y

Sbayti, M., Ghiotti, A., Bahloul, R., Belhadjsalah, H., Bruschi, S.: Finite element analysis of hot single point incremental forming of hip prostheses. In: MATEC Web of Conferences (2016)

A CAD Model for the Tolerancing of Non-rigid Parts Assemblies

Anis Korbi[1(✉)], Mehdi Tlija[1], Borhen Louhichi[2], and Abdelmajid Benamara[1]

[1] Mechanical Engineering Laboratory (LGM), National Engineering School of Monastir (ENIM), University of Monastir, Monastir, Tunisia
anis.korbii@gmail.com, tlija.mehdi@gmail.com, benamara.enim@gmail.com

[2] Mechanical Laboratory of Sousse (LMS), National Engineering School of Sousse (ENISo), University of Sousse, Sousse, Tunisia
louhichi.ca@gmail.com

Abstract. The study of the mechanical assembly behavior during the design stage requires the modeling of components with manufacturing defects. However, the integration of tolerances in the CAD model, only with geometrical and dimensional errors without taking into account several factors such as the deformations of parts during the assembly task, generates an important difference between the digital and the real models. Thus, a new model for the tolerancing of assemblies in CAD is developed considering the deformations of parts under loads. The realistic assembly configurations are obtained basing on the parts with defects. The dimensional and geometrical defects of parts are represented in the CAD modeler basing on the worst-case tolerancing approach and the Small Displacement Torsor (SDT) parameters. The Finite Elements (FE) computation is executed with realistic part models. Sub-algorithms are developed to update the mechanical joints between couples of Non-Rigid/Non-Rigid (NR/NR) components. The tolerance analysis is performed basing on the obtained realistic assembly. A case study is used to show the highlights of the proposed method.

Keywords: CAD · Geometrical and dimensional defects · Deformations · FE · Realistic assembly · Tolerance analysis

1 Introduction

In the Digital Mock-Up (DMU), the modelling of the assembly components with manufacturing deviations as well as deformations caused by external loads, are two important tasks for the control of the product Functional Requirement (FR) during the design step. Several research works validated the need to consider the tolerances in the calculation and simulation phases of the product (Tsai 2007). However, the realistic modelling of the assembly behaviour during operation requires not just the consideration of geometrical and dimensional tolerances but also the deformations of non-rigid

M. Kharrat et al. (Eds.): ICAMEM 2019, LNME, pp. 152–159, 2020.
https://doi.org/10.1007/978-3-030-52071-7_22

components caused by various loads types. Thus, Korbi et al. developed a tolerance analysis model considering the above defects. This model considers the contact between Rigid/Rigid (R/R) and Non-Rigid/Rigid (R/NR) parts (Korbi et al. 2018). The above model presented limitations especially with the neglect of the relationships between both of Non-Rigid/Non-Rigid (NR/NR) parts during the tolerance analysis step. In this context, the proposed tolerancing method improves the previous tolerance analysis models by taking into account the (NR/NR) joints between realistic cylindrical and planar parts during the simulation of the realistic assembly with defects.

The paper is organised as follows. First, a literature review of tolerance analysis approaches of assemblies with non-rigid components is presented. The steps of the new proposed tolerancing model are detailed in Sect. 3. Sub-algorithms to update mechanical joints between couples of Non-Rigid/Non-Rigid (NR/NR) realistic parts are presented. The conclusion and future woks are illustrated at the end.

2 State of the Art

Several approaches are proposed in the literature to consider the parts deformations during the tolerancing of mechanical products. In this context, Liu et al. presented a tolerance analysis method of flexible sheets metal assemblies using the Method of Influence Coefficient (MIC) (Liu et al. 2007). This method is applied to estimate the deformations effect of sheets metal on initial tolerance specifications. Söderberg et al. developed an approach to determine the effect of the spot-welding position variations on the geometrical behaviour of sheet metal assemblies (Söderberg et al. 2012). Mazur et al. proposed a tool for the tolerance analysis and synthesis of mechanism subjected to external and internal loads (Mazur et al. 2011). Jayaprakash et al. established a tolerance design method for assemblies considering deformations dues to temperature and inertia (Jayaprakash et al. 2014). Pierre et al. developed a method based on torsorial formulations to model the relationship between the thermal effects and the geometrical deviations of assembly parts (Pierre et al. 2009). Zhang et al. established a method for the tolerance analysis of annular surfaces considering form defects and the local surface deformations dues to external loads (Zhang et al. 2018).

The most of approaches already presented in the literature allow to consider the deformations of non-rigid parts during the tolerancing process. Some methods such as (Liu et al. 2007; Söderberg et al. 2012) can be applied just in the case of assembly composed of sheets metal and are not valid for thick parts. The other methods are focused on the use of mathematical formulations, statistical tools and FE computations without a realistic geometric modelling of assembly components on the CAD software. Thus, this paper presents a tolerance analysis model of non-rigid parts assemblies using the CAD modeller.

3 The Tolerancing Approach of Non-rigid Parts Assemblies

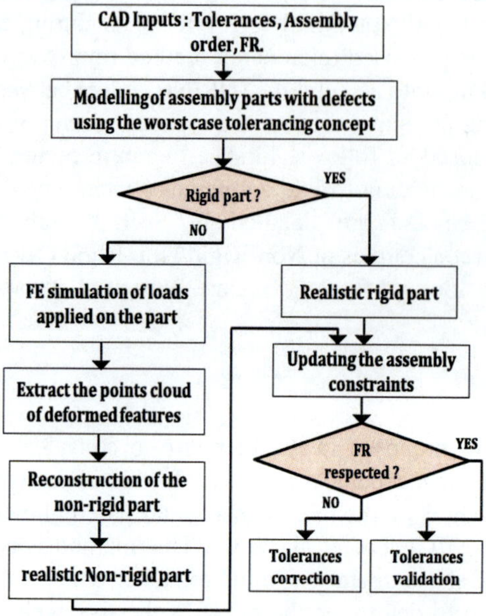

Fig. 1. Flow chart of the developed model

A new tolerance analysis model is proposed in this paper to overcome the limitations of the existed tolerancing approaches of non-rigid parts assemblies. The developed model considers simultaneously the manufacturing defects of parts caused by machining errors as well as the deformations of non-rigid components. The above deformations considered in this paper are caused by external loads. The steps of the developed method are presented in the flow chart of figure (Fig. 1).

The realistic parts with defects are determined using two methods: The Numerical Perturbation Approach (NPA) allows to model the parts with dimensional defects and the worst case tolerancing concept is applied to obtain the assembly components with geometrical deviations (Korbi et al. 2018). An automated method based on the API of Solidworks®, CATIA V5 and MS Excel as well as the FE computation is used to reconstruct the deformed parts through the construction of reference triangular surfaces. The above surfaces are created from the triangulation vertices and elements (Korbi et al. 2018). After the reconstruction of realistic cylindrical and planar components with defects, the initial mating constraints between parts are updated to obtain the final realistic assembly. In the case of planar parts, the initial coincident constraint, between NR/NR planar faces f1 and f2 (Co: f1 & f2) of Non rigid components PR1 and PR2 respectively, is redefined while respecting an Objective Function of the Assembly (OFA) as described in (Korbi et al. 2018). The OFA contains the Assembly Sequence Direction (ASD), contact type between parts, no interferences condition, etc. In the realistic configuration of parts with defects, the constraint Co: f1 & f2 is substituted by one of the following constraints (Fig. 2):

- Coincident constraint (Co: $F1_{OBB}$ & 3 V): $F1_{OBB}$ is a face created from three non-aligned vertices of the realistic face F1 of f1 and considered as the closest points to the non-rigid realistic face F2 of P2 (F2 is the realistic configuration of f2).
- Coincident constraint (Co: $F1_{OBB}$ & 2 V): 2 V are two aligned vertices of F1 considered as the closest points to F2.
- Coincident constraint (Co: $F1_{OBB}$ & V): V is a vertex of F1 considered as the closest point to F2.

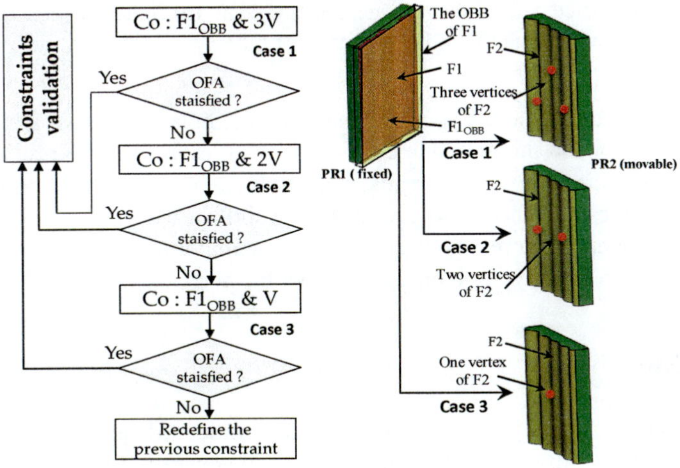

Fig. 2. Sub-algorithm to update the mating constraints between realistic (NR/NR) planar parts

In the realistic configuration, the mating constraints between couples of non-rigid cylindrical parts (hole and shaft) are updated and defined as follows:

- Coincident constraint (Co: L'h & L's) between the realistic axes L'h and L's (Fig. 3(b)).
- Coincident constraint (Co: L's & Vertex of L'h) between L's and a vertex of L'h (Fig. 3(c)).

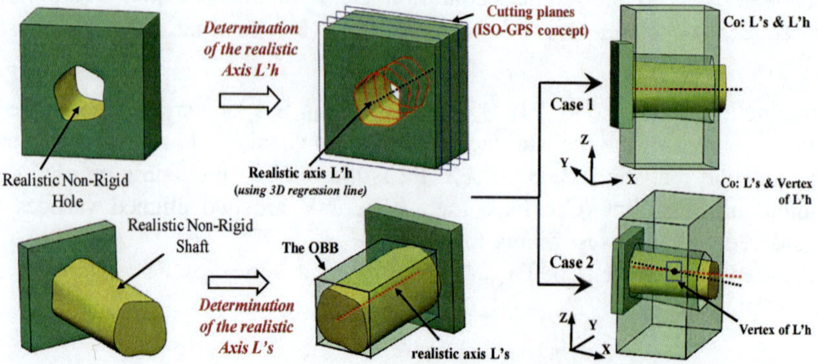

Fig. 3. (a) Reconstructed axes of the realistic hole and shaft, (b) coincident constraint (Co: L'h & L's), (c) coincident constraint (Co: L's & Vertex of L'h)

The realistic axes L's and L'h of the shaft and the hole (Fig. 3(a)) are constructed using the OBB tool and the 3D Square fitting method respectively (Korbi et al. 2018).

After the construction of the realistic assembly, the FR is evaluated. If FR is respected, then the initial tolerances values are valid. In case FR is not satisfied, an algorithm for tolerances correction is used (The most important value in the assembly is selected to be corrected by applying an iterative method).

4 Case Study

In this section, an industrial case study is used in order to show the validity of the proposed tolerancing model. A grinding machine is considered and composed of the following parts with defects (Fig. 4):

- The shaft (1), is considered as non-rigid with coaxiality defect Tco = 0.4 mm.
- The shaft support (2) considered as rigid with perpendicularity defect Tpe = 0.3 mm.
- The other assembly parts are supposed to be rigid without defects.

The FR of the assembly consists of respecting the gap between the wheel (3) and the protective cover (4) of the grinding machine (J1 and J2 > 0).

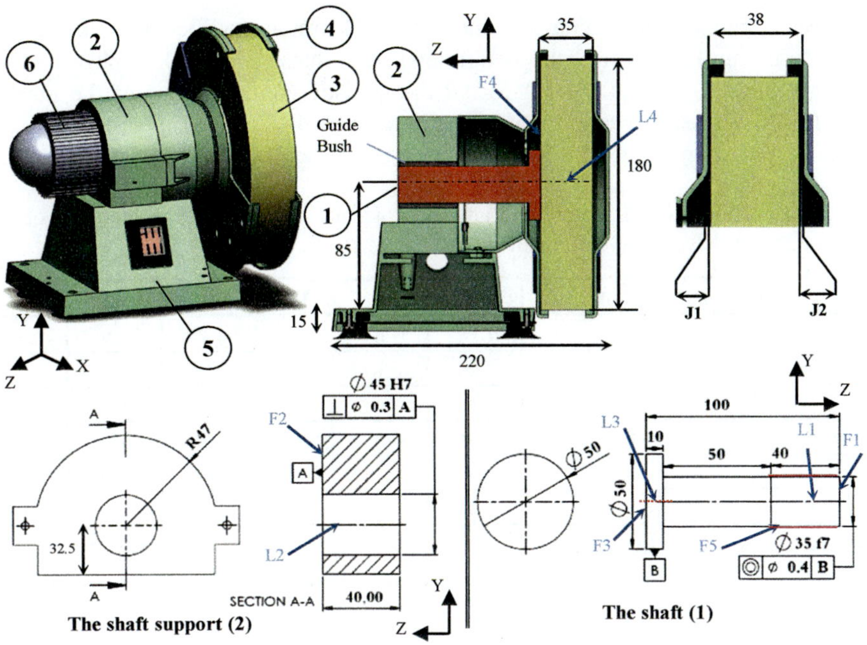

Fig. 4. Drawing of the grinding machine

The Assembly Order (AO) and constraints in the nominal and the 4th realistic configurations of the assembly are presented in Table 1.

Table 1. Assembly Order (AO) and constraints.

AO	Nominal constraints	6th Realistic constraints
1	Co: L1 & L2	Co: L'1 & L'2
2	Co: F1 & F2	Co: Vertex of F1 & F2
3	Co: L3 & L4	Co: L3 & L4
4	Co: F3 & F4	Co: F3 & F4

Co: Coincident constraints, L'I the realistic axis of Li (i = 1 to 2)

Table 2. Boundary conditions and loads.

Features	Boundary conditions/ Loads
Face F1	Fixed
Face F5	Fixed
Face F3	Force F = 120 N (–Y Dir)
–	–

Part (1) material: Steel

The FE computation is used to simulate the deformations of the deformable shaft (1). In this case, the boundary conditions and loads are show in Table 2.

- **Tolerance Analysis results:**

By considering the initial tolerance specifications and the deformations of the non-rigid shaft (1), the tolerance analysis shows that the FR of the mechanism is not satisfied for some realistic configurations (Table 3). After the using of the algorithm of tolerances correction, the tolerance value Tco = 0.4 mm is modified and will be equal to T'co = 0.3 mm. As result, the new tolerances values (T'co = 0.3 mm and Tpe = 0.3 mm) allow to respect the FR for all the realistic configurations of the machine (J1 & J2 > 0) (6 realistic configurations are presented in Table 3).

Table 3. The control of FR before and after the tolerances correction: consideration of parts with geometrical defects and deformations.

	Before tolerances correction			After tolerances correction		
Realistic configuration	**J1**	**J2**	**FR**	**J1**	**J2**	**FR**
1	0.98	0	X	0.32	0.74	✓
2	0	2.63	X	0.15	2.55	✓
3	1.88	0	X	0.21	0,69	✓
4	2,51	0	X	2.6	0.22	✓
5	1,12	0	X	0.76	0.21	✓
6	0	2.78	X	0.45	1.97	✓

✓ : **FR respected; X: FR not respected**

5 Conclusion

A model for the tolerance analysis of CAD assemblies, considering the deformations of deformable parts, is presented. The components with geometrical and dimensional defects are obtained using the worst case concept and de SDT parameters. The non-rigid parts subjected to external mechanical loads are reconstructed using the FE computation. The final realistic assembly is obtained through the redefinition of the mating constraints between realistic parts. The FR is checked and controlled at the end to validate the accuracy of the initial tolerance values. The proposed model is automated using the API of Solidworks® and assists mechanical engineers for the tolerance analysis of mechanisms during an early stage of the product life cycle: design step. Additional work will be required in the future for the consideration of form defects with the developed tolerancing method.

References

Korbi, A., Tlija, M., Louhichi, B., BenAmara, A.: CAD/tolerancing integration: a new approach for tolerance analysis of non-rigid parts assemblies. Int. J. Adv. Manuf. Technol. (2018). https://doi.org/10.1007/s00170-018-2347-5

Jayaprakash, G., Thilak, M., SivaKumar, K.: Optimal tolerance design for mechanical assembly considering thermal impact. Int. J. Adv. Manuf. Technol. **73**(5–8), 859–873 (2014). https://doi.org/10.1007/s00170-014-5845-0

Tsai, J.C.: Stiffness variation of compliant devices due to geometric tolerancing. In: Proceedings of CIRP Conference on Computer Aided Tolerancing, Erlange, Germany (2007)

Pierre, L., Teissandier, D., Nadeau, J.P.: Integration of thermo-mechanical strains into tolerancing analysis. Int. J. Interact. Des. Manuf. **3**(4), 247–263 (2009)

Mazur, M., Leary, M., Subic, A.: Computer Aided Tolerancing (CAT) platform for the design of assemblies under external and internal forces. Comput. Aided Des. **43**(6), 707–719 (2011)

Söderberg, R., Wärmefjord, K., Lindkvist, L., Berlin, R.: The influence of spot weld position variation on geometrical quality. CIRP Ann. Manuf. Technol. **61**(1), 13–16 (2012)

Liu, S.C., Hu, S.J., Woo, T.C.: Tolerance analysis for sheet metal assemblies. J. Mech. Des. **118** (1), 62–67 (2007)

Zhang, Z., Liu, J., Ding, X., Shao, N.: Tolerance analysis of annular surfaces considering form errors and local surface deformations. In: Proceedings of 15th CIRP Conference on Computer Aided Tolerancing, Milan, Italy (2018)

A Developed Static Model for Tool Deflection in Ball-End Milling

Rami Mallek[1(✉)], Moez Smaoui[1], Maher Baili[2], Gilles Dessein[2], and Zoubeir Bouaziz[1]

[1] Laboratory of Applied Fluid Mechanics, Process Engineering and Environment, National School of Engineering of Sfax, Sfax, Tunisia
rami.mallek9@gmail.com
[2] Laboratory of Production Engineering, National School of Engineering of Tarbes, Tarbes, France

Abstract. Accuracy of the workpiece geometry is negatively affected by several sources of errors during milling operation, the most important one is tool deflection which will result in the deviation of the tool path and cutting force is the most significant factor responsible for this error.

A static model of tool deflection in ball-end milling is proposed to predict the unwanted displacements along X, Y and Z, results are then considered in the compensation strategy by tool path modification in order to avoid failure of the tool.

The tool and the tool holder are modeled as a multi-section cantilever beam with a concentrated cutting force load and the cross-section method is used to faithfully determine the deflection, the analytical model shows a good agreement with the finite element method (FEM) which is the proposed method employed to estimate the deflection of the tool in the three directions.

Predicted displacements of the tool are also compared against experimental measurements and a good correlation is found between measured and predicted deflection leading to the validation of the static model proposed for ball-end milling.

Keywords: Tool deflection · Ball-end milling · Static model · Cutting force

1 Introduction

The accuracy of the workpiece geometry is negatively affected by several sources of errors during milling operation, the most important one is tool deflection which will result in the deviation of the tool path. For this reason, it is very important to predict these deflections generated by cutting forces.

According to A.-P. Xu et al. (2003*)* the deflection of the cutting tool is gathered from 2 parts of cutting force one is the static caused by material removal, the other is the dynamic which is generated by vibration. Under stable cutting parameters, the dynamic effect can be neglected and only the static deflection is considered since it is the most significant part.

M. Kharrat et al. (Eds.): ICAMEM 2019, LNME, pp. 160–167, 2020.
https://doi.org/10.1007/978-3-030-52071-7_23

M.A. Salgado et al. (2005) investigates the stiffness of the machine-tool, tool-holder and the tool, and his results show that stiffness of the flexible tool is 5–7 times lower than that of tool-holder. Comparing results from analytical or FEM with experimental values, important differences have been noticed, exceeding 50% in some cases so taking into account the tool-holder deflection can improve the results. It was found also that in the case of ball-end milling, the application of force point load at the tool tip is almost the same as the application of a distributed load along cutting flute (Layegh et al. 2013).

Mamedov et al. (2013) presents a validated force and deflection models where the cutter is considered as a multi cross-sectional cylinder with effective diameter to provide an easy calculation of the end mill deflection, this effective diameter is approximately 80% of the cutter diameter as shown by Kops and Vo (1990).

2 Tool Deflection Estimation by Analytical Method AM

To calculate the tool deflection, the cutter (tool + tool-holder) is modeled as a four-step flexible cantilever beam as shown in Fig. 1 and the cutting force is simulated as a concentrated load at the tool tip. Geometric parameters are mentioned in Table 1 and Young's modulus E = 207000 MPa.

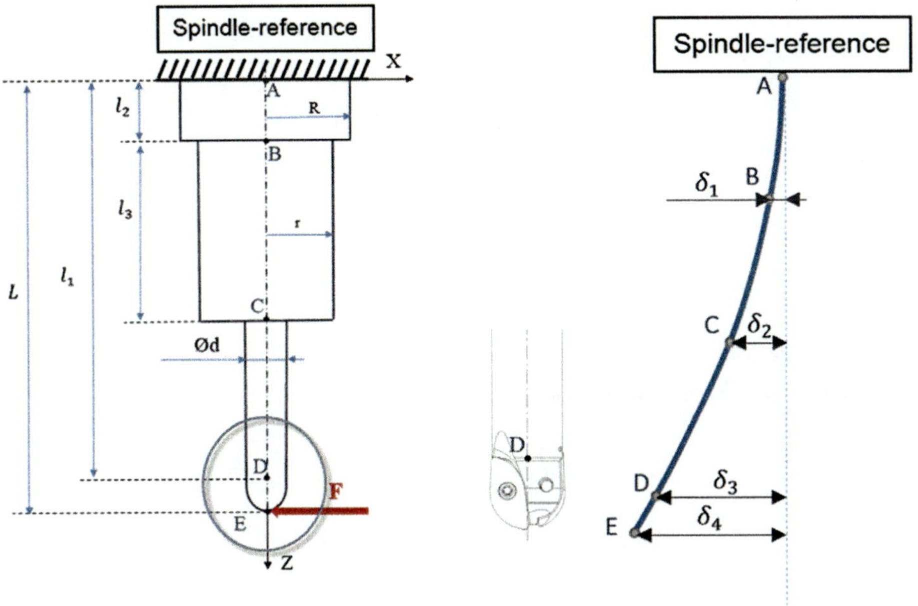

Fig. 1. Deflection model of the cutting tool

Table 1. Geometric parameters of the cutter (Tool + Tool-holder)

Dimensions	Value (mm)	Dimensions	Value (mm)
l_2	26	l_1	178
l_3	89	overhang length L	195
Tool-holder radius R	31.5	Tool diameter d	16
Tool-holder radius r	24		

The bending moment is given by the following expression:

$$M_f = -F.(z - L) \tag{1}$$

Knowing the bending moment (M_f), the deflection δ can be calculated by applying double integration of the following equation:

$$\delta'' = -\frac{M_f}{E.I_{GZ}} \tag{2}$$

a) **Between A and B:**

$$E.I_{p1}.\delta_1'' = F.z - F.L \tag{3}$$

$$E.I_{p1}.\delta_1 = \frac{F}{6}.z^3 - \frac{F.L}{2}.z^2 + C_1.z + C_2 \tag{4}$$

At point A (z = 0), we have $\delta_1' = 0$ and $\delta_1 = 0$ then $C_1 = C_2 = 0$

$$\delta_1(z) = \frac{F}{6.E.I_{p1}}.z^3 - \frac{F.L}{2.E.I_{p1}}.z^2 \tag{5}$$

b) **Between B and C: (Tool-holder)**

$$E.I_{p2}.\delta_2'' = F.z - F.L \tag{6}$$

$$E.I_{p2}.\delta_2 = \frac{F}{6}.z^3 - \frac{F.L}{2}.z^2 + C_3.z + C_4 \tag{7}$$

At point B ($z = l_2$), we have $\delta_1' = \delta_2'$ and $\delta_1 = \delta_2$

$$\delta_2(z) = \frac{F}{6.E.I_{p2}}.z^3 - \frac{F.L}{2.E.I_{p2}}.z^2 + \frac{1}{E.I_{p2}}.\left(\frac{I_{p2}}{I_{p1}} - 1\right).\left(\frac{F.l_2^2}{2} - F.L.l_2\right).z + \frac{1}{E.I_{p2}}.\left(\frac{I_{p2}}{I_{p1}} - 1\right).\left(\frac{F.L.l_2^2}{2} - \frac{F.l_2^3}{3}\right) \tag{8}$$

c) **Between C and D: (Tool shank)**

$$E.I_o.\delta_3'' = F.z - F.L \tag{9}$$

$$E.I_o.\delta_3 = \frac{F}{6}.z^3 - \frac{F.L}{2}.z^2 + C_5.z + C_6 \tag{10}$$

At $l_4 = l_2 + l_3$, we have $\delta_2'(l_4) = \delta_3'(l_4)$ and $\delta_2(l_4) = \delta_3(l_4)$

$$\delta_3(z) = \frac{F}{6.E.I_o}.z^3 - \frac{F.L}{2.E.I_o}.z^2 + \left[\begin{array}{c}\left(\frac{1}{E.I_{p1}} - \frac{1}{E.I_o}\right).\left(\frac{F.l_4^2}{2} - F.L.l_4\right) \\ + \left(\frac{1}{E.I_{p1}} - \frac{1}{E.I_{p2}}\right).\left(\frac{F.l_2^2}{2} - F.L.l_2\right)\end{array}\right].z + \left(\frac{1}{E.I_{p2}} - \frac{1}{E.I_o}\right).\left(\frac{F.L.l_4^2}{2} - \frac{F.l_4^3}{3}\right) + \left(\frac{1}{E.I_{p1}} - \frac{1}{E.I_{p2}}\right).\left(\frac{F.L.l_2^2}{2} - \frac{F.l_2^3}{3}\right) \tag{11}$$

d) **Between D and E: (cutting length)**

$$E.I_s.\delta_4'' = F.z - F.L \tag{12}$$

The nonuniform geometry of the cutting part is considered in this model by using the cross-section method to faithfully calculate the second moment of inertia as shown by A.M. Khorasani et al. (2016)

From the 3D tool model, using Solid Works software, measurements of the second moment of inertia are taken by doing cross-section every 0.2 mm along tool axis between D and E (the spherical part). Figure 2 and 3.

Fig. 2. Cutting tool cross section

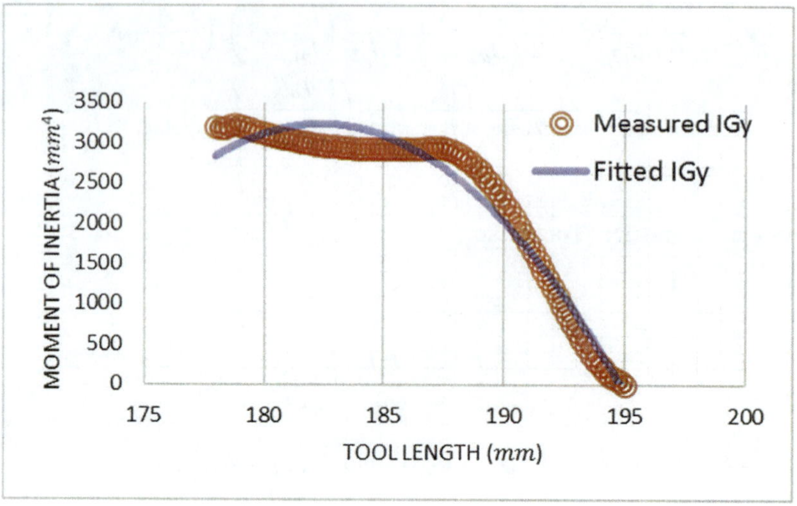

Fig. 3. Moment of inertia variation along tool axis

Between D and E the expression of the moment of inertia is given by the equation $I_{Gy} = -20,901 \times z^2 + 7624,2 \times z - 692056$ with the square of the correlation coefficient $R^2 = 0,9565$.

In this case applying double integration is impossible, δ_4 is solved with Matlab. The maximum deflection, applying 1000 N load at the tip of the tool, is 288,6 μm.

To validate this result, FEM3D analysis and experimental measurement of deflection have been established.

3 Validation Proceedings

To check the correlation of the improved analytical model, two methods of prediction are proposed, numerical and experimental.

3.1 FEM3D Analysis

The cutter (tool + the tool-holder) is modeled by the finite elements method using Abaqus software with respect to the geometric parameters. The cutter is regarded as an elastic cantilever beam loaded at the tip with Young's modulus E = 207,000 MPa and Poisson's ratio $\gamma = 0.33$.

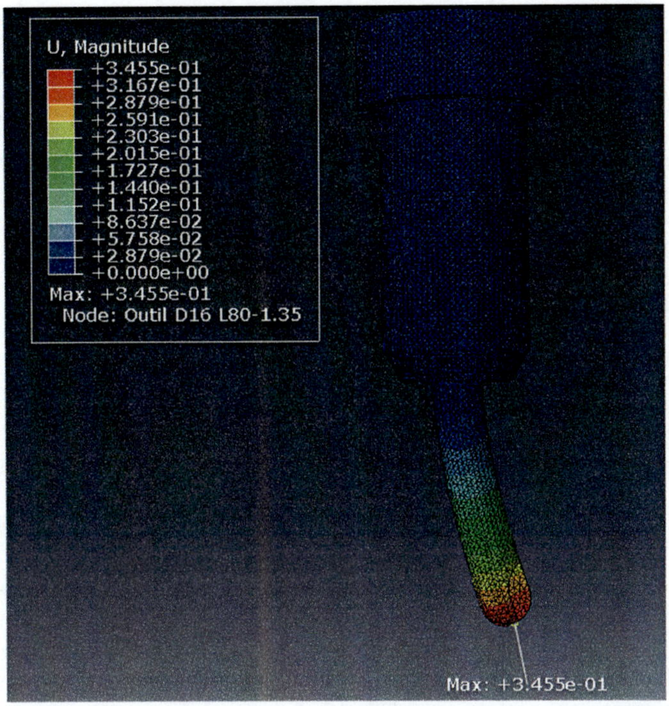

Fig. 4. Deformed shape and maximum deflection for F = 1000 N

Deflection gives a linear behavior with force and for 1000 N load along X $\delta = 0.3455$ mm Fig. 4.

For further investigation of tool deflection, simulations are done in the 3 directions X, Y and Z varying F from 0 to 1000 N.

3.2 Experimental Analysis

To identify the real amount of the tool deflection, an inverse method was used. In a milling machine, the tip of the tool is tangent to a workpiece fixed to a dynamometer sensor then imposed displacement of the table by 0.001 mm are applied and the force is measured by Kistler as shown by M. Smaoui et al. (2011)

Linear behavior was observed and for 1000 N load, $\delta = 0.3436$ mm.

4 Result and Discussion

With regard to the old two-step cantilever beam model in a previous work by the authors (M. Smaoui et al. 2011), the new AM shows an improvement of 13% and the error between AM and experiment results is 16%. This small error is coming from taking into account the real geometry of the tool and the accurate estimation of the moment of inertia using the cross-section method proposed.

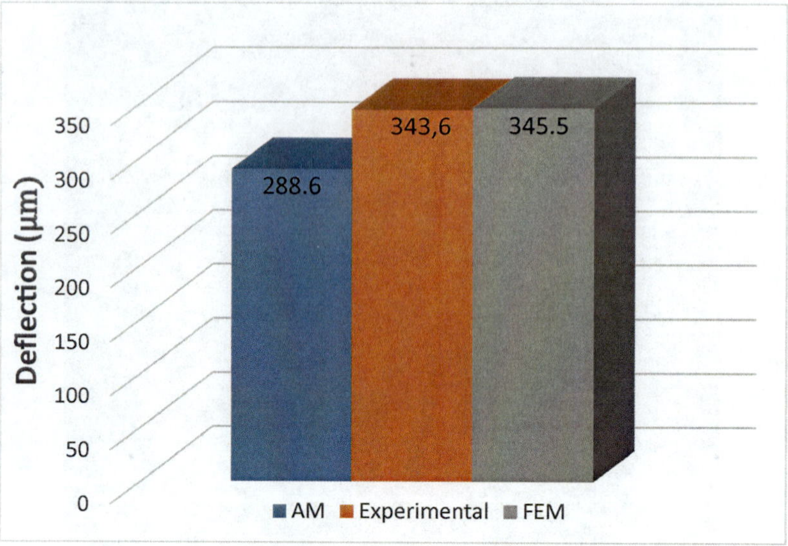

Fig. 5. Tool deflection results for F = 1000 N

From another hand, FEM3D results show good agreement with experimental measurements with an error of less than 1% Fig. 5. This accurate result comes from considering the exact geometry of the cutter including the tool-holder part.

Since the deflection in the Z direction is not negligible and the good correlation between the FEM with the experimental results, the FEM3D is the chosen method of prediction so the static tool deflection model is given by the following expression:

$$\begin{cases} \delta x = \delta x_{Fx} + \delta x_{Fy} + \delta x_{Fz} = 3,455.10^{-4} \times Fx + 3,717.10^{-6} \times Fy + 7,707.10^{-7} \times Fz \\ \delta y = \delta y_{Fx} + \delta y_{Fy} + \delta y_{Fz} = 3,689.10^{-6} \times Fx + 3,431.10^{-4} \times Fy + 4,54.10^{-7} \times Fz \\ \delta z = \delta z_{Fx} + \delta z_{Fy} + \delta z_{Fz} = 4,689.10^{-6} \times Fx + 4,474.10^{-5} \times Fy + 9,784.10^{-6} \times Fz \end{cases} \tag{13}$$

5 Conclusion

In this research, two improved methods for tool deflection were proposed. First the AM by taking into account the tool-holder in the geometric model and carefully estimating the second moment of inertia. Second, in the FEM with less than 1% error compared to experimental results leading to a developed 3D static model for tool deflection Eq. (13).

References

Xu, A.-P., et al.: Simulation and experimental investigation of the end milling process considering the cutter flexibility. Int. J. Mach. Tools Manuf. **43**, 283–292 (2003)

Salgado, M.A., et al.: Evaluation of the stiffness chain on the deflection of end-mills under cutting forces. Int. J. Mach. Tools Manuf **45**, 727–739 (2005)

Layegh, S., Lazoglu, I., Erdim, H.: Tool deflection in five-axis miling. In: International Conference and Exhibition on Design and Production of Machines/Dies/Molds, June 2013

Mamedov, A., et al.: Machining forces and tool deflections in micro milling. Procedia CIRP **8**, 147–151 (2013)

Kops, L., Vo, D.: Determination of the equivalent diameter of an end mill based on its compliance. CIRP Ann. Manuf. Technol. **39**(1), 93–96 (1990)

Khorasani, A.M., et al.: An improved static model for tool deflection in machining of Ti–6Al–4V acetabular shell produced by selective laser melting. Measurement **92**, 534–544 (2016)

Smaoui, M., et al.: Simulation of the deflected cutting tool trajectory in complex surface milling. Int. J. Adv. Manuf. Technol. **56**, 463–474 (2011)

Thermal Sciences and Renewable Energy

Effect of Steam Explosion and Torrefaction Treatments of Wood Chips on the Heating Value of Pellets

Safa Arous[1(✉)], Mariem Mharssi[1], Hassine Bouafif[2], Besma Bouslimi[1], Chedly Bradai[3], and Ahmed Koubaa[1]

[1] Université du Québec en Abitibi-Témiscamingue, 445 Boulevard de l'université, Rouyn-Noranda, QC, Canada
{Safa.arous,Mariem.mharssi,Besma.bouslimi2, Ahmed.koubaa}@uqat.ca

[2] Centre Technologique des Résidus Industriels, 433, Boul. du Collège, Rouyn-Noranda, QC J9X 0E1, Canada
Hassine.Bouafif@cegepat.qc.ca

[3] École Nationale d'ingénieurs de Sfax, Route de la Soukra km 4, 3038 Sfax, Tunisie
Chedly.Bradai@enis.rnu.tn

Abstract. The surplus of wood chips from sawmills is increasing due to the reduction in the demand for newsprint paper. Thus, this study aims to promote the use of wood chips for bioenergy by producing high-quality pellets. More specifically, this project aimed to evaluate the impact of various heat treatment approaches on the heating value of wood pellets made from black spruce (BS), jack pine (JP) and balsam fir (BF). The studied heat treatments were steam explosion and torrefaction. The resulting energy pellets were characterized in terms of density, high heat value (HHV) and durability. The heat treatment improved pellet HHV. Regardless of the type of wood, the HHV of torrefied pellets is significantly higher (27.6–31.1 MJ/kg) compared to steam exploded pellets (21.9–22.7 MJ/kg) and untreated pellets (18–19 MJ/kg). However, as a result of the deterioration of the hydroxyl groups due to torrefaction, the durability of the torrefied pellets is low compared to that of the steam-exploded pellets. Irrespective of the treatment used, JP pellets, with a maximum of 98.9%, had the best durability.

Keywords: Wood pellets · Torrefaction · Steam explosion · High heat value

1 Introduction

The substitution of paper products with digital communications has contributed to a steady decline in the growth of the pulp and paper industry. North American newspaper consumption has decreased by more than 70% between 2000 and 2019 (Government of Quebec 2016). Spruce, pine and fir (SPF) are the main species used for lumber and pulp and paper production. In 2014, SPF accounted for 72% of woody material consumed by pulp and paper mills, mainly in the form of chips (Government of Quebec 2016).

M. Kharrat et al. (Eds.): ICAMEM 2019, LNME, pp. 171–178, 2020.
https://doi.org/10.1007/978-3-030-52071-7_24

As a result, the decline in demand for paper products led to a surplus of wood chips in sawmills and wood processing plants. For example, chip consumption in pulp and paper mills decreased from 6293.2 to 4825.2 oven-dry metric tons between 2007 and 2017 (Salmon 2017). Finding other uses for these residues has become a priority to maintain sawmill profitability and competitivity.

Technological developments are moving towards the exploitation of wood by-products through the creation of a new range of products. In particular, wood chips can be used for the production of oriented strand board (OSB), insulation boards and wood-polymer composites. Since biomass is one of the world's largest energy resources, bioenergy production can be another alternative for wood chip recovery, especially from SPF wood. SPF wood accounts for about 75% of the raw material used for cogeneration and energy production in Quebec (Government of Quebec 2016).

Wood residues pose a challenge to energy producers because of handling difficulties as well as combustion problems. For example, untreated raw wood contains too much moisture and has a relatively low energy density (Lipinsky et al. 2002; Stelte et al. 2011). To develop renewable energies, several methods can be used to overcome these disadvantages. Namely, heat treatments such as steam explosion and torrefaction improve the hydrophobic properties and heating value of raw material (Kosov et al. 2014). In addition, mechanical treatments such as pelletization promote the densification and homogeneity of the final product. The combination of these two types of treatment leads to the creation of fuel with a high heating value (HHV), better handling properties and storage, and which can be used in thermal, electrical or domestic applications. Note that the heating value of torrefied wood pellets can reach 19–23 MJ/kg even though the heating value of raw wood pellets does not exceed 18 MJ/kg (Kosov et al. 2014). The main objective of this study was to investigate the influence of two types of thermal treatment (torrefaction and steam explosion) on the pellet properties (density, HHV and durability) of three softwood species (black spruce, jack pine, and balsam fir).

2 Materials and Methods

2.1 Raw Wood Material

Three softwood species were used for this study: black spruce (BS), balsam fir (BF), and jack pine (JP). The choice of these species was mainly based on their abundance in the province of Quebec. The wood chips were supplied by Laval University, Quebec City and Blanchet Materials Inc., a Sawmill located in Amos, Quebec, Canada. The wood chips had a moisture content that ranged from 11 to 35%, on a wet basis.

2.2 Torrefaction

Torrefaction is a thermal treatment that improves the physical and chemical biomass properties for bioenergy applications (Pach et al. 2002). During the process, raw biomass is heated under an inert or nitrogen atmosphere at a temperature between 200 and 300 °C (Chen and Kuo 2011). In this study, torrefaction was carried out using the

CarbonFX fast pyrolysis technology developed by Airex Energy Inc. (Bécancour, Qc, Canada). The wood chips from each species were milled using a 12 mm hammer mill grate (Industrial grinder, Schutte Buffalo, Hammer Mills, Buffalo, NY, USA). They were then torrefied at three temperatures: 315, 400 and 454 °C for a few seconds with a low level of oxygen.

2.3 Steam Explosion Process

The steam explosion was performed at the laboratory of the University of Sherbrooke. First, the chips were impregnated with water (solid ratio: liquid = 1:8) at ambient temperature for a duration of 16 h until fibre saturation. The impregnated biomass was then pressed at 100 psi to remove excess water. The moisture content of impregnated chips was around 60%, the recommended moisture for the steam explosion process. After 5 min of pressing, the chips were transferred into the preheated steam treatment reactor at 220 °C. The chips were then cooked for 6 min with saturated steam. The treated samples were stored in a plastic bag to retain their moisture. For each steam explosion test, a solution containing lignocellulose was filtered through a synthetic filtration cloth and the exploded fibers were pressed for 5 min at 80 psi to remove excess water. The material was then kept in the freezer to avoid further degradation.

2.4 Pelletization

The pelletization of untreated and treated wood chips was carried out using a KAHL granulator (Amandus Kahl, Reinbek). Each class of wood chips was conditioned to reach 20% moisture content. For the torrefied wood chips, 15% of pyrolytic lignin was added as a binder. A die geometry ratio of 1:5 was used to ensure good material compaction.

2.5 Pellet Properties

The durability or abrasion test was carried out according to the ISO17831 standard. Pellet durability was calculated from the ratio of the weight of recovered pellets to the initial weight (500 g) after 10 min of rotation in a rectangular box at 50 rpm (Belhadef 2016).

The measurement of the higher heating value (HHV) of pellets determines their efficiency during combustion. The HHV of the pellets was measured on 1 g pellet samples using an isoperibol calorimetric bomb model 6400 (Parr Instrument, USA 2010). The wet-based heating value was measured in three replicates for each treatment.

Pellet density was measured with an Accupyc 1330 helium pycnometer. The pellets were introduced into a 12.08 cm^3 cylindrical cell. The pellet density of each treatment was measured in three replicates.

3 Results and Discussion

3.1 Effect of Treatment on Pellet Density

The density of pellets made from the steam-exploded samples is significantly higher than that of the untreated samples for the three studied species (Fig. 1). For example, the density of the pellets made from balsam fir increases slightly from 1395 to 1406 kg/m^3 with the steam explosion pretreatment. In contrast, the pellets made from torrefied wood present a lower density compared to untreated wood pellets and steam-exploded pellets. This may be due to the irregularity of rigid particles and the lack of hydrophilic functional groups during the torrefaction treatment (Hu et al. 2016). Regardless of the species used, the particle density also decreases with increasing torrefaction temperature. The density of jack pine pellets decreases from 1438 to 13453 kg/m^3 as the torrefaction temperature increases from 315 to 400 °C. These results are consistent with previous findings (Manouchehrinejad et al. 2018, Wang et al. 2017). Jack pine torrefied pellets present the highest density regardless of the torrefaction temperature. This can be attributed to the higher lignin content of Jack pine, which results in a stronger adhesion (Hu et al. 2016, Manouchehrinejad and Mani 2018).

3.2 Effect of Heat Treatment on Pellet Durability

Figure 2 shows the durability values of untreated and treated pellets (by both torrefaction and steam explosion) of the three studied species. The durability of the pellets is significantly affected by the species and thermal treatment. Regardless of the species used, the steam exploded pellets present the highest durability, with values as high as 98.9%. However, an increase in the torrefaction temperature decreases the durability. The durability of untreated and steam exploded BF pellets was 97.7 and 98.5%, respectively. The durability of torrefied BF pellets at 315, 400 and 454 °C was 85, 83 and 84%, respectively. This is attributed to the degradation of the hemicellulose structure and the removal of extractives and volatile compounds during the torrefaction process (Stelte et al. 2011). Despite the decreasing trend in the durability of torrefied BS pellets is similar to that of BF torrefied pellets, the JP torrefied pellets have a good durability of 91% at 315 °C. Indeed, cross-linking and poly-condensation of lignin in the pores formed by thermal degradation can lead to good mechanical strength of torrefied pellets (Manouchehrinejad and Mani 2018). Also, the softening of the pyrolytic lignin can lead to good adhesion of torrefied wood fibers despite the severe conditions of torrefaction.

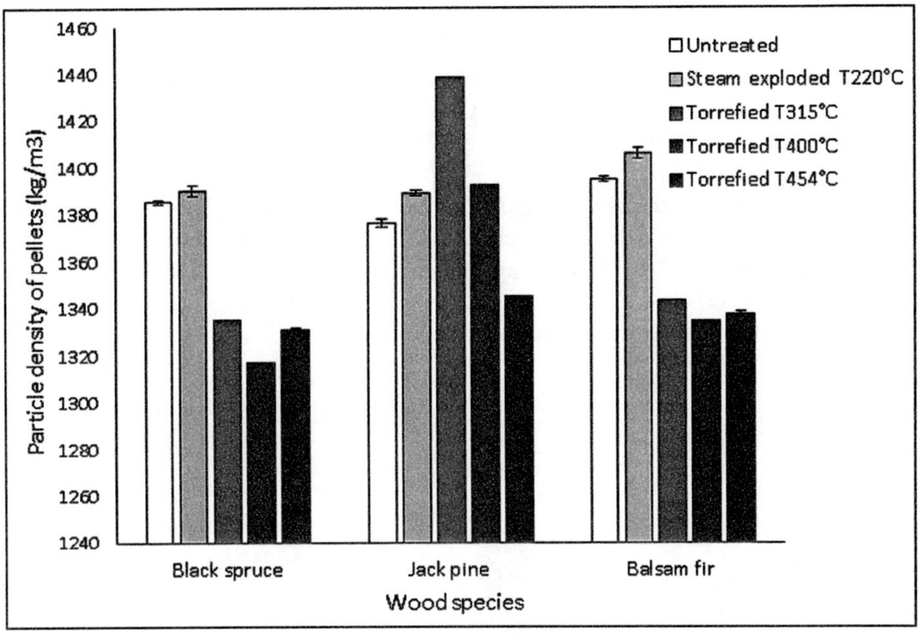

Fig. 1. Effect of thermal treatment and wood species on pellet particle density

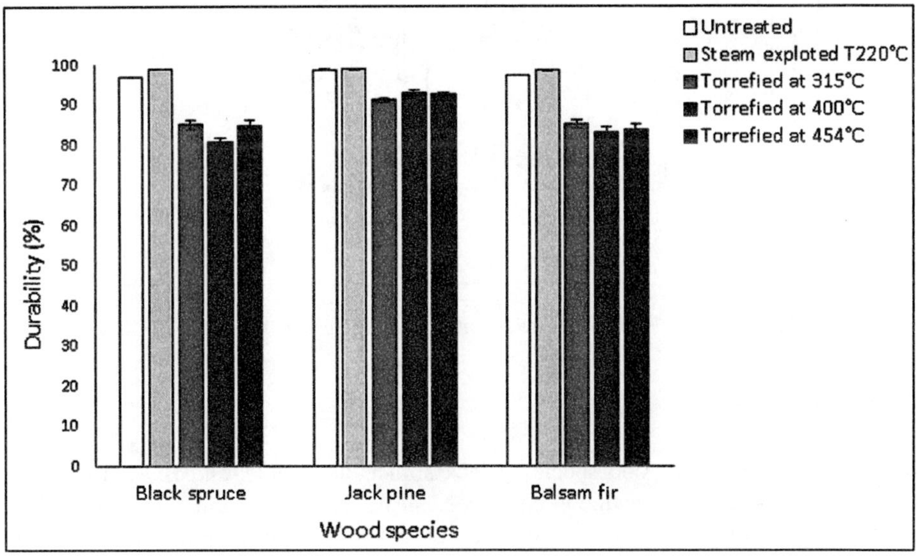

Fig. 2. Effect of thermal treatment and wood species on pellet durability

3.3 Effect of Heat Treatment on Pellet HHV

The higher heating values (HHV) of untreated and treated pellets at different temperatures are presented in Fig. 3. Species and heat treatment affect pellet HHV. However, the temperature has a more pronounced effect on the HHV of the pellets than does species. Torrefaction improves pellet HHV to a greater extent than steam explosion. The HHV of the untreated pellets ranges from 18.5 to 19.3 MJ/kg. The steam explosion treatment improves HHV which reaches 22 to 22.7 MJ/kg, while the HHV of torrefied pellets ranges from 28.8 to 31.1 MJ/kg, which corresponds to an increase of almost 60%. Azargohar et al. (2019) attributed this increase to moisture reduction and hemicellulose degradation during the torrefaction process.

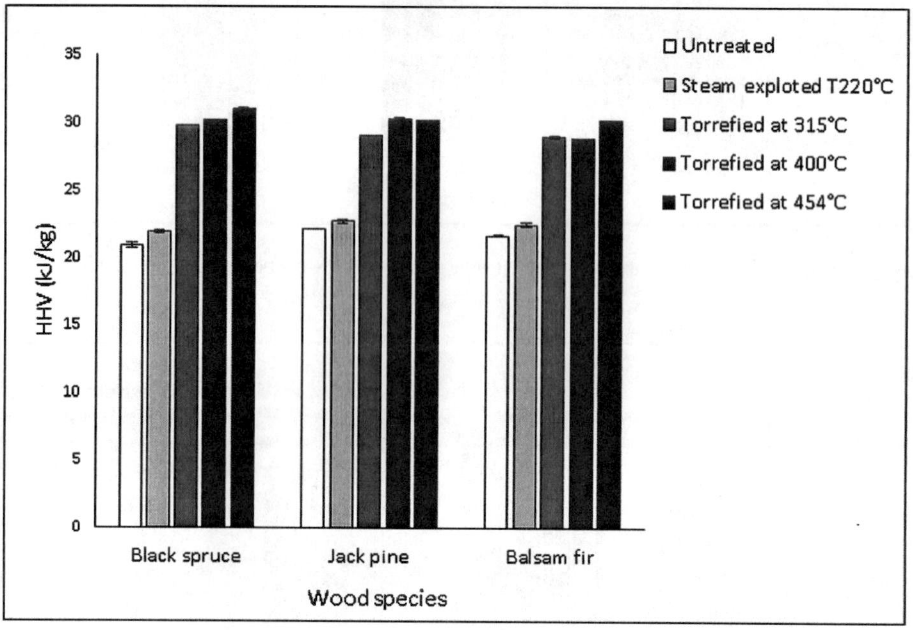

Fig. 3. Effect of thermal treatment and wood species on pellet HHV

4 Conclusions

In this study, three softwood species (Black spruce, Jack pine, and Balsam fir) were treated using two heat treatment processes: steam explosion and torrefaction. The effects of each treatment were evaluated on the HHV, durability index and density of the pellets. The following conclusions can be drawn from this study:

- The density of pellets made from steam-exploded wood is higher than that of pellets made from torrefied and untreated wood.

- The steam explosion process slightly increases pellet HHV compared to that of untreated wood. These improvements were attributed to the softening and the migration of lignin to the fiber surface following the steam explosion process and to the degradation of hemicellulose.
- Torrefaction significantly improved pellet HHV from 18–19 MJ/kg for untreated pellets to 27.6–31.1 MJ/kg for torrefied pellets.
- Stem explosion improved the durability of the pellets while torrefaction decreased this property. Torrefaction causes the reduction of hydroxyl groups, thus fibre-to-fibre bonding.
- Regardless of the heat treatment used, Jack pine pellets show the best performance in terms of HHV and durability compared to the two other species.

Acknowledgments. The authors wish to thank the Natural Sciences and Engineering Research Council of Canada, the Canada Research Chair Program, the Canada Foundation for Innovation and the Centre Technologique des Résidus Industriels (CTRI) for their financial support.

References

Azargohar, R., Soleimani, M., Nosran, S., Bond, T., Karunakaran, C., Dalai, A.K., et al.: Thermo-physical characterization of torrefied fuel pellet from co-pelletization of canola hulls and meal. Ind. Crops Prod. **128**, 424–435 (2019)

Belhadef, W.: Développement de granules énergétiques améliorées à base de bouleau blanc et de graines de canola. Université du Québec en Abitibi-Témiscamingue (2016)

Chen, W.-H., Kuo, P.-C.: Torrefaction and co-torrefaction characterization of hemicellulose, cellulose, and lignin as well as torrefaction of some basic constituents in biomass. Energy **36** (2), 803–811 (2011)

Gouvernement du Québec. Compétitivité de l'industrie forestière au Québec, budget 2016–2017 (2016). http://www.budget.finances.gouv.qc.ca-/budget/2016-2017/fr/documents/Foret.pdf. Consulted on 13 janvier 2019

Haldar, A., Mahadevan, S.: Probability, Reliability and Statistical Methods in Engineering Design. Wiley, New York (2000)

Hu, Q., Yang, H., Yao, D., Zhu, D., Wang, X., Shao, J., et al.: The densification of bio-char: effect of pyrolysis temperature on the qualities of pellets. Biores. Technol. **200**, 521–527 (2016)

Johnson, E.A., Wojtkiewicz, S.F., Bergman, L.A., Spencer Jr., B.F.: Observations with regard to massively parallel computation for monte carlo simulation of stochastic dynamical systems. Int. J. Non-Linear Mech. **32**(4), 721–734 (1997)

Kareem, A., Hsieh, C.C., Tognarelli, M.A.: Frequency-domain analysis of offshore platform in Non-Gaussian seas. ASCE J. Eng. Mech. **124**(6), 668–683 (1998)

Kirkner, D.J., Spencer, Jr, B.F.: Monotonic loading of brittle materials: a stochastic damage model. In: 7th ASCE Specialty Conference on Probabilistic Mechanics and Structural Reliability, Worcester, Massachusetts, 7–9 August 1996, Proceedings, pp. 354–357 (1996)

Kosov, V.V., Sinelshchikov, V.A., Sytchev, G.A., Zaichenko, V.M.: Effect of torrefaction on properties of solid granulated fuel of different biomass types. High Temperature **52**(6), 907–912 (2014). https://doi.org/10.1134/S0018151X14060170

Lam, P.S., Sokhansanj, S., Bi, X., Lim, C.J., Melin, S.: Energy input and quality of pellets made from steam-exploded Douglas fir (Pseudotsuga menziesii). Energy Fuels **25**(4), 1521–1528 (2011)

Lipinsky, E.S., et al.: Enhanced wood fuels via torrefaction. Fuel Chem. Div. Preprints **47**(1), 408–410 (2002)

Manouchehrinejad, M., Mani, S.: Torrefaction after pelletization (TAP): analysis of torrefied pellet quality and co-products. Biomass Bioenerg. **118**, 93–104 (2018)

Manouchehrinejad, M., van Giesen, I., Mani, S.: Grindability of torrefied wood chips and wood pellets. Fuel Process. Technol. **182**, 45–55 (2018)

Pach, M., et al.: Torrefied biomass a substitute for wood and charcoal. In: 6th Asia-Pacific International Symposium on Combustion and Energy Utilization (2002)

Salmon, D.: Ressources et industries forestières du Québec: Portrait statistique (2017). https://mffp.gouv.qc.ca/wp-content/uploads/portrait-statistique-2017.pdf. Consulted on 8 mai 2019

Stelte, W., et al.: Pelletizing properties of torrefied spruce. Biomass Bioenerg. **35**(11), 4690–4698 (2011)

Wang, Q., Han, K., Gao, J., Li, H., Lu, C.: The pyrolysis of biomass briquettes: effect of pyrolysis temperature and phosphorus additives on the quality and combustion of bio-char briquettes. Fuel **199**, 488–496 (2017)

A CFD Investigation of a Turbulent Flow in a Corrugated Plate Heat Exchanger

Sirine Chtourou(✉), Hassene Djemel, Mohamed Kaffel, and Mounir Baccar

CFDTP, National School of Engineers of Sfax (ENIS), 3038 Sfax, Tunisia
sirine.chtourou@enis.tn, hassenedjemel@yahoo.fr, kaffel_moh@yahoo.fr, mounir.baccar@hotmail.fr

Abstract. This paper presents a three dimensional analysis with the computational fluid dynamics CFD code of a turbulent flow through a plate heat exchanger with corrugated wall. For the calculation, the standard k-ε model with "wall-functions" is employed with Reynolds number Re range of 400, 900, 1000.

This paper describes the effect of rib height on the thermal performance of the PHE. To simulate the flow through a corrugated plate and discover the flow characteristics and structure, the commercial CFD software Ansys Fluent R15.0 is used. In order to simplify the simulation, the computational domain was considered as one corrugated plate with chevron angle $\beta = 45°$. The geometry is obtained by the Computed aided design software Solidworks. It consists of chevron plate with different rib height h of 3, 5, 7 mm and a constant chevron angle $\beta = 45°$. The hot water enters the fluid domain with temperature of 60 °C through the corrugated wall which has a constant temperature of 20 °C. The contours of temperature, velocity and turbulence intensity were presented and analyzed numerically.

Keywords: CFD code · Plate heat exchangers · K-ε · Ansys Fluent R15.0 · Solidworks 2017

1 Introduction

The plate heat exchanger (PHE) was invented by Dr. Richard Seligman in 1923, which revolutionized the methods of heating and indirect cooling of fluids. Dr. Richard Seligman founded the VPA in 1910 as Aluminum Plant & Vessel Company Limited. Shah and Wanniarachchi (1991), Kays and London (1998) found that the PHE have the advantage of offering a large exchange surface per unit volume and allow very small temperature differences to be achieved while simultaneously exchanging heat between several fluids. Today, they are widely used in the alimentary sector (Hesselgraves 2001).

Focke et al. (1985), with visualization techniques, investigated experimentally the effect of chevron angle β on the thermal performance of the plate heat exchangers PHE. The chevron angle β not only influence the heat of transfer rate and the pressure drop but also the flow structures. They studied with β range of [0° 30° 45° 60° 72° 90°] and Reynolds number Re range of 100–500. They found that when $\beta < 60°$ the fluid flows

M. Kharrat et al. (Eds.): ICAMEM 2019, LNME, pp. 179–186, 2020.
https://doi.org/10.1007/978-3-030-52071-7_25

along the furrows and for β equal to 80°, it flows zig-zag. They demonstrated that the increase of chevron angle leads to the increase of heat transfer and friction factor f. The maximum of friction factor f and Coburn factor j are obtained with β of 80°. Moreover, Kanaris et al. (2004) investigated the flow characteristics and the heat transfer in the plate heat exchangers with chevron angle of 60°. They carried out numerical simulations for Reynolds number range Re 400–1400 with 870000 elements. With the CFD code, SST (Shear Stress Transport) is chosen for the analysis. They found a good agreement between the numerical results and the experimental results. They developed new correlation of Nusselt number Nu and the friction factor f of a turbulent flow cross corrugated. In addition, Kanaris et al. (2006) studied numerically and experimentally the heat of transfer in the plate heat exchanger with three corrugated plates. It is a continuation of previous works. The analysis is carried out with SST turbulence model for Reynolds number Re range 200–2400. The computational domain was composed of inlet, outlet and the distribution zones which are supposed smooth. They measured with infrared camera the different temperature in the outlet and the inlet of each fluid. Therefore, they determined the pressure drop for each channel. They estimated an error of friction factor f and Nusselt number Nu about 10% of the experimental data. They demonstrated that this error is caused by the effect of corrugation geometry that increases the loses charge. They found that the temperature was uniform in the transversal section of the channel.

Therefore, until this day, many previous works demonstrated that using the chevron form in the plate heat exchanger provides the best thermal performance.

In the performance prediction, recently, the computational fluid dynamics CFD is become more important. In order to study the flow characteristics and the heat transfer enhancement in a corrugated plate heat exchanger, a numerical simulation is carried out. It is based on the Finite Volume Method FVM.

2 Computational Domain

In order to simplify the simulation, we choose to work with one corrugated plate. Paras et al. (2002) used flat plate with one side with chevron angles in their computational analysis. (Kanaris et al. 2004) indicates that the plate of the PHE has dimensions as height h = 5 mm, length L = 200 mm and width W = 110 mm In fact, the computational domain is a rectangular duct with an of height h = 10 mm, length L = 200 mm and width W = 110 mm. In our analysis case, only the fluid domain as a three dimensional computational domain is developed by the commercial CAD software Solidworks 2017 (Fig. 1). Kanaris et al. (2004) assumed the distribution zones, the inlet and the outlet as smooth zones. All dimensions are indicated in Table 1.

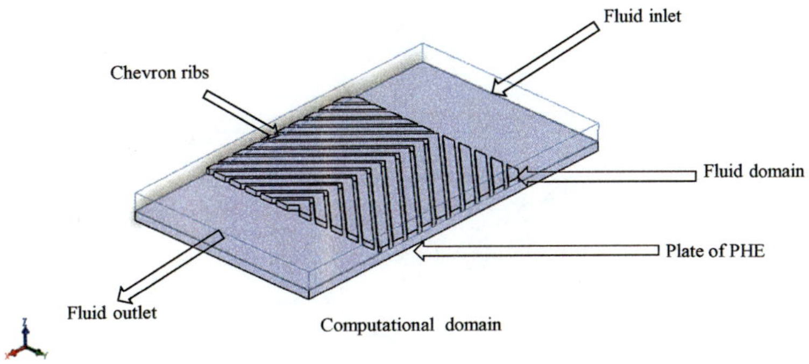

Fig. 1. Different part of the computational domain

Table 1. Computational domain dimensions

	Plate dimensions	Fluid domain dimensions
Height h (mm)	5	10
Length L (mm)	200	200
Width W (mm)	110	110

3 Boundary Conditions

At the entrance of the channel, a constant velocity and temperature are assumed as boundary conditions. And a pressure outlet is applied in the output of the channel. The water as the working fluid enters with temperature of 60 °C and velocity depends the Reynolds number Re. The fluid is supposed incompressible. The flow is assumed turbulent and steady through the corrugated plate which is heated with temperature T of 20 °C (Kanaris 2004; Kanaris et al. 2006). The other walls are adiabatic as indicates the Fig. 2.

4 Mathematical Formulation and Meshing Grid

The Finite Volume Method FVM is the key of the CFD numerical. The Navier Stokes equations are employed to solve the flow characteristics and the fluid phenomenon. In Cartesian tensor system, the governing equations (continuity, momentum, energy) can be written as Djemel et al. (2017),

Continuity equation:

$$\frac{d\rho}{dt} + div\left(\rho\vec{V}\right) = 0 \tag{1}$$

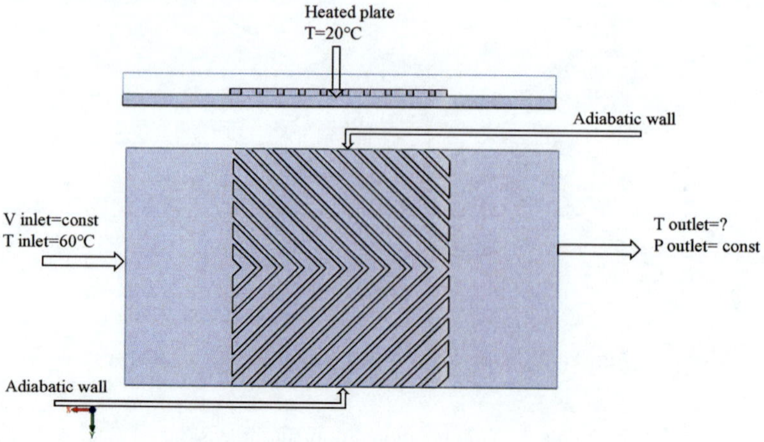

Fig. 2. Boundary condition of the computational domain

Momentum equation:

$$\frac{d(\rho\vec{V})}{dt} + div\left(\rho\vec{V}\otimes\vec{V} + P\bar{\bar{I}}\right) = div\left(\bar{\bar{\tau}} + \overline{\overline{\tau_R}}\right) \tag{2}$$

Energy equation:

$$\frac{d(\rho E)}{dt} + div\left(\rho E\vec{V}\right) = div\left[\left(P\bar{\bar{I}} + \bar{\bar{\tau}} + \overline{\overline{\tau_R}}\right)\vec{V} - \vec{q} - \vec{q_t}\right] \tag{3}$$

Equation of kinetic energy and its dissipation rate:

$$\frac{d(\rho k)}{dt} + div\left(\rho\vec{V}k\right) = div\left[\left(\mu + \frac{\mu_t}{\sigma_k}\right)\nabla k\right] + \overline{\overline{\tau_R}} : \nabla\vec{V} - \rho\varepsilon \tag{4}$$

$$\frac{d(\rho\varepsilon)}{dt} + div\left(\rho\vec{V}\varepsilon\right) = div\left[\left(\mu + \frac{\mu_t}{\sigma_\varepsilon}\right)\nabla\varepsilon\right] + C_1\frac{\varepsilon}{k}\overline{\overline{\tau_R}} : \nabla\vec{V} - C_2\rho\frac{\varepsilon^2}{k} \tag{5}$$

Where: $\vec{V}$ is the velocity vector, p is the pressure, E is the energy, k is the kinetic energy and ε is the dissipation rate. τ is shear tensor, I is the identity matrix. μ is dynamic viscosity, μt is turbulent viscosity and ρ is density (constant as incompressible fluid), C1, C2, σk, and σε are the standard constants of the k–ε turbulence model. The standard k-ε model with "wall-functions" was employed for the analysis (Paras et al. 2002). After a gird independence study, an unstructured tetrahedral mesh was performed with 818629 elements which is the most suitable with the minimum error and time simulation. The simple algorithm is used for the pressure velocity coupling (Pantakar 1980) and the second order upwind scheme (Yadav and Bhagoria (2014)/ Karmare and Tikekar (2010)) are used in order to solve the conservation of mass, momentum and energy. The convergence residuals are maintained 10e-3 for both continuity, momentum and 10e-6 for energy.

5 Numerical Results

The Flow Characteristics

For a constant Reynolds number Re of 400, velocity and temperature, turbulence intensity and turbulence kinematic energy contours are presented in the mid plan as indicated in Fig. 3. They present the flow charactristics with a constant chevron angle β of 45° and a constant rib height h of 3 mm. The Fig. 3a reveals the temperature distribution in the mid plan of turbulent flow over a corrugated plate. The presence of the chevron shapes leads to the improvement of cooling. Indeed, the temperature decreases from the inlet to the outlet. Near the walls the temperature is decreased also. Besides, the velocity distribution in the mid plan is mentioned in Fig. 3b. Behind the walls, the velocity is minimum and increases progressively over the corrugations. By examining the Fig. 3b, we can see the existence of two zones. The distribution zone (at inlet and outlet) where the velocity is evaluated lowly. The transfer zone is characterized by gradual increase in the velocity. Beyond that the fluid enters the corrugated zone the turbulence intensity and the turbulence kinetic energy increases progressively then decreases slowly in the outlet zone as indicated in Fig. 3c and 3d. The turbulence of the fluid is intensified in the area of the chevron.

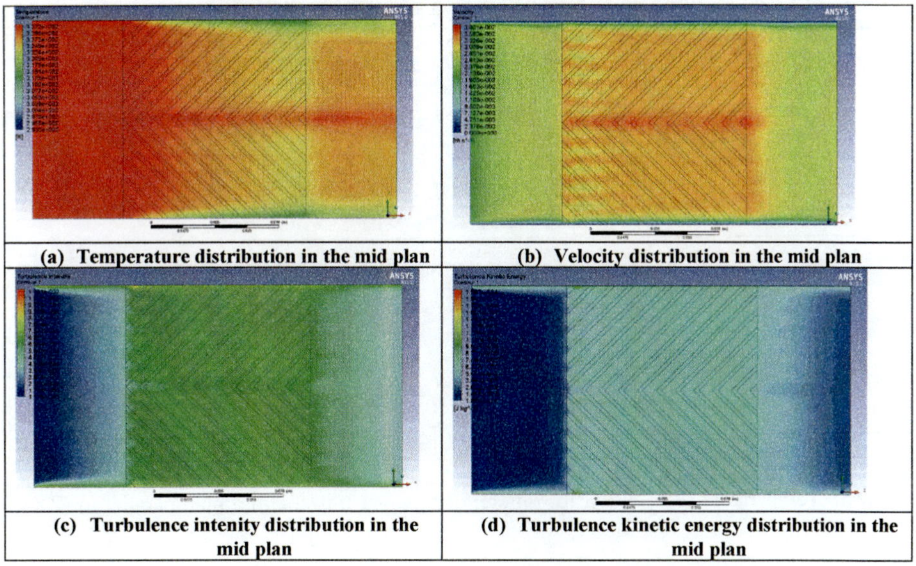

(a) Temperature distribution in the mid plan | **(b) Velocity distribution in the mid plan**

(c) Turbulence intenity distribution in the mid plan | **(d) Turbulence kinetic energy distribution in the mid plan**

Fig. 3. Temperature, velocity, turbulence intensity and turbulence kinetic energy distribution in the mid plan for Re = 400

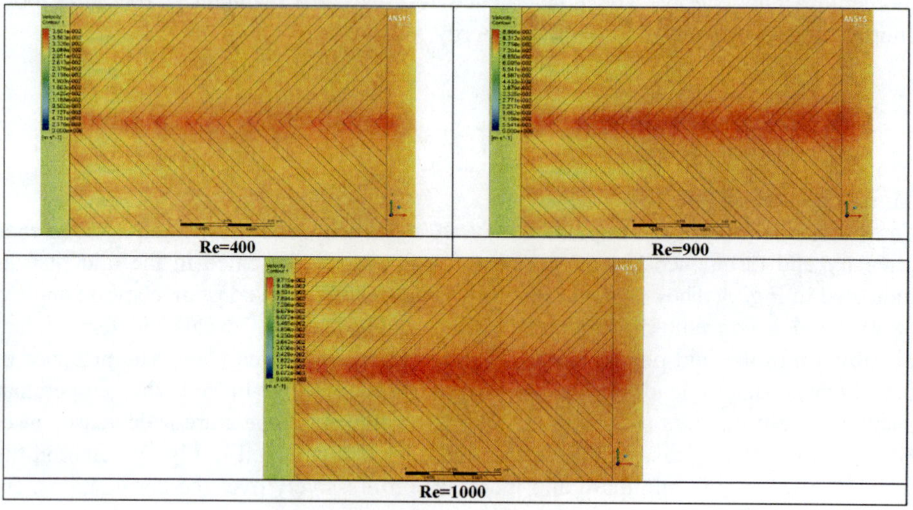

Fig. 4. Velocity contours for different Reynolds number Re in the mid plan

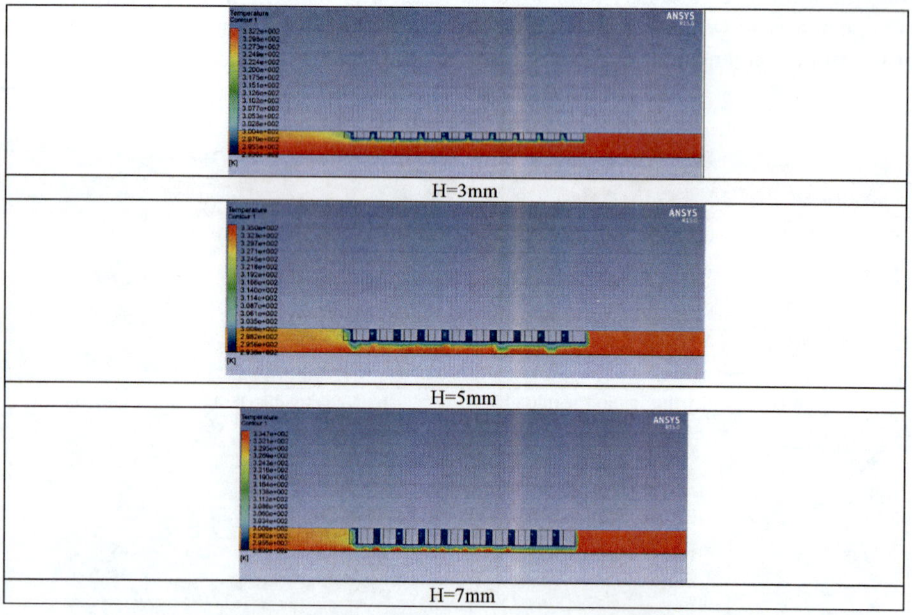

Fig. 5. Temperature distribution for different rib height h in the mid plan parallel to (ZX)

The Effect of the Reynolds Number Re

The Fig. 4 describes the effect of Reynolds number Re on the velocity distribution over a corrugated plate. The velocity contours are located in the mid plane parallel to (XY) plan for a constant rib height h of 3 mm. They show that the increase of Reynolds number Re, causes the increase of velocity cross the flow. In fact, the turbulence is increased and the laminar sublayer is decreased. Therefore, the heat transfer is ameliorated.

The Effect of the Rib Height H For a constant Reynolds number Re of 400, the effect of the rib height h range of 3/5/7 mm on the heat transfer enhancement is presented in the Fig. 5. It presents the temperature distribution for various rib height configurations located in the mid plan parrallel to (ZX) plan. The Fig. 5 reveals the temperature distribution in the corrugated plate. In the inlet the temperature is maximum and at the outlet, it is minimum. Indeed the temperature is minimum and located behind the corrugation shapes. Therefore, the presence of chevron shapes improves cooling and ameliorates the heat of transfer.

6 Conclusion

In this work the hydrodynamic and thermal behavior of the hot fluid inside a corrugated plate heat exchanger has been numerically simulated. It has been shown that the presence of the chevron corrugation in the heated plate of the PHE ameliorates the heat of transfer and its thermal performance. The rib height h has an influence on the flow characteristics. In fact, the temperature in the output of the channel is decresed and the cooling is improved. Hence, the best heat transfer is occurated with rib height h of half of height of the fluid domain.

References

Focke, W.W., Zachariades, J., Olivier, I.: The effect of the corrugation inclination angle on the thermohydraulic performance of plate heat exchangers. Int. J. Heat Mass Transf. **28**(8), 1469–1479 (1985)

Shah, R.K., Wanniarachchi, A.S.: Plate heat exchanger design theory. In: Buchlin, J.M. (ed.) Industrial Heat Exchangers. Karman Institute Lecture Series, no. 991-04. Rhode-Saint-Genese, Belgium (1991)

Kays, W.M., London, A.L.: Compact Heat Exchangers, 3rd edn. Krieger Publishing Company, Melbourne (1998)

Hesselgraves, J.E.: Compact Heat Exchangers: Selection, Design and Operation, 1st edn. Pergammon, New York (2001)

Kanaris, A.G., Mouza, K.A., Paras, S.V.: Designing novel compact heat exchangers for improved efficiency using a CFD code. In: lst International Conférence “From Scientfific Computing to Computational Engineering”, Athène, 8–10 Septembre 2004 (2004)

Kanaris, A.G., Mouza, K.A., Paras, S.V.: Flow and heat transfer prédiction in a corrugated plate heat exchanger using a CFD code. Chem. Eng. Technol. **29**(8), 923–930 (2006)

Paras, S.V., Kanaris, A.G., Mouza, A.A., Karabelas, A.J.: CFD code application to flow through narrow channels with corrugated walls. In: 15th International Congress of Chemical and Process Engineering (2002)

Patankar, S.V.: Numerical Heat Transfer and Fluid Flow. McGraw-Hill, New York (1980)

Yadav, A.S., Bhagoria, J.L.: A CFD based thermo-hydraulic performance analysis of an artificially roughened solar air heater having equilateral triangular sectioned rib roughness on the absorber plate. Int. J. Heat Mass Transf. **70**, 1016–1039 (2014)

Karmare, S.V., Tikekar, A.N.: Analysis of fluid flow and heat transfer in a rib grit roughened surface solar air heater using CFD. Sol. Energy **84**, 409–417 (2010)

Djemel, H., et al.: Three-dimensional numerical study of the performance of the intercooler equipped with vortex generators. Heat Transf. Res. **48**(8), 715–740 (2017)

CFD Simulation of CO_2 Adsorption onto Activated Carbon for Gas Separation and Storage Applications

Skander Jribi[1(✉)], Boutheina Zallama[2], and Takahiko Miyazaki[3]

[1] Laboratory of Electro-Mechanical Systems, The Engineers National School of Sfax, University of Sfax, Sfax, Tunisia
skander.jribi@gmail.com

[2] LESTE Laboratory, National Engineering School of Monastir, University of Monastir, 5000 Monastir, Tunisia
zallama.boutheina@live.fr

[3] Interdisciplinary Graduate School of Engineering Sciences, Kyushu University, Fukuoka 816-8580, Japan
miyazaki.takahiko.735@m.kyushu-u.ac.jp

Abstract. The increase in global energy demand, the scarcity of fossil fuels resources, climate change and global warming are undeniable realities. In this energy context, the implementation of concrete measures in favor of greater energy efficiency systems becomes urgent. Among these systems, the integration of environmental friendly, low energy adsorption technologies is gaining more interest in many chemical, residential and industrial applications such as adsorption refrigeration systems, quality treatment and energy storage processes. Pressure swing adsorption and temperature swing adsorption are among the promising techniques for CO_2 capturing and separation from exhaust gases. In this study, corrected adsorption rates of CO_2 onto activated carbon powder were investigated using computational fluid dynamics simulation. The modified linear driving force (mLDF) model was used as the adsorption kinetics equation by fitting of experimental data with isothermal assumption. Then the adsorbent layer temperature was estimated with CFD simulation which allowed to adjust the diffusion time constants for accurate performance investigation of CO_2 adsorption cooling, storage and separation applications.

Keywords: Carbon dioxide · Activated carbon · CFD simulation · Adsorption rate

1 Introduction

The selection of adsorbent-refrigerant pair is critical for designing thermal and pressure swing adsorption processes. The knowledge of the sorption characteristics of these pairs is crucial for assessing the reactor performance. Multiple studies have considered the sorption kinetics effect of activated carbons-CO_2 pairs. Recently, Singh et al. (2018) proposed an investigation of adsorption isotherms and kinetics of CO_2 on indigenous coconut shell based activated carbon at different temperatures using volumetric

M. Kharrat et al. (Eds.): ICAMEM 2019, LNME, pp. 187–193, 2020.
https://doi.org/10.1007/978-3-030-52071-7_26

method. Adsorption kinetics data were fitted with Fickian Diffusion and Linear Driving Force models. Sarker et al. (2017) proposed the carbon dioxide adsorption isotherm and kinetics on six adsorption materials, namely, three different types of zeolites, carbon molecular sieve and two activated carbons. Similar kinetic measurements were driven by Zhang et al. (2010) and Shen et al. (2010). Even if, in most of the above-mentioned investigations, the linear driving force was commonly adopted for fitting of the adsorption rates, the fact of assuming isothermal assumption has revealed in some cases real limitations in interpreting and analyzing kinetic behavior.

This study is conducted to be continuity to the previous experimental work of Jribi et al. (2017) where the present study proposes corrected adsorption rate of CO_2 onto Maxsorb III by means of CFD simulation overcoming the common practice of neglecting the heat of adsorption.

2 CFD Model

In the present work, the Computational Fluid Dynamics method is proposed to simulate the adsorption kinetics of carbon dioxide (CO_2) onto activated carbon adsorbent Maxsorb III at 30 °C and 520 kPa. Simulations were carried using Fluent software integrated in AnsysWorkbench R19.2.

2.1 Geometry

The geometry illustrated in Fig. 1 consists of a cylindrical basket filled with the adsorbent material namely activated carbon powder Maxsorb III. Axial revolution is considered. Thus, the model is reduced to 2D-axisymmetric simulation.

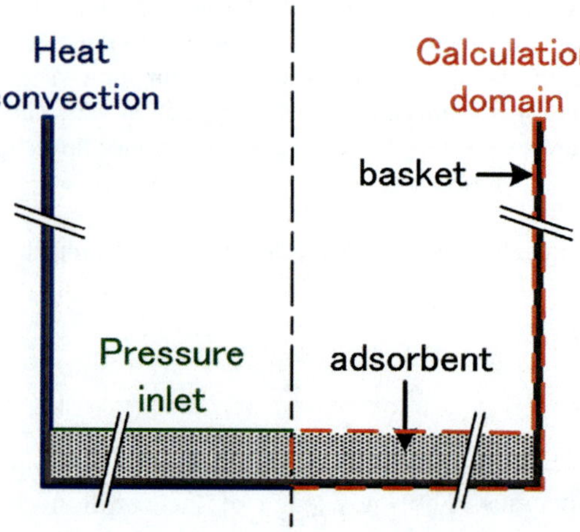

Fig. 1. Schematic of geometry and boundary conditions used in the simulation

2.2 Mathematical Model

The governing equations for fluid flow and heat and mass transfer consists of mass, energy and momentum conservation equations. Further details about the mathematical model can be found in (Jribi et al. 2016).

2.3 Boundary Conditions

2.3.1 Inlet Condition

The CO_2 is charged in measuring chamber by a sharp pressure jump of 520 kPa from 0 Pa initially. More details about the procedure are found in (Jribi et al. 2017).

2.3.2 Heat Transfer Correlation

The basket wall exchanges heat by convection with the surrounding CO_2 gas. The average natural convection heat transfer coefficient at different temperatures and pressures is evaluated for enclosed spaces as 5 $W.m^{-2}.K^{-1}$ (Holman 2010).

3 Results and Discussion

In this paper, we present the results of CFD simulation of transient CO_2 adsorption onto activated carbon at 30 °C and inlet pressure of 520 kPa. The experimental data have been measured by thermogravimetric apparatus (Jribi et al. 2017).

Figure 2 shows the measured pressure and uptake using the thermogravimetric apparatus. The pressure attains almost equilibrium within few seconds. The instantaneous adsorption uptake increases sharply at starting of adsorption and attains the equilibrium uptake within 10 min.

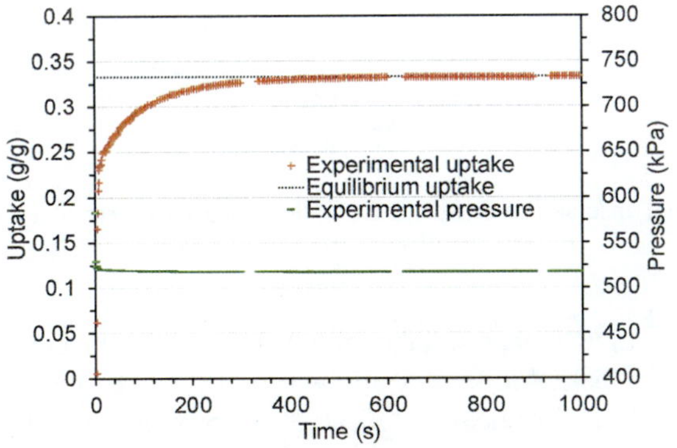

Fig. 2. Transient adsorption uptake of CO_2 onto Maxsorb III at 30 °C

Jribi et al. (2017) also fitted the instantaneous uptake with modified linear driving force (mLDF) model given by the following expression.

$$\frac{dq}{dt} = k^{\frac{q}{q^*}} (q^* - q) \quad (1)$$

Where k is the mass transfer coefficient correlated by:

$$k = A\,exp\left(-\frac{E_a}{RT}\right) \quad (2)$$

A and E_a denote pre-exponential factor and activation energy, respectively. The equilibrium adsorption uptake (q^*) is evaluated using Dubinin-Astakhov (D-A) adsorption isotherm equation (Jribi et al. 2017).

It is worthy to note that the fittings were carried out with isothermal assumption that is the heat of adsorption is neglected and assuming constant temperature during adsorption (T = constant).

As shown in Fig. 3, the isothermal mLDF model matched the experimental data. The diffusion time constant is k_{0m} = 5.06 10^{-3} s^{-1}. Considering the activation energy E_a = 5254.76 J.mol^{-1}, the pre-exponential factor is then A = 0.046 s^{-1}. The root mean square deviations (RMSD) was 2.46.

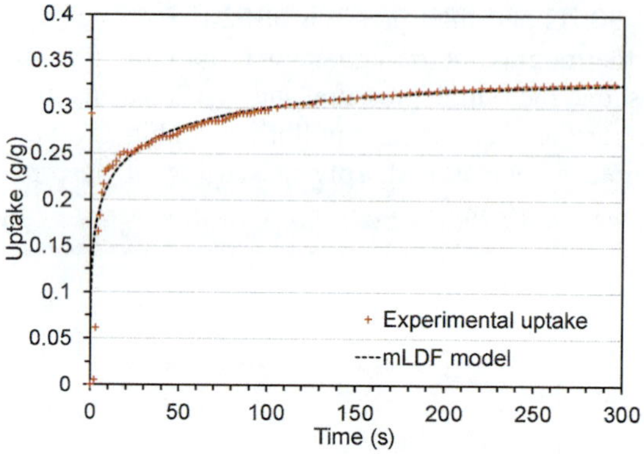

Fig. 3. Experimental instantaneous uptakes (+) fitted with isothermal mLDF model (dashed line)

Although the mLDF model suitably fitted the adsorption uptake, the isothermal assumption used can make further deviations from calculated models. In fact, a small temperature change due to release of heat of adsorption can cause a considerable deviation of calculated uptake using isothermal assumption comparing to experimental data (Jribi et al. 2016).

In this study, the adsorption of CO_2 onto Maxsorb III was simulated using the same conditions as in the thermogravimetric apparatus to evaluate the effect of heat of adsorption on the established adsorption kinetics model.

Figure 4 shows that the simulated uptakes were lower than those calculated from mLDF model using the isothermal assumption as the heat of adsorption is not taken into account in the energy conservation equation of the mathematical model. The RMSD values increased from 2.46 for calculated values to 4.46 for simulated values.

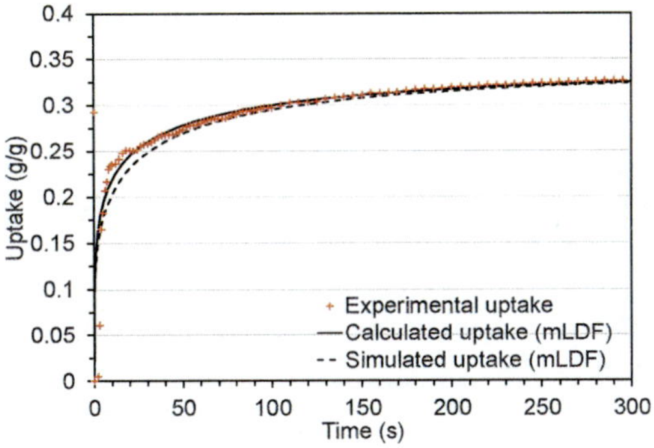

Fig. 4. Experimental vs. calculated and simulated uptakes using mLDF model.

The average temperature change of adsorbent layer is presented in Fig. 5. It shows that the adsorbent temperature increases sharply at starting of adsorption then decreases due to convection cooling from the basket wall. The temperature increase was about 5.3 °C from 30 °C initially which corresponds to a decrease in the simulated equilibrium uptake of 10.3%. Therefore, the simulated uptakes were lower than those calculated with isothermal mLDF model.

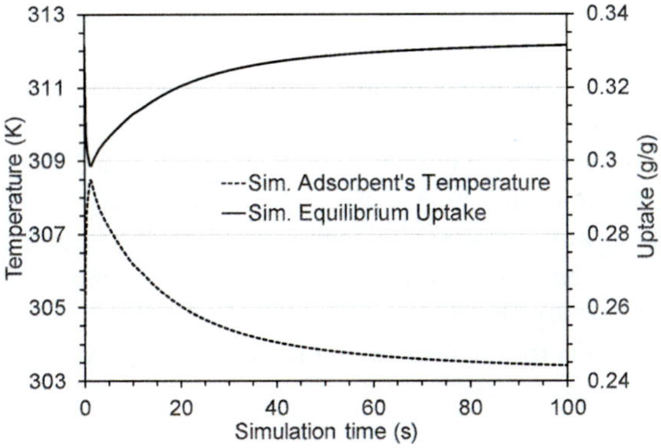

Fig. 5. Simulated average adsorbent layer temperature (dashed line) and equilibrium uptake (line)

Using the simulated temperature and equilibrium uptake profiles to fit the experimental data to mLDF equation, we get the non-isothermal parameters of the adsorption rate equation which are $A = 0.06\ s^{-1}$ and $E_a = 5245.76\ J.mol^{-1}$. This means that the corrected adsorption rate equation is higher of about 31.7% than that with isothermal assumption. Both the calculated and the simulated uptakes matches those measured experimentally as shown in Fig. 6.

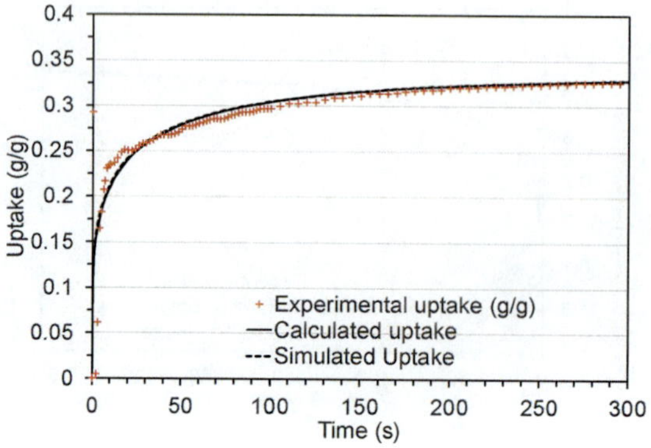

Fig. 6. Calculated (line) and simulated (dashed line) uptakes using the non-isothermal mLDF model vs. experimental data

4 Conclusion

Adsorption kinetics of carbon dioxide onto activated carbon were investigated using computational fluid dynamic simulation. Transient adsorption uptake measured experimentally with thermogravimetric apparatus were first fitted with modified linear driving force (mLDF) model with isothermal assumption (T = 30 °C) and charging pressure of 520 kPa from 0 Pa initially. However, the simulated uptakes were lower than those measured experimentally and the RMSD values increased from 2.46 for calculated values with isothermal mLDF model to 4.46 for simulated values. The release of heat of adsorption caused a temperature increase which yielded a decrease in the simulated equilibrium uptake of about 10.3%. The simulated temperature was taken into consideration in fitting the non-isothermal mLDF and the corresponding adsorption rate parameters were $A = 0.06\ s^{-1}$ and $E_a = 5245.76\ J.mol^{-1}$. The simulated uptake with the corrected adsorption kinetics equation matched correctly the experimental data. The developed adsorption rate model allows accurate performance investigation of CO_2 adsorption storage and separation applications.

References

Holman, J.P.: Heat Transfer-Natural Convection Systems, Chap. 7, pp. 327–378, 10th edn. McGraw-Hill, New York (2010)

Jribi, S., Miyazaki, T., Saha, B.B., Pal, A., Younes, M., Koyama, S., Maalej, A.: Equilibrium and kinetics of CO2 adsorption onto activated carbon. Int. J. Heat Mass Transf. **108**, 1941–1946 (2017)

Jribi, S., Miyazaki, T., Saha, B.B., Koyama, S., Maeda, S., Maruyama, T.: Corrected adsorption rate model of activated carbon–ethanol pair by means of CFD simulation. Int. J. Refrig **71**, 60–68 (2016)

Sarker, A.I., Aroonwilas, A., Veawab, A.: Equilibrium and kinetic behaviour of CO2 adsorption onto zeolites, carbon molecular sieve and activated carbons. In: 13th International Conference on Greenhouse Gas Control Technologies, Lausanne, Switzerland, GHGT-13, 14–18 November 2016 (2017), Energy Procedia **114**, 2450–2459

Shen, C., Grande, C.A., Rodrigues, A.E.: Adsorption equilibrium and kinetics of CO_2 and N_2 on activated carbon beads. Chem. Eng. J. **160**, 398–407 (2010)

Singh, V.K., Kumar, E.A., Saha, B.B.: Adsorption isotherms, kinetics and thermodynamic simulation of CO_2-CSAC pair for cooling application. Energy **160**, 1158–1173 (2018)

Zhang, Z., Zhang, W., Chen, X., Xia, Q., Li, Z.: Adsorption of CO_2 on zeolite 13X and activated carbon with higher surface area. Sep. Sci. Technol. **45**, 710–719 (2010)

Enhancement of Fixed-Wing Space Drone Performance Through Thermoelectric Power Generation

Devyn Rice[1], Samah Ben Ayed[2](✉), Stephen Johnstone[3], and Abdessattar Abdelkefi[1]

[1] Department of Mechanical and Aerospace Engineering, New Mexico State University, Las Cruces, NM 88003, USA
[2] Department Engineering Technology and Surveying Engineering, New Mexico State University, Las Cruces, NM 88003, USA
benayed@nmsu.edu
[3] Los Alamos National Laboratory, Los Alamos, NM 87545, USA

Abstract. The thermal effects of wing color for drones performing in the Martian atmosphere are investigated during the summer and winter seasons. The optimal wing color configuration is identified in each season. This study focuses on two useful consequences of the thermal effects of wing color: the skin friction drag reduction and the thermoelectric generation of power. According to its color, every side has a certain temperature affecting the local skin friction drag. Investigations of various configurations have shown that the thermal effect on the wing boundary layer is insignificant because of the low atmospheric pressure. However, the large temperature difference the top and the bottom of the wing has a great potential of thermoelectric power generation. The maximum temperature differences between the top and bottom surfaces for the summer and winter seasons are, respectively, 65 K and 30 K. The drag reduction and the power generation via thermoelectric generators both contribute to enhancing the endurance of drones. Future drone designs will benefit from the increased endurance through optimizing the wing color configuration.

Keywords: Aerodynamics · Mars · Drones · Thermoelectric power · Wing color

1 Introduction

Over the past few decades, various government and private programs have been developed for space exploration. Among the high interest exploration targets is the planet Mars. NASA has performed many missions in the past and will continue with new mission in the future including the upcoming Mars 2020 mission [1]. In the past few years, the concept of using drones for planetary exploration has gained traction and many design concepts have been developed [2]. To design a drone suitable for Mars exploration, it is important to understand the Martian atmosphere and other properties for flight. Due to the extreme conditions and challenges associated with flying in the Martian atmosphere, maximizing the efficiency of a drone is essential [2, 3]. Drones

M. Kharrat et al. (Eds.): ICAMEM 2019, LNME, pp. 194–199, 2020.
https://doi.org/10.1007/978-3-030-52071-7_27

require high endurance to be able to perform many defined missions for planetary exploration. The endurance of a drone is primarily limited by the available energy [4].

Researchers have explored various techniques to decrease and manage the required power for drones. An example is to propose new design methods for drones, which improve their aerodynamic and mechanical efficiencies and consequently reduce the power consumption. Therefore, design methods enhance the performance and efficiency of drones which will extend their endurance and improve their utility in complex environments. Reduction of drag can be considered as one of the main factors during the design process to conserve energy for the mission. In other words, drag reduction of fixed wing space drones is crucial to flight efficiency on Mars [5]. This study will provide information on the surface temperature of a wing during different seasons and for different color configurations. An optimal wing color configuration will be determined to achieve maximum drag efficiency. Also, due to a possible temperature difference between the top and bottom wing surfaces, an investigation into the application of thermoelectric generators (TEGs) will be performed. Thermoelectric generators are composed of semiconductor material that produces thermoelectric modules. These modules are between ceramic wafers and then connected to an electrical load resistance. Electrical power is generated from the temperature difference as a result that can be explained through the Seebeck effect. The specific type of TEG considered for this study is the BiTe which is considered suitable for low-temperature heat recovery applications [6].

2 Thermal Modeling

To understand the thermal effects of different wing colors, a thermal analysis is performed using an energy balance and including the influences of solar energy absorption, convection heat loss, and radiative heat loss:

$$P_{solar} = P_{convective} + P_{radiative} + P_{conductive} \tag{1}$$

The various sources of energy acting on the wing can be seen in Fig. 1. Determining the various fluxes in the energy balance requires information from the Martian atmosphere. In this study, the ambient temperature, pressure, wind speed, and other

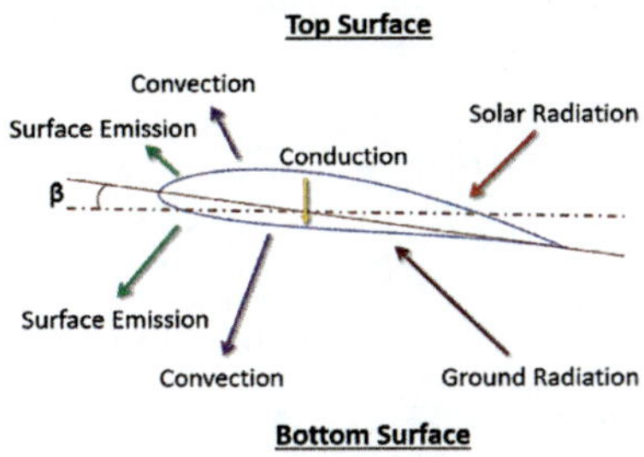

Fig. 1. Heat flux on a wing.

atmospheric properties are obtained from the dataset collected by Viking Lander 1 and the entry probe (VL1) which is located at 22.37 N latitude and 47.97 W longitude on the surface of Mars.

The top and the bottom wing surfaces experience different heat fluxes which can be determined by the view factors, surface temperatures, surface absorptance, and surface emissivity. The total energy balance equation for both the top and the bottom surfaces of the wing are given in Eqs. 2 and 3, respectively [7].

$$\alpha_t \cdot G_b \cdot \cos\theta + \alpha_t \cdot F_d^t \cdot G_{dh} + \alpha_t \cdot al \cdot F_{al}^t \cdot (G_{bh} + G_{dh}) - h \cdot (T_{s-top} - T_{amb}) - \varepsilon_t \cdot \sigma \cdot F_d^t \cdot \left(T_{s-top}^4 - T_{sky}^4\right) - \varepsilon_t \cdot \sigma \cdot F_{al}^t \cdot \left(T_{s-top}^4 - T_{grd}^4\right) - \frac{K \cdot A}{L} \cdot \left(T_{s-top} - T_{s-bottom}\right) = 0 \tag{2}$$

$$\alpha_b \cdot F_d^b \cdot G_{dh} + \alpha_b \cdot al \cdot F_{al}^b \cdot (G_{bh} + G_{dh}) - h \cdot (T_{s-bottom} - T_{amb}) - \varepsilon_b \cdot \sigma \cdot F_d^b \cdot \left(T_{s-bottom}^4 - T_{sky}^4\right) - \varepsilon_b \cdot \sigma \cdot F_{al}^b \cdot \left(T_{s-bottom}^4 - T_{grd}^4\right) - \frac{K \cdot A}{L} \cdot \left(T_{s-bottom} - T_{s-top}\right) = 0 \tag{3}$$

In the energy balance, we denote by: the absorptivity α, direct beam irradiance on a surface G_b, solar incidence angle θ, view factor from the top surface to the sky F_d^t, diffuse irradiance on a horizontal surface G_{dh}, albedo al, view factor from the top surface to the ground F_{al}^t, direct beam irradiance on a horizontal surface G_{bh}, heat transfer coefficient h, top surface temperature T_{s-top}, ambient temperature T_{amb}, emissivity ε, Stephen-Boltzmann constant σ, sky temperature T_{sky}, ground temperature T_{grd}, atmospheric thermal conductivity K, wing surface area A, wing chord length L, bottom surface temperature $T_{s-bottom}$, view factor from the bottom surface to the sky F_d^b, and the view factor from the bottom surface to the ground F_{al}^b. The subscripts *t* and *b* denote top or bottom wing surface.

The thermal boundary layer effects on the skin friction drag are determined using Eq. 4 [8], where the wingspan ws, is 3.5 m and the chord length, L, is 0.22 m. To determine the viscosity μ, Sutherlands formula, which can be found in Eq. 5, is utilized [9]. T_0 and μ_0 are the reference temperature and viscosity, respectively, which are 293.14 K and 1.48×10^{-5} kg/m · s for the Martian atmosphere [10]. The Sutherlands constant C, for CO_2 is 240 K [10] and T_s is the wing surface temperature. The density ρ is determined using ideal gas law considering CO_2 in Eq. 6. The pressure P is found from the information collected by VL1.

$$D = 0.664 \cdot ws \cdot \sqrt{\rho \cdot \mu \cdot u^3 \cdot L} \tag{4}$$

$$\mu = \mu_0 \cdot \left(\frac{T_0 + C}{T_s + C}\right) \cdot \left(\frac{T_s}{T_0}\right)^{3/2} \tag{5}$$

$$\rho = \frac{P}{192.1R \cdot T_s} \tag{6}$$

Equations 2 and 3 are used to determine the surface temperature for both the bottom and top wing surfaces for black and white wing colors as presented in Fig. 2. During the summer and winter seasons, the wing surface experiences the highest temperatures when a black color is applied to the top surface. The lowest wing temperatures occur when white paint is used on the top surface. As expected, due to the high solar absorption during summer, the overall temperatures on the wing surfaces are higher during summer compared to winter. The total drag of the combined wing color configurations is plotted in Figs. 3(a) and 3(b) for summer and winter, respectively. The four different wing color combinations are as follows: black on top and black on bottom (black-black), black on top and white on bottom (black-white), white on top and black on bottom (white-black), and white on top and white on bottom (white-white). It follows from Fig. 3 that the drag difference is insignificant.

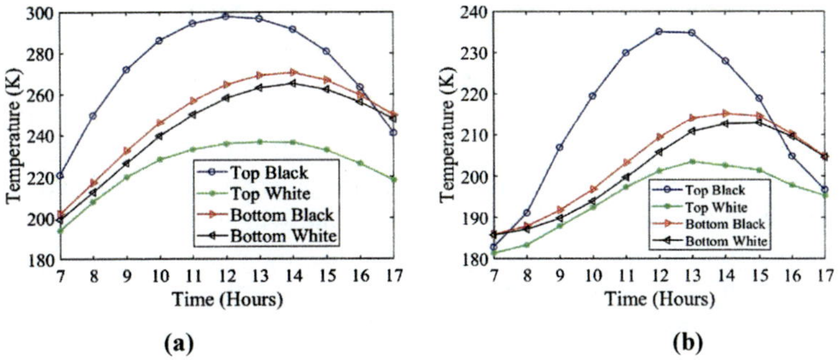

Fig. 2. Surface temperature for (a) summer and (b) winter.

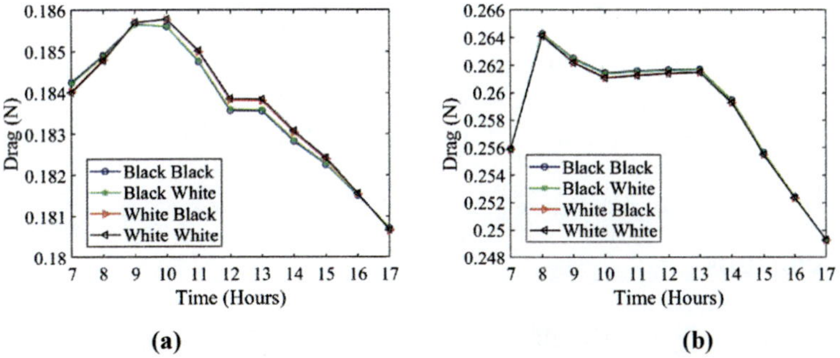

Fig. 3. Surface drag for (a) summer and (b) winter in the Martian atmosphere.

3 TEGs for a Martian Drone

The basis of a thermoelectric generator is the ability to convert heat energy into electrical power when there exists a temperature difference between two surfaces. The power produced from the TEG is the difference between the heat supply P_h and heat removal P_c which is also equal to the product of the voltage across the external load resistance V and the current I. The energy balance can be expressed as:

$$P_{TEG} = P_h - P_c = V \cdot I \tag{7}$$

The derived equation for the electrical harvested power is written as [11]:

$$P_{TEG} = R_L \cdot \frac{\alpha^2 \cdot (T_h - T_c)^2}{(R_i - R_L)^2} \tag{8}$$

It follows from Eq. 8 that the electrical power is dependent on the Seebeck coefficient α, the internal resistance R_i, the external resistance R_L, and the temperature difference between the hot surface T_h and the cold surface T_c. Depending on the specific TEG and the manufacturer, the coefficients will differ.

An analysis is performed to determine the harvested power over a range of values for the external resistance, Seebeck coefficient, and internal resistance.

In Table 1, the optimal power generation along with the maximum temperature difference and the time of day for each color configuration for both summer and winter is given. The optimal wing color configuration that generates the most harvested power is the black-white configuration for both summer and winter. This is due to having the highest temperature difference. The white-white configuration has the lowest power generation and the lowest temperature difference. Overall, the TEG performs better during the summer than during the winter.

Table 1. Optimal power comparison: time, temperature and power

Color	Summer	Winter
Black-White	(10:00, 46.3 °C, 4.3 W)	(11:00, 36.1 °C, 2.6 W)
Black-Black	(10:00, 38.4 °C, 2.9 W)	(11:00, 30.5 °C, 1.8 W)
White-Black	(15:00, 32.6 °C, 2.1 W)	(15:00, 11.6 °C, 0.3 W)
White-White	(15:00, 27.2 °C, 1.5 W)	(15:00, 27.2 °C, 1.5 W)

4 Conclusions

To optimize the design of an efficient drone capable of flying in the Martian atmosphere, a study was performed to analyze the thermal effects of wing color on the heated boundary layer of a wing. This study provided information on the optimal wing color configuration during different seasons on Mars. The drag does not significantly change as a result of the color configuration. However, due to the high temperature differences on the top and bottom wing surfaces that occur on Mars, TEGs are capable of considerable power generation.

References

1. https://mars.nasa.gov/programmissions/missions/future/mars2020/, November 2019
2. Hassanalian, M., Rice, D., Abdelkefi, A.: Evolution of space drones for planetary exploration: a review. Prog. Aerosp. Sci. **97**, 61–105 (2018)
3. Hassanalian, M., Rice, D., Johnstone, S., Abdelkefi, A.: Performance analysis of fixed wing space drones in different solar system bodies. Acta Astronaut. **152**, 27–48 (2018)
4. Abdelkefi, A., Ghommem, M.: Piezoelectric energy harvesting from morphing wing motions for micro air vehicles. Theor. Appl. Mech. Lett. **3**(5), 052004 (2013)
5. Hassanalian, M., Rice, D., Abdelkefi, A.: Aerodynamic performance analysis of fixed wing space drones in different solar system bodies. In: 2018 AIAA Aerospace Sciences Meeting, p. 1533 (2018)
6. Hsu, C.T., Huang, G.Y., Chu, H.S., Yu, B., Yao, D.J.: Experiments and simulations on low-temperature waste heat harvesting system by thermoelectric power generators. Appl. Energy **88**(4), 1291–1297 (2011)
7. Matz, E., Appelbaum, J., Taitel, Y., Flood, D.J.: Solar cell temperature on mars. J. Propul. Power **14**(1), 119–125 (1998)
8. Bird, R.B., Stewart, W.E., Lightfoot, E.N.: Transport Phenomena. Wiley, New York (2006)
9. Smits, A.J., Dussauge, J.P.: Turbulent Shear Layers in Supersonic Flow. Springer, Heidelberg (2006)
10. Wei, T., Xiaofeng, Y., Yewei, G., Yanxia, D.: Aerodynamic prediction and performance analysis for mars science laboratory entry vehicle. World Acad. Sci. Eng. Technol. Int. J. Mech. Aerosp. Ind. Mech. Manuf. Eng. **9**(6), 1040–1045 (2015)
11. Abdelkefi, A., Alothman, A., Hajj, M.R.: Performance analysis and validation of thermoelectric energy harvesters. Smart Mater. Struct. **22**(9), 095014 (2013)

Simulation of Cellulose Pyrolysis Kinetics

Amal Masmoudi(✉), Moez Hammami, and Mounir Baccar

National School of Engineers of Sfax, Mechanical Department (CFDTP), University of Sfax, Soukra Street Km 3.5, Bp 1173, 3038 Sfax, Tunisia
masmoudiamal@hotmail.fr, hammamimoez2003@yahoo.fr, mounir.baccar@enis.rnu.tn

Abstract. By definition, the controlled pyrolysis of cellulose is the physical and chemical decomposition of organic compound under the action of heat and in the absence of oxygen. Understanding this phenomenon firstly involves identifying the reaction mechanisms and the kinetic parameters involved in the thermal degradation of cellulose. The diversity of results from the literature reflects the difficulty of explaining these complex kinetics related to pyrolytic degradation of such large biological macromolecules. This study consists in determining the effect of different temperatures and kinetic parameters which influence the mass Damköhler number on the pyrolysis of cellulose in a perfectly agitated reactor that ensures the uniformity and homogenization of the medium. The results of our simulation show that higher conversion is obtained for high values of mass Damköhler number which increases with temperature. We have also noticed that the conversion rate of cellulose pyrolysis process depends on the mass Damköhler imposed by the reaction model.

Keywords: Pyrolysis · Conversion rate · Simulation

1 Introduction

Cellulose being the principal compound of most ligno-cellulosic materials, it is not surprising that the study of cellulose pyrolysis takes a lot of interest, as attested by the large number of articles and works devoted to it (Chen et al. 2019; Gao et al. 2019). Historically, the first multi-reaction mechanism proposed is that of Broido and Weinstein (Broido and Weinstein 1972). A slightly endothermic first step between 200 and 280 °C leads to the formation of anhydro-cellulose, a molecule that will be degraded beyond 280 °C (exothermic reaction) into charcoal and volatile. At the same time, an endothermic de-polymerization reaction produces the tars competitively. A few years later, Shafizadeh et al (Shafizadeh and Bradbury 1979) confirmed the fundamentals described in the Broido model and propose a slightly modified mechanism. The initial reaction, characterized at low temperature by an acceleration of the loss of mass leads to the formation of an active cellulose characterized by a degree of polymerization approximately 10 times lower than the initial cellulose. Various studies conducted on this compound show that from 200 °C, intrinsic modifications may occur. Subsequently, this model will be simplified with the elimination of intermediate steps (formation of anhydro-cellulose or active cellulose). This mechanism is often

M. Kharrat et al. (Eds.): ICAMEM 2019, LNME, pp. 200–207, 2020.
https://doi.org/10.1007/978-3-030-52071-7_28

proposed as the most serious interpretation of cellulose thermal degradation (Várhegyi et al. 1994). Other authors have formulated similar patterns of cellulose degradation involving competitive and consecutive reactions. As a result, a single first-order reaction would be sufficient to describe the thermal degradation of cellulose (Zaror and Pyle 1986; Antal et al. 1998). However, other authors discussed and defended the multi-reaction scheme (Milosavljevic and Suuberg 1995).

Thus, this study aims to determine the temporal evolution of the pyrolytic degradation of cellulose at higher temperatures for different reaction models and different kinetic parameters by highlighting the influence of the mass Damköhler number.

2 Kinetics of Thermal Degradation of Cellulose

Understanding the pyrolysis phenomena firstly involves identifying the reaction and determining the kinetic parameters involved in the thermal degradation of the cellulose.

The kinetic values recorded in the literature are widely dispersed (Table 1) and confirm the difficulty in explaining the reaction mechanisms. In order to explain this dispersion, some authors have advanced the concept of "kinetic compensation effect" based on a compensating change of the pre-exponential factor A and the variation of the activation energy E (Zsakó 1976; Chornet and Roy 1980). In general, the pyrolysis of cellulose is frequently described by a single irreversible reaction of order 1:

$$\text{Cellulose} \xrightarrow{K} \upsilon_t \text{Tar} + \upsilon_g \text{Gas} + \upsilon_c \text{Char}$$

The reaction rate constant K is deduced from the Arrhenius equation:

$$K(T) = A\,\text{Exp}\,(-E/RT) \tag{1}$$

Where A, E, R and T are respectively the frequency or pre-exponential factor, the activation energy, the perfect gas constant and temperature.

3 Numerical Model

A numerical investigation of the evolution of the cellulose pyrolysis is performed under different kinetic models. The reactor is supposed perfectly agitated (batch reactor) which is characterized by the uniformity and homogenization of the medium. Thus, the conservation equation of the species involved in the thermo-chemical conversion is expressed in term of its dimensionless concentration and is given as follow:

$$\frac{\partial C_i^*}{\partial t^*} = \pm Da_{m_j}\, exp(E^*) C_j^* \tag{2}$$

With: $C_i^* = \frac{C_i}{C_R}$ is the dimensionless concentration of the ith species, $t^* = 2\pi Nt$ is the dimensionless time, Da_{m_j} is the mass Damköhler number and $C_j^* = \frac{C_j}{C_R}$ is the reactant

Table 1. Different mechanisms of cellulose pyrolysis and Mass Damköhler number at different temperatures

Model	Description	Mass Damköhler number at T = 600 K	Mass Damköhler number at T = 700 K	Mass Damköhler number at T = 800 K	Reference
Model 1	Cellulose → Tar	$Da_m = 2.58\ 10^{-3}$	$Da_m = 1.98\ 10^{-2}$	$Da_m = 9.11\ 10^{-2}$	(Barooah and Long 1976)
Model 2–1	Cellulose → 0.92 Volatiles + 0.08 Char	$Da_m = 4.77\ 10^{-3}$	$Da_m = 2.60\ 10^{-1}$	$Da_m = 5.21$	(Lewellen et al. 1977)
Model 2–2		$Da_m = 2.14\ 10^{-3}$	$Da_m = 3.80\ 10^{-2}$	$Da_m = 3.29\ 10^{-1}$	(Tabatabaie-Raissi et al. 1989)
Model 2–3		$Da_m = 1.61\ 10^{-3}$	$Da_m = 3.80\ 10^{-1}$	$Da_m = 3.23$	(Ding et al. 2018)
Model 3	Cellulose → 0.65 Volatiles + 0.35 Char	$Da_m = 1.64\ 10^{-3}$	$Da_m = 2.94\ 10^{-2}$	$Da_m = 2.55\ 10^{-1}$	(Zaror and Pyle 1986)
Model 4	Cellulose → Tar → 0.65 Volatiles + 0.35 Char	$Da_{m1} = 8.84\ 10^{-4}$ $Da_{m2} = 7.85\ 10^{-4}$	$Da_{m1} = 2.18\ 10^{-2}$ $Da_{m2} = 1.01\ 10^{-2}$	$Da_{m1} = 2.41\ 10^{-1}$ $Da_{m2} = 6.86\ 10^{-2}$	(Zaror and Pyle 1986)
Model 5	Cellulose → Anhydrocellulose → 0.65 Volatiles + 0.35 Char → Tar	$Da_{m1} = 2.80\ 10^{-5}$ $Da_{m2} = 8.84\ 10^{-5}$ $Da_{m3} = 1.73\ 10^{-3}$	$Da_{m1} = 3.59\ 10^{-3}$ $Da_{m2} = 5.0310^{-2}$ $Da_{m3} = 7.17\ 10^{-1}$	$Da_{m1}=1.37\ 10^{-1}$ $Da_{m2} = 5.85$ $Da_{m3} = 65.78$	(Broido and Weinstein 1972)
Model 6	Cellulose → Active Cellulose → Tar → 0.65 Volatiles + 0.35 Char	$Da_{m1} = 2.04\ 10^{-2}$ $Da_{m2} = 1.89\ 10^{-3}$ $Da_{m3} = 6.09\ 10^{1}$	$Da_{m1} = 2.13\ 10^{1}$ $Da_{m2} = 5.46\ 10^{-1}$ $Da_{m3} = 4.89\ 10^{3}$	$Da_{m1} = 3.92\ 10^{3}$ $Da_{m2} = 38.29$ $Da_{m3} = 1.31\ 10^{5}$	(Shafizadeh and Bradbury 1979)

species for the jth reaction, $E^{*} = \frac{E}{RT_0}$ is the dimensionless activation energy, C_R is the reactant concentration and N is the agitation velocity.

4 Initial Conditions

The initial conditions considered in numerical simulation are:

- At $t^{*} = 0$: the reactant is introduced to the reactor with a dimensionless concentration equal to 1.
- At $t^{*} = 0$: the dimensionless concentration of the product is equal zero.

5 Results and Discussion

The results of our numerical simulations are presented in this section. All the figures illustrate the evolution of the species concentration profiles with time. The results of simulations indicate that the cellulose degradation is almost complete for high mass Damköhler numbers and with increasing the temperature. In fact, the mass Damköhler number value increases with the augmentation of the temperature. Besides, for some models, the cellulose pyrolysis process necessitates more high temperature to achieve cellulose conversion (Figs. 1, 3, 5 and 6). In fact, these cases can be explained by the fact that the corresponding mechanisms have low activation energies and therefore low mass Damköhler number values. Moreover, we can explain this phenomena by the fact that for low values of the activation energy, the quantity of energy required is not sufficient for the completion of the reaction and for this reason the temperature must be increased to obtain a better conversion. For the majority of the studied models, it was noted for the pyrolysis process that the reactions with the highest activation energies are the fastest reactions (Figs. 2, 4, 7 and 8). Especially, the two last models (Figs. 7 and 8) have the highest mass Damköhler number and that's why the cellulose conversion is almost complete for a really low dimensionless time less than one. However, for the same reaction model, we find that the mass Damköhler number does not increase with the activation energy (model 2–1 (Fig. 2) and model 2–3 (Fig. 4)) and hence the yield of the products. While the conversion rate is faster for the largest mass Damköhler number and the latter increases with the increase of temperature which explains that the pyrolysis reaction is highly dependent on temperature and this justifies the fact that the reaction constant (K) depends only on the temperature.

In conclusion, a better conversion of cellulose is determined for large values of activation energy while the most influencing parameter is the mass Damköhler number which increases with increasing the temperature to obtain high products yields.

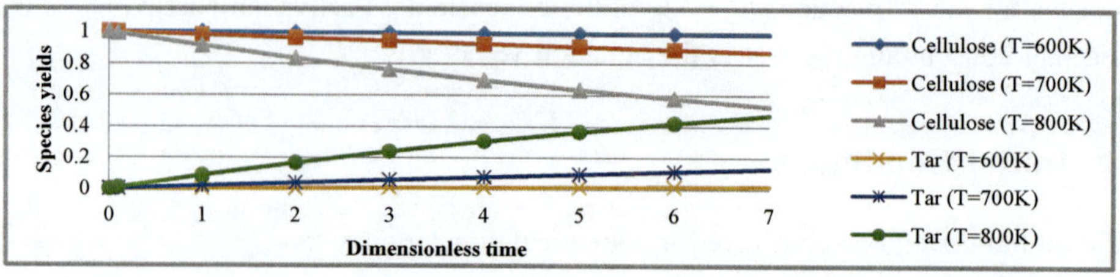

Fig. 1. Species yields profiles evolution at different temperatures for model 1

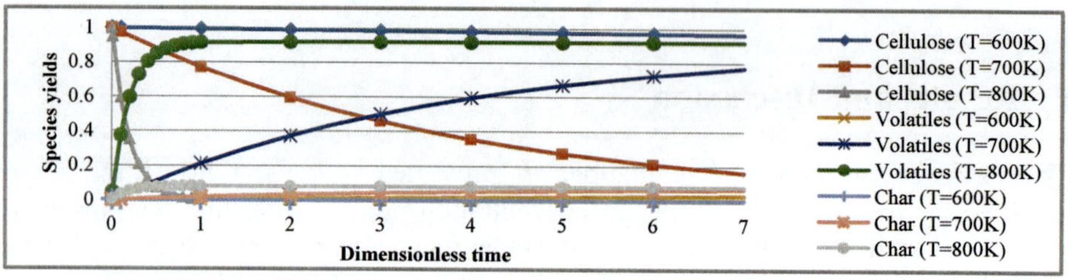

Fig. 2. Species yields profiles evolution at different temperatures for model 2–1

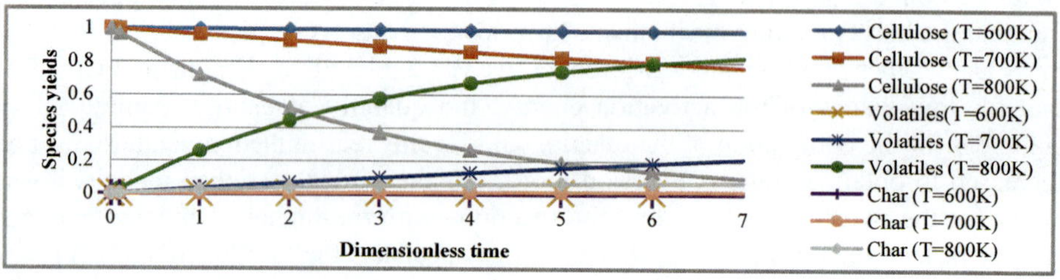

Fig. 3. Species yields profiles evolution at different temperatures for model 2–2

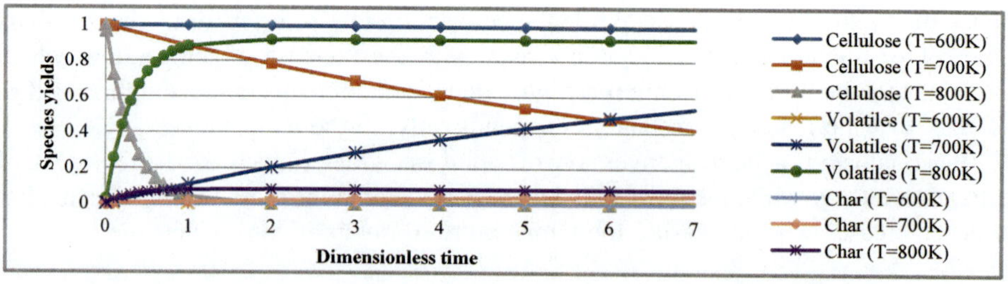

Fig. 4. Species yields profiles evolution at different temperatures for model 2–3

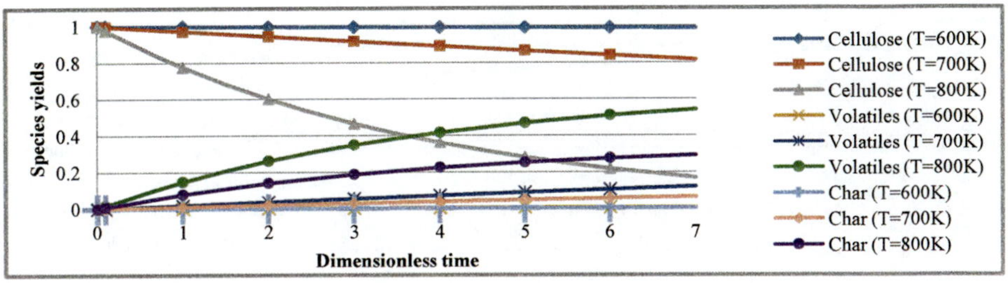

Fig. 5. Species yields profiles evolution at different temperatures for model 3

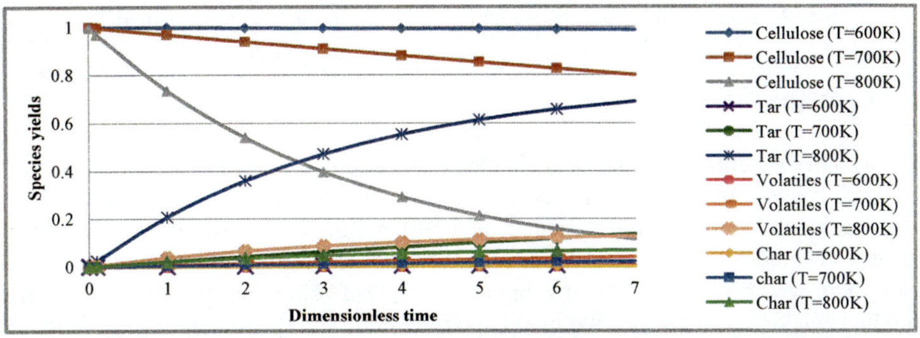

Fig. 6. Species yields profiles evolution at different temperatures for model 4

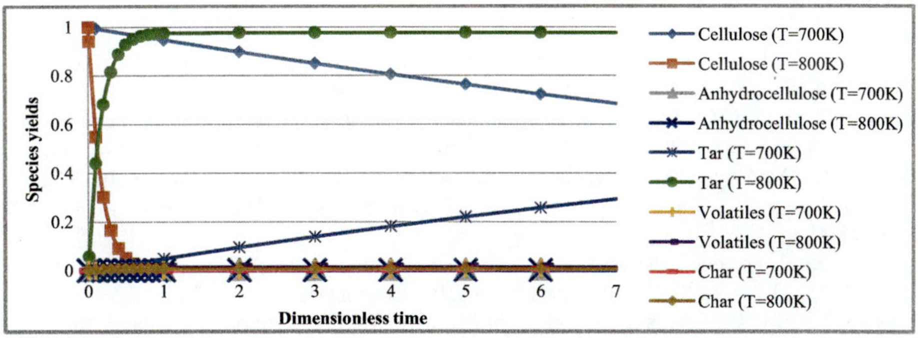

Fig. 7. Species yields profiles evolution at different temperatures for model 5

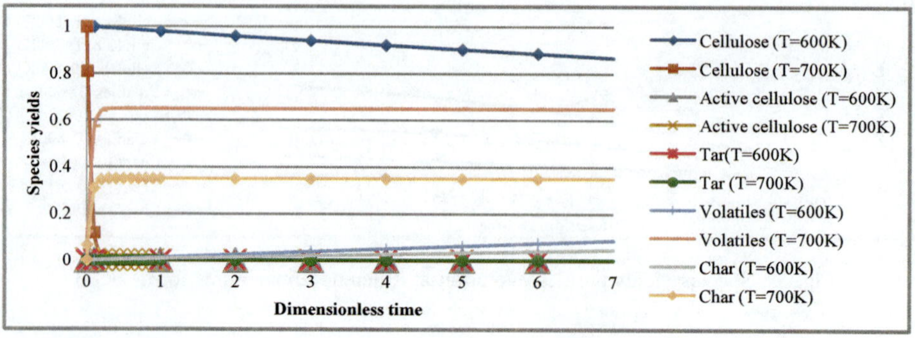

Fig. 8. Species yields profiles evolution at different temperatures for model 6

6 Conclusion

Pyrolysis represents a promising technology of cellulose wastes degradation. In this research paper, a numerical simulation of the temporal evolution of cellulose pyrolysis was studied for different kinetic models under different temperatures. High temperature leads to increase the mass Damköhler number and accordingly the cellulose conversion. The reaction model offering higher mass Damköhler number induces higher products yields. As a perspective, we propose to study the spatiotemporal evolution of cellulose pyrolysis in a tubular reactor with helical ribbon heated through the wall.

References

Antal, M.J., Várhegyi, G., Jakab, E.: Cellulose pyrolysis kinetics: revisited. Ind. Eng. Chem. Res. **37**(4), 1267–1275 (1998). https://doi.org/10.1021/ie970144v

Barooah, J., Long, V.: Rates of thermal decomposition of some carbonaceous materials in a fluidized bed. Fuel **55**(2), 116–120 (1976). https://doi.org/10.1016/0016-2361(76)90007-7

Broido, A., Weinstein, M.: Low temperature isothermal pyrolysis of cellulose. In: Thermal Analysis, pp. 285–296. Birkhäuser, Basel (1972). https://doi.org/10.1007/978-3-0348-5775-8_25

Chen, L., et al.: Products distribution and generation pathway of cellulose pyrolysis. J. Clean. Prod. **232**, 1309–1320 (2019). https://doi.org/10.1016/j.jclepro.2019.06.026

Chornet, E., Roy, C.: Compensation effect in the thermal decomposition of cellulosic materials. Thermochim. Acta **35**(3), 389–393 (1980). https://doi.org/10.1016/0040-6031(80)87140-1

Ding, Y., et al.: The effect of chemical reaction kinetic parameters on the bench-scale pyrolysis of lignocellulosic biomass. Fuel **232**, 147–153 (2018). https://doi.org/10.1016/j.fuel.2018.05.140

Gao, Z., et al.: Comparative study on the pyrolysis of cellulose and its model compounds. Fuel Proces. Technol. **193**, 131–140 (2019). https://doi.org/10.1016/j.fuproc.2019.04.038

Lewellen, P.C., Peters, W.A., Howard, J.B.: Cellulose pyrolysis kinetics and char formation mechanism. In: International Symposium on Combustion, vol. 16, no. 1, pp. 1471–1480 (1977). https://doi.org/10.1016/s0082-0784(77)80429-3

Milosavljevic, I., Suuberg, E.M.: Cellulose thermal decomposition kinetics: global mass loss kinetics. Ind. Eng. Chem. Res. **34**(4), 1081–1091 (1995). https://doi.org/10.1021/ie00043a009

Shafizadeh, F., Bradbury, A.G.W.: Thermal degradation of cellulose in air and nitrogen at low temperatures. J. Appl. Polym. Sci. **23**(5), 1431–1442 (1979). https://doi.org/10.1002/app.1979.070230513

Tabatabaie-Raissi, A., Mok, S.-L.W., Antal, M.J.: Cellulose pyrolysis kinetics in a simulated solar environment. Ind. Eng. Chem. Res. **28**, 595–608 (1989). https://pubs.acs.org/sharingguidelines

Hiby, J.W.: Mixture in Turbulent Tube Flow. Clarendon Press, Oxford

Várhegyi, G., Jakab, E., Antal, M.J.: Is the Broido-Shafizadeh model for cellulose pyrolysis true? Energy Fuels **8**(6), 1345–1352 (1994). https://doi.org/10.1021/ef00048a025

Zaror, C.A., Pyle, D.L.: Competitive reactions model for the pyrolysis of lignocellulose: a critical study. J. Anal. Appl. Pyrol. **10**(1), 1–12 (1986). https://doi.org/10.1016/0165-2370(86)85015-X

Zsakó, J.: The kinetic compensation effect. J. Therm. Anal. **9**(1), 101–108 (1976). https://doi.org/10.1007/BF01909271

Systems and Dynamics

Free Vibration Analysis of Torsional Line Shafts of Four-Stroke Engine

Abdelouahab Rezaiguia[1(✉)], Salah Guenfoud[1], and Debra F. Laefer[2]

[1] Applied Mechanics of New Materials Laboratory and Department of Mechanical Engineering, University of 8 May 1945-Guelma, Guelma, Algeria
a.rezaiguia@gmail.com,
rezaiguia.abdelouahab@univ-guelma.dz
[2] Center for Urban Science and Department of Civil Engineering, New York University, New York, USA
debra.laefer@nyu.edu

Abstract. The free vibration analysis of torsional line shafts is the preliminary step for any torsional forced vibration system and, in certain cases, can be sufficiently accurate. This work introduces a stream-lined approach for free vibration analysis of the torsional line of a four-stroke engine with four cylinders (operating range 600–5500 rpm). The equivalent torsional model, calculated by the British International Combustion Engine Research Association empirical method, has eighteen degrees of freedom. On the basis of some simplifying hypothesis, the equations of motion are obtained using Lagrange's formalism. Then the Rutishuser method is used to calculate natural frequencies and mode shapes. The main objective is to determine particularly dangerous crankshaft rotational speeds, as well as the magnitudes of the torsional vibrations of the different disks during resonance. The adaptation example of the motor shaft line shows that for the investigated configuration a speed of 710 rpm (importance 2.36 for harmonic 2), is a particularly dangerous rotational speed for the entire operating range of the motor.

Keywords: Torsional free vibrations · Vehicle engine · Critical speed · Rutishauser method

1 Introduction

The torsional vibrations of reciprocating engines result from the application of periodic combustion forces in the cylinder and inertial forces associated with rotation and articulation members such as the crankshaft, the camshaft, and the connecting link. Other excitations result from the eccentric rotation of the trunnion supports and the steering wheel. At each moment, the radial component of the force exerted on the piston is transmitted along the connecting rod and is applied to the crankpin. This force tends to pull the crankpin and induces an instantaneous load on the bearing support. On the other hand, the tangential component of the force applied to the piston causes rotation of the crankshaft. This action causes a torsion torque on the crankshaft causing torsional vibrations. According to Drezig and Holzweissig (2010), in a four-stroke engine, the fundamental frequency of the applied torque can coincide with the half of rotational speed of the crankshaft and its harmonics are at full or half orders of the same speed.

M. Kharrat et al. (Eds.): ICAMEM 2019, LNME, pp. 211–217, 2020.
https://doi.org/10.1007/978-3-030-52071-7_29

Multiple studies have been conducted to investigate torsional vibrations of various engines and tree lines (branched and unbranched). For example, Pasricha and Hashim (2006) studied the influence of cyclic variation on rotational inertia of the moving crank linkage connecting rod of a single-cylinder diesel engine on torsional vibration amplitudes of this latter. Their numerical results showed that the cyclic variation of the inertia of the motor of the reciprocating parts widens the zone of the rotational instability speeds of the motor, where large amplitudes were observed. Thus dangerous vibrations can be caused by a secondary resonance phenomenon in marine diesel engines. The regions of instability were demonstrated both analytically and experimentally. Previously, Boysal and Rahnejat (1997) analyzed torsional vibrations of a nonlinear, multi-mass, single cylinder internal combustion engine. Their model included all rigid mass inertias, support bearings, seals, couplings and connections between the various components of the motor, as well as damping. Their results provided histories of the dynamic response of all the inertial elements (e.g. piston, flywheel, connecting rod). The model also output crankshaft torsional vibrations and synchronous and asynchronous whirl motions of the sliding bearings. In the frequency domain tested, there was agreement with generally known experimental spectra, as well as being consistent with the analytical solution presented by several authors. Notably, Yuwen (2004) studied torsional vibration of multi-branch and multi-junction systems. An analytical model and method of predicting speed-related excitation frequencies of complex rotating systems was proposed. Damping and excitation forces for multi-branch torsional vibration systems were also discussed. Lubomír and Pavel (2017) investigated torsional vibration analysis of crank train with low friction losses and Mendes et al. (2008) conducted modeling and experimental validation of the crankshaft torsional vibration phenomenon in the internal combustion of a six-cylinder diesel engine. Related work by Liu et al. (2019) analyzed a single-cylinder reciprocating compressor considering the coupling effects of torsional vibration why Burdzik et al. (2016) analyzed bending-torsional vibrations for diagnostics.

2 Diagram and Equivalent Model of the Motor Line Shafts

Figure 1 shows a shaft line of a four-stroke spark ignition, four-cylinder automobile engine. The brake is of an electric Schenck type. The connection between the motor and the brake is an elastic coupling. Figure 2 shows the equivalent torsional model of the kinematic scheme, calculated by the BICERA method (Blanc 2000). Polar torsional inertia J_i (kgm^2) and torsional rigidity k_i (Nm/rad) of shafts of the equivalent model are shown in Table 1.

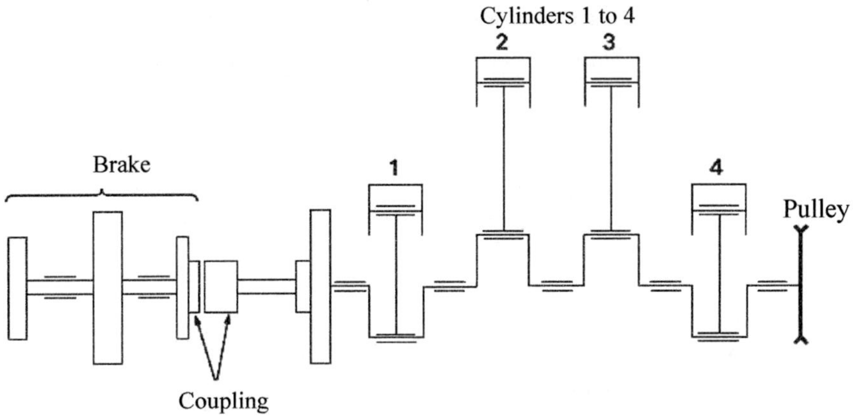

Fig. 1. Kinematic scheme of the automobile engine

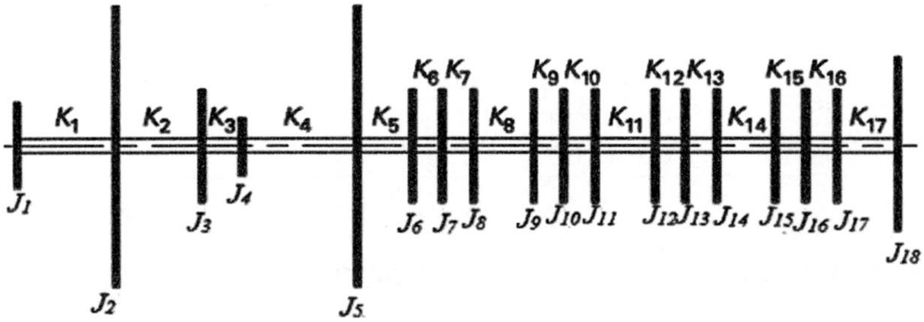

Fig. 2. Equivalent torsional model to study torsional free vibrations

3 Equations of Free Vibrations

The devised model is conservative. Specifically, the deformations are treated as proportional to the forces, thereby producing linear vibrations. As such, only torsional deformations are taken considered. The gyroscopic effect (i.e. the axis of rotation is the main inertia) is neglected, along with the flexural vibrations of shafts. Consequently, the centers of gravity of the discs are on the axis of rotation. Bearings are considered as pivotal connections, therefore rigid. Each disc J_i is associated with a variation of the harmonic angular displacement θ_i (rad). Applying Lagrange equations, the equations of motion of the model (Fig. 2) are (1):

$$[M]\left\{\ddot{\theta}\right\} + [K]\{\theta\} = 0 \tag{1}$$

with [M] the diagonal square inertia mass matrix and [K] is symmetrical square torsional stiffness matrix.

Table 1. Inertias and equivalent torsional stiffness's of the line shaft.

Disc number	Inertia (kgm^2)	Shaft number	Rigidity (Nm/rad)
1	0.98×10^{-2}	1	0.285×10^{6}
2	**0.14**	2	0.285×10^{6}
3	0.1475×10^{-1}	**3**	$\mathbf{0.21 \times 10^{4}}$
4	0.25×10^{-2}	4	0.205×10^{5}
5	**0.1476**	5	0.4302×10^{7}
6	0.343×10^{-2}	6	0.9124×10^{7}
7	0.15×10^{-2}	7	0.9124×10^{7}
8	0.145×10^{-2}	8	0.22147×10^{7}
9	0.1175×10^{-2}	9	0.9124×10^{7}
10	0.15×10^{-2}	10	0.9124×10^{7}
11	0.345×10^{-2}	11	0.22147×10^{7}
12	0.345×10^{-2}	12	0.9124×10^{7}
13	0.15×10^{-2}	13	0.9124×10^{7}
14	0.1175×10^{-2}	14	0.22147×10^{7}
15	0.145×10^{-2}	15	0.9124×10^{7}
16	0.15×10^{-2}	16	0.9124×10^{7}
17	0.343×10^{-2}	17	0.22147×10^{7}
18	0.5×10^{-2}	–	–

4 Natural Frequencies and Natural Mode Shapes

For single tree lines, with one or more branches at each tree end, the Rutishauser numerical method for quasi-triangular matrices is adopted. This method has the advantage of being precise and quickly convergent, enabling calculation of all natural frequencies of a conservative system. Also, this method only requires a small number of memory registers for its implementation and rapidly converges.

5 Free Vibrations Results

This first study is necessary for any study of forced torsional vibrations and in some cases may be sufficient. The results of the first eight natural frequencies are shown in Table 2. The rigid mode is ignored. The first mode shape of torsional vibration of the shaft line of the motor is shown in Fig. 3. Analysis of this figure shows that the vibration node is located at the elastic coupling of low rigidity (see Table 1). This is logical, because the coupling is located between two large disks inertia J_2 (brake) and J_5 (flywheel). However, in general, elastic couplings are mounted near disks of large inertia, and the vibrations are almost always located in the couplings.

Table 2. Natural frequencies of the tree line of Fig. 2.

Mode number	1	2	3	4	5	6	7	8
Frequency [Hz]	23.65	480.90	730.83	862.09	893.68	2382.26	4113.04	5048.73

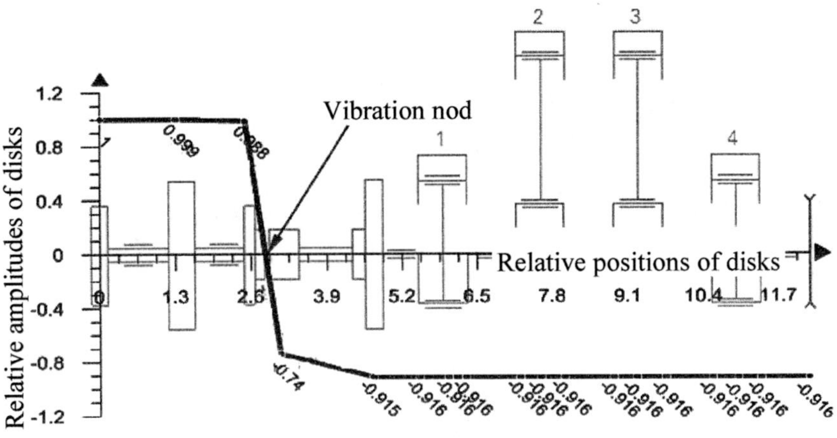

Fig. 3. First vibration mode shape.

Figure 4 shows the modal deformation of the second vibration mode shape. The analysis of this vibration mode shows two nodes: one near the coupling and one near the flywheel. A large amplitude value corresponding to disc J_4 is observed, which corresponds to the half-coupling. So significant elastic energy is stored in the coupling. Table 3 ranks the dangerous harmonics, critical speeds, and amplitudes of the corresponding harmonic couples resulting from the Fourier series de-composition of the couple and the importance of the harmonics for the first two modes. The results demonstrate particularly dangerous rotation speeds for the engine are equal to:

- for the first mode 710 rpm (importance 2.36, harmonic 2);
- for the second mode 3608 rpm (importance 0.0417, harmonic 8), as well as 4811 rpm (importance 0.097, harmonic 6).

Use of importance allows selection of truly dangerous harmonics and limits the calculations of forced vibrations to only dangerous harmonics (Table 3 and Fig. 5).

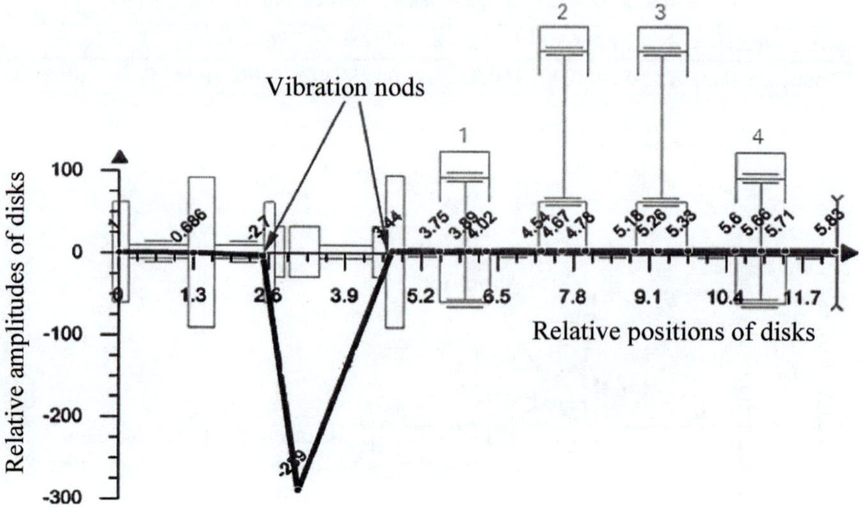

Fig. 4. Second vibration mode shape.

Table 3. Dangerous harmonics, critical speeds, amplitudes of harmonic couples and the importance of harmonics for the first two modes

	Mode 1				Mode 2							
	f = 23.65 Hz				f = 480.90 Hz							
Dangerous harmonic ranks	0.5	1	1.5	***2***	5.5	***6***	6.5	7	7.5	***8***	8.5	9
Critical speeds (rpm)	2840	1420	946	***710***	5248	***4811***	4441	4124	3849	***3608***	3396	3207
Amplitudes of the harmonics couples (Nm)	59.22	63.96	47.91	***30.57***	5.64	***4.78***	3.77	3.12	2.55	***2.06***	1.67	1.57
Harmonic's importance	0.001	0.0002	0.0007	***2.36***	0.011	***0.097***	0.0077	0.0012	0.0049	***0.0417***	0.0032	0.006

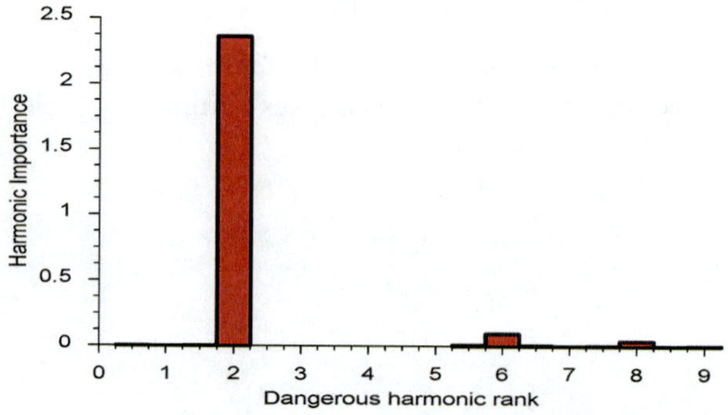

Fig. 5. Importance of dangerous harmonics

6 Conclusions

The analysis of free vibration of torsional line shafts of a four-stroke engine generated the following conclusions:

- for first mode shapes, the vibration nodes are almost always located in the couplings, because these are generally mounted near the disks of large inertia;
- knowledge of the natural frequencies and possible excitation sources defines dangerous rotating speeds within the engine's normal operating range;
- calculation of harmonic torque magnitudes enables defining and ranking t potentially dangerous rotating speeds [e.g. 710 rpm (importance 2.36 and harmonic 2 for first mode for the engine studied herein)];

Identification of dangerous operating speeds enables the undertaking of targeted modifications to the shaft line, redefining the safe range of operating range, or reducing the amplitudes of the torsional vibrations. Future work may examine forced vibration by explicitly considering absolute damping in the equations of motion of the model.

References

Drezig, H., Holzweissig, F.: Dynamics of Machinery, Theory and Applications. Springer, Heidelberg (2010)

Pasricha, M.S., Hashim, F.M.: Effect of the reciprocating mass of slider-crank mechanism on torsional vibrations of diesel engine systems. AJSTD **23**(1), 71–81 (2006)

Boysal, A., Rahnejat, H.: Torsional vibration analysis of a multi-body single cylinder internal combustion engine model. Department of Mechanical and Manufacturing Engineering, University of Bradford, West Yorkshire, UK (1997)

Yuwen, Y.: Analysis of multi-branch torsional vibration for design optimization. Graduate theses, West Virginia University, USA (2004)

Lubomír, D., Pavel, N.: Torsional vibration analysis of crank train with low friction losses. J. VibroEng. **19**(8), 5691–5701 (2017)

Mendes, A.S., Meirelles, P.S., Zampieri, D.E.: Analysis of torsional vibration in internal combustion engines: modelling and experimental validation. Proc. Inst. Mech. Eng. J. Multi-body Dyn. **222**(2), 155–178 (2008)

Liu, S., Li, W., Shuai, Z., Chen, M.: Vibration analysis of a single cylinder reciprocating compressor considering the coupling effects of torsional vibration. J. Shock Vibr. **2019**, 1–9 (2019)

Burdzik, B., Pankiewicz, J., Wądołowski, M.: Analysis of bending-torsional vibrations for diagnostics. Diagnostyka **17**(4), 79–84 (2016)

Blanc, H.: Dynamique des rotors en torsion: répartition de l'inertie et de la raideur (Dynamics of torsional rotors: distribution of inertia and stiffness), Technical Engineer, BM 5 122 (2000)

Stability Analysis of One Degree of Freedom System Equipped with Friction Vibration Absorber

Aymen Nasr(✉), Charfeddine Mrad, and Rachid Nasri

Laboratory of Applied Mechanics and Engineering (LMAI), National School of Engineers of Tunis (ENIT), University of Tunis El Manar (UTM), BP 37, Le Belvedere, 1002 Tunis, Tunisia
aymen.nasr@outlook.com, {charfeddine.mrad, Rachid.Nasri}@enit.rnu.tn

Abstract. In this paper, the free vibration response of a linear one degree of freedom system equipped with a friction vibration absorber is analyzed. The friction between the two systems is modeled using Coulomb friction law. The effect of the friction parameters on the stability of the linear one degree of freedom system is studied. First, around the equilibrium point, the local stability is analyzed using the Routh-Hurwitz criterion: the discontinuous non-smooth description of the friction force is approximated by a tangent hyperbolic continuous smooth function. Second, the global stability is analyzed using the Lyapunov direct method. The effect of negative viscous damping and the limit cycle behavior are considered

Keywords: Vibration absorber · Friction contact · Stability analysis

1 Introduction

Friction vibration absorber is widely used as a passive device to control vibration of mechanical systems (Nasr et al. 2018). Friction force, defined as the tangential force between two bodies in contact represents the dissipative mechanism of energy. The Coulomb friction law is the first model used to describe the friction force. This approach is widely used to represent a rigid friction contact.

Mechanical systems with friction contact are considered as discontinuous systems of Filippov-type (Leine and Nijmeijer 2004). This discontinuity can be approximated by the tangent hyperbolic smooth function and many stability analysis techniques can be used. The stability of system around the equilibrium point can be studied by the Routh-Hurwitz method. This method is frequently used for nonlinear systems (Berger and Sinou 2011) and (Yu and Yuan 2001). It can be concluded on the stability of systems without calculating the eigenvalues of the linearized Jacobin matrix. In addition, the direct method of Lyapunov allows concluding on the global stability of mechanical systems (Rouche et al. 1977). Aleksandrov et al. (2013) used this method to study the stability of a system with two degrees of freedom with a switched force. Leine and Van de Wouw (2008) studied the stability of the equilibrium points of

M. Kharrat et al. (Eds.): ICAMEM 2019, LNME, pp. 218–224, 2020.
https://doi.org/10.1007/978-3-030-52071-7_30

mechanical systems in the presence of friction contact; mechanical energy is used as a function of Lyapunov to study stability.

In this paper, a one degree of freedom system equipped with a friction vibration absorber is considered. A general formulation is elaborated to study stability of the vibratory system. The stability of system in presence of negative damping is also addressed.

2 Vibratory Model and Motion Equations

The primary system, illustrated in Fig. 1, is modeled by the mass m_1, the linear stiffness k_1 and the linear viscous damping c_1. The vibration absorber is of mass m_2 and of stiffness k_2. The Coulomb law represents friction force as a function of friction coefficient μ, normal applied load N and sign of relative velocity between the two systems.

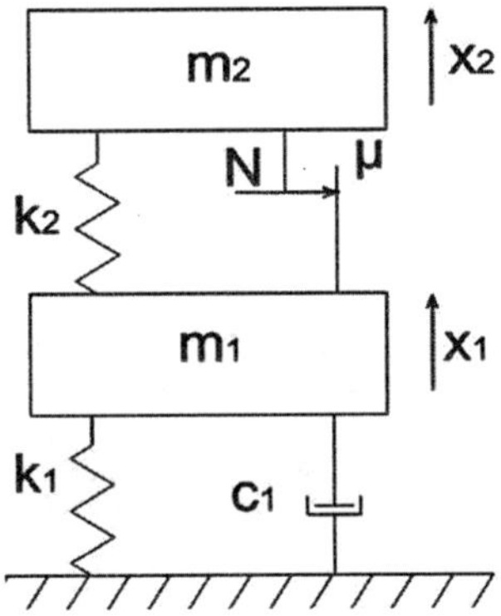

Fig. 1. Principal system with friction vibration absorber

The equations of motion can be written as

$$\begin{array}{c} m_1\ddot{x}_1 + k_1x_1 + c_1\dot{x}_1 + k_2(x_1 - x_2) + \mu N sign(\dot{x}_1 - \dot{x}_2) = 0 \\ m_2\ddot{x}_2 - k_2(x_1 - x_2) - \mu N sign(\dot{x}_1 - \dot{x}_2) = 0 \end{array} \quad (1)$$

The following quantities are used to transform equations of motion to the non-dimensional form; $y_1 = \frac{x_1}{x_s}$, $y_2 = \frac{x_2}{x_s}$, $x_s = \frac{N}{k_1}$, $\omega_1 = \sqrt{\frac{k_1}{m_1}}$, $\omega_2 = \sqrt{\frac{k_2}{m_2}}$, $\omega_a = \frac{\omega_2}{\omega_1}$, $\xi_1 = \frac{c_1}{2\omega_1 m_1}$ $r = \frac{m_2}{m_1}$ and $\tau = \omega_1 t$.

The non-dimensional equations of motion are written as

$$\begin{array}{c} \ddot{y}_1 + y_1 + 2\xi_1\dot{y}_1 + r\omega_a^2(y_1 - y_2) + \mu sign(\dot{y}_1 - \dot{y}_2) = 0 \\ \ddot{y}_2 - \omega_a^2(y_1 - y_2) - \frac{\mu}{r} sign(\dot{y}_1 - \dot{y}_2) = 0 \end{array} \quad (2)$$

3 Stability Analysis

A nonlinear system is stable, if under an initial perturbation, the trajectories of motion converge to an equilibrium point. In addition, the stability can be examined using the Lyapunov direct method; if the derivative of Lyapunov function is negative, the system is stable. A function V is considered as Lyapunov function if $V(y_e) = 0$ and $V(y) > 0$. y_e is the equilibrium point of vibratory system. The equilibrium point is represented by a vector of static displacement because velocity and acceleration are equal to zero. It is written as

$$y_e = \alpha(0, 0, 1, 0) \text{ with } |\alpha| \leq -\mu \tag{3}$$

As a function of the kinetic and potential non-dimensional energies, Lyapunov function is given by

$$V = y_1^2 + r\omega_a^2(y_1 - y_2)^2 + \dot{y}_1^2 + r\dot{y}_2^2 \tag{4}$$

And

$$\dot{V} = -4\xi_1\dot{y}_1^2 - 2\mu|\dot{y}_1 - \dot{y}_2| \tag{5}$$

Regarding Eq. (4), it is clearly seen that the proposed function verifies the condition of Lyapunov direct method, mentioned in Eq. (3). As shows Fig. 2, if viscous damping ratio ξ_1 is positive, $\dot{V}$ is strictly negative and the system is globally stable at the equilibrium point. But, for a negative damping ratio, the stability behavior of the system cannot be interpreted; $\dot{V}$ has negative and positive signs depending on the system parameters.

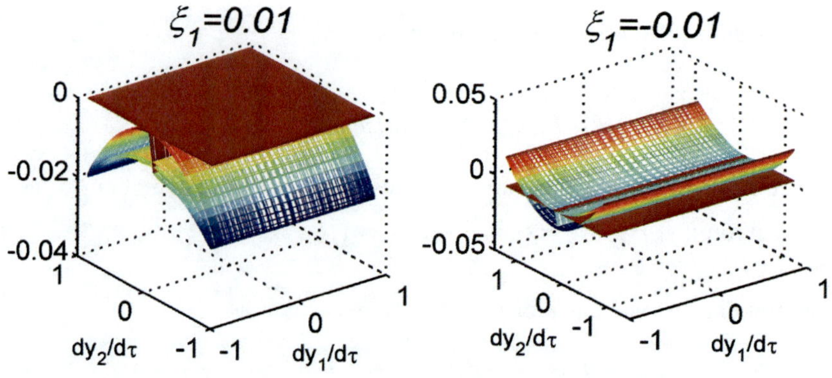

Fig. 2. $\dot{V}$ for positive and negative system viscous damping, $\mu = 0.2$

For negative viscous damping, the stability of equilibrium point can be examined using Routh-Hurwitz criteria. The system is stable if the criteria of Routh-Hurwitz are positive. Stability is investigated by the following approach;

- Calculating the Jacobian matrix,
- Determining the characteristic polynomial
- Calculating the Routh-Hurwitz criteria.

The sign function is a discontinuous function at zero velocity and it is not differentiable. This discontinuity makes the calculation of Jacobian matrix impossible. So, $sign(\dot{y}_1 - \dot{y}_2)$ can be approximated by $\tanh(\beta\dot{y}_1 - \beta\dot{y}_2)$ where β is equals to 10^6. Jacobian matrix of the system is written as follows

$$J_R = \begin{bmatrix} 0 & 1 & 0 & 0 \\ -1 - r\omega_a^2 & -2\xi_1 - \mu\beta & r\omega_a^2 & \mu\beta \\ 0 & 0 & 0 & 1 \\ \omega_a^2 & \frac{\mu\beta}{r} & -\omega_a^2 & \frac{\mu\beta}{r} \end{bmatrix} \tag{6}$$

Figure 3 shows the variation of stability as function of friction and principal mass viscous damping. The system is stable at the equilibrium point if viscous damping is positive. If viscous damping is negative, an unstable zone appears, and the stability is dependent on the system parameter.

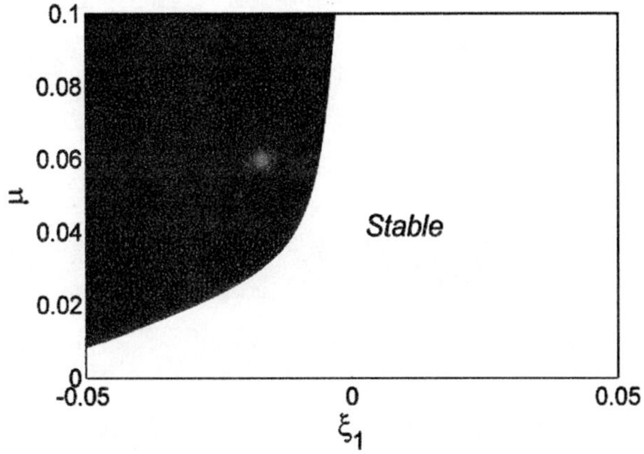

Fig. 3. Stability region function of friction coefficient and system viscous damping

The resolution of differential equations shows that there is a limit cycle where both masses vibrate in a harmonic way. This cycle is a stable limit cycle; whatever the initial condition applied to the main mass, the trajectories of the motion converge at this cycle (Figs. 4 and 5).

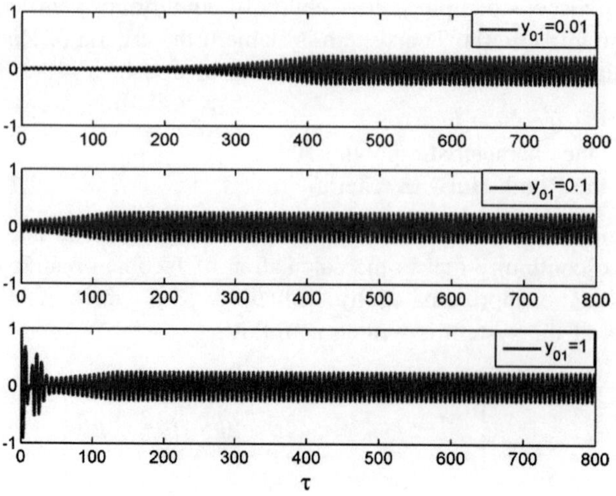

Fig. 4. Responses of principal system for $\mu = 0.02$

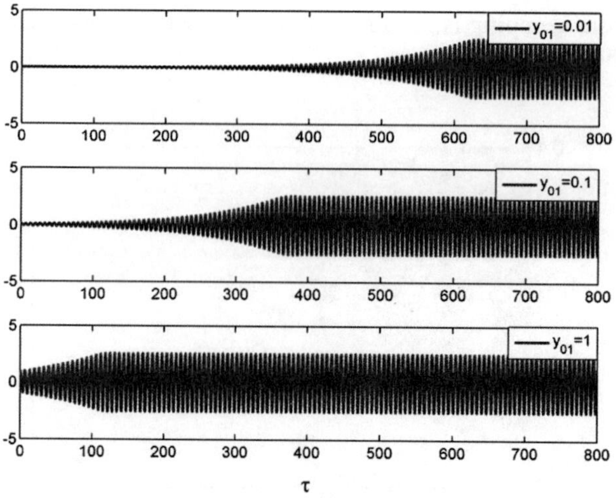

Fig. 5. Responses of principal system for $\mu = 0.2$

The two examples in Figs. 4 and 5 show the effect of the initial conditions on the response of the main mass with two levels of the friction coefficient. First, the increase in the imposed initial displacement accelerates the convergence of the vibrations to a stable limit cycle. Second, the maximal amplitude of vibration has monotonous behavior with respect to friction coefficient; it increases with the increase of the coefficient of the friction.

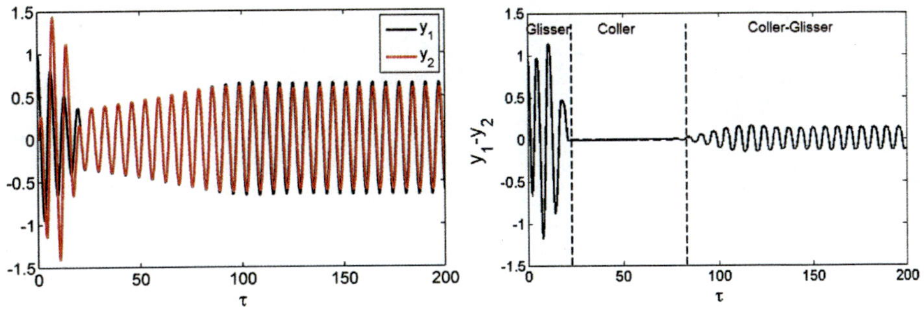

Fig. 6. Reponses of system and absorber/stick-slip behavior as a function of y_1-y_2

This behavior is due to the presence of the dissipative force of the friction. From the results shown in Fig. 6, both masses are in pure slip and vibration amplitude decreases. Then the two masses stick together and the relative displacement y_1-y_2 becomes zero, in this case the vibratory system is a system with a one degree of freedom with negative viscous damping, it is an unstable system and the amplitude of the vibrations amplifies. Finally, the increase reaches a steady state where the two masses vibrate harmonically; the energy provided in the vibratory system by viscous damping is dissipated by friction damping.

4 Conclusion

In this works, the stability of one degree of freedom system equipped with friction vibration absorber is examined. The friction contact between the two masses is modeled by Coulomb friction force. The stability analysis is established using Routh-Hurwitz criterion and the direct Lyapunov function. The non-smooth sign function was approximated by the smooth continuous tangent hyperbolic function.

The results show that the stability depends on the behavior of the system in the sticking state. The system is globally stable, so locally, if the viscous damping dissipates the energy. In addition, if the viscous damping pumps the energy to the mechanical system, the stability depends on the contact parameters.

References

Aleksandrov, A.Y., Lakrisenko, P.A., Platonov, A.V.: Stability analysis of nonlinear mechanical systems with switched force fields. In: 21st Mediterranean Conference on Control and Automation (MED) Control & Automation (MED), pp. 628–633 (2013)

Berger, S., Sinou, J.J.: Stability analysis for a self-exciting mechanism with friction using interval computations. Int. J. Veh. Struct. Syst. **3**(1), 4–20 (2011)

Leine, R., Nijmeijer, H.: Dynamics and Bifurcations of Non-Smooth Mechanical Systems. Springer, Heidelberg (2004)

Leine, R.I., van de Wouw, N.: Stability properties of equilibrium sets of nonlinear mechanical systems with dry friction and impact. Nonlinear Dyn. **51**(4), 551–583 (2008)

Nasr, A., Mrad, C., Nasri, R.: Friction tuned mass damper optimization for structure under harmonic force excitation. Struct. Eng. Mech. **65**(6), 761–769 (2018)

Rouche, N., Habets, P., Laloy, M.: Stability Theory by Liapunov's Direct Method. Springer, New York (1977)

Yu, P., Yuan, Y.: The simplest normal forms associated with a triple zero eigenvalue of indices one and two. Nonlinear Anal.: Theory Methods Appl. **47**(2), 1105–1116 (2001)

Materials

Study of Fe-W-P Coating Performance Under Scratch Test: Modeling and Experiments

F. Zouch(✉), A. Bahri, and K. Elleuch

Laboratory of Materials Engineering and Environment (LGME), National Engineering School of Sfax, University of Sfax, B.P.W. 1173, 3038 Sfax, Tunisia
fatma.zouch@enis.tn, amirl9t@yahoo.fr, Khaled.Elleuch@enis.rnu.tn

Abstract. Fe-W-P coating was deposited on stainless steel 316 L using electrodeposition process. An experimental and numerical evaluation of the adhesion strength between this coating and the stainless steel was studied with a scratch test using Rockwell indenter and progressive load from 0 to 165 N. Scanning electron microscopy and 3D profilometry were used to determine the morphology of the scratch track and the indenter penetration, respectively. The main experimental result shows the propagation of cracks inside and outside of the scratch track which increase with the increasing of the normal applied load. Moreover, a pile-up phenomenon was detected in the border. The obtained numerical results are in good agreement with the experimental one. In fact, a stress concentration was observed in the center and the border of the track due to the pile-up phenomenon which is the origin of the creation of the crack network. Besides, the cross sectional numerical profile in the maximum penetration area show clearly a pile-up phenomenon. A good agreement was found when comparing the numerical and the experiemental results prouve the robustness of the model.

Keywords: Fe-W-P · Stainless steel 316 L · Scratch test · Numerical evaluation

1 Introduction

Barite (Barium sulphate, $BaSO_4$) is the most abundant and economically important barium mineral produced in the worldwide (Pieper et al. 2013; Focus on Pigments 2010; Kallel et al. 2017). In particular, barite powder which presents an important product used in petroleum industry (Kallel et al. 2017). In fact, in order to obtain the barite powder, the obtained barite rocks extracted from the mines follow an important crushing procedure. Meanwhile, during crushing step, the crusher undergoes a catastrophic wear mechanism due to the severity of the work conditions such as high impact velocity, particles shape as mentioned in previous study (Kallel et al. 2017). Many authors have proposed several coatings in order to resolve the wear damage (Bahri et al. 2016; Sriraman et al. 2006). Rare those who have take into account their adhesion as a principal key of their mechanical performances. Based on some research studies

M. Kharrat et al. (Eds.): ICAMEM 2019, LNME, pp. 227–232, 2020.
https://doi.org/10.1007/978-3-030-52071-7_31

(Zouch et al. 2018; Vernickaite et al. 2015), Fe-W-P are known by their important mechanical and tribological performance, namely, the hardness, Young's modulus and the wear rate. However, to the author knowledge, no research found about the coating-substrate adhesion. In the present study, Fe-W-P layer has been deposited on 316 L stainless steel using electrodeposition procedure. Both numerical and experimental scratch tests have been conducted to evaluate the adhesion strength. The obtained results are compared and discussed.

2 Materials and Methods

Fe-W-P coatings were deposited by electrodeposition process. Samples of stainless steel 316 L with dimensions of $40 \times 40 \times 1$ mm^3 were used. Specimens were mechanically polished, degreased and activated in 20% sulfuric acid and washed with distilled water. After the preparation of the substrate, a prior layer of nickel was deposited in order to enhance the adhesion of the coating. Then, Fe-W-P alloy were electrodeposited with pH = 8 and J = 30 mA/cm^2. In order to investigate the adhesion between Fe-W-P and stainless steel 316 L, a scratch test has been performed using Rockwell type C indenter. The load has been applied progressively from 0 N to 165 N.

Scanning electron microscope (SEM, Joel JSM 5410) and 3D (WYKO NT1100) profilometry were used to determine the morphology and the depth of the scratch track, respectively.

A finite element model has been elaborated using a commercial nonlinear FE code ABAQUS. The indenter has been modeled as a rigid body. The substrate has been described with an elasto-plastic behavior. Moreover, the coating has been described with purely elastic behavior. The used parameters are illustrated in the Table 1.

Table 1. Coating and substrate mechanical properties (Zhou et al. 2004; Bradai 2006).

Stainless steel 316 L								
E (GPa)	**υ**	**Yield strain**	0	0.01	0.02	0.03	0.04	0.05
190	0.29	**Yield stress (MPa)**	300	484	553	592	615	621
Ni								
E (GPa)	**υ**							
190	0.3							
Fe-W-P								
E (GPa)	**υ**							
184.53	0.3							

3D continuum elements C3D8R have been used to model the stainless steel as substrate, Ni layer and Fe-W-P. There are 105984 nodes and 88200 finite elements for the substrate and for the Ni and Fe-W-P there are (47104 nodes and 33075 elements) for each layer, as shown in Fig. 1. The model has been constrained on the bottom surface of the stainless steel substrate. The load has been applied as a displacement in both X and Y directions.

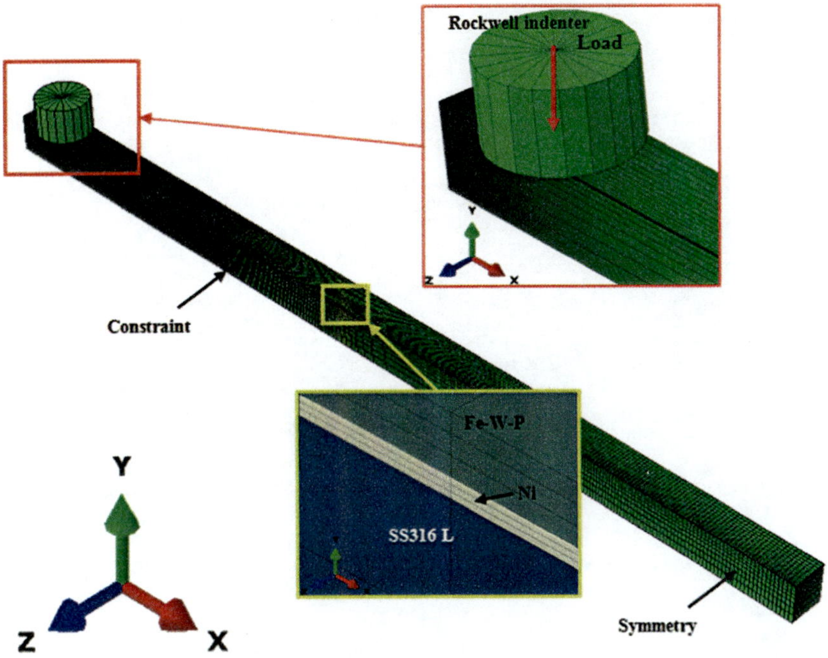

Fig. 1. Mesh configuration and boundary conditions.

3 Result and Discussion

During scratch test, the hard Rockwell indenter will sink into the surface at the first time after applying the normal load. Besides, the indenter moves from one side to another with the increase of load. As a consequence, a scratch path was formed on the surface of the specimen. As shown in the Fig. 2, the displaced material which could be accumulated at the border of the scratch track may create a pile-up phenomenon. Moreover, both the penetration and the width of the track increase with the increase of the normal load. Several cracks inside and outside of the track were detected, it is important to note that the crack initiation start from 12 N of the normal applied load.

Figure 3a shows the evolution of the friction coefficient during scratch test under progressive load from 0 N to 165 N. It is observed that the friction coefficient reaches its maximum at the beginning at the first contact indenter-coating. During the displacement of the indenter, the friction coefficient increases from 0.2 to approximately 0.42. Moreover, the analysis of the curve reveals that the friction coefficient increases sharply with the increase of the normal load. It can be seen some fluctuations in the curve which become more important when the friction coefficient increases indicating creation of several cracks in the track.

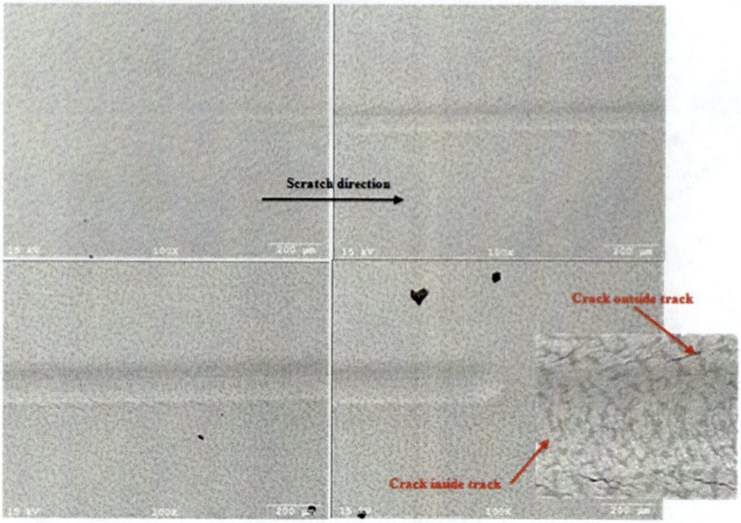

Fig. 2. SEM photo of scratch track of the Fe-W-P coating.

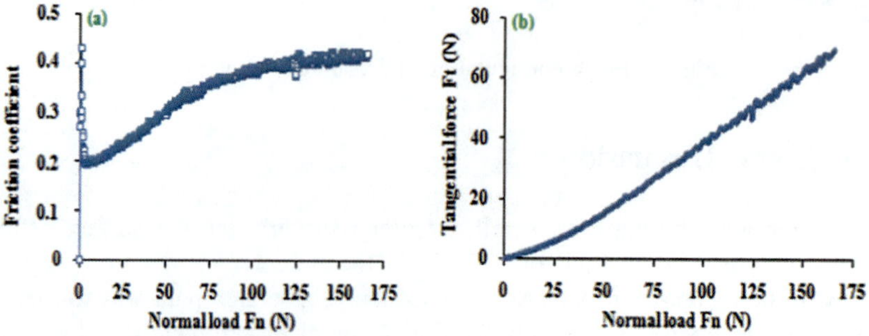

Fig. 3. Comparison of the evolution in the a) Friction coefficient and b) tangential force of the scratch track during progressive load from 0 N to 165 N under scratch test of Fe-W-P coating deposited on 316 L.

The evolution of the tangential load Ft of the Fe-W-P as a function of the normal load was shown in Fig. 3b. It was remarked that Ft increases with the increase of the applied load. Some fluctuations were detected in particular with the highest load from approximately 130 N to 165 N which confirm the increase of the crack propagation in the scratch track.

Figure 4 illustrates a comparaison between the experimental and the numerical obtained results during scratch test. The penetration-displacement curves (Fig. 4(a)) show a good accordance between experimental and numerical profile. The maximum penetrations reach approximately 8 μm under a maximum load of F = 165 N. It can be seen some clear fluctuations in the experimental curve which indicate the creation of a crack network already mentioned in the previous section. A cross sectional view of the

Von Mises stress distribution throughout the layers and the substrate (Fig. 4(b)) reveals a stress concentration in the center and the border of the track for both Ni and Fe-W-P. The numerical results consolidate the experimental one.

The cross sectional profiles have been extracted in the maximum penetration area (Fig. 4(c)), the comparaison between those curves shows a good agreement. The main results show a pile-up phenomenon in both experimental and numerical curves which prouve the accuracy of the numerical model.

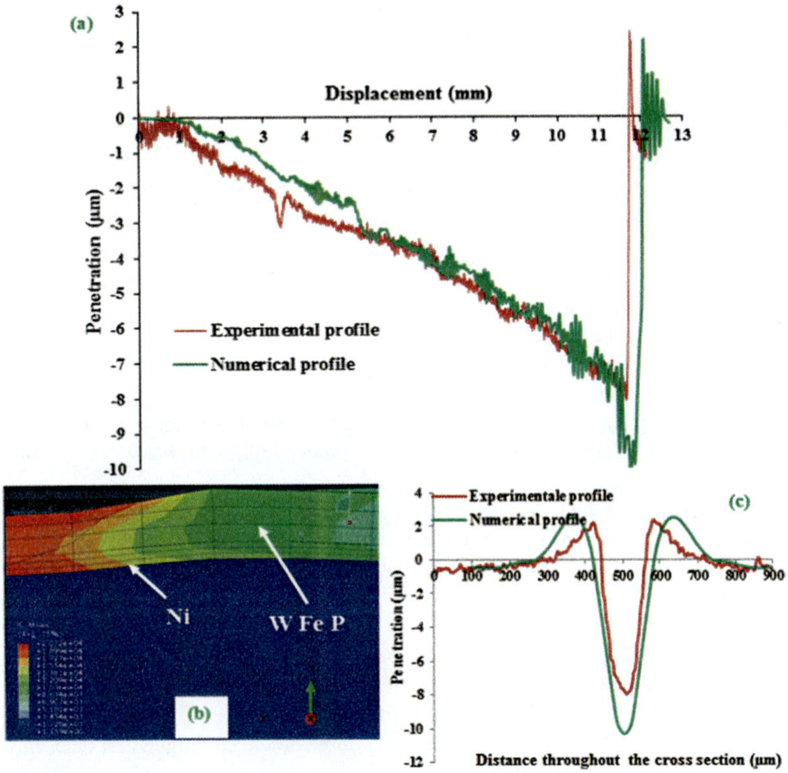

Fig. 4. Experimental and numerical comparison: a) Indenter penetration-displacement, b) Cross section throughout coating and substrate and c) penetration profile throughout cross section track.

4 Conclusion

In this paper, FE model has been performed to analyze the adhesion strength between Fe-W-P and the stainless steel 316 L using scratch test. The stainless steel 316 L was described with elasto-plastic material properties and both Ni and Fe-W-P were considered with elastic behavior. A good accordance was found between experimental and numerical results. The experimental results reveal several cracks propagation inside and outside the scratch track starting from F = 12 N. The intersection of the different cracks creates a crack network. The numerical model results permit localize the stress concentration area and to understand an advantage the damage phenomenon.

In a future work, the numerical model will be developed in order to take into account the damage criteria of the Ni and the Fe-W-P layers. Moreover, a numerical study of the thickness and the friction coefficient effect during scratch process on the layer damage will be conducted.

Acknowledgments. This work was supported by the ministry of higher education and scientific research Tunisia. The authors would like to thank Professor Mustafa Ürgen, from Istanbul Technical University (Turkey), for his help in carrying out SEM observation and scratch test.

References

Pieper, M., Kutelova, Z., Aman, S., Tomas, J.: Modeling of baryte batch grinding in a vibratory disc mill. Adv. Powder Technol. **24**, 229–234 (2013)

World demand for filler-grade barytes approaches 1 M tonnes/y, Focus on Pigments 4 (2010)

Kallel, M., Zouch, F., Antar, Z., Bahri, A., Elleuch, K.: Hammer premature wear in mineral crushing process. Tribol. Int. **115**, 493–505 (2017)

Bahri, A., Kaçar, E., Akkaya, S.S., Elleuch, K., Ürgen, M.: Wear protection potential of TiN coatings for 304 stainless steels used in rotating parts during olive oil extraction. Surf. Coat. Technol. **304**, 560–566 (2016)

Sriraman, K.R., Ganesh Sundara Raman, S., Seshadri, S.K.: Synthesis and evaluation of hardness and sliding wear resistance of electrodeposited nanocrystalline Ni–Fe–W alloys. Mater. Sci. Technol. **22**, 14–20 (2006)

Zouch, F., Antar, Z., Bahri, A., Elleuch, K., Ürgen, M.: Tribological study of Fe-W-P electrodeposited coating on 316 L stainless steel. J. Tribol. **140**, 1–9 (2018)

Vernickaite, E., Antar, Z., Nicolenco, A., Kreivaitis, R., Tsyntsaru, N., Cesiulis, H.: Tribological and corrosion properties of iron-based alloys. In: Proceedings of the International Conference BALTTRIB (2015)

Zhou, Z.M., Zhou, Y., Yang, C.S., Chen, J.A., Ding, G.F., Ding, W., Wang, M.J., Zhang, Y.M.: The evaluation of Young's modulus and residual stress of nickel films by microbridge testings. Meas. Sci. Technol. **15**, 2389–2394 (2004)

Bradai, S.: Etude de l'effet de l'équibiaxialité sur la durée de vie en fatigue de l'acier inoxydable austénitique, Doctoral thesis (2006)

Corrosion Resistance Enhancement of AISI 304 Stainless Steel by Deep Rolling Treatment

Khouloud Gharbi(✉), Naoufel Ben Moussa, and Nabil Ben Fredj

Laboratoire de Mécanique, Matériaux et Procédés (LR99ES05), ENSIT, Université de Tunis, 5 Avenue Taha Hussein Montfleury, 1008 Tunis, Tunisia
gharbi_khouloud90@hotmail.fr, emailnaoufel@gmail.com, benfredjnabil@gmail.com

Abstract. As one of severe plastic deformation (SPD) techniques, the deep rolling treatment was carried out after machining to enhance the surface integrity (microstructure, grain size, residual stress, etc.) of many materials. Recrystallized surface layers containing submicron-scaled grains could be produced during deep rolling process due to severe plastic deformation. These ultra-fine grained materials showed improved mechanical properties (fatigue, wear resistance, corrosion resistance, etc.). In this work, deep rolling process was performed to enhance the corrosion behavior of Stainless Steel type AISI 304 in severe environments like seawater. Findings of this study depicts that deep rolling generates an ultra-fine grain structure, multi-directional mechanical twins, strain-induced martensite in the surface layers. In addition, the application of deep rolling produced high and deep compressive residual stress distributions. As a matter of fact, the corrosion rate was reduced by 53% for deep rolled sample comparatively to machined one. Thus, the enhancement of AISI 304 corrosion behavior results from microstructural changes, deformation-induced martensite and compressive residual stresses.

Keywords: AISI 304 · Deep rolling · Ultra-fine grains · Martensite · Corrosion

1 Introduction

Among the most extensively used grade in manufacturing of chemical components as well as food and beverage equipments is AISI type 304 austenitic Stainless Steel (SS). However, this grade is mainly susceptible to localized corrosion and Stress Corrosion Cracking (SCC) in spite of its high resistance to corrosion as reported in the work of Park et al. (1974). In fact, it is known that localized corrosion resistance is particularly influenced by surface conditions of materials as shown in the work of Rustandi et al. (2017). Thus, mechanical treatments are increasingly used to enhance the surface integrity (microstructure, hardness, residual stress, etc.) of many materials, which efficiently enhance their corrosion behavior in severe environments. The introduction of surface compressive stresses by peening was shown to sufficiently improve the AISI 304L corrosion resistance as investigated in the work of Badran et al. (2008). Pyre et al. (2000) concluded that an appreciable enhancement in surface properties of laser peened AISI 316L specimens led to better corrosion behavior as compared to that generated by

M. Kharrat et al. (Eds.): ICAMEM 2019, LNME, pp. 233–239, 2020.
https://doi.org/10.1007/978-3-030-52071-7_32

shot peening. Many investigations were interested on the effect of strain-induced martensite on the austenitic stainless steels' mechanical characteristics and its further impact on their corrosive behavior. Lee et al. (2009) showed that ultrasonically peened AISI 304 workpieces, having better quality-finished surface and more strain-induced martensite, exhibited a high localized and general corrosion resistance behavior compared to the as-received specimens. Toppo et al. (2012) denoted the significance of grain refinement in the passive film stability and highlighted the fact that thermo-mechanical surface treatment of AISI 304, including sand blasting with a subsequent laser surface heating, resulted in better uniform and pitting corrosion resistance. Thus, the formation of refined grains, strain-induced martensite and compressive residual stresses in the near-surface regions are considered as the key factors for enhancing the corrosive behavior of many materials. Ben Moussa et al. (2018) reported that the control of deep rolling parameters enhances the surface quality, the work hardening and the compressive residual stresses, leading to an improvement of the initial and the residual fatigue life of AISI 304 SS components working under low strain cyclic loading. The purpose of this work is studying the impact of deep rolling treatment on AISI 304 mechanical properties and its further influence on the corrosion resistance.

2 Experimental Work

In this work, all experimental tests were performed on AISI type 304 SS. Tables 1 and 2 report the chemical composition as well as the mechanical properties of this grade. The as-received material was subjected to stress-relief annealing in order to remove the effects resulting from previous manufacturing processes. As performed in the work of Ben Moussa et al. (2018), the workpieces were firstly heated to 1050 °C for 60 min, and then air-cooled. The annealed structure is shown in Fig. 1 in which grain size was about 50 ± 10 μm.

Table 1. Chemical composition (weight %) of AISI 304 SS

C	Si	Mn	Cr	Ni	Mo	Cu	N	F
0.05	0.41	1.14	18	9	0.193	18.04	0.004	Balance

The workpieces were turned on a CNC lathe using a cutting insert VBMT 16 04 04-MF 2015 at a machining feed of 0.125 mm/rev, a cutting speed of 150 m/min and a cutting depth of 0.4 mm as recommended by SANDVIK.

Deep rolling (DR) experiments were conducted after turning without removing the workpiece from the lathe chuck to avoid the roundness error. A specially designed and manufactured DR tool (Fig. 2) was mounted on the lathe's tool post. A spherical roller in AISI D2 steel, with a curvature radius of 8 mm, freely rotates along with the workpiece during DR. The tool was pressed into the workpiece at a depth of penetration of 0.1 mm. During the process, the feed rate was 0.06 mm/rev, the speed was fixed to 100 rev/min and the operation was done at 4 passes.

Table 2. Mechanical properties of AISI 304 SS at 20 °C

E (GPa)	Rp0.2 (MPa)	Rm (MPa)	Toughness (J/cm2)	A (%)
193	320	615	270	58

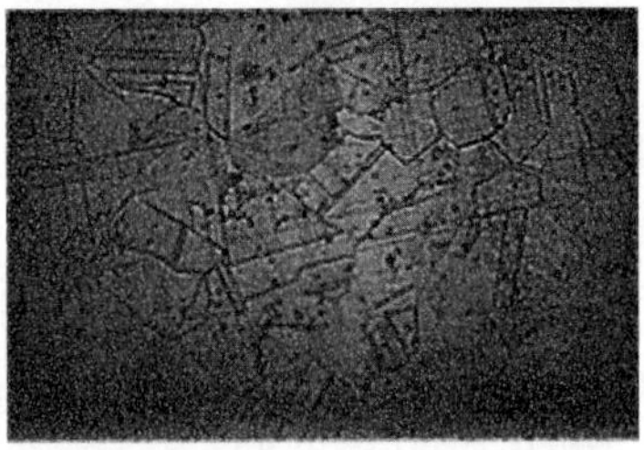

Fig. 1. Austenitic structure of the annealed AISI 304 SS

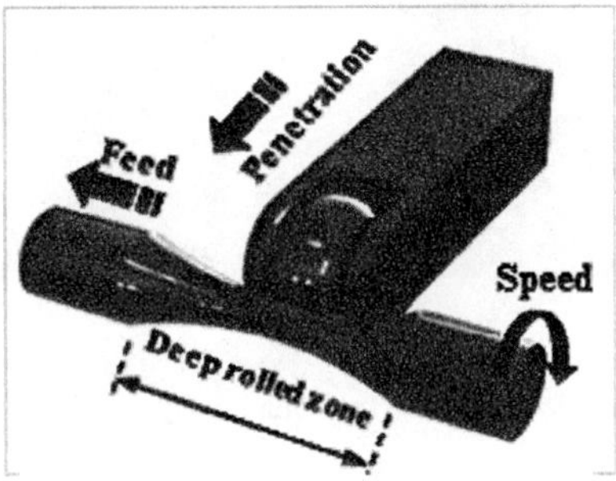

Fig. 2. Deep rolling set-up

In this work, 4 test samples were cut from machined workpieces as well as deep rolled ones. One AISI 304 sample was used for each experimental test. For metallurgical characterization, specimens were first polished using silicon carbide emery papers up to 2400 grit finish. After that, oxalic acid solution was applied for electrolytic etching in order to reveal the surface grain structure.

Optical microscopy were performed to examine the microstructure of the turned and deep rolled samples. Microstructural changes were assessed by X-ray diffraction using a Philips PANalytical goniometer with Co Kα radiation (λ = 1.79 Å). A set-X diffractometer was used to measure the residual stress distributions by means of ψ tilt X-ray method. Corrosion behavior was studied by means of potentiodynamic anodic polarization test in synthetic seawater using a PGZ301 Voltalab 40 potentiostat at a temperature of 25 °C. In this work, artificial seawater was prepared in the laboratory according to the standard ASTM D 1141-98 and its pH was calibrated to 8.30 using 0.1 mol.L^{-1} of NaOH solution.

3 Results and Discussions

3.1 Microstructure

The cross section micrograph reveals a deformed structure resulting from machining characterized by slip lines localized in the stretched grains. The bulk region was free of plastic deformation and the grain boundaries were clearly noticeable with an average size of 50 ± 10 μm (Fig. 3a). Figure 3b shows the microstructure near the surface of deep rolled sample. After DR, the grain size was reduced to submicron scale near the surface top layer. A refined layer in which grain boundaries were no longer observable and an intermediate layer with the pattern of sweeping grains with an average size of about 17 ± 5 μm were observed in deep rolled surface. High density of intersecting twins was noticed above the interface (between the top surface and the transition layer) due to plastic deformation, whereas no twinning was seen in the bulk material.

(a) (b)

Fig. 3. Cross section micrograph of (a) machined and (b) deep rolled AISI 304 (magnification ×1000)

3.2 Phase Changes

Figure 4 shows the surface XRD patterns of turned and deep rolled specimens in which only γ-austenite peaks appeared for machined sample. However, the XRD spectra of deep rolled sample depicts both γ-austenite and α'-martensite peaks. These results agree with the density of multi-directional mechanical twins in the plastic deformed area generated after deep rolling (Fig. 3b). In this case, the deformation localized in the slip bands results in plastic-induced martensite formation, which appears as thin laths in the austenitic grains.

3.3 Residual Stress

The near-surface residual stress levels were measured in longitudinal (σ_{11}) and transversal (σ_{22}) directions. Tensile residual stresses in both directions were obtained for AISI 304 machined surface. The measured levels were σ_{11} = 339 ± 20 MPa and σ_{22} = 476 ± 20 MPa, respectively. The biaxial residual stress profiles produced by DR process are illustrated in Fig. 5. The surface residual stress in the longitudinal direction (σ_{11} = −383 MPa) was much higher than that in the transversal direction

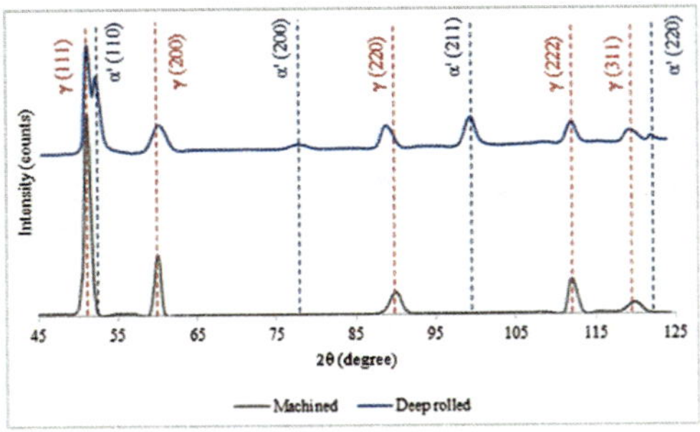

Fig. 4. Phase analysis of AISI 304 processed surfaces

corresponding to the roller loading direction ($\sigma_{22} = -149$ MPa). After DR application, both high and deep peaks of compression were noticed. These results point out the fact that this mechanical treatment induces high plastic deformation in near-surface layers.

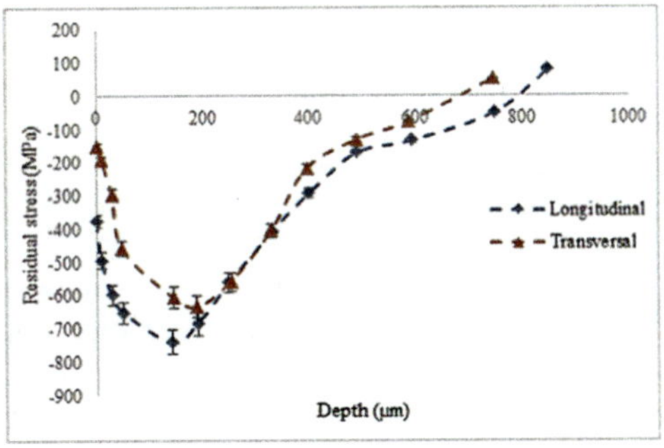

Fig. 5. Residual stress profiles generated after the application of deep rolling

3.4 Corrosion Behavior

Figure 6 depicts the potentiodynamic polarization curves used to determine corrosion potential (Ecorr), corrosion current density (Icorr) and corrosion rate values of each processed workpiece. The results were calculated by means of Tafel extrapolation method (Table 3). After DR, the corrosion current was decreased and the corrosion potential increased. Lee et al. (2009) showed that lower values of Icorr and likewise, ennobled values of Ecorr indicate higher resistance to localized and general corrosion. Thus, the deep rolled specimen showed a better corrosion behavior with regard to the machined sample and the corrosion rate was considerably reduced from 1.822×10^{-3} to 0.855×10^{-3} mm/year.

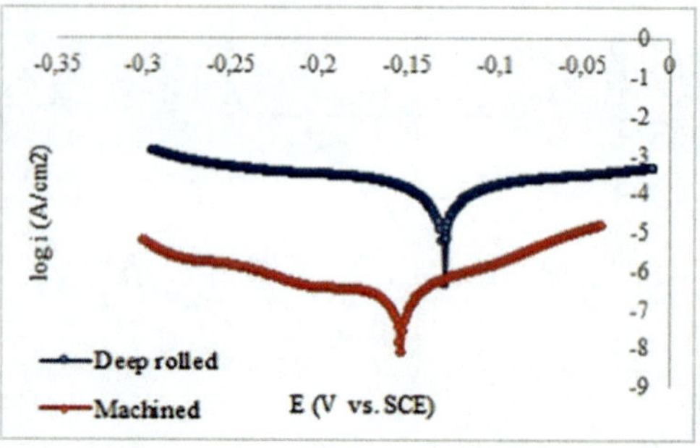

Fig. 6. Potentiodynamic polarization responses in synthetic seawater

Table 3. Potentiodynamic polarization test results of the machined and deep rolled AISI 304

Specimen	Icorr (mA/cm^2)	Ecorr (mV/SCE)	Corrosion rate (mm/year)
Machined	140	−153.305	1.822×10^{-3}
Deep rolled	68	−128.050	0.855×10^{-3}

4 Conclusion

In this investigation, the application of DR produced an ultra-fine grained structure, multi-directional intersecting twins as well as deformation-induced martensite in the near-surface layers due to high plastic deformation rate. Besides, the deep rolled surface was characterized by both high and deep compressive residual stress fields. The measured corrosion rate for machined sample was decreased by 53% after the application of DR. Consequently, it can be stated that the surface integrity modifications (microstructure, grain size and residual stress) generated after DR led to the improvement of AISI 304 corrosive behavior in seawater.

References

Park, P.S., Yoon, S.R.: A study on the stress corrosion cracking of AISI 304 stainless steel. J. Corros. Sci. Soc. Korea **3**(2), 77–88 (1974)

Rustandi, A., et al.: The use of cyclic polarization method for corrosion resistance evaluation of austenitic stainless steel 304L and 316L in aqueous sodium chloride solution. Int. J. Mech. Eng. Robot. Res. **6**, 512–518 (2017)

Badran, O., Kloub, N., Musa, A.-T.: The effect of shot peening and polishing on the pitting corrosion resistance of stainless steel. Am. J. Appl. Sci. **5**, 1397–1402 (2008)

Peyre, P., et al.: Surface modifications induced in 316L steel by laser peening and shot-peening. Influence on pitting corrosion resistance. Mater. Sci. Eng., A **280**(2), 294–302 (2000)

Lee, H.-S., et al.: Influence of peening on the corrosion properties of AISI 304 stainless steel. Corros. Sci. **51**(12), 2826–2830 (2009)

Toppo, A., et al.: Enhancement of corrosion resistance of type 304 stainless steel through a novel thermo-mechanical surface treatment. J. Mater. Eng. Perform. **22**, 632–639 (2012)

Ben Moussa, N., Gharbi, K., Chaieb, I., Ben Fredj, N.: Improvement of AISI 304 austenitic stainless steel low-cycle fatigue life by initial and intermittent deep rolling. Int. J. Adv. Manuf. Technol. **101**, 435 (2018)

Check for updates

Numerical Simulation of Reciprocating Sliding Test: Effects of Surface Topography on the Wear Behavior

Fatma Elwasli[1(✉)], Slah Mzali[1], Farhat Zemzemi[2], and Salah Mezlini[1]

[1] LGM, Ecole Nationale d'Ingénieurs de Monastir, Université de Monastir, Avenue Ibn Aljazzar, 5019 Monastir, Tunisia
elwaslifatma@gmail.com, salah.mezlini@gmail.com, mzalislah@hotmail.com

[2] LMS, Ecole Nationale d'Ingénieurs de Sousse, BP 264 Sousse Erriadh, 4023 Sousse, Tunisia
fzemzemi@gmail.com

Abstract. In recent years, researchers have shown an increased interest in the investigation of the wear track evolution.

The present work proposes a Finite Element Approach (FEA) to evaluate the impact of the surface properties and the wear resistance of a 5083 aluminum alloy. Reciprocated sliding test was carried out numerically, using Abaqus/Explicit software. Based on python scripts, it was possible to consider the tribological behavior of a real surface profile. Tests were performed with a rigid ball (R = 5 mm) under a normal load of 1 N at a maximum number of sliding cycles of 10 cycles.

During the numerical simulation, the wear track evolution was inspected for different surface topographies. A comparison of the findings with those of experimental studies confirms the validity of the numerical results. This numerical investigation has revealed the significant impact of the surface topography and the number of sliding cycles (*N*) on the wear track evolution.

Keywords: Reciprocating sliding test · Surface topography · Wear behavior · Abaqus/Explicit

1 Introduction

There is a growing body of literature that recognizes the importance of the initial surface roughness in tribological behavior. Researchers have stepped up their efforts to understand the influence of the surface properties on the contact-pair behavior. Based on experimental works, they have shown that the surface topography plays a significant role in controlling the tribological behavior (Sedlaček et al. 2009, 2012; Elwasli et al. 2018a, b).

However, far too little attention has been given to the development of a numerical model simulating the contact behavior of rough surfaces (Ripoll et al. 2011; Lestari et al. 2017). Ripoll et al. (2011) have developed a numerical model simulating a

M. Kharrat et al. (Eds.): ICAMEM 2019, LNME, pp. 240–245, 2020.
https://doi.org/10.1007/978-3-030-52071-7_33

reciprocating sliding test. They have highlighted the significant role of the textured surface on the tangential force.

The main purpose of the current research was to examine the dependence of the wear track evolution on the surface topography. A 3D numerical simulation of a reciprocating sliding test was implemented using Abaqus/Explicit software. The evolution of the wear track of *AA5083* was examined after 10 cycles of reciprocating sliding against a rigid ball.

2 FE Modeling

A simulation of a reciprocating sliding test was carried out on the ABAQUS/Explicit software. The numerical model consists of a rough parallelepiped aluminum sample of $2 \times 2 \times 1\ mm^3$ in dimension and a rigid ball with a radius $R = 5$ mm (Fig. 1). The simulation starts with the application of a normal load $F_n = 1$ N on the rigid ball. Then, the ball moves horizontally with a reciprocating sliding distance $d = 1$ mm. During the test, full constraints were applied for the displacement degrees of freedom at the bottom face of the sample. The maximum number of sliding cycles was fixed at 10.

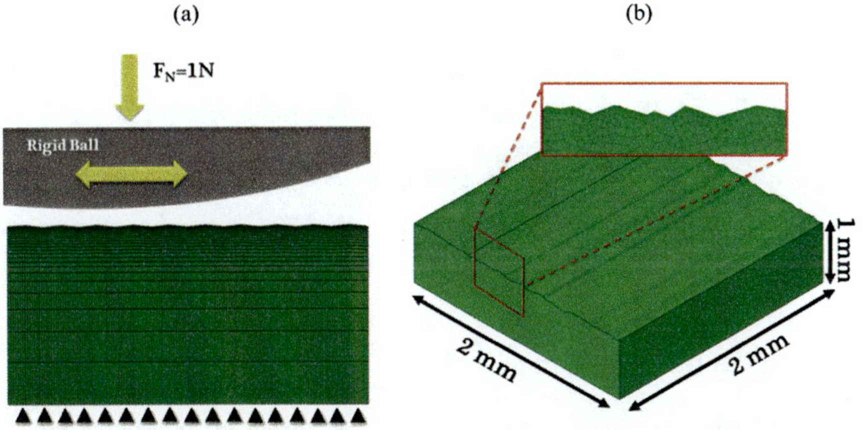

Fig. 1. (a) Geometry, load and boundary conditions, (b) Dimensions and mesh structure.

In an attempt to accurately reproduce the surface irregularities, the aluminum sample was partitioned into three different densities mesh. The mesh was highly refined close to the surface roughness and along the sliding path with 1 μm element size. Upon distancing from the surface, mesh decreases gradually. The geometry was meshed with *164290* brick element, with reduced integration *C3D8R* (Fig. 1-b) elements. This mesh was fixed after a numerical convergence analysis.

To examine the correlation between the wear behavior and the surface properties, three profiles were tested. Based on python script, experimental profiles were investigated. The average roughness R_a, and the mean absolute profile slope Δ_a of each surface were displayed in Table 1.

Table 1. Surface roughness values.

	G_{50}	G_{80}	M
R_a (μm)	4.98	3.31	0
Δ_a (°)	3.9	12.9	0

The isotropic and nonlinear kinematic hardening model was adopted for describing the material behavior of *AA5083* specimen during cyclic sliding. The Johnson–Cook damage criterion was proposed. The detailed constitutive parameters were reported in Table 2.

Table 2. Constitutive model parameters for *AA5083* (Elwasli et al. 2018a).

Mechanical properties					
E [GPa]	ν	ρ [Kg/m^3]			
70	0,3	2700			
Cyclic behavior law coefficient					
K [MPa]	Q [MPa]	B	c [MPa]	γ	
100	103	6.44	30 000	350	
Johnson–Cook damage parameters					
d_1	d_2	d_3	d_4	d_5	U_f^{pl}
0,0261	0,263	−0,349	0,147	16,8	1 10^{-6}

3 Results and Discussion

Figure 2 provides the line profiles of wear tracks obtained by experimental and numerical methods for the surface G_{80}. It can be noticed that the experimental wear track presents a fluctuated aspect compared to the predicted wear track.

The difference between numerical and experimental wear track was thought to be related to the wear debris entrapped at the bottom of the wear track. The discrepancy was estimated at about *8%*. Thus, the results exhibit good concordance.

Figure 3 shows line profiles of the wear track for the surfaces G_{50}, G_{80} and *M*. It can be noticed that the rise in *N* causes the increase of the wear track area. Moreover, the deformation rate decreases as a function of *N*. The first sliding cycle has the highest deformation, followed by only a slight deformation in the next cycles. In fact, with the rise in *N*, the worn zone widens and the plastic deformation becomes slower.

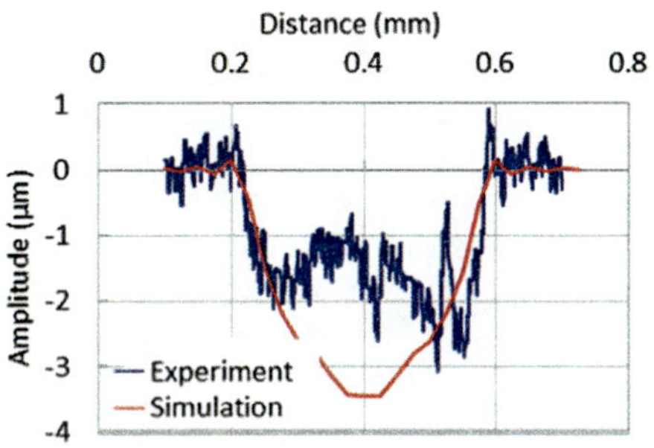

Fig. 2. Numerical and experimental line profiles of wear tracks after 10 sliding for G_{80}.

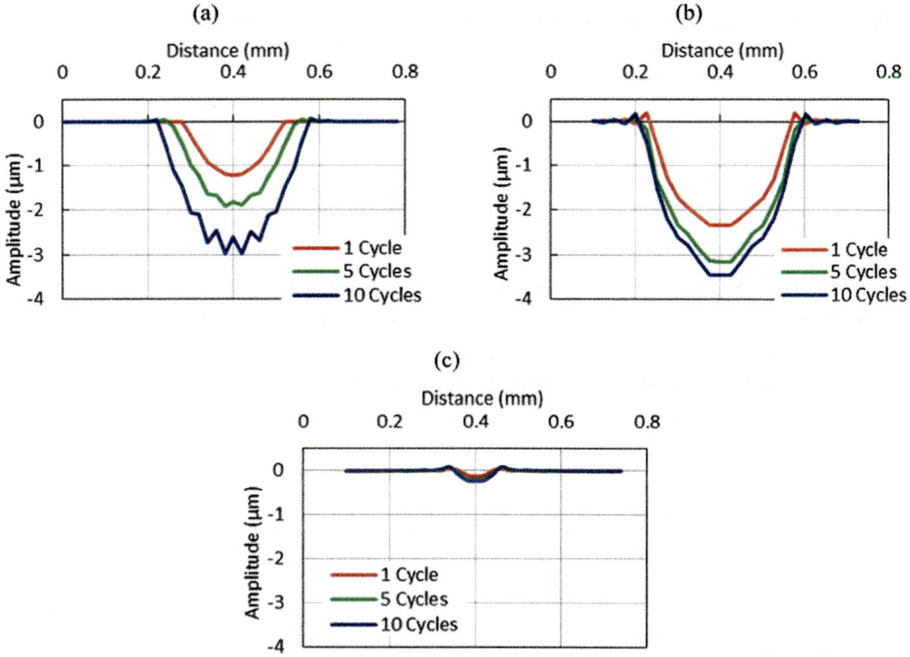

Fig. 3. Line profiles of wear tracks of different aluminum surfaces. (a) G_{50}, (b) G_{80} and (c) M.

The initial surface topography was an important determinant of the wear behavior (Fig. 4). A small wear track was observed when sliding over the smooth surface *M*. Indeed, large real contact area was generated which results in the low contact pressure and thus low plastic deformation. However, only a small contact area was obtained when sliding over rough surfaces, provoking high contact pressure. As a result, larger wear tracks were detected for surfaces G_{50} and G_{80}.

Moreover, the surface G_{80} generates the widest wear track. This result can be attributed to the high value of Δ_a. In fact, high shear stress was generated as a consequence of the

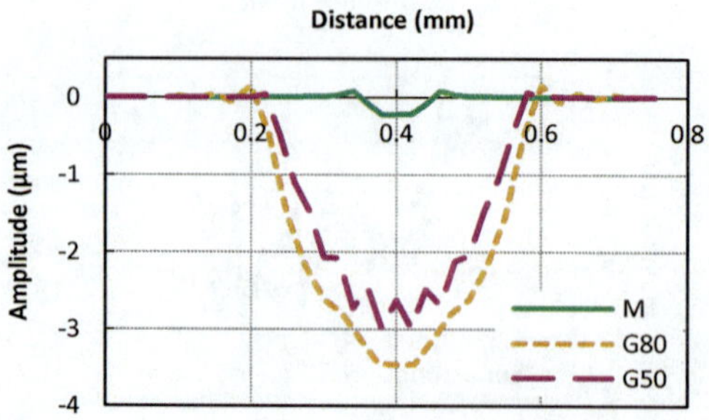

Fig. 4. Line profiles of wear tracks after 10 sliding cycles.

sliding on a surface characterized by a high Δa (Elwasli et al. 2018b). Therefore, failure occurs leading to a larger wear tracks area.

4 Conclusion

The aim of this paper was to examine the significant effect of the initial surface topography on the wear track evolution. The main findings were as follows:

(1) Numerical simulation of a reciprocating sliding test was performed under Abaqus/Explicit software. The proposed models exhibit excellent ability to investigate the wear behavior of rough and smooth surfaces.
(2) During the simulation, the wear track evolution was inspected for different surface topographies. A comparison of the findings with those of experimental studies confirms the validity of the numerical results.
(3) The predicted result has revealed that the wear behavior exhibits a significant sensitivity to R_a and Δ_a.

The increase of R_a and Δ_a provokes the increase of the wear track.

References

Elwasli, F., et al.: Correlation between roughness anisotropy and tribological behavior of AA5083: Experimental and numerical analysis. Proc. Inst. Mech. Eng. Part L J. Mater. Des. Appl. **233**(3), 372–382 (2018a)

Elwasli, F., et al.: Effects of initial surface topography and contact regimes on tribological behavior of AISI-52100/AA5083 materials' pair when reciprocating sliding. Int. J. Mech. Sci. **137**, 271–283 (2018b)

Lestari, W.D., Jamari, J., Bayuseno, A.P.: The Von Mises Stress distribution on the surface of UHMWPE with texture-shaped variation in the presence of normal load and dry sliding contact. In: 7th International Conference on Mechanical and Manufacturing Engineering. American Institute of Physics, pp. 1–6 (2017)

Ripoll, M.R., Podgornik, B., Vižintin, J.: Finite element analysis of textured surfaces under reciprocating sliding. Wear **271**(5–6), 952–959 (2011)

Sedlaček, M., Podgornik, B., Vižintin, J.: Correlation between standard roughness parameters skewness and kurtosis and tribological behaviour of contact surfaces. Tribol. Int. **48**, 102–112 (2012)

Sedlaček, M., Podgornik, B., Vižintin, J.: Influence of surface preparation on roughness parameters, friction and wear. Wear **266**(3–4), 482–487 (2009)

Ni-Ti Superelastic Wire Coupled with Conventional Brackets During Bending Tests: Cross-section Effect Comparison

Aroua Fathallah[1(✉)], Tarek Hassine[1], and Fehmi Gamaoun[1,2]

[1] Laboratory of Mechanics of Sousse, National Engineering School of Sousse, University of Sousse, Sousse, Tunisia
aroua.fathallah.07@gmail.com, tarekhassine007@gmail.com, Fehmi.Gamaoun@insat.rnu.tn
[2] College of Engineering, King Khalid University, Abha, Saudi Arabia

Abstract. Ni-Ti shape memory alloy has become a famous alloy in the dental field. This alloy is used to design orthodontic wire suitable for the alignment stage. The Ni-Ti superelasticity has permitted to the clinicians to ensure an optimal orthodontic displacement during long period. The choice of round or rectangular cross section wire depended on the force delivered by the Ni-Ti wire, since this archwire exhibits a different mechanical behavior compared to an ordinary alloy. Thus, the aim of this work is to examine the effect of varying the cross-sectional dimension of superelastic Ni-Ti wire inserted in three brackets during bending tests which reproduce the most accurately the mechanical response of the wire as it is inserted in the oral cavity. The experimental findings showed that, when studying the cross-section size effect, the engagement loading, and the orthodontic force are dominated by the martensitic transformation process rather than the moment of inertia wherein the pseudo-elastic behavior is revealed. Moreover, the activation and deactivation loads have been affected by the brackets ligated system when comparing the experimental results to the literature ones. This was due to the friction resistance induced by the wire and the braces coupled with the ligatures.

Keywords: Ni-Ti · Superelastic wire · Bending · Cross-section · Brackets effect

1 Introduction

Ni-Ti shape memory alloys are classified as smart materials thanks to their excellent mechanical properties. Besides their ability to recover their initial shape, these alloys are superelastic and they have good corrosion resistance (Mohd et al. 2014). These features were exploited in many fields such as aeronautic and biomedical fields. Because of their ability to apply light and constant load, the Ni-Ti alloys were used as orthodontic archwire during the earlier stage of the orthodontic treatment (Thompson 2000; Zuhanek 2011). This superelastic property is related to the occurrence of the martensitic thermomechanical transformation. These wires are inserted in the mouth with the intention of aligning and leveling the irregular teeth. Otherwise, the Ni-Ti wire

M. Kharrat et al. (Eds.): ICAMEM 2019, LNME, pp. 246–251, 2020.
https://doi.org/10.1007/978-3-030-52071-7_34

are generally bent during this stage. Therefore, many experimental researches have been performed to examine the mechanical behavior of the Ni-Ti alloys during bending tests. The aim of those works was to determine the load generated from the Ni-Ti wire in different situations (Aghili et al. 2015; Gatto et al. 2013; Mallory et al. 2004; Tonner and Waters 1994). Garrec and Jordan (2004) have compared load deflection curves of different square cross-section Ni-Ti wire size, while Fujio et al. (1986) have studied the response of several round section Ni-Ti wire during 3 points bending tests. They have found that increasing the cross-section dimensions in both shapes, permitted to generate higher horizontal load during the activation and the deactivation. These results were obtained through 3 point classic bending test. Whereas, Nucera et al. (2014) have demonstrated that brackets and even their type affect the level of force. Lombardo et al. (2012) were interested in the lower plateau average load of traditional and heat activated round Ni-Ti wire with different cross section dimensions inserted in self-ligating brackets.

The purpose of this work is to investigate the mechanical response of several cross-sectional dimension Ni-Ti superelastic wire subjected to bending test using braces coupled with elastomeric modulus and analyze their load deflection curve to determine the mechanical properties of such alloy when it is engaged in the oral cavity. The result of this study will consist an experimental base to future works.

2 Experimental Procedure

Superelastic Ni-Ti orthodontic archwires with different cross-section dimensions were used. 0.40 and 0.45 mm round section wires and 0.45 * 0.64 mm^2 rectangular section wires were selected for this work. They were cut into straight specimens of 25 mm length and inserted in three 0.55 mm slot conventional brackets attached with elastomeric ligatures. These braces were glued on an acrylic device fabricated to reproduce a vertical displacement of mal-aligned tooth as illustrated in Fig. 1(a). The inter-bracket distance was equal to 7.5 mm.

All the bending tests were realized at room temperature using the flexure device fixed on a uniaxial tensile machine as it is displayed in Fig. 1(b). These bending tests simulate a cycle of loading until 4 mm and then unloading until returning to the initial position. This cycle was carried out at a velocity of 1 mm/min during the charging and the discharging. For each test, new elastomeric modules were used. In order to validate the results, each test was repeated at least three times with new specimen.

3 Results and Discussion

The experimental results are depicted in Fig. 2. This figure represents the load deflection average curves of the three different cross-section wires bent until 4 mm. The three curves demonstrate the same superelastic mechanical behavior during the whole cycle. It was verified that the load deflection curves exhibited a similar behavior compared to the stress strain curves extracted from the tensile tests (Zhang et al. 2010). During the charging, the load rises linearly followed by a horizontal plateau. This step

(a) (b)

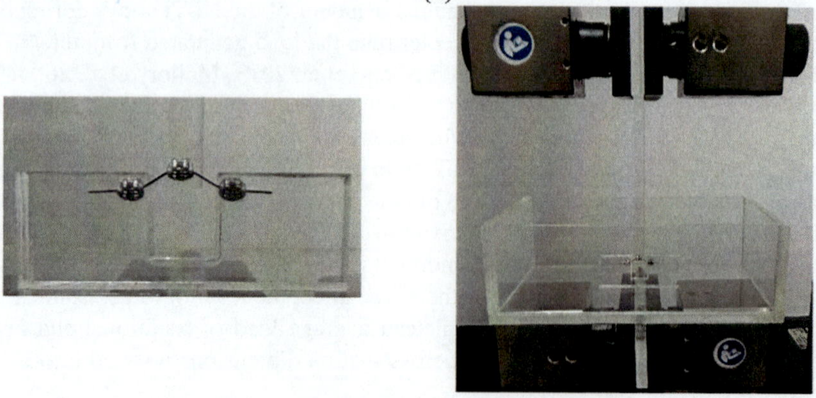

Fig. 1. (a) bending device, (b) flexure device fixed on the tensile machine during bending test

simulates the engagement of the wire in brackets glued on the teeth during the orthodontic treatment. After that, during the discharge, where the wire applies force in the oral cavity, a new lower plateau is displayed. Then, a linear curve is exhibited until the load annulment. In fact, at the room temperature, the Ni-Ti wire is governed by the austenite phase. During the loading, this phase is initially deformed according to the Hooke's law until the beginning of the martensite bands nucleation wherein the non-constant Young modulus provokes the decrease of the bending rigidity. When unloading the wire, the backward transformation from the martensite to the austenite is activated by presenting an inferior force plateau compared with the upper plateau. Then, at a critical force value, the wire recuperates its initial elastic austenite phase.

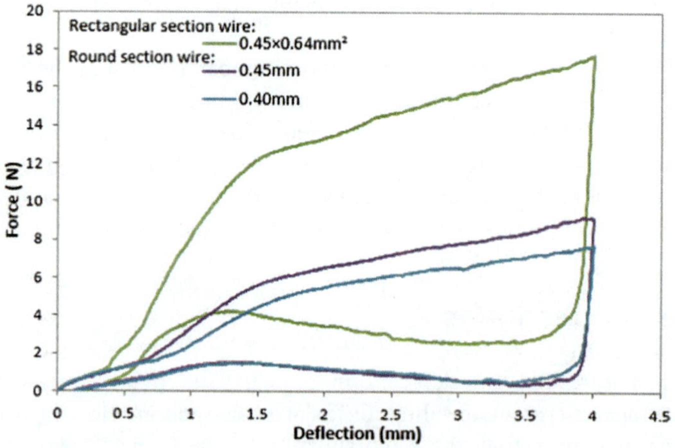

Fig. 2. Load Deflection curves of the superelastic 0.4 and 0.45 round Ni-Ti wire and $0.45 \times 0.64\ mm^2$ rectangular Ni-Ti wire bent until 4 mm

When comparing the different wire cross-section, it was observed that the engagement load increased when rising the area section size and that the Ni-Ti rectangular wire delivered the highest force during the loading as it is illustrated in Fig. 3. During the elastic deformation of the austenite, the difference of the flexure modulus could be due to area moment of inertia which depend on the cross-section size and form. This reason cannot be still available to justify the difference between the plateau forces during the martensitic transformation. Indeed, according to Garrec et al. (2005), this augmentation of load level when varying the wire cross-section is caused by the upsurge of the volume fraction of the transformed martensite. When this latter augment as well as the frictional energy, the stress needed to transform the austenite rises. During the unloading, which is governed by the stored elastic energy, the difference of the load during the transformation process is less noticed between the rectangular and the round wire. In addition, the load deflection curves of the 0.40 and 0.45 mm round wire are almost superimposed despite the difference of the moment of inertia. These results prove that, the higher the cross-section wire is, the greatest is the area of mechanical hysteresis. This latter is evaluated as the difference between the upper and lower plateau loads. This phenomenon was explained by the fact that the more important the volume fraction of martensitic would be, the less the unloading force would be when compared to the loading plateau (Garrec et al. 2005).

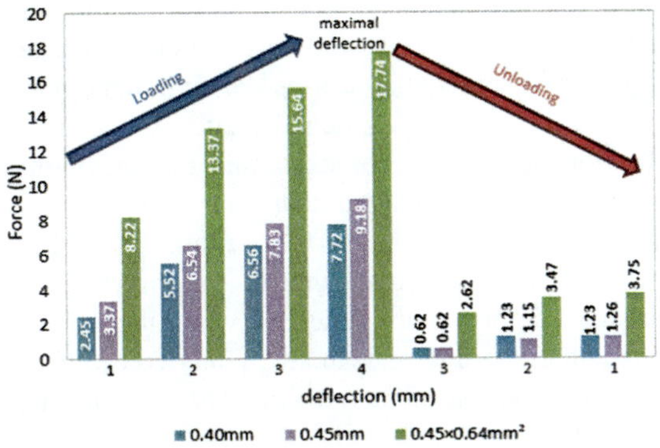

Fig. 3. Load level difference of the three different cross-section wire during loading and the unloading at each 1 mm of deflection

Figure 3 illustrates the forces recorded at each 1 mm of deflection during the charging and the discharging. The 0.4 and 0.45 mm round cross-section wires delivered 5.52 and 6.54 N when they were bent until 2 mm. these loads are higher than the force reported by Fujio et al. (1986). Besides, at a deflection of 3 mm, the round wires exhibited higher load than the 0.4 and 0.45 mm square wire investigated in the work of Garrec and Jordan (2004). The fact that the wires used in the aforementioned researches are not fabricated by the same companies, can be responsible of the load level differences, but this difference is not significantly important as produced by the orthodontic braces. In fact, when the wire is engaged in conventional brackets with elastomeric ligatures, the engagement load is increased because of the friction (Nucera et al. 2014). This latter affects also the initial load delivered at small deformation, when the wire is still austenitic. In other words, the change of slope observed during the elastic deformation of the austenite is due to the increase of the contact between the archwire and the orthodontic attachments. During the deactivation, the wire firstly slips between the brackets, which lead to a sudden decrease of the load. Then, when the backward martensitic transformation is produced, the load initially decreases, then it increases. For instance, at 3 and 1 mm of deflection, the rectangular wire generated respectively loads of 2.62 N and 3.75 N. The raise of this orthodontic load was produced when the wire overcame the friction resistance caused by the brackets ligated system (Nucera et al. 2014).

At the end of the bending tests, the three deflection load curves, shown in Fig. 2, exhibit a residual deflection about 0.15 mm. Indeed, this latter is due to the play between the brackets and the wire.

4 Conclusion

Superelastic Ni-Ti wire behavior was investigated during bending tests using a specific device. This device consists on replacing the 3 points classic bending test by three brackets bending test to examine the effect of these attachments. Furthermore, the cross-section size and shape effect were studied to elucidate the behavior of the Ni-Ti wire when it is engaged in the oral cavity during the alignment phase. The experimental findings were compared to other experimental results found in the literature. It was observed that the friction between the brackets and the orthodontic wire affected the load released during the loading, which simulates the force generated wherein inserting the wire in the brackets. This friction has also changed the level of clinical loads delivered during the unloading. Moreover, the raise of the cross-section size induces a load increase at the small deformation. Whereas, when the pseudo-elastic behavior is exhibited, the loading and unloading plateau are mainly controlled by the strain energy, depending on the volume fraction of martensitic. These results will be used in future work as an experimental reference.

Acknowledgments. This research did not receive any specific grant from funding agencies in the public, commercial, or not-for-profit sectors.

References

Aghili, H., Yasssaei, S., Ahmadabadi, M.N., Joshan, N.: Load deflection characteristics of nickel titanium initial archwires. J Dent. (Tehran) **12**(9), 695–704 (2015)

Fujio, M., Mogi, M., Ohura, Y., Hamanaka, H.: The wire super-elastic property of the Japanese NiTi alloy for use in orthodontics. Am. J. Orthod. Dentofacial Orthop. **90**(1), 1–10 (1986)

Garrec, P., Jordan, L.: Stiffness in bending of a superelastic Ni-Ti orthodontic wire as a function of cross-sectional dimension. Angle Orthod. **74**(5), 691–696 (2004)

Garrec, P., Tavernier, B., Jordan, L.: Evolution of flexural rigidity according to the cross-sectional dimension of a superelastic nickel titanium orthodontic wire. Eur. J. Orthod. **27**, 402–407 (2005). https://doi.org/10.1093/ejo/cji014

Gatto, E., Matarese, G., Di Bella, G., Nucera, R., Borsellino, C., Cordasco, G.: Load – deflection characteristics of superelastic and thermal nickel – titanium wires. **35**, 115–123 (2013). http://doi.org/10.1093/ejo/cjr103

Lombardo, L., Marafioti, M., Stefanoni, F., Mollica, F., Siciliani, G.: Load deflection characteristics and force level of nickel titanium initial archwires. **82**(3), 4–7 (2012). http://doi.org/10.2319/032511-213.1

Mallory, D.C., English, J.D., Powers, J.M., Brantley, W.A., Bussa, H.I.: Force-deflection comparison of superelastic nickel-titanium archwires. 110–112 (2004). http://doi.org/10.1016/j.ajodo.2004.03.012

Mohd, J., Leary, M., Subic, A., Gibson, M.A.: A review of shape memory alloy research, applications and opportunities. Mater. Des. **56**, 1078–1113 (2014). https://doi.org/10.1016/j.matdes.2013.11.084

Nucera, R., Gatto, E., Borsellino, C., Aceto, P., Fabiano, F.: Influence of bracket-slot design on the forces released by superelastic nickel-titanium alignment wires in different deflection configurations. **84**(3) (2014). http://doi.org/10.2319/060213-416.1

Thompson, S.A.: An overview of nickel – titanium alloys used in dentistry. Int. Endod. J. **44**(0), 297–310 (2000)

Tonner, R.I.M., Waters, N.E.: The characteristics of super-elastic Ni-Ti wires in three-point bending: Part I: the effect of temperature. Eur. J. Orthod. **16**, 409–419 (1994)

Zhang, X., Feng, P., He, Y., Yu, T., Sun, Q.: Experimental study on rate dependence of macroscopic domain and stress hysteresis in NiTi shape memory alloy strips. Int. J. Mech. Sci. **52**(12), 1660–1670 (2010). https://doi.org/10.1016/j.ijmecsci.2010.08.007

Zuhanek, S.: Material characteristics of the orthodontic archwires. (18), 301–308 (2011). http://doi.org/10.2507/daaam.scibook.2011.24

A Comparative Investigation of the Tribological and the Mechanical Behavior of Polyester Powder Coatings Filled with Graphite Depending on the Filling Percentage and the Size of the Graphite Particles

Nedia Gafsi[1(✉)], Mohamed Kharrat[1], Maher Dammak[1], Raquel Verdejo[2], Miguel Angel Lopez Manchado[2], and Massimiliano Barletta[3]

[1] Laboratory of Electromechanical Systems (LASEM), National Engineering School of Sfax, University of Sfax, Sfax, Tunisia
gafsi.nedia89@gmail.com, {mohamed.kharrat, maher.dammak}@ipeis.rnu.tn

[2] Instituto de Ciencia y Tecnología de Polímeros (ICTP-CSIC), Spanish National Research Council (CSIC), C/Juan de la Cierva, 3, 28006 Madrid, Spain
r.verdejo@csic.es, lmanchado@ictp.csic.es

[3] Department of Enterprise Engineering, University of Rome Tor Vergata, Rome, Italy
barletta.ing@gmail.com

Abstract. Composite coatings are used in order to reduce the friction and protect contacting surfaces against wear with good adhesive to substrate.

The present study deals with two sizes of solid particles used in composite coatings. As received graphite G1, the mean size is 20 μm and the crushed one, obtained by ball mill, is 10 μm.

The resulting tribological behavior and micromechanical properties were discussed and compared. As expected, the smaller the particle size the greater reduction in the coefficient of the friction and wear rate. The experimental findings show that adding 5 wt% of crushed graphite provide similar COF than when introducing 10 wt% of as received graphite. Coatings scratch behavior is affected by the size and percentage of fillers inside the polyester matrix. In fact, the scratch resistance is observed with polyester coatings filled at 5 wt% G2 and 10 wt% G1.

Keywords: Composite coating · Polyester · Graphite · Tribology · Scratch-test

M. Kharrat et al. (Eds.): ICAMEM 2019, LNME, pp. 252–258, 2021.
https://doi.org/10.1007/978-3-030-52071-7_35

1 Introduction

Tribological characterization of materials deals with friction, wear, scratch resistance and design of interactive surfaces in relative motion. The use of polymer-based materials is increasing in several industries, notably in automotive and aerospace manufacture (Marouen 2017). Coating aluminum mechanical parts with polyester matrix containing graphite particles can be a solution to decrease friction coefficient, wear loss and also to protect environment (Brostown et al. 2003; Glendimar et al. 2019; and Friedrich et al. 2011).

In the present work, polyester powder coatings filled with various content of graphite G1, with 20 μm mean particles size, and graphite G2, with 10 μm mean particles size, had been coated by electrostatic spray (ES) onto aluminum substrate pre-cleaned in an ultrasonic bath and ethanol. We purpose to analyze the effect of graphite particle size on the mechanical properties and the tribological response of the elaborated composite coatings. Friction and wear tests were carried out using a reciprocating a flat tribometer under dry sliding condition in order to evaluate and collate the tribological properties of the prepared composites. The wear mechanisms were identified throw the analyzes of the worn surfaces' morphology. Scratch test were conducted on the different coatings in order to assess the scratch resistance.

2 Materials and Methods

The polyester powder was mixed with the graphite fillers using three filling ratios 0.5 wt%, 5 wt% and 10 wt%. The mixture was coated on the substrates using an ES deposition system and thereafter samples were put in an oven at 180 °C during 10 min for the consolidation. The solid lubricant was as received graphite G1 (20 μm mean particle size) and crushed graphite G2 (10 μm mean particle size). The thickness of the coatings was approximately 130 μm ± 20 μm.

2.1 Sliding Test

The friction and wear analysis of the coatings were analyzed via linear reciprocating tribometer. The normal load applied was 10 N between the flat coating surface and a high chromium steel ball. The length of the wear crack was under ±7.5 mm using 1 Hz frequency.

2.2 Scratch Test

Scratch tests, in a progressive mode, were did on the coating systems using a Micro-Combi Tester with Rockwell indenter, 200 μm tip of radius. The length of the track was 3 mm with an indenter advancing at 1 mm/min. The spacing between two consecutive tracks was minimum 3 mm to scratch the total of the surface simples. Penetration depth (Pd) and residual depth (Rd) were the output of scratch test to evaluate the scratch resistance of the coating.

For sliding and scratch tests, three tests were done for each percentage.

3 Results and Discussion

3.1 Friction and Wear Behavior

Figure 1 present the friction coefficient after 10000 sliding cycles versus the different weight fractions of graphite G1 and G2 in order to explore the effect of weight fractions of G1 and G2 on the friction behavior of the polyester coating. Adding graphite G1 to the polyester matrix at different percentages provide a progressive reduction of the coefficient of friction shown in the work by The friction coefficient decreases from 0.6 for pure polyester to 0.45 for polyester filled with 10 wt% graphite G1. This is consistent with the previous research on the effect of graphite which is an effective lubricant to reduce the coefficient of friction, this is in argument with Shalwan and Yousif (2014) and Jitendra et al. (2016). Also, in case of G2 the crushed fillers are able to decrease the friction coefficient, until 0.42 for polyester filled with 5 wt%, and the decrease is more remarkable in G2 than in the case of G1 (Fig. 1). This is may be due to a better incorporation of crushed fillers into the polyester matrix than as received one, as show Satheesan (2017).

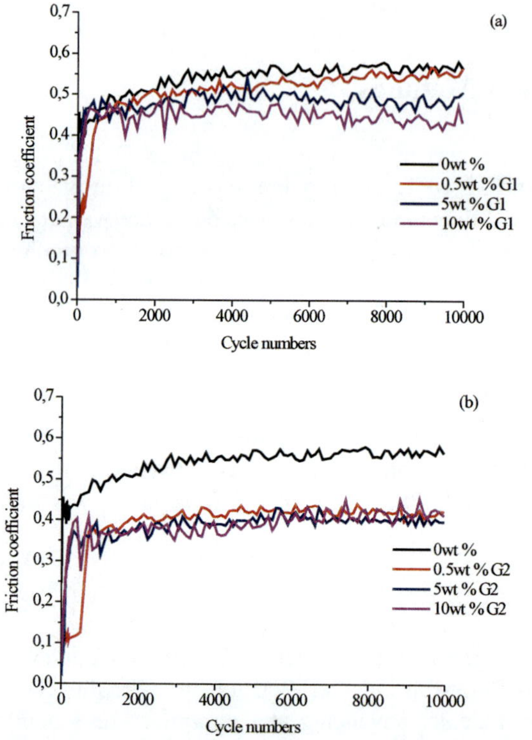

Fig. 1. Friction coefficient versus number of cycles for polyester-graphite composite coatings filled with different weight fractions of graphite: (a) G1 and (b) G2.

Introduced graphite fillers increase the wear resistance, Kenneth (2019). For the two microparticles, at 10 wt%, the wear resistance was higher than 0.5 wt% and 5 wt% (Fig. 2). The graphite fillers increase the wear resistance by reducing the plastic deformation of the coating.

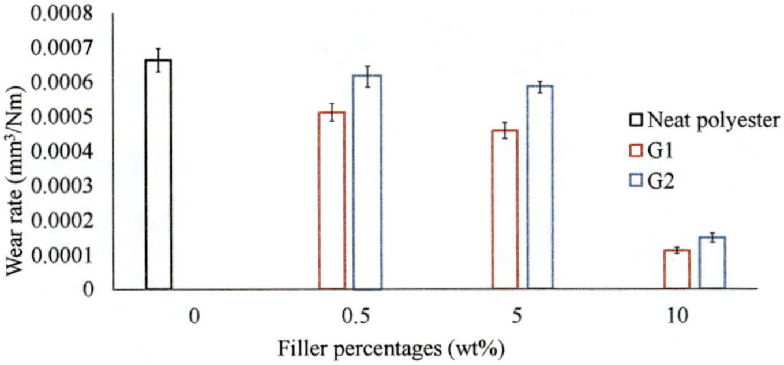

Fig. 2. Wear volume versus percentage of graphite G1 and G2

During the friction test, transfer film (third body) is created. Therefore, the contact change from stain steel/composite coating to transfer film/third body film, which may be softer. Indeed graphite particles accelerate the formation of the third body thanks to their lubricating properties. In addition, microstructures of both third body and transfer films should change with the graphite particles size and this fact can explain the obtained difference in wear resistance.

3.2 Wear Mechanism

The observations of the wear track via SEM after sliding tests were did to investigate the wear mechanisms of coatings filed with different amount of graphite (Fig. 3). For the matrix coating without solid particles, SEM micrograph reveal plastic deformation with the appearance of micro-cracks and micro-grooves formed in the sliding direction located at the wear mark. At 5 wt% of both solid lubricants G1 and G2, the SEM micrographs show a decrease of plastic deformation intensity and micro-cracking extend comparing with pure polyester. At 10 wt% of graphitic lubricants G1 and G2, the contact area becomes smooth with limited plastic deformation. The reduction of wear can be attributed to easy shearing action and good lubricity provided by graphite particles, Basavarajappa (2009). In fact, during the friction test, the contact pressure creates movement of the graphite particles in the sliding direction. These particles (G1 and G2) have a hexagonal structure formed by graphene sheets linked together with weak leaves (Van der Waals forces). May be due to the weakness of the bonding how gives low shear strength and responsible of reduction of the wear.

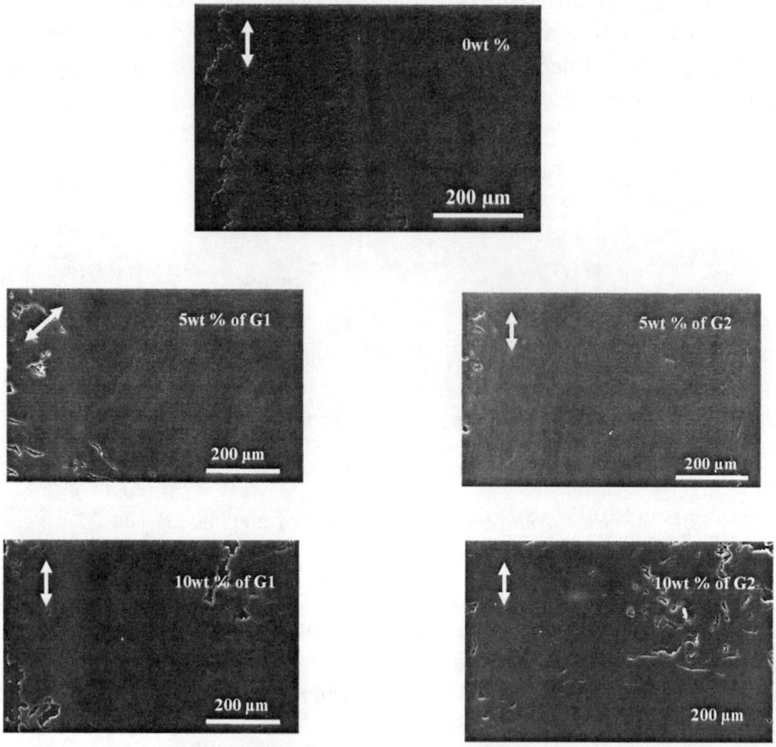

Fig. 3. SEM micrographs of the different composite coatings after 10000 sliding cycles

3.3 Scratch Test Analysis

Figure 4, present the curve of the Pd and Rd of the scratch tests. The reinforcement of the polyester resin by the dispersion of the solid particles G1 and G2 decrease the penetration depths during progressive load scratch tests. For the neat matrix, coating have approximately the same trend and with no significant jumbling it may be from the detachment of the polyester coating. Adding 5 wt% graphite, the trends showed with G1 irregularities at 13 N, although they appeared at 20 N in case of G2. At 10 wt% graphite, we observed the appearance of irregularities at low normal force (10 N) with G2. However, with graphite G1 irregular jumbling was observed at 20 N. In addition, at 10 wt% G1 reduced the penetration depth more than G2.

So, at 5 wt% graphite, the polyester resin filled with G2 limited more the penetration and residual depth during the test then G1. This result may be due to the small size of particles G2 providing a quasi-homogeneous micro-structure of the composite coating since this low filler content.

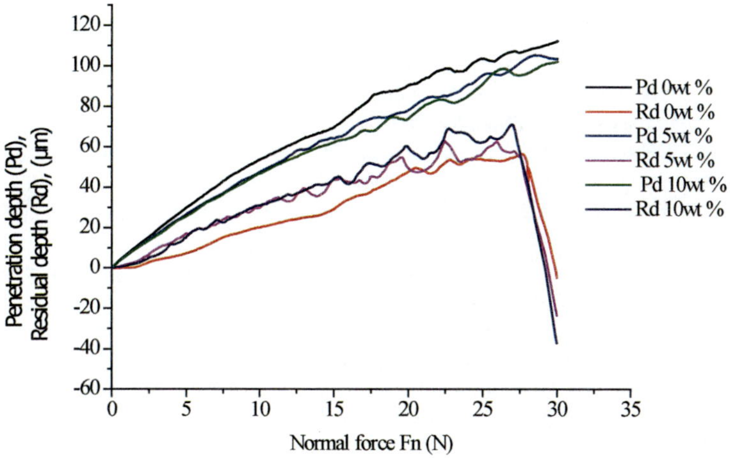

(a)

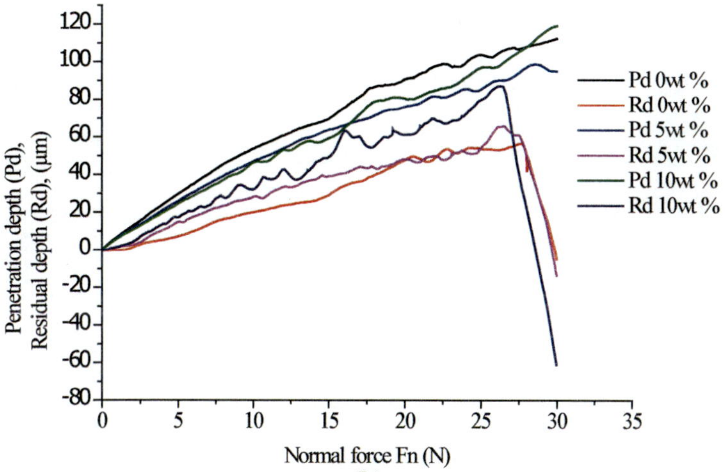

(b)

Fig. 4. Penetration and Residual depth: (a) G1 (b) G2

4 Conclusion

The results showed that the mechanical properties of the composite coatings mainly depend on size of filler particles. It also revealed that the addition 5 wt% of crushed graphite (10 µm) provide a similar friction coefficient than when introducing 10 wt% of as received graphite (20 µm). The wear volume of composite coatings with G1 is lower than with G2. At 5 wt% filler content, composite coating with G2 was more resistant in scratch test. However, at 10 wt% filler content, composite coating with G1 was more resistant.

References

Marouen, H., Xiao, Z., Hung-Jue, S.: Fundamental understanding on scratch behavior of polymeric laminates. Wear **380–381**, 203–216 (2017)

Brostow, W., Deborde, J.-L., Jaklewicz, M., Olszynski, P.: Tribology with emphasis on polymers: friction, scratch resistance and wear. J. Mater. Ed. **25**, 119–132 (2003)

Glendimar, M., Hung-Jue, S.: Scratch behavior of model epoxy resins with different crosslinking densities. Mater. Des. **182**, 107965 (2019)

Friedrich, K., Sue, H.J., Liu, P., Almajid, A.: Scratch resistance of high performance polymers. Tribol. Int. **44**, 1032–1046 (2011)

Shalwan, A., Yousif, B.F.: Influence of date palm fibre and graphite filler on mechanical and wear characteristics of epoxy composites. Mater. Des. **59**, 264–273 (2014)

Jitendra, K.: Friction and wear durability study of epoxy-based polymer (SU-8) composite coatings with talc and graphite as fillers. Wear **362–363**, 199–208 (2016)

Satheesan, B., Mohammed, A.S.: Enhancement of tribological performance of epoxy bulk composites and composite coatings using micro/nano fillers. Polym. Adv. Technol. **28**, 633–644 (2017)

Kenneth, K.A., Adetomilola, V.F., di Nthabiseng, B.M.: Development of aluminium-based composites reinforced with steel and graphite particles: structural, mechanical and wear characterization. Mater. Res. Tech. **8**, 670–682 (2019)

Basavarajappa, S.: Studies on dry sliding wear behaviour of Graphite filled glass epoxy composites. Mater. Des. **30**, 2670–2675 (2009)

Physicochemical, Morphological and Thermal Characterization of Composites Based on Olive Wood Flour

Nesrine Bouhamed[1,2](✉), Slim Souissi[1], Pierre Marechal[2], Mohamed Ben Amar[1], and Olivier Lenoir[2]

[1] LASEM, Université de Sfax, Sfax, Tunisia
Bouhamednesrine34@gmail.com
[2] LOMC, Université Le Havre Normandie, Le Havre, France

Abstract. This study deals with the evaluation of the physicochemical, morphological and thermal properties of a biocomposite material based on polypropylene (PP) reinforced with vegetable fibers olive wood flour (OWF). Initially, the olive wood flour was chemically treated with a coupling agent (amino-silane) and then characterized to provide the modification influx. In a second step, various PP/OWF bio-composites with variable contents of OWF (0, 10, 20 and 30% by weight) were implemented by twin-screw extrusion and then shaped by injection molding. The effect of the OWF content on the one hand and the chemical modification of the latter by adding silane as coupling agent on the other hand, are the two parameters that are the subject of the present sensitivity study. Finally, the characterization results are compared and discussed. The significant influence of the treatment of OWF with the silane coupling agent is demonstrated, clearly demonstrating that the silane promotes the adhesion of the plant-based filler that is OWF with the PP polymer matrix.

Keywords: Physicochemical · Biocomposite · Coupling agent

1 Introduction

The development of a composite material makes it possible to obtain properties that are better than those of the constituents that compose it. Thus, it is imperative to know the composition and the properties of the constituents, as well as the stages of implementation to understand the interaction mechanisms between the constituents. This study aims to focus on this approach and allowed us to determine the physicochemical, thermal and morphological characteristics of the fibers, in this case olive wood flour (OWF), the matrix, polypropylene (PP), and elaborate composite (PP/OWF).

M. Kharrat et al. (Eds.): ICAMEM 2019, LNME, pp. 259–263, 2020.
https://doi.org/10.1007/978-3-030-52071-7_36

2 Materials and Method

2.1 Materials

The polypropylene (PP) used in this study is a standard homopolymer PP H9069 (Braskem, Brazil). FBO is recovered from wood waste generated during agricultural work. It is sieved and stored to protect it from moisture. Two types of FBO were used as reinforcing fillers, untreated (OWF) and treated (OWF) at 3% by weight with a coupling agent [1], amino-propyl-triethoxy-silane (OWFT).

2.2 Filler Characterization

The biomass used in this study consisted of olive wood flour and sieved to obtain finer particles, from 84.6 to 254.9 μm in size. The chemical composition of the dried fibers was determined according to [2]. It was determined by carrying out a series of extractions to isolate the various substances contained in the fibers such as: cellulose, hemicellulose and lignin. The length and aspect ratio of OWF filler was determined by Morfi analyses using JEOL JSM-5400 Morfi equipment

2.3 Elaboration of Bio-composite

In order to eliminate all the moisture absorbed and to avoid the formation of agglomerates, the FBOs were pre-dried at 105 °C for 24 h before extrusion. The PP polymer and the FBO fibers were mixed in a co-rotating twin-screw extruder (DSM Xplore Netherlands) with 3 heating zones. The 10 mm diameter screw was set at a rotational speed of 100 rpm. The mixture is extruded through a cylindrical matrix of diameter 1 mm at a rate of 200 g/h. The rod is cooled, then cut into granules which are in turn dried for 24 h at 60 °C. After extrusion, the dried granules are injection molded using a molding machine (Demag Plastic Group Extra 50–200). The injection is produced at a temperature of 180 °C and a pressure of 1000 bar to produce the bio-composite samples required for the tests, in the form of plates and test pieces. Different charge rates of 0, 20 or 30% have been developed.

2.4 Scanning Electron Microscopy (SEM)

Scanning electron microscopy (SEM) was performed with a JEOL JSM-5400 SEM microscope operating at 15 kV. The specimens were carried out under liquid nitrogen to avoid plastic deformation of the matrix and to obtain a well-defined fiber–matrix interface. Prior to analysis, all samples were coated with a 10 nm layer of gold to avoid sample charging under the electron beam.

2.5 Thermo-Gravimetric Analyzes (ATG)

Thermogravimetric analyzes (ATG) were performed using Perkin Elmer Pyris-1 equipment (USA) under a nitrogen flow up to 800 °C and with a temperature rise rate of 10 °C/min. The objective is to study the thermal stability of our samples (fibers alone, virgin polymer, and composite materials).

3 Results and Discussion

3.1 Morphology and Chemical Composition of Olive Wood Flour

Olive wood flour is similar to a composite material reinforced by longitudinally bonded fibrils of cellulose, lignin and other non-cellulosic materials. Waxes and impurities regularly cover the whole [3, 4]. Figure 1 shows the micrographs obtained by the SEM show morphological transformations of the fibers following chemical treatment with silane The surface of raw olive wood flour is smooth, while that of the treated FBO shows exposed fibrils. The reduction in the level of non-cellulosic material such as lignin, hemicellulose and wax, thanks to the treatment performed, has a positive effect on the roughness of the fiber surface, which increases visibly according to the micrograph [5].

(a) (b)

Fig. 1. SEM images of olive wood flour fibers: (a) untreated OWF, (b) OWF treated

The chemical composition of the OWF. It confirmed the lignocellulosic character of the filler being composed of lignin, hemicelluloses and cellulose, with content about 21.9%, 29.2% and 45.9%, respectively.

3.2 Morphological and Thermal Characterization of Elaborate Bio-composites

By switching from crude fiber-based biocomposite (Fig. 2 (a)) to processed fiber-based fiber (Fig. 2 (b)), the number of these cavities decreases. As a result, the improvement in adhesion between the fibers and the matrix can be distinguished by addition of the coupling agent (the treatment with the silane).

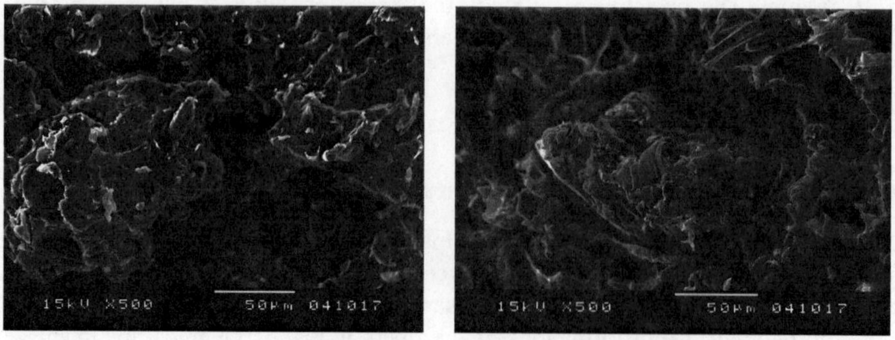

(a) (b)

Fig. 2. SEM images of composite: (a) PP/OWF, (b) PP/OWFT

From the PP matrix. As shown in photo (a) of Fig. 2, the fiber surfaces are smooth and clean, showing poor inter-facial adhesion between the fibers and the matrix. On the other hand, the photo (b) of Fig. 2 shows that for the bio-composite loaded by the treated FBO the fibers are closely linked to the matrix with a surface characterized by a coating of a layer of the polymer matrix (PP). Indeed, it can be seen that the fracture occurred at the cross-section of the fibers themselves. This confirms that there is a transfer of the stresses from the matrix to the load, because of the adhesion to the contact between fibers and matrix. The fibers therefore participate in this case in the energy required for breaking and thus ensure good cohesion of the entire composite material.

These results are in agreement with previous studies which confirmed that the chemical treatment with silane of natural fibers used as reinforcements improves the cohesion between the fibers and the matrix of the corresponding bio-composite materials.

Biocomposites loaded with the raw and treated OWF were analyzed by thermogravimetry (ATG) in order to know the effect of chemical treatment of the load and the rate of reinforcement on the thermal stability of the material.

The results obtained by thermogravimetric analysis (ATG) are presented in Table 1 as a function of degradation temperature (Tmax).

Table 1. Degradation temperatures of raw and processed PP/OWF composites.

Composition	Degradation temperature of PP/OWF	Degradation temperature of PP/OWF
PP/	466.61	466.61
PP/OWF 10%	471.96	466.54
PP/OWF 20%	469.41	469.82
PP/OWF 30%	475.54	472.58

The addition of 10%, 20%, 30% of untreated fibers significantly increases the degradation temperatures respectively by 5, 3, and 9 °C relative to the matrix alone. This effect is comparable with the treated fiber content, for which the temperature varies from 0, 3, and 6 °C with the addition of 10%, 20%, 30% of treated fibers, respectively. In our case, neither the addition of fiber content nor the chemical treatment significantly affected the thermal stability of the PP/OWF composite.

The literature [6, 7] shows that generally the concentration of chemical treatment is the factor that significantly influences the thermal stability of composites because of its ability to remove hemicellulose, lignin and impurities from the fiber. This allows an improvement in the thermal stability of the fibers. However, the modification made to the fibers in our study is not sufficient to affect the thermal stability of the composite in its entirety

4 Conclusion

In this study we characterized the raw olive wood flour and chemically treated with silane. Indeed, the development of new bio composites requires a prior knowledge of the different properties and the composition of fibers.

The SEM micrographs of both types of flours revealed an increase in the roughness of the treated versus untreated fibers due to the removal of non-cellulosic constituents. The chemical analysis reveals that after chemical treatment, the hemicellulose and lignin contents decrease, which causes the increase of the cellulose content in the treated fibers. Secondly, we have characterized morphologically and thermally elaborate bio-composites. SEM images showed good adhesion between the matrix and treated or untreated fibers. Thus, the cohesion at the fiber/matrix interface seems to improve with the chemical treatment.

Thermogravimetric analysis has shown that the thermal stability of bio-composites does not change much for both treated and untreated fibers.

References

1. Arbelaiz, A., Fernandez, B., Cantero, G., Llano-Ponte, R., Valea, A., Mondragon, I.: Mechanical properties of flax fibre/polypropylene composites. Compos. Part A **36**, 1637–1644 (2005)
2. Jayaramudu, J., Guduri, B.R., Rajulu, A.V.: Characterization of new natural cellulosic fabric Grewia tilifolia. Carbohydr. Polym. **79**(14), 847–851 (2010)
3. Fiore, V., Scalici, T., Valenza, A.: Characterization of a new natural fiber from Arundo donax L. as potential reinforcement of polymer composites. Carbohydr. Polym. **106**, 77–83 (2014)
4. Mwaikambo, L.Y., Ansell, M.P.: Chemical modification of hemp, sisal, jute, and kapok fibers by alkalization. J. Appl. Polym. Sci. **84**(112), 2222–2234 (2002)
5. Narendar, R., Dasan, K.P.: Chemical treatments of coir pith: morphology, chemical composition, thermal and water retention behavior. Compos. Part B Eng. **56**, 770–779 (2014)
6. Shukor, F., Hassan, A., Islam, M.S., Mokhtar, M., Hasan, M.: Effect of ammonium polyphosphate on flame retardancy, thermal stability and mechanical properties of alkali treated kenaf fiber filled PLA biocomposites. Mater. Des. (1980–2015) **54**, 425–429 (2014)
7. Manshor, M.R., Anuar, H., Aimi, M.N.N., Fitrie, M.I.A., Nazri, W.B.W., Sapuan, S.M., El-Shekeil, Y.A., Wahit, M.U.: Mechanical, thermal and morphological properties of durian skin fibre reinforced PLA biocomposites. Mater. Des. **59**, 279–286 (2014)

A Comparative Study on the Physical and Mechanical Behavior of AA6082-T6 and AA5083- H116 Aluminum Alloys in Friction Stir Spot Welding

Nasra Hannachi[1,2,5](✉), Ali Khalfallah[1,3], Carlos Leitao[4], and Dulce M. Rodrigues[5]

[1] Mechanical Engineering Laboratory (LGM), ENIM, University of Monastir, Monastir, Tunisia
nasra.hannachi@gmail.com, ali.khalfallah@gmail.com
[2] National School of Engineers of Sousse (ENISO), Sousse, Tunisia
[3] Higher Institute of Applied Science and Technology of Sousse (ISSATSO), University of Sousse, Sousse, Tunisia
[4] CEMMPRE, Department of Mechanical Engineering, University of Coimbra, Coimbra, Portugal
carlos.leitao@dem.uc.pt
[5] ISISE, Department of Mechanical Engineering, University of Coimbra, Coimbra, Portugal
dulce.rodrigues@dem.uc.pt

Abstract. Aluminum alloys are widely used in manufacturing industry due to its beneficial combination of formability and strength. However, these alloys are prone to imperfections when joined using conventional fusion welding technologies. A revolutionary solid-state welding technique, labeled Friction Stir Welding (FSW), was demonstrated to be particularly suited for aluminum alloys joining. This solid-state welding process produces joining by mixing materials under intense plastic deformation and heat generation. To gain insight into the base material flow and heat generation during FSW, 3D coupled thermo-mechanical finite element model (FEM) was developed to reproduce the friction stir spot welding (FSSW) variant of FSW process. The numerical model, which uses an Arbitrary Lagrangian Eulerian (ALE) formulation, was implemented in ABAQUS/Explicit. The constitutive behavior of the base material was modeled using the classical Johnson-Cook model. The developed FE model was used to predict and compare the temperature distributions, forces and torques during FSSW of AA6082-T6 and AA5083-H116 aluminum alloys. Some conclusions on AA6082-T6 and AA5083-H116 aluminum alloys weldability in FSSW will be presented based on the comparative analysis of numerical and experimental results.

Keywords: FSSW · FE model · ALE · AA6082-T6 · AA5083-H116

M. Kharrat et al. (Eds.): ICAMEM 2019, LNME, pp. 264–270, 2020.
https://doi.org/10.1007/978-3-030-52071-7_37

1 Introduction

In the last years, Friction Stir Welding (FSW) gained significant attention as a solid-state joining process industrialized for a wide range of materials. The concept of this joining technique is that, a non-consumable welding tool is rotated, slowly plunged then translated along the joint line between two plates. The combined action of the frictional heat and the plastic deformation in the interface tool/workpiece leads to an increasing temperature which softens the material, thereby, allowing the tool to "stir" the joint surfaces.

During FSW, several variables act to determine the joint properties, including mechanical properties, hardness, fatigue strength, etc. Hence, great efforts have been made to investigate the effect of these variables on the weld quality, in order to maximize productivity and performance. The wide majority of researches, such as John and Iqbal (2019), have been conducted to study the effect of the tool geometry. Results showed that these geometric parameters exert significant effect on the material flow pattern and temperature distribution, thereby influence the microstructural evolution of material. On the other hand, numerous authors such as Costa et al. (2019) stated that the tool rotation rate and the welding speed are the major independent variables that control the FSW process. In fact, a higher tool rotation rate generates higher friction heating and result in more intense stirring material.

However, until now only few studies have reported the effect of material behavior on weldability. It is only possible to emphasize the work presented by Costa et al. (2018) which analyzed the influence of high temperature plastic behavior on aluminum alloy weldability. To the author's knowledge the material factors have not been thoroughly investigated yet. Indeed, the purpose of this paper is to investigate the behavior of AA6082-T6 and AA5083-H116 aluminum alloys during Friction Stir Spot Welding (FSSW), one of FSW derivative, with the aim to identify the influence of heat treatment and material components on weldability. A numerical analysis of the process is conducted and the results are validated by literature.

2 Thermo-Mechanical Modeling

The Finite Element Method (FEM) is the modeling approach adopted in this work. A 3D thermo-mechanical model is developed in the commercial finite element solver ABAQUS/Explicit using the Arbitrary Lagrangian Eulerian (ALE) formulation, the Johnson Cook material law, and the Coulomb's friction Law. In the following, the geometries of the tool and the workpiece as well as the ALE formulation, the material and the friction law are described.

2.1 Geometric Model and Boundary Conditions

The model consists of a rigid H13 steel tool and a deformable workpiece wherein assigned, firstly the AA6082-T6, and secondly the AA5083-H116 materials.

The computational domain used to model the workpiece has a parallelepiped shape measuring $40 \times 40 \times 2\ \text{mm}^3$. A pinless spot welding tool was used due to its ability to create high strength joints in thin aluminum plate without leaving a key- hole (Cox 2014). Figure 1 shows the numerical model developed (Part geometries, dimensions and boundary conditions). A plunge velocity of 1 mm/s and a rotational velocity of 2000 rpm are applied to the tool. The plunge depth is 0.2 mm.

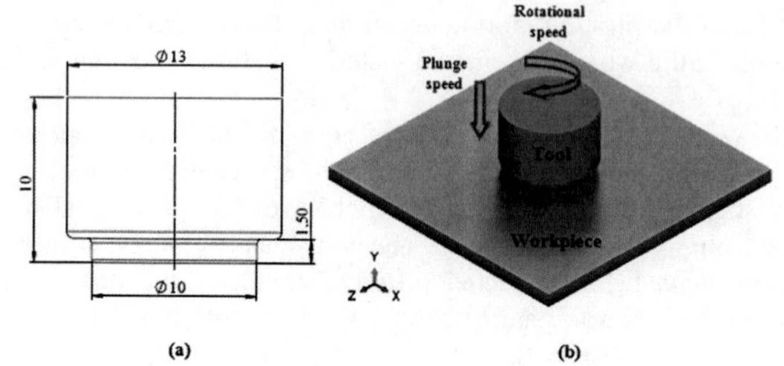

Fig. 1. FSSW process: (a) tool design (b) schematic overview and boundary conditions

A total of 64,582 eight node coupled-temperature displacement, brick elements (C3D8RT) are used to mesh the workpiece. A refined circular mesh is assigned around the tool plunge area in order to improve the accuracy of the solution at the tool-workpiece interface.

2.2 Materials Properties

Welding of aluminum has always been a challenge due to its high thermal conductivity, high coefficient of thermal expansion and high solubility of hydrogen. Hence, in this study, AA6082-T6 (heat-treatable aluminum alloy) and AA5083-H116 (non heat-treatable aluminum alloy) are selected for the FSSW analysis.

For the material behaviour, the constitutive Johnson Cook model (1) is currently used (Lorrain et al. 2009). This model accounts the strain, the strain rate and the temperature dependence of the material behaviour in an uncoupled manner. It is written:

$$\sigma = [A + B(\varepsilon)^n] \times \left[1 + Cln\frac{\dot{\varepsilon}}{\dot{\varepsilon}_0}\right] \times \left[1 - \left(\frac{T - T_{room}}{T_{melt} - T_{room}}\right)\right] \quad (1)$$

where T_{melt} is the melting point of the material, T_{room} the ambient temperature (25 °C), T the effective temperature, A the yield stress, B the strain factor, n the stain exponent, m the temperature exponent, ε the plastic strain, $\dot{\varepsilon}$ the plastic strain rate, $\dot{\varepsilon}_0$ the reference strain rate, and C the strain rate factor. A, B, C, n and m are material/test constants. The material properties of AA6082-T6 and AA5083-H116, considered for simulations, are as per the values taken by Birsan et al. (2011). Chemical compositions as well as material parameters are listed in Table 1, 2 and 3.

The contact between the tool and the workpiece is defined by Coulomb friction law with $\mu = 0.3$.

Table 1. Chemical composition of AA6082-T6 and AA5083-H116 (Matweb)

Chemical composition (%)							
	Si	Fe	Mn	Mg	Ti	Cr	Al
AA6082-T6	0.7–1.3	<=0.50	0.4–1.0	0.6–1.2	<=0.10	<=0.25	95–98
AA5083-H116	<=0.4	<=0.40	0.4–1.0	4.0–4.9	<=0.15	0.05–0.25	92–95

Table 2. Johnson Cook parameters

	A (MPa)	B (MPa)	C	m	n	T_{melt} (°C)	$\dot{\varepsilon}$
AA6082-T6	285	94	0.007	1.31	1.00	582	0.0001
AA5083-H116	167	596	0.001	0.85	0.55	620	0.0001

Table 3. Materials physical and mechanical properties (Matweb)

	AA6082-T6 (workpiece)	AA5083-H116 (workpiece)	H13 steel (Tool)
Young's modulus; E (GPa)	70	71	210
Poisson's ratio; υ	0.33	0.33	0.1
Tensile Strength, Yield (MPa)	285	167	1650
Elongation at Break (%)	10	16	–
Density; ρ (kg/m3)	2700	2650	7800
Conductivity; k (W/mK)	180	120	25
Expansion coefficient; α (C^{-1})	$21.1.10^{-6}$	$23.4.10^{-6}$	$12.6.10^{-6}$
Specific heat; cp (J/kgC)	900	880	500

3 Results and Discussion

3.1 Temperature Results

Figure 2 shows the temperature recorded at the end of the dwelling period for both AA6082-T6 and AA5083-H116 aluminum alloys.

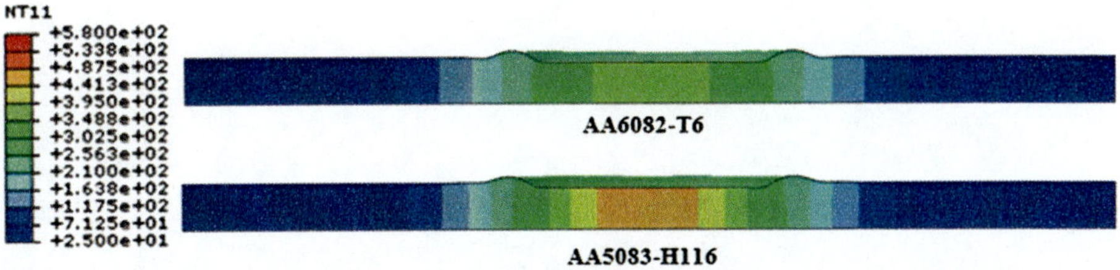

Fig. 2. Thermal distribution in (°C) predicted by FE model along cross-section FSSW joint of the AA6082-T6 and AA5083-H116 aluminum alloys

As seen in Fig. 2, the temperature in the workpiece under the tool is in the range between 300 and 500 °C. These results indicate that the maximum of temperature for both materials still under the melting points (Table 2) which is suitable for defect-free welds.

3.2 Force and Torque Variation

In the current work, the FE model is capable of predicting the axial force and torque generated during the FSSW process. It should be noted that this model was validated by comparing the numerical results to referenced testing performed in ISISE laboratory. Typical plots of force and torque during FSSW are represented respectively in Fig. 3 and Fig. 4.

Maximum force is generated during the plunging phase followed by a gradual reduction throughout the dwelling and retraction of the tool. This evolution is stemming from the high temperature which soft the material. The same logic explains the torque variation during all FSSW phases as shown in Fig. 4. Greater force and torque magnitudes were generated for the AA5083-H16 aluminum alloy. This is certainly due to its higher mechanical properties, namely the yield and the ultimate strength (Table 3), either the content of Magnesium (Table 1) which increases the alloy strength (Yalavarthy et al. 2009).

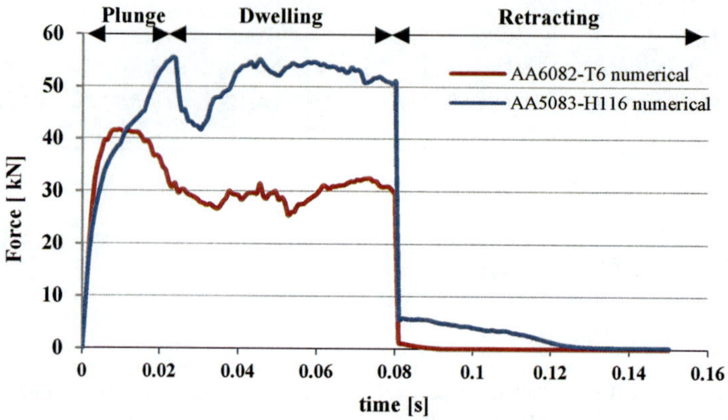

Fig. 3. Axial force variation during FSSW process

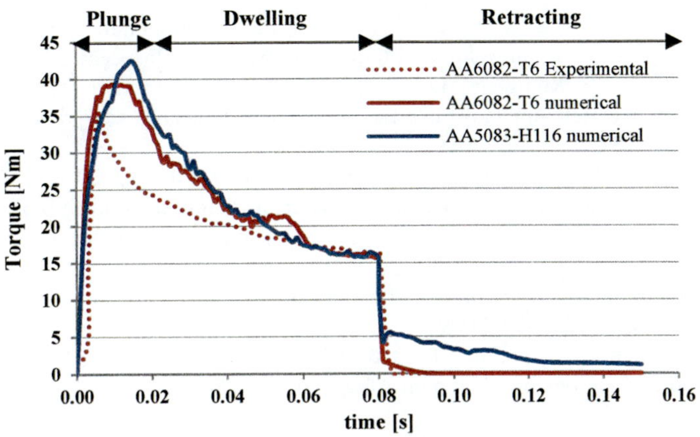

Fig. 4. Torque variation during FSSW process

4 Conclusion

Based on the results presented and discussed above, the main conclusions are cited as following:

- The proposed coupled thermo-mechanical model, using Johnson Cook material law and Coulomb friction law under ABAQUS/Explicit, can estimate the material behaviour during FSSW.
- The proposed FE modeling method comprehended the possibility of widening the effective optimization study on FSSW for a large variety of materials.
- The results obtained by the FE modeling of FSSW would greatly help in getting insight of FSW process since the initial thermo-mechanical conditions appeared in the plunge stage (initial common stage between FSSW and FSW).

Acknowledgments. The research described in this paper was held in the Mechanical Engineering Laboratory LGM of the National School of Engineering of *Monastir, Tunisia* and The Institute for Sustainability and Innovation in Structural Engineering (ISISE), a research unit incorporating Universities of *Coimbra* and *Minho, Portugal*. The technical assistance received from the staff of both laboratories is greatly appreciated. The author *C. Leitão* is supported by the Portuguese Foundation for Science and Technology through SFRH/BPD/93685/2013 fellowship.

References

Haldar, A., Mahadevan, S.: Probability, Reliability and Statistical Methods in Engineering Design. Wiley, New York (2000)

Birsan, D.A.N., Scutelnicu, E., Visan, D.: Behaviour simulation of aluminium alloy 6082-T6 during friction stir welding and tungsten inert gas welding. Recent Adv. Manuf. Eng. 103–108 (2011)

Costa, M.I., Leitao, C., Rodrigues, D.M.: Influence of the aluminium alloy type on defects formation in friction stir lap welding of thin sheets. Soldagem Inspeção **23**(1), 32–42 (2018)

Costa, M.I., Leitão, C., Rodrigues, D.M.: Parametric study of friction stir welding induced distortion in thin aluminium alloy plates: a coupled numerical and experimental analysis. Thin Walled Struct. **134**, 268–276 (2019)

Cox, C.D.: Friction stir spot welding: engineering analysis and design (2014)

John, J., Mohammed Iqbal, U.: Optimization of dissimilar friction stirwelding process parameters of AA5083-H111 and AA6082-T6 by CCD-RSM technique. Adv. Manuf. Process. 49–60 (2019)

Lorrain, O., Zahrouni, H., El Hadrouz, M.: Mechanics of materials and structures. J. Mech. Mater. Struct. **4**(2), 351–369 (2009)

Yalavarthy, H.: Friction Stir Welding Process and Material Microstructure Evolution Modeling in 2000 and 5000 Series of Aluminum Alloy (2009)

Tribological Characterization of Composites Based on Si_3N_4 Ceramics

Amine Charfi[1(✉)], Ismail Dhaou[1], Mohamed Kharrat[1], Mohd Farooq Wani[2], and Maher Dammak[1]

[1] Laboratory of Electromechanical Systems (LASEM), National Engineering School of Sfax, University of Sfax, Sfax, Tunisia
charfi.amin@yahoo.fr, ismaildhaou93@gmail.com, {mohamed.kharrat,maher.dammak}@ipeis.rnu.tn

[2] Tribology Laboratory, Mechanical Engineering Department, NIT Srinagar, Srinagar, India
mfwani@nitsri.net

Abstract. Silicon nitride (Si_3N_4) has excellent mechanical properties even at high temperatures, low inertia, high hardness, very good resistance to thermal shock and excellent friction and wear behavior. For this purpose, silicon nitride is used in various applications such as for welding nozzles, tools for metal transformation, calibration, control, bearing balls and prosthetic components for orthopedic surgery.

This paper presents tribological behavior of two shades of nano-sized silicon nitride ceramics, the first is silicon nitride with 10 wt% of boron nitride (SN-BN) and the second is silicon nitride in β phase (β-SN). Tribological behavior was studied in dry sliding condition, by means of reciprocating ball-on-flat tribometer.

The aim of this work is to study the effect of normal load and the counterface materials on tribological properties of SN-BN and β-SN ceramics. The results have shown that tribological properties of studied materials strongly depends on normal load and counterface material.

Keywords: Silicon nitride · Tribological behavior · Boron nitride · β phasis

1 Introduction

Because silicon nitride has a good thermomechanical property, high thermal conductivity, excellent wear behavior, high hardness and toughness at room and high temperatures, it is used in several applications were the wear and friction phenomenon occur such as in seal rings, rolling balls, knee joints, prosthetic hip… [1–3].

In order to improve the mechanical properties of silicon nitride, we can either introduce addition elements such as boron nitride or transform the phase into a more stable phase, phase, which promotes a better behavior of material. Recent studies showed that addition of boron nitride in silicon nitride makes the cracks course which leads to a reduction of the grain size. The reduction in grain size is accompanied by a reduction in friction and wear coefficients [4]. Fayaz et al. [5] show also that friction

M. Kharrat et al. (Eds.): ICAMEM 2019, LNME, pp. 271–276, 2020.
https://doi.org/10.1007/978-3-030-52071-7_38

coefficient values of SN-BN (5 wt% of boron nitride) increase with load. Dogan and al. showed that volume loss is not directly influenced by the quantity of β phase in silicon nitride [3]. The amount of β phase depends on temperature, heating time, and the type of crucible in which Si_3N_4 particles are heated [6].

In this paper we study the tribological behavior of two shade of silicon nitride ceramic, the first is a mixture of Si_3N_4 with 10 wt% of boron nitride and the second is a silicon nitride in β phase. Tests were performed under dry sliding conditions by means of reciprocating ball-on-flat tribometer for different normal loads. The effect of counterface materials on wear and friction coefficient has also been studied. The results show that friction coefficient increase with normal load for SN-BN and decrease for β-SN. We note also that sliding against SiC ball, allows us to get a lower wear rate than using Si_3N_4 ball.

2 Materials and Experimental Procedure

Tribological tests were carried out with a reciprocating ball-on-flat tribometer as presented in Fig. 1 [7]. Nano-sized silicon nitride with 10 wt% boron nitride and β-Si_3N_4, were fabricated using Spark Plasma Sintering (SPS). Samples were made in cylindrical shape with 15 mm as diameter and 5 mm as thickness. Two counterface materials were used, the first is silicon carbide (SiC) and the second is the silicon nitride (Si_3N_4). Counterface balls have been delivered from Nanoshell company (USA) with a diameter of 10 mm.

During tribological tests, the counterface was kept stationary and the specimen was on a tangential cyclic motion using a crank system driven by an electric motor. The frequency was adjusted by an electronic speed regulator at the beginning of tests and the tangential force gets measured by a load cell placed between the specimen holder and the crank slider system. A data acquisition system stores the output of the load cell. Before each test, the ball and the specimen surfaces get cleaned with ethanol than dried. The test conditions were given in Table 1.

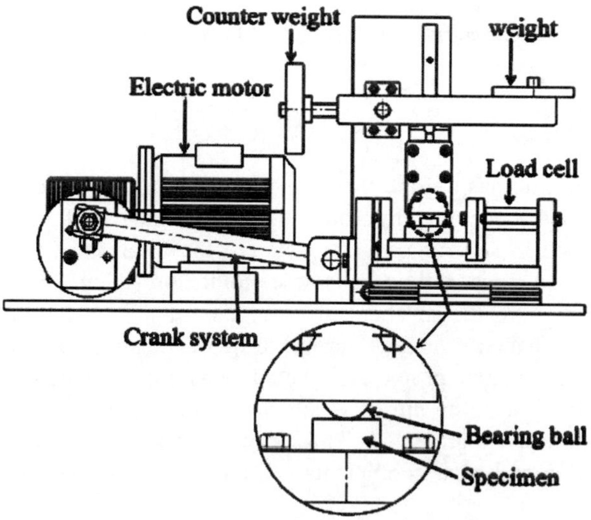

Fig. 1. Reciprocating ball-on-flat tribometer

Table 1. Experimental conditions

Parameters	Experimental conditions
Normal load (N)	22 and 46
Displacement amplitude (mm)	±5
Frequency (hz)	1
Temperature (°C)	20–25
Humidity (%)	50–60
Test duration (min)	120

3 Results

Figures 2-a and 2-b represent respectively the variation of friction coefficient versus cycles number for β-SN and SN-BN materials using SiC ball as counterface. As shown in figures, for both shades and whatever the normal load, friction coefficient experiences an initial decrease to reach a steady state toward the end of the test. According to Fig. 2-a, for β-SN ceramic material, friction coefficient is equal to 0.27 for 22N and 0.25 for 46N. We deduce that friction coefficient values slightly decrease with the increase of load. From Fig. 2-b, we deduce that friction coefficients for SN-BN are equals to 0.4 for 22N as normal load and 0.95 for 46N. In that case, friction coefficient increases with load. This behavior is different to that observed for pure silicon nitride; this is due to the added amount of boron nitride to ceramic material. The same result was found by Fayaz et al. for silicon nitride with 5 wt% boron nitrides [5]. Wei et al. have shown that 10 wt% boron nitrides added to silicon nitride promotes a porous structure of material. These pores lead to stress concentration and hence microcracks are easily produced by friction [4]. This explains the very high friction coefficient for SN-BN for 46N as normal load.

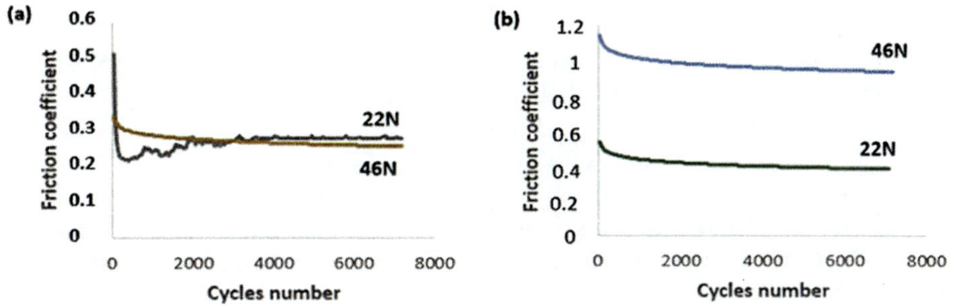

Fig. 2. Evolution of friction coefficient against cycles number for different normal loads (SiC counterface): (a) β-SN; (b) SN-BN

The worn surfaces of β-SN and SN-BN, for different loads, are illustrated respectively in Figs. 3 and 4. From these figures, we note that wear tracks of SN-BN material, for the two tested normal loads, were the smaller and smoother with some micro grooves. The higher amount of debris has been observed in the wear tracks of β-SN material. We note also, for two studied materials, that abrasion marks were parallel to the sliding direction. We conclude that adhesive wear is the main wear mechanism.

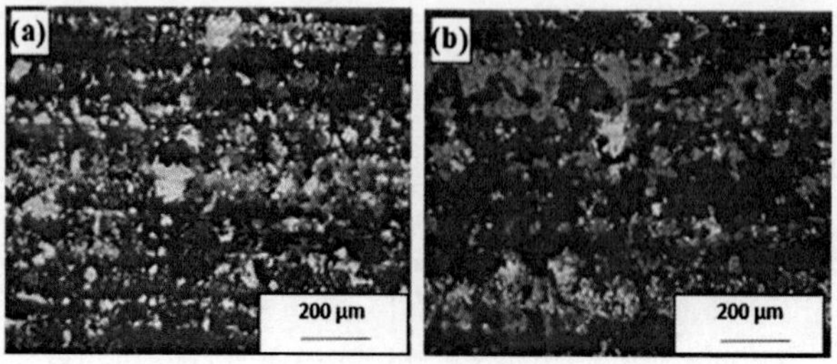

Fig. 3. Optical micrographs of wear track for β-SN for different loads (SiC counterface): (a) 22N; (b) 46N

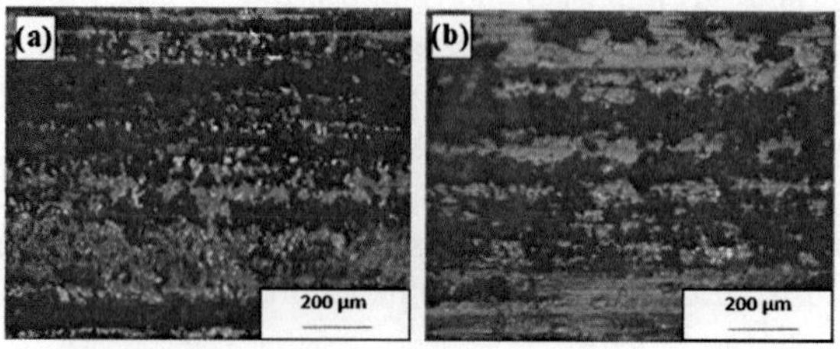

Fig. 4. Optical micrographs of wear track for SN-BN for different loads (SiC counterface): (a) 22N; (b) 46N

Table 2 present the depths, the volume loss and the wear rate (Kw) for SN-BN and β-SN for different loads with SiC ball as counterface.

$$\mathbf{Kw}\left(\mathbf{mm^3/mN}\right) = \mathbf{V/(L.Fn)} \tag{1}$$

From this table and according to volume loss and wear rate values, we conclude that SN-BN is more wear resistant than β-SN. This is due to the effect of Boron nitride that used as a solid lubricant.

Figure 5 shows the influence of counterface materials on friction coefficient values. From this figure, we note that the lowest friction coefficient value, for SN-BN and β-SN material, were obtained with SiC as counterface. This is due to mechanical properties and hardness difference between the two ball materials.

Table 2. Volume loss and wear rates for SN-BN and β-SN materials

	Normal load (N)	Depth (± 0.2 μm)	Volume loss (± 2 10^{-6})	K_W (± 3 10^{-9})
SN-BN	22	2.13	1.49 10^{-5}	10^{-8}
SN-BN	46	3.53	3.53 10^{-5}	1.1 10^{-8}
β-SN	22	5.84	4.1 10^{-5}	2.7 10^{-8}
β-SN	46	5.6	5.84 10^{-5}	1.7 10^{-8}

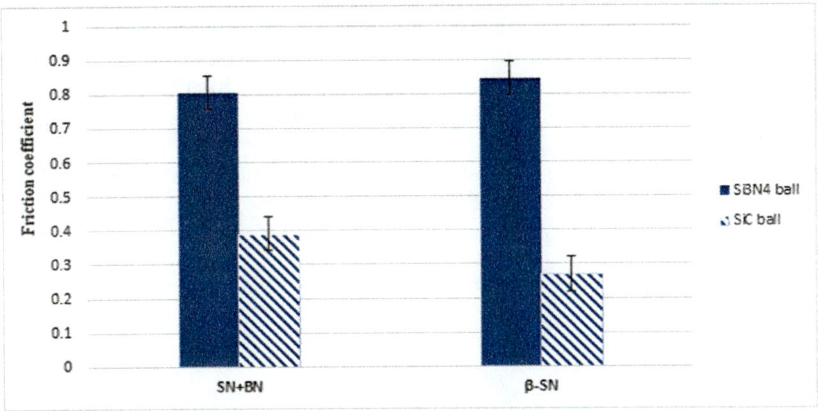

Fig. 5. Influence of counterface on friction coefficient for different studied material with 22N as normal load

4 Conclusion

This research paper is interested to study the tribological behavior of two different ceramic materials, the first is silicon nitride in beta phases, while the second is silicon nitride that added with 10 wt% of boron nitrides. The results allow to conclude the following:

- Boron nitride and β phases have a notable influence on the tribological behavior of silicon nitride.
- Friction coefficient increase with normal load for SN-BN and slightly decrease for β-SN.
- For 46N as normal load, friction coefficient of SN-BN is nigh to 1. This friction coefficient value is unacceptable by the most part of sliding contact.
- For both ceramic materials and whatever the normal load, the friction coefficient is lower with SiC ball as counterface than obtained with Si_3N_4 ball.

Acknowledgment. This work is developed by means of research grant under the Tunisia/India agreement on science and technology cooperation

References

1. Rutkowski, P., Stobierski, L., Zientara, D., Jaworska, L., Klimczyk, P., Urbanik, M.: The influence of the graphene additive on mechanical properties and wear of hot-pressed Si3N4 matrix composites. J. Eur. Ceram. Soc. **35**, 87–94 (2015)
2. Ghalme, S., Mankar, A., Bhalerao, Y.: Integrated Taguchi-simulated annealing (SA) approach for analyzing wear behaviour of silicon nitride. J. Appl. Res. Technol. **15**, 624–632 (2018)
3. Dogan, C.P., Hawk, J.A.: Microstructure and abrasive wear in silicon nitride ceramics. Wear **250**, 256–263 (2001)
4. Wei, D., Meng, Q., Jia, D.: Mechanical and tribological properties of hot-pressed h-BN/Si3N4 ceramic composites. Ceram. Int. **32**, 549–554 (2006)
5. Fayaz, S.D., Wani, M.F.: High temperature tribological characterization of Nano silicon nitride + 5% Boron nitride ceramic composite. IJAET 2093–2101 (2011)
6. Hirata, T.: Synthesis of b-Si3N4 particles from a-Si3N4 particles. J. Eur. Ceram. Soc. **20**, 1191–1195 (2000)
7. Trabelsi, M., Kharrat, M., Dammak, M.: On the friction and wear behaviors of PTFE based composites filled with MoS2 and/or bronze particles. Trans. Indian Inst. Met. **69–5**, 1119–1128 (2015)

The Effect of the Loading Path History on the Fracture Loci

Meriem Nouira[1,2](✉), Marta Oliveira[3], Ali Khalfallah[1,3,4], José Alves[5], and Luís Menezes[3]

[1] Laboratoire de Génie Mécanique, Ecole Nationale d'Ingénieurs de Monastir, University of Monastir, Av. Ibn El Jazzar, 5019 Monastir, Tunisia
nouirameriem93@gmail.com

[2] Ecole Nationale d'Ingénieurs de Sousse, University of Sousse, 4003 Sousse, Tunisia

[3] CEMMPRE, Department of Mechanical Engineering, University of Coimbra, Polo II, Pinhal de Marrocos, 3030-788 Coimbra, Portugal
{marta.oliveira, ali.khalfallah, luis.menezes}@dem.uc.pt

[4] Institut Supérieur des Sciences Appliquées et de Technologie de Sousse, University of Sousse, 4003 Sousse, Tunisia

[5] CMEMS, Department of Mechanical Engineering, University of Minho, Campus de Azurém, 4800-058 Guimarães, Portugal
jlalves@dem.uminho.pt

Abstract. Theoretical analysis and experimental studies have shown that the stress triaxiality η is one of the most important factor that characterizes the ductile fracture of metallic materials. However, there is another variable that should be involved in fracture prediction models, which is Lode angle parameter ξ. In this study, two uncoupled phenomenological models are used for predicting the ductile fracture of a cast A356 aluminium alloy. These models define the equivalent strain to fracture in function of variables that characterize the stress state. The first considers only the stress triaxiality, while the other takes into account the mean pressure and the Lode angle parameter. Two groups of experimental tests were considered: (i) tensile tests on smooth and notched round bars and (ii) tensile plane strain tests on flat grooved plates. The numerical results show that both stress triaxiality and Lode angle parameter evolve during the tests. Therefore, it is necessary to understand the impact of this variation on the calibration of the parameters of uncoupled ductile fracture models. Note that the aim is to be able to predict the fracture under non-proportional, complex loading conditions. In order to assess the effect of the evolution of the stress triaxiality on the prediction of ductile fracture, two strategies are proposed for calibrating a 2D fracture *locus*, for a wide range of positive stress triaxiality, named direct and inverse, since the first is based on the initial value of the stress triaxiality, while the second uses the average. The numerical results show that if ductile fracture occurs before the onset of localized deformation, the stress triaxiality and the Lode angle present quite constant values, although there is a small variation in the transition between the elastic and the plastic regime. However, when ductile fracture occurs after the onset of localized deformation, these variables can present a strong variation, which should be taken into account in the calibration of the 2D fracture *loci*.

M. Kharrat et al. (Eds.): ICAMEM 2019, LNME, pp. 277–290, 2020.
https://doi.org/10.1007/978-3-030-52071-7_39

Keywords: Stress states · Uncoupled models · Fracture *locus* · Loading history

1 Introduction

Several ductile fracture criteria have been proposed in literature, in order to predict the occurrence of this phenomenon for metallic materials. These criteria are based on models that can be classified into two groups: uncoupled and coupled. For coupled models, the influence of damage is included in the constitutive equations. This class of models includes, for instance, the Lemaitre model (see e.g. (Hambli 2001) and the Gurson model (see e.g. (Fesich et al. 2008). In fact, the later one lead to the most popular coupled model, commonly refer as the Gurson, Tvergaard and Needleman model (see e.g. (Rahimidehgolan et al. 2017). However, the calibration of the parameters for coupled models can be a complex task. In this context, many researchers advocate the use of the uncoupled models, since the hardening law and the yield criterion parameters can be identified without taking into account the influence of damage. This class of models includes, for instance, the Johnson-Cook model (see e.g. (Johnson and Cook 1985), the Wilkins model (see e.g. (Wilkins et al. 1980) and the ones proposed by Wierzbicki and his collaborators (see e.g. (Mae et al. 2008); (Xue 2007); (Xue and Wierzbicki 2009)).

It is consensual that, at the microscopic scale, ductile fracture results from void nucleation, followed by growth and coalescence. In this context, several authors assume that ductile fracture is strongly dependent on the stress triaxiality (Johnson and Cook 1985); (Rice and Tracey 1969); (Mae et al. 2008); (Mae et al. 2009), since it influences the void growth rate. In this context, a decrease of the equivalent strain to fracture is commonly observed for increasing positive values of stress triaxiality. However, other authors state that the stress triaxiality is not enough to enable an accurate prediction of fracture. These authors developed uncoupled fracture models that define the equivalent strain to fracture as a function of the stress triaxiality, the Lode angle parameter or the mean pressure. Moreover, these models were used to investigate the sensitivity of the fracture loci to the Lode angle parameter as well as to the mean pressure, for a wide range of stress states (Lou et al. 2012); (Bai and Wierzbicki 2010) and (Xue and Wierzbicki 2008).

In this work, the effect of the loading path history on the fracture *loci* is analyzed for stress states characterized by a positive value of stress triaxiality. In this context, the calibration of the fracture *locus* is performed for positive values of stress triaxiality, using two different strategies. The first one considers the initial value of the stress triaxiality, while the second takes into account its evolution with the plastic deformation. The material selected is a cast A356 aluminium alloy for which experimental results are available in literature, for tensile tests performed on two types of specimens: smooth and notched round bars; and tensile plane strain tests on flat grooved plates, with different radii.

2 Ductile Fracture Criteria

In this work, two uncoupled ductile fracture models are used to predict the equivalent strain to fracture, for a wide range of positive values of stress triaxiality. Before presenting the models, it is important to define the equivalent stress $\bar{\sigma}$ (von Mises), the mean pressure p, and the stress triaxiality η, as follows:

$$\bar{\sigma} = \sigma_{\mathrm{eq}} = \sqrt{\frac{1}{2}}\sqrt{\left[(\sigma_1 - \sigma_2)^2 + (\sigma_2 - \sigma_3)^2 + (\sigma_3 - \sigma_1)^2\right]}, \tag{1}$$

$$p = -\sigma_{\mathrm{m}} = -\frac{1}{3}\mathrm{tr}(\sigma) = -\frac{1}{3}(\sigma_1 + \sigma_2 + \sigma_3), \tag{2}$$

$$\eta = -\frac{p}{\sigma_{\mathrm{eq}}} = \frac{\sigma_{\mathrm{m}}}{\sigma_{\mathrm{eq}}}, \tag{3}$$

where σ_1, σ_2 and σ_3 are the principal components of the Cauchy stress tensor.

In the model proposed by Mae and Wierzbicki, 2008, the equivalent strain to fracture is predicted using the following relations (Mae et al. 2008):

$$\left\{\begin{array}{l} \bar{\varepsilon}_{\mathrm{f}} = \frac{D_1}{1+3\eta}, \quad -\frac{1}{3} \leq \eta \leq 0 \\ \bar{\varepsilon}_{\mathrm{f}} = \bar{\varepsilon}_{\mathrm{f,t}} + \left(\bar{\varepsilon}_{\mathrm{f,t}} - \bar{\varepsilon}_{\mathrm{f,s}}\right)(3\eta - 1), \quad 0 \leq \eta \leq \frac{1}{3} \\ \bar{\varepsilon}_{\mathrm{f}} = D_2 e^{D_3\eta} + D_4, \quad \eta \geq \frac{1}{3} \end{array}\right\}. \tag{4}$$

This model considers that the equivalent strain to fracture is a monotonic function of the stress triaxiality, with a different expression for three branches. It involves a total of six parameters, i.e. D_1, D_2, D_3, D_4, $\bar{\varepsilon}_{\mathrm{f,s}}$ and $\bar{\varepsilon}_{\mathrm{f,t}}$, which need to be identified. Nevertheless, in order to assure the continuity between the three branches defined in Eq. (4), thee six parameters are not independent. In fact, $\bar{\varepsilon}_{\mathrm{f,s}}$ corresponds to the equivalent strain to fracture for the shear test and $\bar{\varepsilon}_{\mathrm{f,t}}$ is the equivalent strain to fracture for the uniaxial tensile test.

In the case of the model proposed by Xue and Wierzbicki, 2008, the ductile fracture envelope assumes the following form (Xue and Wierzbicki 2008)

$$\bar{\varepsilon}_{\mathrm{f}} = \varepsilon_{\mathrm{f}_0} \cdot u_\theta(\theta_l) \cdot u_p(p) \tag{5}$$

where $\varepsilon_{\mathrm{f}_0}$ is the reference fracture strain, corresponding to zero mean stress, and the functions $u_p(p)$ and $u_\theta(\theta_l)$ are introduced todescribe the dependence on the hydrostatic pressure and on the Lode angle, respectively. $u_p(p)$ adopts the following logarithmic form:

$$u_p(p) = \left\{\begin{array}{l} 1 - \mathrm{q}\,\log\left(1 - \frac{p}{\mathrm{p_{lim}}}\right), \quad p \geq \mathrm{p_{lim}}\left(1 - \exp\left(\frac{1}{\mathrm{q}}\right)\right) \\ 0, \quad p \leq \mathrm{p_{lim}}\left(1 - \exp\left(\frac{1}{\mathrm{q}}\right)\right) \end{array}\right\}, \tag{6}$$

where q is the shape parameter of the pressure sensitivity function and p_{lim} is the limit pressure, below which no damage occurs. The second kind of Lode angle function is expressed as follows

$$u_\theta(\theta_l) = \gamma + (1-\gamma)\left(\frac{6|\theta_l|}{\pi}\right)^k, \tag{7}$$

where γ presents the ratio of the Lode angle function and k is the exponent. The azimuthal angle θ_l on an octahedral plane, in the principal stress space, can be defined by

$$\theta_l = -\frac{1}{3}\sin^{-1}\left(\frac{27J_3}{2\bar{\sigma}^3}\right) \tag{8}$$

where J_3 is the third invariant of the deviatoric stress tensor. The Lode parameter is commonly defined as

$$\xi = \frac{27J_3}{2\bar{\sigma}^3}, \tag{9}$$

such that its range corresponds to $\xi \in [-1, 1]$. However, the range for the Lode angle is $\theta_l \in \left[-\frac{\pi}{6}, \frac{\pi}{6}\right]$ ($\theta_l = 0$ (pure shear test); $\theta_l = -\frac{\pi}{6}$ (uniaxial tensile test)). This model assumes that the equivalent strain to fracture depends on the Lode angle parameter θ_l and on the hydrostatic pressure p (see Eq. (6)). In this case, it is also necessary to six calibrate parameter: ε_{f_0}, q, p_{lim}, γ, k and m. The parameter m is a damage exponent that will be detailed in the following section.

A set of experimental tests was carried out to calibrate both model parameters under a wide range of stress states (Mae et al. 2007), leading to the values presented in Tables 1 and 2.

Table 1. Mae and Wierzbicki, 2008 model fracture parameters for A356

D_1	D_2	D_3	D_4
0.2733	0.1417	−1.545	0.0

Table 2. Mae and Wierzbicki, 2008 model fracture parameters for A356

ε_{f_0}	p_{lim}	q	γ	k	m
1.20	800.0 MPa	1.5	0.6	1.0	1.0

When using uncoupled ductile fracture models, the prediction of ductile failure for non-monotonous strain paths can be carried out using a damage parameter D, such that (Xue and Wierzbicki 2008)

$$D = \int_{0}^{\bar{\varepsilon}_p} f\left(\bar{\varepsilon}_{\mathrm{p}}, \bar{\varepsilon}_{\mathrm{f}}\right) d\bar{\varepsilon}_{\mathrm{p}} = 1, \tag{10}$$

where $\bar{\varepsilon}_{\mathrm{p}}$ refers to the equivalent plastic strain and $\bar{\varepsilon}_{\mathrm{f}}$ is the equivalent strain to fracture. When the damage parameter D reaches the critical value ($D = 1$), fracture occurs. In the case of Mae and Wierzbicki, 2008 model (Mae et al. 2008), $f\left(\bar{\varepsilon}_{\mathrm{p}}, \bar{\varepsilon}_{\mathrm{f}}\right)$ is defined by the ratio between the equivalent plastic strain and the equivalent strain to fracture. For Xue and Wierzbicki, 2008 model (Xue and Wierzbicki 2008), the rate of damage accumulation $\dot{D}$ is expressed by the following expression:

$$\dot{D} = \mathrm{m}\left(\frac{\bar{\varepsilon}_{\mathrm{p}}}{\bar{\varepsilon}_{\mathrm{f}}}\right)^{\mathrm{m}-1} \frac{\bar{\varepsilon}_{\mathrm{p}}}{\bar{\varepsilon}_{\mathrm{f}}}. \tag{11}$$

In this work, the damage exponent m is considered to be equal to one, i.e. the damage function is linear as stated also for the Mae and Wierzbicki, 2008 model. In the next section, it will be shown that this assumption will have a strong impact in the equivalent strain to fracture predicted by the Xue and Wierzbicki, 2008 model.

3 Numerical Simulations: Results and Discussion

All the tests were numerically simulated using the in house-code DD3IMP (Menezes 1998); (Oliveira et al. 2008); (Neto et al. 2014) to analyze the evolution of the stress triaxiality with the equivalent plastic strain. The results were obtained for the two uncoupled models, i.e. Mae and Wierzbicki, 2008 (Mae et al. 2008) and Xue and Wierzbicki, 2008(Xue and Wierzbicki 2008), for the cast A356 aluminium alloy. Two groups of classical specimens were used to predict the ductile fracture: smooth and notched round bars, with R = 4 mm and R = 12 mm; and flat grooved plates with different groove radii, namely R = 1.59 mm, R = 3.97 mm and R = 12.7 mm. All flat grooved specimens have the same thickness T = 1.6 mm. Figure 1 presents the geometries and the dimensions of the considered specimens.

The CAD files of the specimen geometries were imported to GID pre-processor, which was used to generate the corresponding three-dimensional finite element meshes. The models use hexaedral linear solid finite elements (eight node) combined with a selective reduced integration (SRI) technique. The work-hardening behaviour is described by the Swift law:

$$Y = C(\varepsilon_0 + \bar{\varepsilon})^{\mathrm{n}}, \tag{12}$$

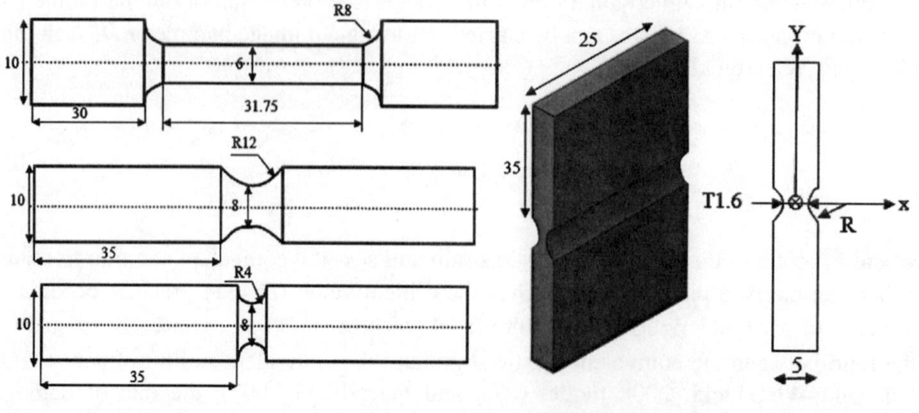

Fig. 1. Geometries and dimensions of smooth and notched round bars specimens (left) and flat-grooved plates (right)

where Y is the flow stress and C, ε_0 and n are material parameters, such that $Y_0 = C\varepsilon_0^n$ defines the material yield stress for the simple tensile test. Table 3 presents the material parameters used to describe the elastoplastic behaviour of the material (Mae et al. 2007).

Table 3. Material parameters for a cast A356 aluminium alloy

Isotropic work-hardening law			Isotropic elastic parameters	
Y_0 (MPa)	C(MPa)	n	υ	E(GPa)
200.167	556.06	0.2010	0.33	86

The evolution of the stress triaxiality and the Lode angle parameter during the test were analysed, for the Gauss point (GP, or integration point) for which damage is predicted to occur first. Figures 2 and 3 present the evolutions of these variables, in function of the equivalent plastic strain, as predicted by the Mae and Wierzbicki, 2008 model and the Xue and Wierzbicki, 2008, respectively. In these figures, it can be seen that both models predict very different values for the equivalent strain to fracture, i.e. Xue and Wierzbicki, 2008 model envisages values approximately 10 times higher than that obtained by Mae and Wierzbicki, 2008 model. This is related to the assumption of a linear accumulation of damage for the Xue and Wierzbicki, 2008 (see Eq. (10)). Nevertheless, this allows to visualize the evolution of the stress triaxiality and the Lode angle parameter for small values of equivalent plastic strain (Fig. 2) and larger ones (Fig. 3). The fact that the evolution may different for both models, even for small values of equivalent plastic strain, results from the fact that the analysis is performed for the GP for which fracture is predicted to occur first. Since the fracture location may not be the same for both models, the evolutions can be different.

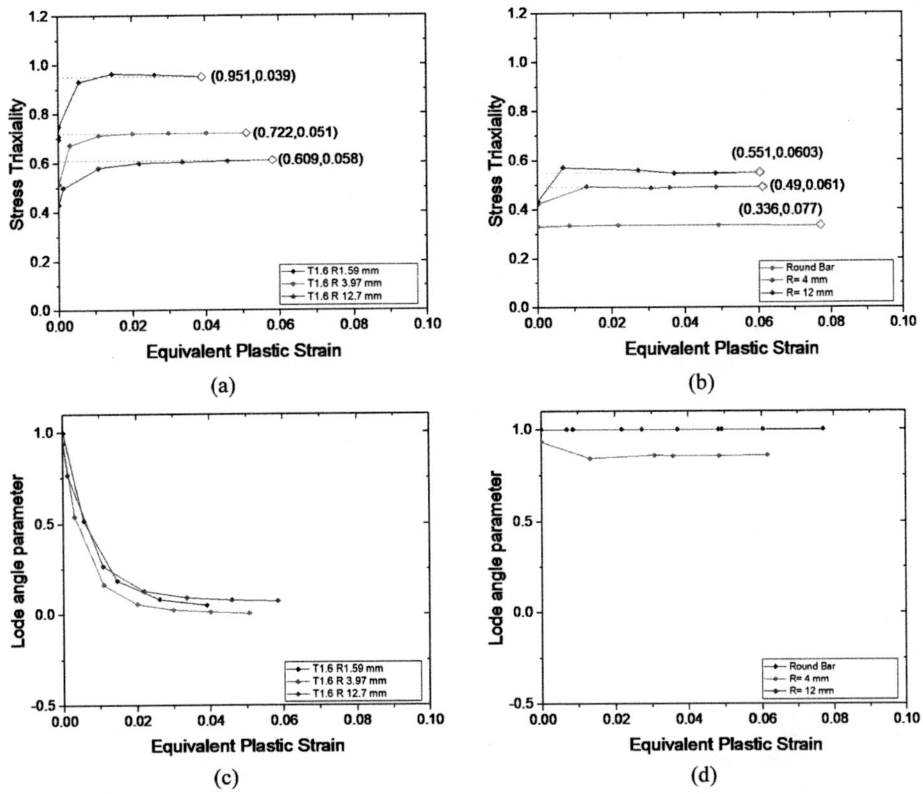

Fig. 2. Evolution of the stress triaxialityand the Lode angle parameter during the test, considering the equivalent strain to fracture predicted by Mae and Wierzbicki, 2008 model: (a), (b) Flat grooved plates and (c), (d) Smooth and notched round bars

For the smooth and notched round bars, both variables show a slightly variation at the beginning of test, but when the equivalent plastic strain is higher than 0.04, the stress state seems to remain almost constant (see Fig. 2). Nevertheless, for large strains a wider variation is obtained, particularly for the smooth round bar. For the flat grooved plates, the stress triaxiality presents relatively constant values during the tests. However, the Lode angle parameter shows a quite large variation from the beginning ($\xi = 1$) to the end of tests ($\xi = 0$), even for values of equivalent plastic strain lower than 0.08 (see Fig. 2). Thus, it can be concluded that, it is very difficult to define a specimen geometry that undergoes a constant loading path until fracture, i.e. the stress state is not monotonous.

The parameters of ductile fracture models are commonly evaluated based on numerical simulation results, to enable the proper evaluation of the variables that characterize the stress state. Therefore, the next step in the current work is to evaluate the impact of taking, or not, the evolution of these variables in the calibration procedure of the 2D fracture *locus,* in the space of equivalent strain to fracture and stress triaxiality. Nevertheless, since the fracture initiation is controlled by the stress state,

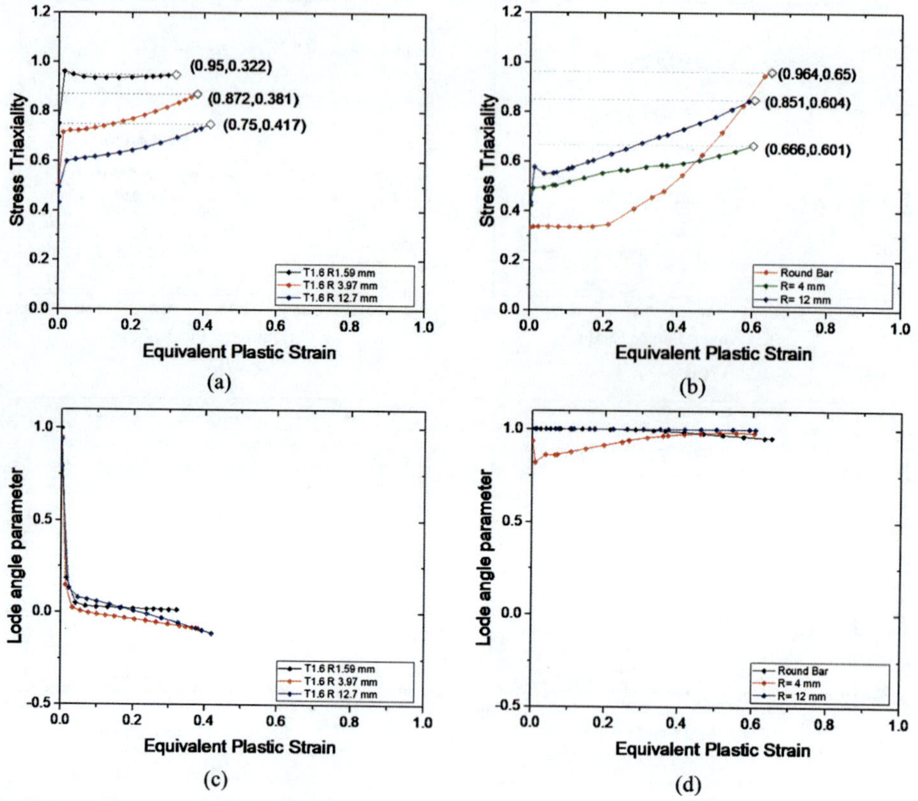

Fig. 3. Evolution of the stress triaxiality and the Lode angle parameter during the test, considering the equivalent strain to fracture predicted by Xue and Wierzbicki, 2008 model: (a), (b) Flat grooved plates and (c), (d) Smooth and notched round bars

the evolutions of both parameters η and ξ were analysed. Two calibration strategies were used, which are summarized as follows:

- The first strategy considers only the initial value of the stress triaxiality. Therefore, this method is referred as a Direct method. In fact, several authors indicate that it is preferable to select tests for which the stress triaxiality remains constant, such that the initial and final values are similar (e.g. Hosford–Coulomb model in (Papasidero et al. 2015), or the McClintock model in (Bai and Wierzbicki 2015).
- The second strategy is based on the average value of the stress triaxiality, defined by the sum of the values of the stress triaxiality, from the first step until the failure one, divided by the number of steps, such as

$$\eta_{\text{average}} = \frac{1}{n}\sum_{i=1}^{n} \eta_i. \tag{13}$$

Since this method requires a first analysis of the numerical test, it is referred as Inverse method. Note that the same procedure can also be applied to the Lode angle parameter, but was not considered in this work.

Table 4 summarizes the results collected from the numerical simulations performed with both models, i.e. Mae and Wierzbicki, 2008 (Mae et al. 2008) and Xue and Wierzbicki, 2008 (Xue and Wierzbicki 2008). The analysis of this table shows that there can be considerable differences between the initial and average values for both ductile fracture models. However, as previously mentioned, the range of equivalent plastic strain to fracture attained by both models is completely different. Thus, although the initial values of stress triaxiality and Lode angle parameter are similar for both models, there are high differences in the average value. This can be observed especially for the Lode angle parameter, and in particular for the flat grooved plates. According to some previous studies, the Lode angle parameter cannot be analysed at the beginning of the test, close to the transition between the elastic and the plastic regime, because it relates the damage plasticity model with ductile fracture (Bai et al. 2009); (Gao et al. 2009); (Zhuang et al. 2016). Thus, many researchers assume that, for smooth and notched round bars, the Lode angle parameter presents a value close to one, while it has a value close to zero for flat grooved plate specimens.

Table 4. Summary of the numerical values of the equivalent plastic strain to fracture and the initial and average values of stress triaxiality and Lode angle parameter for both models

Models	Tests	Initial η	Initial ξ	Average η	Average ξ	$\bar{\varepsilon}_f^p$
Mae and Wierzbicki, 2008	*Smooth and Notched round bars*					
	Smooth bar	0.331	0.999	0.335	0.999	0.077
	R = 4 mm	0.622	0.935	0.478	0.869	0.061
	R = 12 mm	0.434	0.999	0.535	0.999	0.060
	Flat grooved plates					
	T1.6 R1.59 mm	0.696	0.999	0.817	0.648	0.039
	T1.6 R3.97 mm	0.515	0.948	0.633	0.461	0.051
	T1.6 R12.7 mm	0.430	0.940	0.544	0.412	0.058
Xue and Wierzbicki, 2008	*Smooth and Notched round bars*					
	Smooth bar	0.331	0.999	0.500	0.988	0.650
	R = 4 mm	0.422	0.935	0.550	0.927	0.601
	R = 12 mm	0.434	0.999	0.657	0.999	0.604
	Flat grooved plates					
	T1.6 R1.59 mm	0.696	0.999	0.899	0.231	0.322
	T1.6 R3.97 mm	0.515	0.948	0749	0.083	0.381
	T1.6 R12.7 mm	0.430	0.940	0.632	0.119	0.417

The equivalent plastic strain at fracture values numerically predicted by either model (see Table 4) were fitted by an exponential function of the stress triaxiality, as follows.

$$\bar{\varepsilon}_{\mathrm{f}}^{\mathrm{p}} = D_1 e^{D_2 \eta}. \tag{14}$$

Note that the values calibrated for parameters D_1 and D_2 depend on the value considered for the stress triaxiality, η. These parameters were identified by minimizing an objective function, $f_{\mathrm{Objective}}$, defined as follows

$$\underset{(D_1,D_2)}{Min}\left(f_{\mathrm{Objective}}\right) = \underset{(D_1,D_2)}{Min} \sum_{i=1}^{\mathrm{N}} \left|\bar{\varepsilon}_{\mathrm{f},i\,\mathrm{Num}} - \bar{\varepsilon}_{\mathrm{f},i\,\mathrm{Cal}}\right|, \tag{15}$$

where $\bar{\varepsilon}_{\mathrm{f},i\,\mathrm{Num}}$ is the equivalent plastic strain to fracture obtained by the numerical simulation and $\bar{\varepsilon}_{\mathrm{f},i\,\mathrm{Cal}}$ is the one calculated by the exponential function defined in Eq. 15. N is the total number of tests that in this case presents a value of six, for each model.

Table 5 presents the values identified for D_1 and D_2 for all the cases. A comparison between the predicted 2D fracture *locus* obtained using the inverse and direct methods is shown in Fig. 4, for both models. As indicated in Table 5, for the Xue and Wierzbicki, 2008 the calibration was performed assuming constant values of the Lode angle parameter of 1.0 (valid for smooth and notched round bars) and 0.0 (for flat grooved plate specimens). Therefore, Fig. 4 (b) shows a set of results labelled with 0 that correspond to $\xi = 0$, while the label 1 is added for $\xi = 1$.

Table 5. Calibrated parameters

Model		Direct Method		Inverse Method	
		D_1	D_2	D_1	D_2
Mae and Wierzbicki, 2008	–	0.132	–1.79	0.124	–1.402
Xue and Wierzbicki, 2008	$\xi = 0$	0.632	–0.97	0.783	–0.982
	$\xi = 1$	0.836	–0.76	0.785	–0.423

Figure 4 also presents the equivalent plastic strain to fracture values (open black circles) predicted by the Mae and Wierzbicki, 2008 and the Xue and Wierzbicki, 2008 models, which were the ones used as input for calibrating Eq. (14). For the Mae and Wierzbicki, 2008, these values are coincident with the model predictions for a constant value of η (continuous black line in Fig. 4 (a), labelled as "Input DD3IMP"). The figure also shows that this model prediction is very similar to the fitting obtained using the average value of the η, while there is a slight deviation when only the initial value is considered. This confirms that the Mae and Wierzbicki, 2008 model parameters were calibrated taking into account the average value of η. Moreover, the figure highlights that η remains almost constant during the test, after a small change in the transition between the elastic and the elastoplastic regimes. Thus, the use of the initial value of η does not really reflect the majority of the test conditions. Finally, it should be mentioned that the equivalent plastic strain to fracture values shown in this figure correspond to the experimental ones for the A356 alloy (Mae et al. 2008).

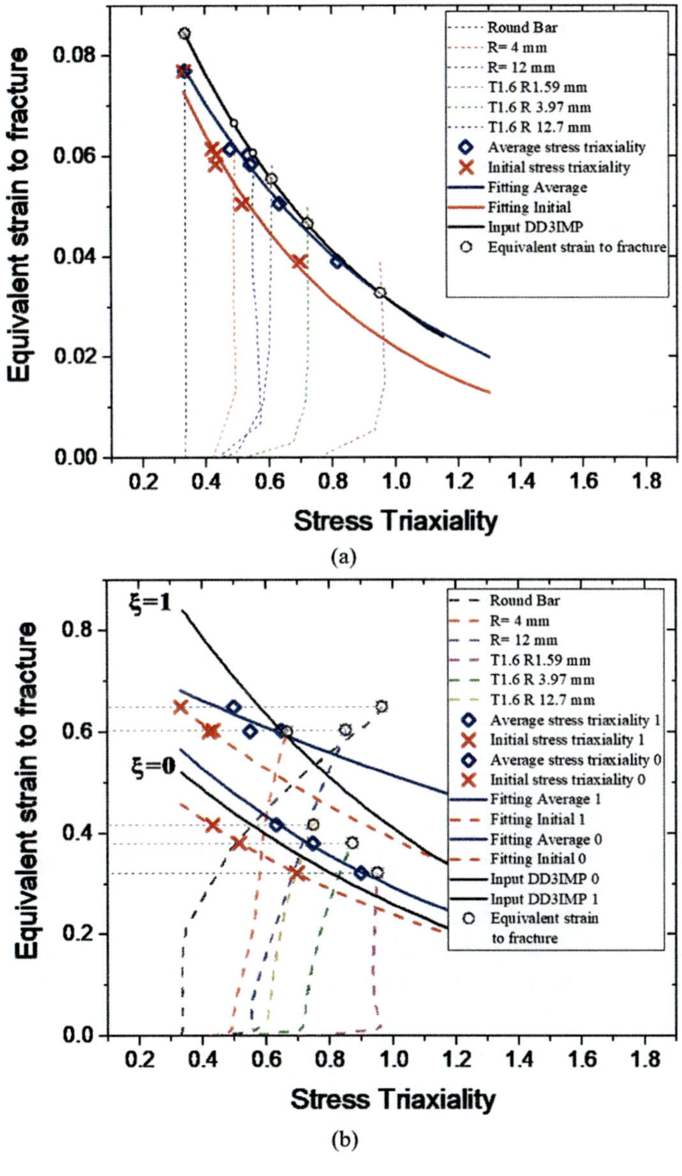

Fig. 4. Comparison of the fracture locus calibrated using the two proposed strategies for both models: (a) Mae and Wierzbicki, 2008 and (b) Xue and Wierzbicki, 2008 (the series labelled 0 correspond to while label 1 is for).

As previously mentioned, the assumptions adopted in this work for the Xue and Wierzbicki, 2008 clearly lead to an overestimation of the equivalent plastic strain to fracture. However, in the context of this work, it allows the analysis of test conditions that involve a strong change of the stress state with the increase of the equivalent plastic strain. The strong change of the stress state during the test justifies the fact that the equivalent plastic strain to fracture values (open black circles) predicted by the Xue and Wierzbicki, 2008 models, are no longer coincident with the model predictions for a constant value of η (black lines in Fig. 4 (b), labelled as "Input DD3IMP"). It also explains why the fracture *loci* constructed using the direct and inverse strategies are so different. In this context, it should be mentioned that the impact is higher for $\xi = 1$, i.e. smooth and notched round bars, than for $\xi = 0$. This seems to result from the fact that the selected tests cover a smaller range of η values for $\xi = 1$ than for $\xi = 0$.

The importance of the stress triaxiality value considered in the calibration procedure has been previously reported in literature by other authors. The effect of the loading paths used to calibrate the fracture *locus* was studied for a structural steel (Weldox 460 E) using notched axisymmetric specimens by (Hopperstad et al. 2003). Further studies, on 1045 and DH36 steels have also shown that the use of the initial and average values of stress triaxiality has an impact on the fracture *locus* (Bai et al. 2009). Figure 4 confirms the effect of the assumptions assumed to characterize the loading path to fracture on the fracture *loci*.

4 Conclusion

In this work, two strategies are used to calibrate the fracture *locus* in order to evaluate the impact of the stress triaxiality value considered in the procedure on the equivalent plastic strain to fracture. The equivalent plastic strain to fracture values consider where obtained using two uncoupled fracture models, i.e. Mae and Wierzbciki, 2008 (Mae et al. 2008) and Xue and Wierzbciki, 2008. The loading path to fracture is defined by the evolution of the equivalent plastic strain to fracture as a function of the stress triaxiality and the Lode angle parameter. Two series of tensile tests on flat grooved plates and round bars specimens were used to construct the fracture *locus* for a wide range of positive values of the stress triaxiality. The numerical results show that it can be difficult to define a specimen geometry that guarantees a monotonic evolution of the stress state, i.e. that the stress triaxiality and the Lode angle parameter remain constant duri the test, up to fracture. Therefore, the calibration of the 2D fracture *loci* needs to take into account this evolution in order to enable a proper prediction of ductile fracture.

References

Bai, Y., Teng, X., Wierzbicki, T.: On the application of stress triaxiality formula for plane strain fracture testing. J. Eng. Mater. Technol. **131**(2), 021002 (2009). https://doi.org/10.1115/1.3078390

Bai, Y., Wierzbicki, T.: A comparative study of three groups of ductile fracture loci in the 3D space. Eng. Fract. Mech. **135**, 147–167 (2015). https://doi.org/10.1016/J.Engfracmech.2014.12.023

Barros, P.D., Neto, D.M., Alves, J.L., Oliveira, M.C., Menezes, L.F.: DD3IMP, 3D fully implicit finite element solver: implementation of CB2001 yield criterion. Rom. J. Tech. Sci. Appl. Mech. **60**(1–2), 105–136 (2015)

Neto, D.M., Oliveira, M.C., Menezes, L.F., Alves, J.L.: Applying Nagata patches to smooth discretized surfaces used in 3D frictional contact problems. Comput. Meth. Appl. Mech. Eng. **271**, 296–320 (2014). https://doi.org/10.1016/J.Cma.2013.12.008

Fesich, T., Mohan, P., Marzougui, D., Kan, C.: A study of the Gurson damage model and numerical simulation of ductile failure in LS-DYNA. 7. LS-DYNA Anwenderforum, pp. 1–12 (2008). http://www.Dynamore.Se/En/Resources/Papers/Forum08/Dokumente/B-III-01.Pdf

Gao, X., Zhang, T., Hayden, M., Roe, C.: Effects of the stress state on plasticity and ductile failure of an aluminum 5083 alloy. Int. J. Plast **25**(12), 2366–2382 (2009). https://doi.org/10.1016/J.Ijplas.2009.03.006

Hambli, R.: Comparison between Lemaitre and Gurson damage models in crack growth simulation during blanking process. Int. J. Mech. Sci. **43**(12), 2769–2790 (2001). https://doi.org/10.1016/S0020-7403(01)00070-4

Hopperstad, O.S., Børvik, T., Langseth, M., Labibes, K., Albertini, C.: On the influence of stress triaxiality and strain rate on the behaviour of a structural steel. Part I. Experiments. Eur. J. Mech. A/Solids **22**(1), 1–13 (2003). https://doi.org/10.1016/S0997-7538(02)00006-2

Johnson, G.R., Cook, W.H.: Fracture characteristics of three metals subjected to various strains, strain rates, temperatures and pressures. Eng. Fract. Mech. **21**(1), 31–48 (1985). https://doi.org/10.1016/0013-7944(85)90052-9

Lou, Y., Huh, H., Lim, S., Pack, K.: New ductile fracture criterion for prediction of fracture forming limit diagrams of sheet metals. Int. J. Solids Struct. **49**(25), 3605–3615 (2012). https://doi.org/10.1016/J.Ijsolstr.2012.02.016

Oliveira, M.C., Alves, J.L., Menezes, L.F.: Algorithms and strategies for treatment of large deformation frictional contact in the numerical simulation of deep drawing process, vol. 11831 (2008). https://doi.org/10.1007/S11831-008-9018-X

Mae, H., Teng, X., Bai, Y., Wierzbicki, T.: Calibration of ductile fracture properties of a cast aluminum alloy. Mater. Sci. Eng., A **459**(1–2), 156–166 (2007). https://doi.org/10.1016/J.Msea.2007.01.047

Mae, H., Teng, X., Bai, Y., Wierzbicki, T.: Comparison of ductile fracture properties of aluminum castings: sand mold vs. metal mold. Int. J. Solids Struct. **45**(5), 1430–1444 (2008). https://doi.org/10.1016/J.Ijsolstr.2007.10.016

Mae, H., Teng, X., Bai, Y., Wierzbicki, T.: Ductile fracture locus of AC4CH-T6 cast aluminium alloy. Comput. Mater. Sci. Surf. Eng. **2**(2), 100–105 (2009)

Menezes, L.F.: Three-Dimensional Numerical Simulation of the Deep-Drawing Process Three-Dimensional Numerical Simulation of the Deep-Drawing Process Using Solid ® Nite Elements, **0136**(July), 99–106 (1998). https://doi.org/10.1016/S0924-0136(99)00345-3

Papasidero, J., Doquet, V., Mohr, D.: Ductile fracture of aluminum 2024-T351 under proportional and non-proportional multi-axial loading: Bao-Wierzbicki results revisited. Int. J. Solids Struct. **69–70**, 459–474 (2015). https://doi.org/10.1016/J.Ijsolstr.2015.05.006

Rahimidehgolan, F., Majzoobi, G., Alinejad, F., Sola, J.: Determination of the constants of GTN damage model using experiment, polynomial regression and Kriging methods. Appl. Sci. **7** (11), 1179 (2017). https://doi.org/10.3390/App7111179

Rice, J.R., Tracey, D.M.: On the ductile enlargement of voids in triaxial stress fields. J. Mech. Phys. Solids **17**(3), 201–217 (1969). https://doi.org/10.1016/0022-5096(69)90033-7

Wilkins, M.L., Streit, R.D., Reaugh, J.E.: Cumulative-Strain-Damage Model of Ductile Fracture: Simulation and Prediction of Engineering Fracture Tests. Livermore, CA (1980). https://doi.org/10.2172/6628920

Xue, L.: Damage accumulation and fracture initiation in uncracked ductile solids subject to triaxial loading. Int. J. Solids Struct. **44**(16), 5163–5181 (2007). https://doi.org/10.1016/J.Ijsolstr.2006.12.026

Xue, L., Wierzbicki, T.: Ductile fracture initiation and propagation modeling using damage plasticity theory. Eng. Fract. Mech. **75**(11), 3276–3293 (2008). https://doi.org/10.1016/J.Engfracmech.2007.08.012

Xue, L., Wierzbicki, T.: Aluminum alloy 2024-T351 using damage plasticity theory, **1**(2), 267–304 (2009)

Zhuang, X., Wang, T., Zhu, X., Zhao, Z.: Calibration and application of ductile fracture criterion under non-proportional loading condition. Eng. Fract. Mech. **165**(October), 39–56 (2016). https://doi.org/10.1016/J.Engfracmech.2016.08.021

Characterization of CrAlN Films Synthesized by DC Reactive Magnetron Sputtering

K. Aouadi[1(✉)], C. Nouveau[2], A. Besnard[2], B. Tlili[3], A. Montagne[4], and M. Chafra[1]

[1] Applied Mechanics and Systems Research Laboratory, Tunisia Polytechnic School, University of Carthage, BP 743, 2078 La Marsa, Tunisia
aouadikhalil@hotmail.com, chafra_moez@yahoo.fr
[2] Arts et Metiers Institute of Technology, LaBoMaP, Rue Porte de Paris, 71250 Cluny, France
{corinne.NOUVEAU, aurelien.BESNARD}@ensam.eu
[3] El-Manar Ecole Nationale d'Ingénieurs de Tunis, LR-11-ES19, Laboratoire de Mécanique Appliquée et Ingénierie (LR-MAI), Université de Tunis, Tunis 1002, Tunisia
tlilii_brahim@yahoo.fr
[4] Arts et Metiers Institute of Technology, MSMP, 8, Boulevard Louis XIV, 59046 Lille Cedex, France

Abstract. CrAlN films were synthesized by DC reactive magnetron sputtering. The influence of the aluminium content on the microstructure, hardness and the friction coefficient of the CrAlN films have been investigated. Results indicate that all CrAlN coatings present a columnar and dense structure. Also, CrAlN coatings present a better hardness than CrN coating. The increase of Al rate from 3.5 to 11% is accompanied by a decrease of hardness from 28 to 24 GPa and then it increases for an Al rate of 24%. The addition of Al to the CrN system causes an increase of the friction coefficient and the wear resistance.

Keywords: CrAlN coatings · Hardness · Friction coefficient · Wear

1 Introduction

For the past few years, chromium nitride coatings are widely used in various applications and especially for the protection of cutting tools. This is due to their good wear resistance, high hardness, and good oxidation and corrosion resistance (Kong et al. 2011; Jianxin et al. 2012; Shan et al. 2014). However, the incorporation of Al in the cubic lattice of CrN system improves their properties and especially enhances its hardness. Indeed; Wang and Nie (Wang and Nie 2014) studied the effect of annealing temperature on CrAlN and CrN coatings properties. They confirm that CrAlN coating exhibits a better thermal stability than that of CrN. Indeed, at 900 °C, CrAlN film presents a strong XRD intensity and it exhibits a higher hardness than that of CrN. The increase of the Al content can lead to the change of the structure from cubic to hexagonal (Reiter et al. 2005). Also, Nouveau et al. (Nouveau et al. 2012) deposited CrN and CrAlN films by DC magnetron sputtering. They varied Al rate and revealed

M. Kharrat et al. (Eds.): ICAMEM 2019, LNME, pp. 291–298, 2020.
https://doi.org/10.1007/978-3-030-52071-7_40

that CrAlN coating hardness reaches a value of 35 GPa however the CrN coating hardness is about 18 GPa. Likewise, CrAlN films are known by their good oxidation resistance above 1000 °C which is due to the formation of a dense oxide layer of (Al, $Cr)_2O_3$ (Lin et al. 2008). Additionally, the incorporation of Al in CrN system improves it corrosion resistance. Ding et al. (Ding et al. 2008) compared the corrosion resistance of CrAlN and TiAlN films. They confirm that both coatings exhibit a good corrosion resistance while the CrAlN showed a better one. To study CrAlN coatings and to improve their properties many parameters were investigated. Lv et al. (Lv et al. 2012) deposited CrAlN coating by mid-frequency unbalanced magnetron sputtering system. They studied the effect of the duty cycle on the properties of CrAlN films. They confirm that coatings present cubic crystalline structure.

In the present work, a series of CrAlN coatings were deposited on stainless steel and silicon substrate using a DC reactive magnetron sputtering technique. The Al content is investigated. The microstructure, hardness and tribological properties of the CrAlN coatings were studied.

2 Experimental Procedures

CrAlN films were synthesized on silicon wafer and stainless steel (90CrMoV8) samples (20×20 mm^2 and 5 mm thick) using a Dc reactive magnetron sputtering technique. The deposition time was fixed to 4 h. The pressure during deposition was 0.5 Pa and the substrate temperature was held at 300 °C. The substrate bias voltage was held at -500 V. The flow rates of Ar and N_2 were 68.8 and 33.3 sccm respectively. To deposit CrAlN coatings a chromium and aluminum targets with 99.95% was used. All this parameters are optimized in a previous work (Aouadi et al. 2019). To study the influence of Al content on the properties of CrAlN films, different power were applied to the Al target (from 500 to 2800 W). Table 1 summarizes the CrAlN deposition conditions.

Table 1. CrAlN deposition conditions

Deposition parameter	Nominal value
Substrate bias voltage (-V)	500
Temperature during deposition	300
Argon flow rate (sccm)	68.8
Nitrogen flow rate (sccm)	33.3
Power of the Cr target (w)	1500
Power of the Al target (W)	500, 1000, 1500, 2200, 2800
Holder rotation speed (rpm)	3
Time of deposition (H)	4

The cross-section and film thicknes of obtained films was characterized by SEM field emission (JEOL JSM 7610F). The rotative ball-on-disk test is used to determine the friction coefficient: The sliding speed was 3 cm/s, the sliding distance is about 100 m and the applied load was 5 N. An alumina ball (Al_2O_3) with a diameter of 6 mm is used as counterpart. After friction sliding, wear volume was calculated by profilometre profiles. The coating hardness and the Young's modulus were measured by nano-indentation test using MTS XP Nano indenter. To determine hardness and Young's modulus, Rahmoun's model is used (Rahmoun et al. 2009). The adhesion tests were realized by means of a micro-scratch tester (Scratch Tester Millennium 200). The indenter is a Rockwell spherical diamond and equipped with an acoustic emission detector. For each CrAlN film, three tests were carried out.

3 Results and Discussion

3.1 Microstructure and Morphology

Figure 1 shows the SEM cross section observation of CrAlN films deposited under different Al rates.

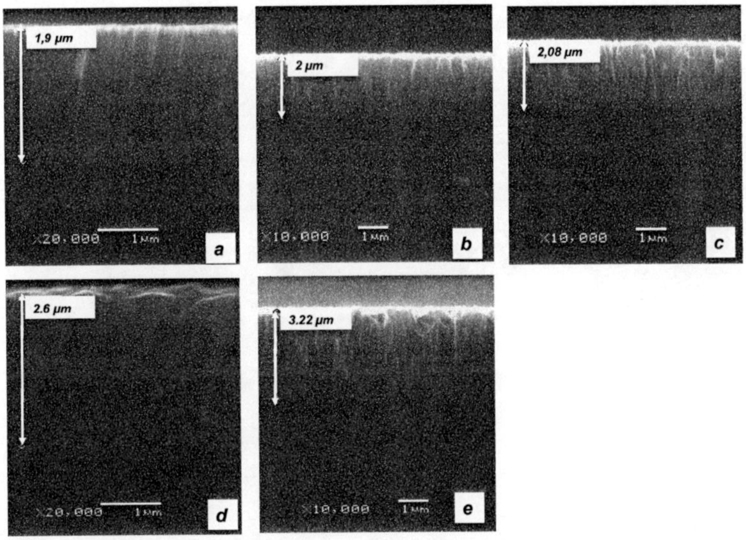

Fig. 1. SEM images of cross-section of coatings with different content of Al: (a) 3.5% (b) 7% (c) 11% (d) 19% (e) 24%

All CrAlN coatings exhibit a dense and a strong columnar structure. The columns are perpendicular to the substrate surface. The SEM images show that the size of the columns increases with the increase of the Al rate. This is ascribed to the increase of the grain size which increases with the increase of the Al rate. Likewise, it observed that the interface between the substrate and the coating doesn't represent any defects. This plays a crucial role to enhance the adhesion of films and consequently the improvement of their mechanical properties, resistance to wear and friction. Also, it's well seen that the film thickness increases by increasing Al content to 19 and 24% at. This is explained by the increase in the deposition rate for these two films.

3.2 Hardness

The hardness and Young's modulus of CrAlN films as a function of Al content are presented in Fig. 2.

From this figure, it can be seen that the addition of Al rate to the Cr-N binary system improves its hardness and Young modulus. Indeed, all CrAlN coatings exhibit a higher hardness and Young's modulus than that of CrN and it varies respectively between 22 and 30 GPa and 244 and 347 GPa. Similar values have been obtained in other works (Barshilia et al. 2006). The addition of 3.5% at. of Al permits to increase the hardness from 22 to 28 GPa. This improvement in hardness can be attributed to several factors such as the decrease of the grain size, the small inter-atomic distance of CrAlN films and the solution hardening (Barshilia et al. 2006). By increasing the Al rate up to 11 at.%, the hardness and the Young modulus slightly decrease and reach respectively 24 and 287 GPa. This can be explained by the relaxation of residual stresses. When the Al rates increase to 19 and 24 at.%, hardness and Young's modulus increase and are maximum, they reach respectively 30 and 347 GPa for 24% at. of Al. This is due to the formation of a solid solution that results from the substitution of Cr atoms by Al.

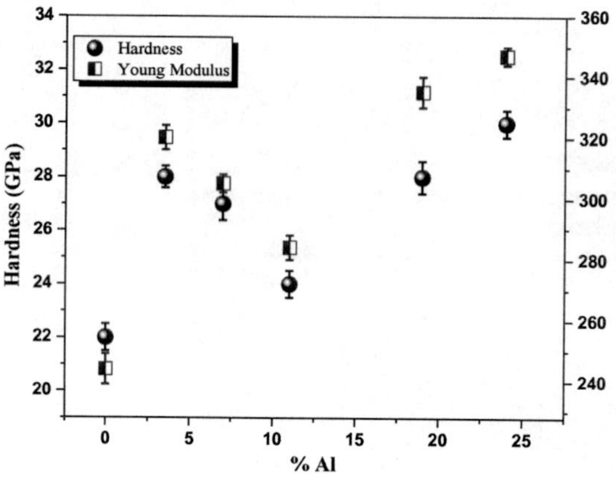

Fig. 2. Hardness and Young modulus of CrAlN films under different Al content

3.3 Friction Coefficient

The variation of the friction coefficient versus the sliding distance for the CrAlN films with different Al rates is shown in Fig. 3.

The addition of Al to the binary system (Cr-N) increases its friction coefficient. Indeed, the friction coefficient of CrN is about 0.68 and remains lower than that of CrAlN films. These results coincide with the work of Lin et al. (Lin et al. 2006). Ding and Zeng (Ding and Zeng 2005) explained this by the effect of the high roughness of CrAlN compared to that of CrN. Also, this increase in the friction coefficient is due to the AlN phase that exists in the CrAlN film. Indeed, the AlN phase has a coefficient of

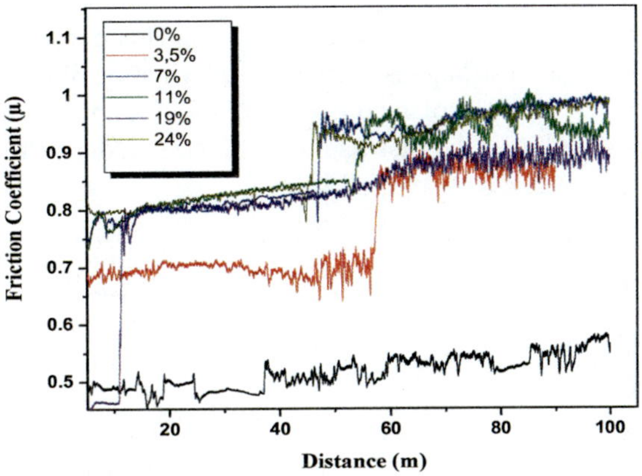

Fig. 3. Friction coefficient of CrAlN coatings

friction higher than that of CrN (Bobzin et al. 2007). According to Lin et al. (Lin et al. 2006), the increase of the friction coefficient of CrAlN films compared to that of CrN is ascribed to the increase of the oxidation resistance. They consider that the oxide layer formed in the surface of CrN coating acts as a lubricant because of its low shear strength, resulting in a low coefficient of friction. The friction coefficient of CrAlN films varied between 0.7 and 0.8. Similar values were obtained in other works (Wang et al. 2010) (Aihua et al. 2012). The addition of a small amount of Al of 3.5% at. to the binary system Cr-N causes a slight increase in the friction coefficient up to 0.7. This can be explained by the low presence of the AlN phase in this film. By adding a quantity of Al between 7 and 24% at., the friction coefficient is almost stable (about 0.8) and the increase in the Al content hasn't any effect. Similar results were found by Wang et al. (Wang et al. 2010). At the end of the test, all coatings exhibit an increase of friction coefficient and it reaches that of the substrate. Thus, the CrAlN layer is delaminated. It can be observed that CrAlN coating with 3.5% at. of Al has the best wear resistance and that with 24% at. possesses the worst one.

3.4 Wear Volume and Critical Load

The adhesion of thin films to cutting tools is an important parameter that affects their lifetime. Thus, it's imperative to perform a scratch-test to determine the adhesion of CrAlN films.

Figure 4 shows the evolution of Lc_2 as a function of Al rate in the layer. Here, the adhesion is represented by the resistance to spallation (Lc_2). The critical load (Lc_2) of CrAlN films varies between 49 and 64 N. Indeed, by adding 3.5% at. Al, Lc_2 increases up to 64 N. By increasing Al rate to 7% at., Lc_2 decreases to 54 N and then stabilizes for higher Al contents. The reduction in the resistance to spallation with the increase in Al content is attributed to its microstructure (Hua et al. 2006). Indeed, the CrAlN film with an 3.5% of Al exhibits the densest structure (Fig. 1). This dense structure has strong inter-columnar bonds, thus a strong resistance to the formation of inter-columnar cracks. Also, Li et al. (Li et al. 2007) explained this trend of adhesion by the evolution of the grain size. They observed that the increase of grain size leads to the decrease of Lc_2.

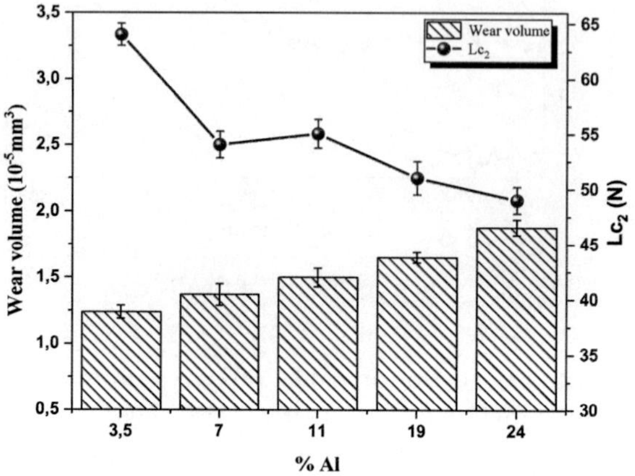

Fig. 4. Critical load and wear volume of CrAlN coatings under different Al content

After the ball-on-disk tests, the wear volume is calculated and it presents in Fig. 4. The increase of the Al %, is accompanied by a slight increase in the wear volume. Indeed, it varies from 1.2 10^{-5} mm^3 for an A1 rate of 3.5% at. to 1.9 10^{-5} mm^3 when the Al rate is about 24%. This increase in wear volume is related to the decrease in hardness, adhesion and microstructure density. Nevertheless, from Fig. 3 the hardness decreases by increasing the percentage of Al from 3.5 up to 11 at.%. Then, it increases for 19 and 24% at. However, the critical load Lc_2 decreases with increasing Al content in the layer (Fig. 4). So, the increase in wear volume is due to the decrease in adhesion. Shew and Huang (Shew and Huang 1995) reported that mechanical properties and chemical stability have a great influence on the wear resistance of films. As the oxidation temperature of CrAlN films is high (≈ 1000 °C), it is always higher than the ball/deposit contact temperature. Thus, the wear behavior of CrAlN films is controlled by the mechanical properties and not by the chemical stability.

4 Conclusion

The aim of this work was to investigate the effect of Al content on the properties of CrAlN films obtained by DC reactive magnetron sputtering. The above results show that the addition of Al rate affects the CrN properties. CrAlN coating hardness was increased with the increase of Al rate. The maximum hardness is about 30 GPa obtained in the coating with 24% of Al content. The friction coefficient of CrAlN films was superior than that of CrN and is about 0.8. CrAlN coatings present a good behavior of adhesion which reaches 64 N. However the increase of Al rate in the layer of CrAlN films is accompanied by a decrease of Lc_2. CrAlN film with 3.5% rate of Al exhibits the best wear resistance. The increase of Al% rate leads to the decrease of the wear resistance.

Acknowledgments. The authors are grateful to the Regional Council of Burgundy for its financial support, Mr. Denis LAGADRILLERE for the SEM observations. Also, authors would like to think Pr. Alain IOST and Dr. Alberto MEJIAS for their help in nano-indentation and scratch test measurements.

References

Kong, Q., Ji, L., Li, H., Liu, X., Wang, Y., Chen, J., Zhou, H.: Composition, microstructure, and properties of CrN_x films deposited using medium frequency magnetron sputtering. Appl. Surf. Sci. **257**, 2269–2274 (2011)

Jianxin, D., Fengfang, W., Yunsong, L., Youqiang, X., Shipeng, L.: Erosion wear of CrN, TiN, CrAlN, and TiAlN PVD nitride coatings. Int. J. Refractory Metals Hard Mater. **35**, 10–16 (2012)

Shan, L., Wang, Y., Li, J., Chen, J.: Effect of N_2 flow rate on microstructure and mechanical properties of PVD CrN_x coatings for tribological application in seawater. Surf. Coat. Technol. **242**, 74–82 (2014)

Wang, L., Nie, X.: Effect of annealing temperature on tribological properties and material transfer phenomena of CrN and CrAlN coatings. J. Mat. Eng. Perform. **23**(2), 560–571 (2014)

Reiter, A., Derflinger, V., Hanselmann, B., Bachmann, T., Sartory, B.: Investigation of the properties of $Al_{1-x}Cr_xN$ coatings prepared by cathodic arc evaporation. Surf. Coat. Technol. **200**, 2114–2122 (2005)

Nouveau, C., Tlili, B., Aknouche, H., Benlatreche, Y., Patel, B.: Comparison of CrAlN layers obtained with one (CrAl) or two targets (Cr and Al) by magnetron sputtering. Thin Solid Films **520**(7), 2932–2937 (2012)

Lin, J., Mishra, B., Moore, J., Sproul, W.: A study of the oxidation behavior of CrN and CrAlN thin films in air using DSC and TGA analyses A study of the oxidation behavior of CrN and CrAlN thin films in air using DSC and TGA analyses. Surface Coat. Technol. **202**, 3272–3283 (2008)

Ding, X., Tan, A., Zeng, X., Wang, C., Yue, T., Sun, C.: Corrosion resistance of CrAlN and TiAlN coatings deposited by lateral rotating cathode arc. Thin Solid Films **516**, 5716–5720 (2008)

Lv, Y., Ji, L., Liu, X., Li, H., Zhou, H., Chen, J.: The structure and properties of CrAlN fi lms deposited by mid-frequency unbalanced magnetron sputtering at different substrate bias duty cycles. Surf. Coat. Technol. **206**, 3961–3969 (2012)

Aouadi, K., Tlili, B., Nouveau, C., Besnard, A., Chafra, M., Souli, R.: Influence of substrate bias voltage on corrosion and wear behavior of physical vapor deposition CrN coatings. J. Mater. Eng. Perform. **28**(5), 2881–2891 (2019)

Rahmoun, K., Iost, A., Keryvin, V., Guillemot, G., Sari, N.: A multilayer model for describing hardness variations of aged porous silicon low-dielectric-constant thin films. Thin Solid Films **518**(1), 213–221 (2009)

Barshilia, H.C., Selvakumar, N., Deepthi, B., Rajam, K.S.: A comparative study of reactive direct current magnetron sputtered CrAlN and CrN coatings. Surf. Coat. Technol. **201**, 2193–2201 (2006)

Lin, J., Mishra, B., Moore, J., Sproul, W.: Microstructure, mechanical and tribological properties of $Cr_{1\text{-}x}Al_xN$ films deposited by pulsed-closed field unbalanced magnetron sputtering (P-CFUBMS). Surf. Coat. Technol. **201**, 4329–4333 (2006)

Ding, X., Zeng, X.: Structural, mechanical and tribological properties of CrAlN coatings deposited by reactive unbalanced magnetron sputtering. Surf. Coat. Technol. **200**, 1372–1376 (2005)

Bobzin, K., Lugscheider, E., Nickel, R., Bagcivan, N., Kramer, A.: Wear behavior of $Cr1_{\text{-}x}Al_xN$ PVD-coatings in dry running conditions. Wear **263**, 1274–1280 (2007)

Wang, L., Zhang, G., Wood, R., Wang, S., Xue, Q.: Fabrication of CrAlN nanocomposite films with high hardness and excellent anti-wear performance for gear application. Surf. Coat. Technol. **204**(2010), 3517–3524 (2010)

Aihua, L., Jianxin, D., Haibing, C., Yangyang, C., Jun, Z.: Friction and wear properties of TiN, TiAlN, AlTiN and CrAlN PVD nitride coatings. J. Refractory Metals Hard Mater. **31**(2012), 82–88 (2012)

Hua, M., Ma, H., Li, J., Mok, C.K.: Tribological behaviours of patterned PVD TiN spot coatings on M2 steel coated with different bias voltages. Surface Coat. Technol. **200**(11), 3612–3625 (2006)

Li, T., Li, M., Zhou, Y.: Phase segregation and its effect on the adhesion of Cr–Al–N coatings on K38G alloy prepared by magnetron sputtering method. Surf. Coat. Technol. **201**(2007), 7692–7698 (2007)

Shew, B., Huang, J.: The effects of nitrogen flow on reactively sputtered Ti-A1-N films. Surf. Coat. Technol. **71**(1995), 30–36 (1995)

Mechanical Characterization of a Composite Sandwich Core Under Shear Stress Based on a Torsion Test

Karim Mharsi[1,3], Pascal Casari[1], Amira Sellami[3], Jamal Fajoui[1], and Mohamed Kchaou[2,3](✉)

[1] Institut de recherche en génie civil et mécanique (GeM), UMR, (CNRS 6183), Université de Nantes, Centrale Nantes, 58, Rue Michel Ange-BP 420, 44600 Saint-Nazaire Cedex, France
[2] Department of Industrial Engineering, College of Engineering, University of Bisha, Bisha, Saudi Arabia
kchaou.mohamed@yahoo.fr
[3] Laboratory of Electromechanical Systems (LASEM), National Engineering School of Sfax, University of Sfax, 3038 Sfax, Tunisia

Abstract. The development of modern technologies requires the use of materials with high mechanical properties and lightest weight structure. With their low density, high strength, high rigidity and excellent durability, the sandwich structures are one of the best materials to meet these requirements. Despite of these advantages, sandwich structure is rather sensitive to failure by shear load. Previous work has shown that standard shear characterization tests of a sandwich structure (ASTM C273, Standard Test Method for Shear Properties of Sandwich Core Materials) exhibits shear stress concentrations at the edges, which negatively affect the results of this test. To solve this issue, this research work proposes a new test methodology to determine shear proprieties using a torsion test. A custom test rig has been developed in order to exert a torsional moment on a cylindrical specimen made of PVC foam. Based on the experimental results, this new test method provides a more precise characterization of the sandwich core materials since it eliminates the phenomenon of concentration of shear stresses on the edges found in the standard test.

Keywords: Sandwich structure · Shear test · Foam PVC · Torsion

1 Introduction

Sandwich structures are distinguished from other materials such as metals, ceramics and polymers by the fact that they consist of:

- Two skins, or face sheets; made of relatively high strength materials.
- A core material; much thicker with low density.

The core and the two face sheets are mechanically separable (Fig. 1). The mechanical properties of the sandwich structure can, therefore, be modified by varying the proportions or the properties of its constituents. These adjustments make it possible

M. Kharrat et al. (Eds.): ICAMEM 2019, LNME, pp. 299–305, 2020.
https://doi.org/10.1007/978-3-030-52071-7_41

to obtain the desired properties for applications which require a very high stiffness/mass ratio. Under solicitations, the core undergoes shear loading and the skins takes traction and compression loadings [1].

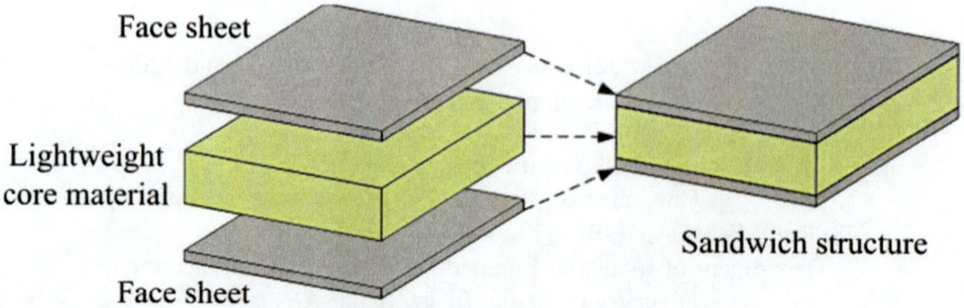

Fig. 1. Sandwich structure [1]

The determination of the mechanical properties of these structures is essential to predict their behavior in specific environments and solicitations [2].

Several researchers [3, 4] performed bending tests under static and cyclic loading conditions in order to identify the mechanical flexural properties of the sandwich material. This bending test was carried out according to ASTM C393 and D790 [5, 6].

Yet, this research work aims to identify the shear properties of the core of the sandwich structure, because during the service of a sandwich structure, the shear stresses are significant at the level of the core. ASTM C273 is one of the most widely used standardized tests to meet this need. The Fig. 2 shows the geometry of the specimen and the used technique.

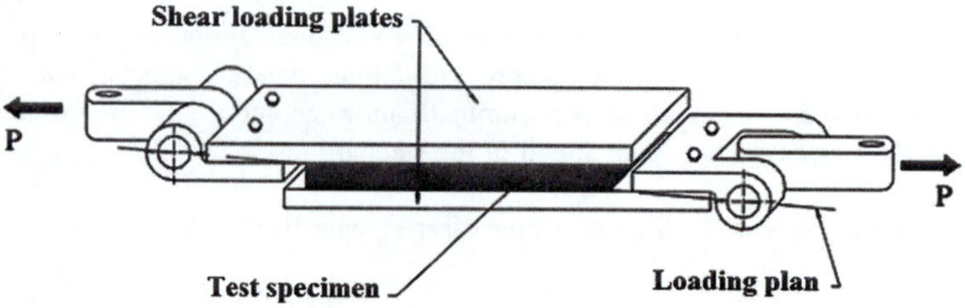

Fig. 2. Shear tests of Sandwich Core Materials: Standard C273 [7]

Experimental and numerical tests were performed for the ASTM C273 test. As the results show below (Fig. 3), the experimental test and the numerical simulation of the ASTM C273 test reveal that the shear stresses are concentrated at the edge of the specimen. For a static load, the concentration of shear stresses may not have much influence on the results, but with a cyclic load, it will be more influential.

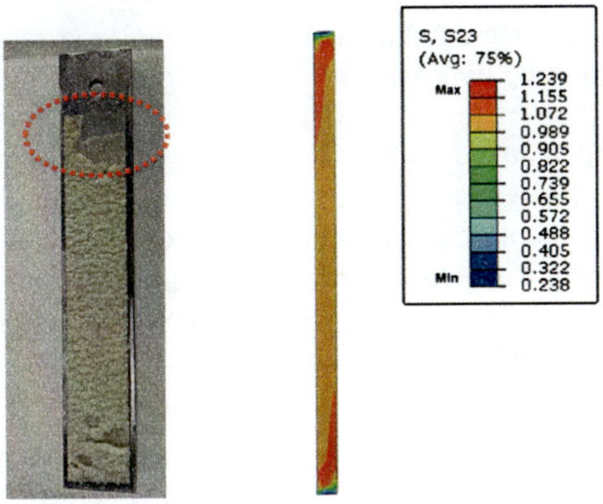

a. Experimental specimen b. Numerical simulations results

Fig. 3. S23 shear stresses concentrated at the edge of the sample of the ASTM C273 standard specimen [8]

Therefore, it is necessary to propose a new shear test methodology which eliminate stress concentration at the edges, gives the same results as the standard during a static load and improves the results of fatigue characterization.

2 Torsional Rig Shear Test

A new test rig has been designed in order to determine the mechanical shear properties of the core of the sandwich structure. The assembly applies a torque to a cylindrical PVC foam test specimen (Fig. 4).

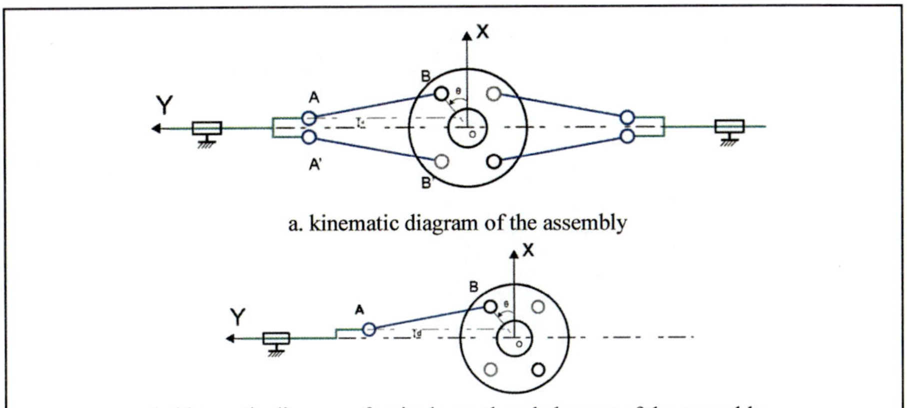

a. kinematic diagram of the assembly

b. kinematic diagram of a single crank rod element of the assembly

Fig. 4. Kinematic diagram of the assembly

It can be considered as a combination of two crank-rod mechanisms mounted in opposition in order to have two opposite torsion moments. These two moments produce shear at the section of the cylindrical specimen. The kinematic diagram of the assembly is presented in the Fig. 5-a. To simplify, we isolated a single crank rod element (Fig. 5-b).

As a hypothesis, we consider that the system's efficiency is equal to 1, the crank turns at a constant speed and the inertia is negligible. A simplified relationship is established giving the torque C as a function of the force F from the equality of the powers consumed and supplied.

$$C\omega = FV \tag{1}$$

The torque applied to the test piece can be expressed according to the geometric parameters as follows:

$$M_t = 4 \times C = 4F \times \left(R \left(\cos\theta + \frac{\sin\theta(\cos\theta + \varepsilon)}{\sqrt{\left(\frac{L}{R}\right)^2 - (\cos\theta + \varepsilon)^2}} \right) \right) \tag{2}$$

with; $d = \varepsilon \times R$.

For small displacements, we can get a simplified expression of the maximum torsional stress.

$$\tau_{max} = \frac{R}{\pi(R^4 - r^4)} \times M_t \tag{3}$$

$$\tau_{max} = 2F \frac{R^2}{\pi(R^4 - r^4)} \times \cos\theta \tag{4}$$

The following figure shows the torsion test rig assembled and mounted on the traction machine with its CAD design on Solidworks (Fig. 5).

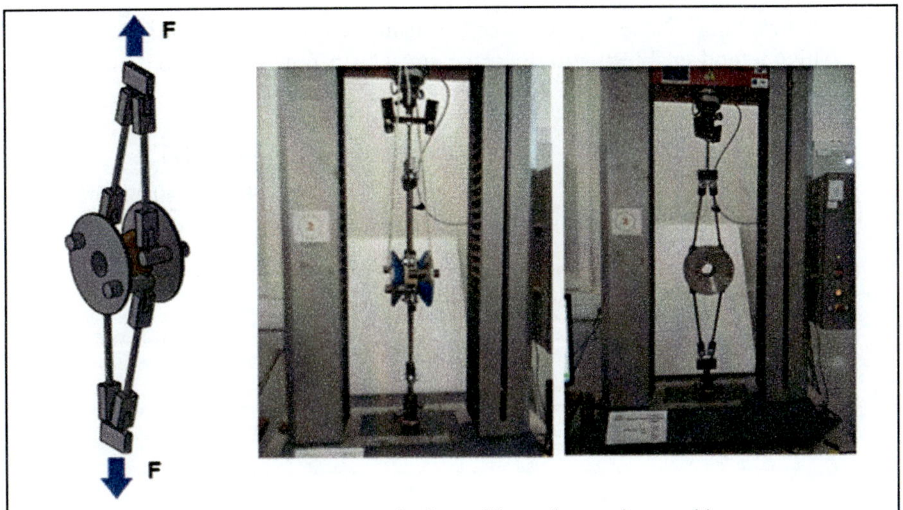

a. Solidworks design

b. Assembly on the traction machine

Fig. 5. Torsion assembly

3 Results and Discussion

The test has been done on the traction machine using this new assembly. The values of the applied force F has been recorded in order to trace the evolution of the maximum torsional stress during the test (Fig. 6).

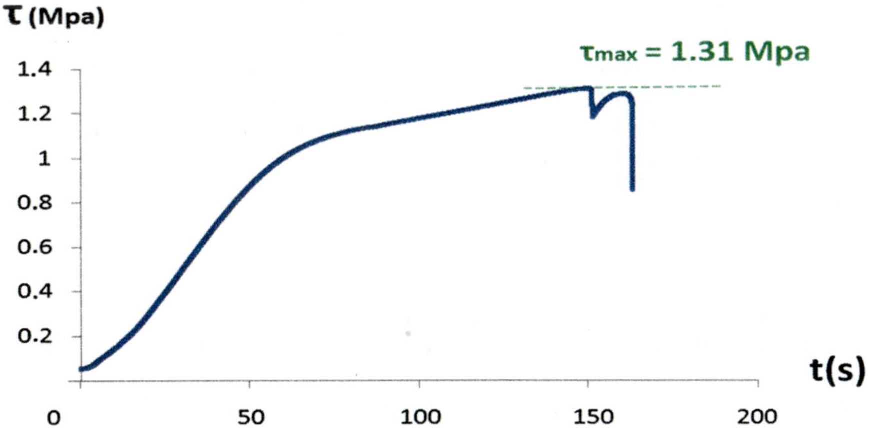

Fig. 6. Evolution of the maximum torsional stress during the test

The deformation values were recorded using digital image correlation (DIC), Fig. 7. This method enables us to draw the stress as a function of the deformation.

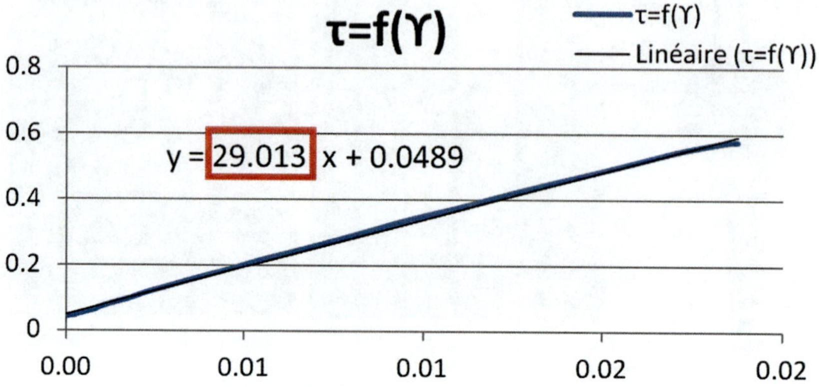

Fig. 7. Deformations according to shear stresses

The shear modulus G of the PVC specimen can be deduced graphically. It presents the slope of the curve $\tau = f(\Upsilon)$ in the elastic response region. The value found matches well to that mentioned in the Divinycell H100 foam data sheet used in this test.

4 Conclusion

The phenomenon of shear stress concentration at the edges of the shear characterization test (ASTM C273) may not be very influential for static characterization. But for a fatigue characterization, this phenomenon will influence the results considerably. The proposed new method relies on a torsion test using a cylindrical specimen which, according to the preliminary results found, avoided this unwanted phenomenon.

As a perspective, this method will be used equally for fatigue characterization test using H100 PVC foam as core material.

References

1. Camille, B.: Development and optimization of a formable sandwich sheet. Materials and structures in mechanics [physics.class-ph]. Ecole Polytechnique X. English. fftel-00691246 (2012)
2. Chemami, A.: Etude de la Performance des Matériaux Composites Sandwichs - Application en Fatigue. Doctoral thesis, Annaba (2012)
3. Rocca, S.V., Nanni, A.: Mechanical characterization of sandwich structure comprised of glass fiber reinforced core. In: Composites in Construction–Third International Conference (2005)
4. Alila, F.: Analyse de la tenue en fatigue de structures composites sandwich. Doctoral thesis, University of Nantes (2017)

5. ASTM C393: Flexural Properties of Sandwich Constructions, American Society for Testing and Materials (2000)
6. ASTM D790: Standard Test Methods for Flexural Properties of Unreinforced and Reinforced Plastics. American Society for Testing and Materials, Philadelphia (2002)
7. ASTM C273: Standard Test Method for Shear Properties of Sandwich Core Materials, American Society for Testing and Materials (2000)
8. Mharsi, K., Sellami, A., Kchaou, M., Fajoui, J., Casari, P., Alila, F., Jacquemin, F.: Development of a new numerical method for the mechanical characterization of a composite sandwich material under shear solicitation. In: 4th Tunisian Congress of Mechanics COTUME 2018, Hammamet, Tunisia, 13–15 October 2018 (2018)

Study of the Wear Behavior of HDPE-Molybdenum Disulphide Composites

A. Salem(✉), M. Guezmil, W. Bensalah, and S. Mezlini

Université de Monastir, École Nationale d'Ingénieurs de Monastir (ENIM), Laboratoire de Génie Mécanique (LGM), LR99ES32, 5019 Monastir, Tunisia
amirasalem0302@gmail.com, walidbensalah@gmail.com, salah.mezlini@gmail.com, manelguezmil@yahoo.fr

Abstract. This work investigates tribological study of high-density polyethylene-molybdenum disulphide (HDPE-MoS_2) bio-composites via an alternative pin-disk tribotester in the dry conditions. The stainless steel pin (M30NW) was used as a counter-part. The obtained HDPE powder after crushing was blended with MoS_2 particles with various proportions (0, 2, 4, 6, 10 and 25 wt%). HDPE composite specimens were elaborated with thermal-compression molding press. The coefficient of friction of the HDPE composite decreased considerably after reinforcing with MoS_2 for all loadings. The specific wear rate was improved by over 40% for up to a loading of 6 wt% MoS_2. It was found that the utilization of molybdenum disulphide improves tribological behavior of the elaborated composites in respect of the unfilled HDPE and the optimal content was about 4 wt%. In fact, after the wear test the optimized percent exhibited a decrease of 30% for the coefficient friction and 55% for the specific wear rate.

Keywords: HDPE-MoS_2 composite · Friction coefficient · Specific wear rate

1 Introduction

Polyethylene is widely exploited in diverse industrial domains in view of its advantages and great performances. High density polyethylene (HDPE) is a biomaterial which had been employed as an acetabular cup for joint prosthesis, also it can be easily molded (Fouad and Elleithy 2011; Zhil et al. 2009).

In order to ameliorate the tribological behavior of HDPE, many researchers have used numerous fillers (graphite, NTC, graphene, PTFE, Al_2O_3, talc, MoS_2…) (Fouad and Elleithy 2011; Panin et al. 2015; Chang et al. 2015; Senatov et al. 2015; Suñer et al. 2014).

Molybdenum disulphide (MoS_2) was characterized with a high biocompatibility (Appel et al. 2016; Wang et al. 2015). Moreover, it has been used as an additive in order to boost the tribological characteristics of different polymers for common mechanical applications (Lansdown 1999; Pettarin et al. 2010).

In the present work, the tribological properties (friction coefficient & specific wear rate) of HDPE-MoS_2 bio-composites versus metallic pin (M30NW) in the dry conditions was investigated using pin-on-disk tribotester.

M. Kharrat et al. (Eds.): ICAMEM 2019, LNME, pp. 306–311, 2020.
https://doi.org/10.1007/978-3-030-52071-7_42

2 Experimental Process

2.1 Materials

HDPE-MoS_2 composite disks (Ø = 30 mm and e ≈ 8 mm) was elaborated with a thermo-compression molding. M30NW stainless steel pin with hemi-spherical shape (Ø = 10 mm) given by C2F implants company-France was used as a counter-face for the tribotester (Fig. 1). Table 1 illustrates the pin's chemical composition.

Table 1. M30NW pin's chemical composition.

Elements (%)	M30NW
C	≤ 0.06
Cr	21
Ni	9.5
Mn	4
Mo	2.2
N	0.4

2.2 Techniques

2.2.1 Specimens Preparation

The high density polyethylene pellets (Ø = 3 mm) crushed and a powder form was obtained (Ø = 315 µm). The HDPE was blended with distinct ratios (2, 4, 6, 10 and 25 wt%) of commercial MoS_2 powder (Ø = 0.75 µm) (SAMARO Company). Eventually, the composite was shaped with a thermal-compression molding press below 18 MPa as a pressure, at 135 °C during 4 h. Finally, the mold cold at air environment, composites disks were extracted and used in tribotester (Fig. 1).

2.2.2 Tribological Test (Tribotester)

The wear behaviors of HDPE-MoS_2 composites was inquiry via straight pin-disk tribotester at ambient environment (Fig. 1). Before tribological test, the surfaces refined to a roughness of 0.21 ± 0.05 µm for the disks and of 0.06 ± 0.01 µm for the pin. During tests, the polymer disk kept animated with an alternative motion of the table and the metallic pin was secured on the holder. Table 2 shows the wear tests' experimental parameters.

2.2.3 Characterization Technics

In order to quantify the wear surface of the disks, the 2D profilometry (Taylor Hobson Surtronic S116 profilometer) was employed. The multiplication of the scanned area by the

Table 2. Wear tests' experimental parameters.

Parameters	Values
Normal force	20 N (Borruto 2010)
Contact pressure	58 MPa
Slip speed	30 mm/s
Total sliding distance	450 m
Stroke length	30 mm (Widmer et al. 2001)

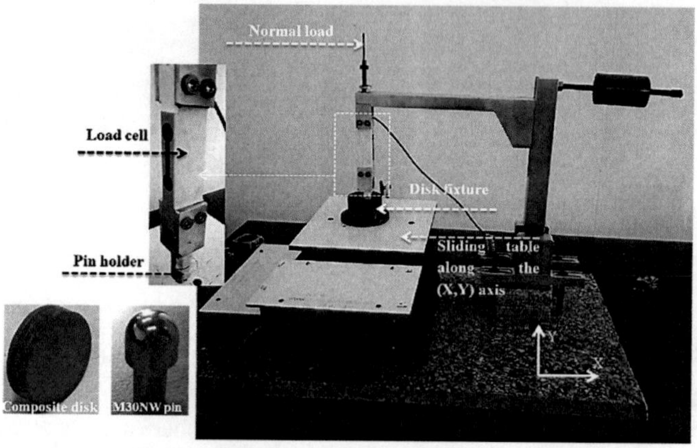

Fig. 1. Photography of pin-disk tribotester.

length of wear track allows the mean wear volume ΔV (mm^3) to be achieved. The specific wear rate K was given following the mathematical expression (Huang et al. 2013):

$$K\ (mm^3/N.m) = \frac{\text{Average volume loss}}{\text{Normal load} \times \text{Total sliding distance}}$$

3 Results and Discussion

3.1 Friction Comportment

Figure 2 displays the variability of friction coefficient of HDPE-MoS_2 composite disks versus metallic pins under dry condition.

The employment of MoS_2 as a filler reduces the friction coefficient notably with 4 wt%. The decrease was almost the same for loading of 4, 6 and 10 wt%, it was about 30% in dealing with the unload HDPE. In contrast, the friction coefficient boosts for loading of 25 wt%. It conducts to a reduction of 14% with regard to the unload HDPE. The enhancing of friction coefficient with an important content of the load could be due to the agglomeration of MoS_2 particles with HDPE matrix. This phenomena was examined with different polymer composites (Ben Difallah et al. 2014; Lansdown 1999; Pettarin et al. 2010; Zhu et al. 2006).

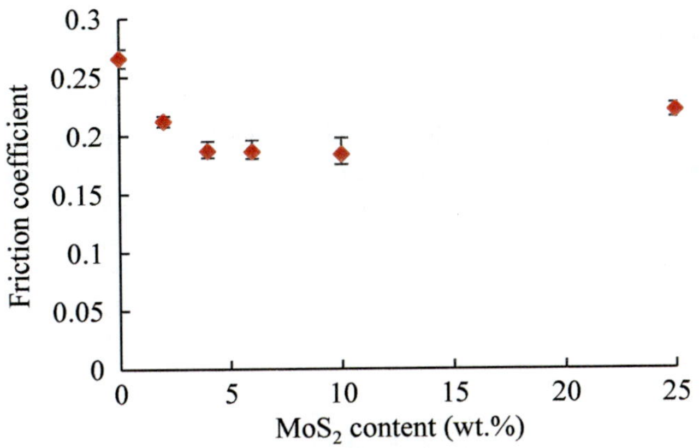

Fig. 2. Variability of HDPE-MoS_2 composites' friction coefficient vs metallic pins.

3.2 Wear Comportment

Figure 3 mentioned the evolution of the specific wear rate of HDPE-MoS_2 composites after wear tests versus metallic pins.

The incorporation of the filler to the polymer matrix conducts to the decay of the specific wear rate which obtain 55% for 4 wt% of MoS_2 in respect of the unload HDPE. In contrast, the wear rate increases with a high content of MoS_2 to achieve 74% for loading of 25 wt% with reference to the unfilled HDPE. A similar finding were mentioned by Pettarin et al. (2010).

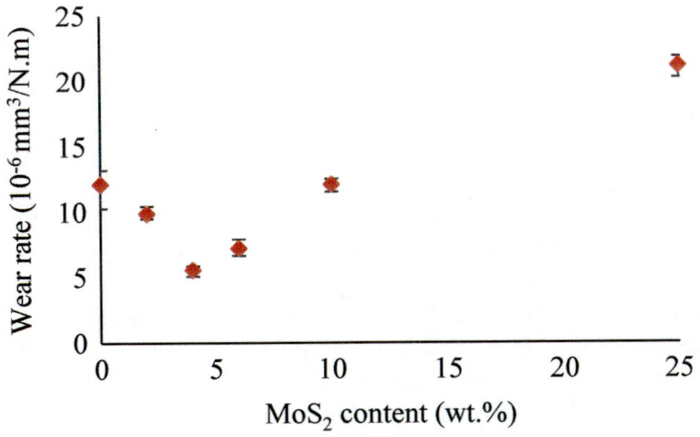

Fig. 3. Evolution of HDPE-MoS_2 composites' specific wear rate vs metallic pins.

4 Conclusion

The use of MoS_2 particles as a filler into HDPE matrix with different percentages (2, 4, 6, 10, and 25 wt%) under dry conditions reduces the friction coefficient. The declines was around 30% for 4 wt% of MoS_2 (0.18) as regard to the unfilled HDPE (0.26). An increase of the specific wear rate was found with loading of 25 wt% of molybdenum disulphide. The specific wear rate of the optimal content of MoS_2 (4 wt%) represents 55% compared to the unload HDPE.

References

Appel, J.H., Li, D.O., Podlevsky, J.D., Debnath, A., Green, A.A., Wang, Q.H., Chae, J.: Low cytotoxicity and genotoxicity of two-dimensional MoS_2 and WS_2. ACS Biomater.-Sci. Eng. **2**, 361–367 (2016)

Borruto, A.: A new material for hip prosthesis without considerable debris release. Med. Eng. Phys. **32**, 908–913 (2010)

Ben Difallah, B., Kharrat, M., Dammak, M., Monteil, G.: Improvement in the tribological performance of polycarbonate via the incorporation of molybdenum disulfide particles. Tribol. Trans. **57**(5), 806–813 (2014)

Fouad, H., Elleithy, R.: High density polyethylene/graphite nano-composites for total hip joint replacements: processing and in vitro characterization. J. Mech. Behav. Biomed. Mater. **4**, 1376–1383 (2011)

Huang, J., Qu, S., Wang, J., Yang, D., Duan, K., Weng, J.: Reciprocating sliding wear behavior of alendronate sodium-loaded UHMWPE under different tribological conditions. Mater. Sci. Eng., C **33**(5), 3001–3009 (2013)

Lansdown, A.R.: Molybdenum Disulphide Lubrication. Elsevier, Amesterdam (1999)

Panin, S.V., Kornienko, L.A., Nguen Suan, T., Ivanova, L.R., Korchagin, M.A.: Wear resistance of composites based on hybrid UHMWPE–PTFE matrix: mechanical and tribotechnical properties of the matrix. J. Frict. Wear **36**(3), 249–256 (2015)

Chang, B.P., Akil, H.M., Nasir, R.B., Khan, A.: Optimization on wear performance of UHMWPE composites using response surface methodology. Tribol. Int. **88**, 252–262 (2015)

Pettarin, V., Jose, M., Felhös, D., Karger-kocsis, J., Maria, P.: Changes in tribological performance of high molecular weight high density polyethylene induced by the addition of molybdenum disulphide particles. Wear **269**, 31–45 (2010)

Senatov, F.S., Kopylov, A.N., Yu Anisimova, N., Kiselevsky, M.V., Maksimkin, A.V.: UHMWPE-based nanocomposite as a material for damaged cartilage replacement. Mater. Sci. Eng., C **48**, 566–571 (2015)

Suñer, S., Bladen, C.L., Gowland, N., Tipper, J.L., Emami, N.: Investigation of wear and wear particles from a UHMWPE/multi-walled carbon nanotube nanocomposite for total joint replacements. Wear **317**, 163–169 (2014)

Wang, S., Li, K., Chen, Y., Chen, H., Ma, M., Feng, J., Zhao, Q., Shi, J.: Biocompatible PEGylated MoS2 nanosheets: controllable bottom-up synthesis and highly efficient photothermal regression of tumor. Biomaterials **39**, 206–217 (2015)

Widmer, M.R., Heuberger, M., Vörös, J., Spencer, N.D.: Influence of polymer surface chemistry on frictional properties under protein-lubrication conditions: implications for hip-implant design. Tribol. Lett. **10**(1), 111–116 (2001)

Zhil, T.V., Oliveira, M.S.A., Ferreira, J.A.F.: Relative influence of injection molding processing conditions on HDPE acetabular cups dimensional stability. J. Mater. Process. Technol. **209**, 3894–3904 (2009)

Zhu, P., Wang, X., Wang, X.D., Huang, P., Shi, J.: Tribology performance of molybdenum disulphide reinforced thermoplastic polyimide under dry and water lubrication conditions. Ind. Lubr. Tibol. **58**, 195–201 (2006)

Microstructural and Mechanical Characterization of a Baby Diaper

Basma Ajmi[1,3], Mohamed Kchaou[2,3(✉)], Amilcar Ramalho[1], Amira Sellami[3], Antonio J. Gamez[4], and Nabil Bouzayani[5]

[1] FCTUC-Department of Mechanical Engineering-788, Univ Coimbra, Rua Luis Reis Santos, 3030 Coimbra, Portugal
ajmi.basma2@gmail.com, amilcar.ramalho@dem.uc.pt

[2] Department of Industrial Engineering, College of Engineering, University of Bisha, Bisha, Saudi Arabia
kchaou.mohamed@yahoo.fr

[3] Laboratory of Electromechanical Systems (LASEM), National Engineering School of Sfax, University of Sfax, 3038 Sfax, Tunisia
amira_sellami@hotmail.com

[4] School of Engineering, Univ Cadiz, 10 Av. Universidad de Cádiz, 11519 Puerto Real, Cadiz, Spain
antoniojuan.gamez@uca.es

[5] Hygiene Products Company, HYPCOM, Industrial Zone Agareb Road Mahres Km1, 3000 Sfax, Tunisia
nbc@gnet.tn

Abstract. Baby diaper, despite their evolution, still caused serious problems after using on the baby's skins, as diaper rash or seat dermatitis. To avoid these problems, understanding of the structure of the material in contact with the baby's skin and its mechanical behavior is necessary. Therefore, the study of the non-wovens fabrics of the top-sheet is compulsory. Our aim is to relate mechanical properties of the tissues to their microstructure. To achieve that goal, the microscopic observation have been performed to study the composition and the structure of the textile, particularly the distribution of the fibers orientation. Also, tensile tests in both longitudinal and transversal directions and compression tests have allowed to mechanically characterize some mechanical properties, as stiffness. Results show that longitudinal directions are favored comparing to the transversal ones and that actual baby diapers are more rigid than others under study.

Keywords: Mechanical behavior · Baby diaper · Top-sheet · Non-woven fabric

1 Introduction

Previous research has shown that the problem caused to the baby by the use of the hygienic textile is very serious [1]. Although, baby's skin is much more sensitive than the adults, there are not much specific research about its mechanical characteristics.

M. Kharrat et al. (Eds.): ICAMEM 2019, LNME, pp. 312–319, 2020.
https://doi.org/10.1007/978-3-030-52071-7_43

Baby diaper is fabricated by a number of elements [2] as shown in Fig. 1, which are divided into two categories:

- Textiles: top-sheet, leakage barriers, acquisition layer, cellulose and back-sheet
- Polymer : the Super Absorbent Polymer (SAP)

Three different types of baby diaper were considered:

- Classic baby diaper, which contains 30% SAP and 70% of cellulose
- Actual baby diaper that contains 45% of SAP and 55% of cellulose
- SAP paper baby diaper is a different type we didn't talk about SAP and cellulose

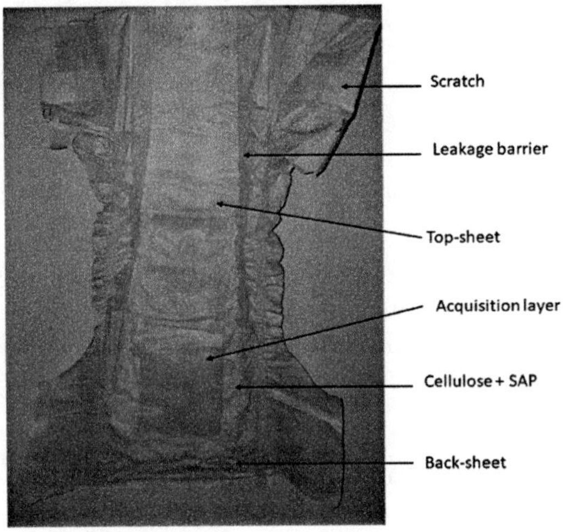

Fig. 1. Composition of baby diaper

However, very few publications are available in the literature that discuss the issue of the mechanical behaviour of the non-wovens fabric.

The top-sheet used in this diaper is a non-wovens fabric. These textiles are engineered fabrics manufactured from a set of disordered fibers collected together by localized melting or by the use of chemicals. Simple complication of fibres is also possible by means of mechanical pressure using water or cold calendaring. Thermal bonding is the most commonly used technique for their manufacture [3]. In thermally calender-bonded non-wovens, a fibrous web is passed through a hot calendar with an embossed pattern. Bonding mainly occurs at raised areas bond points other parts of the web, which were not in contact with the hot engraved pattern, remained unaffected and form the permeable part called fibre matrix that acts as a link between the bond points.

The main objective of this work is the mechanical characterization of the top-sheet and the different types of baby diaper.

2 Microstructural Characterization of the Top-Sheet

In order to have an idea about existing privileged directions in the top-sheet of the baby diaper, a microscopic observation is done in the first step. Figure 2 shows the result of the structure of thermally bonded non-woven fabric. It is a combination of continuous and discontinuous regions.

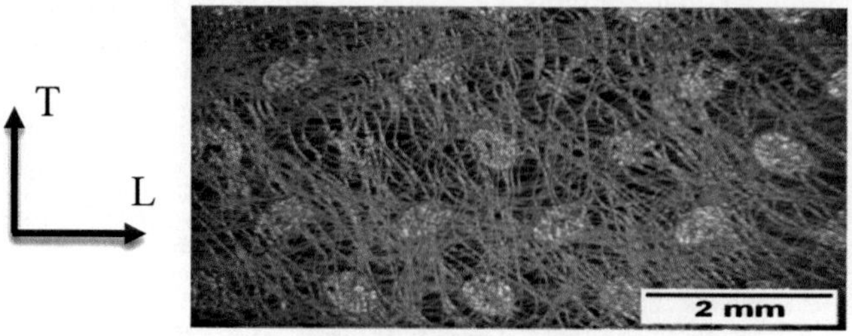

Fig. 2. Microscopic observation of non-wovens fabrics

The microscopic observation of the non-wovens fabrics in Fig. 2 is not sufficient to conclude about the specific direction of the constituent fibers of the top-sheet. In this case, to study the mechanical characterization of the top-sheet, the tensile test in both Longitudinal and Transversal directions is necessary.

3 Mechanical Characterization of the Top-Sheet

Previous study [4] analyses and compares various aspects of non-wovens fabrics. Nevertheless, the research was focalized on the study of the mechanical behavior of reinforced non-woven fabrics. Aiming to study their mechanical behavior, tensile tests are performed in two orthogonal directions: longitudinal and transversal using the parameters as follows. For the tensile test of the textile fabric, the speed displacement used in [4] shouldn't be applied in this case of test because of the sensitivity of this fabric. The strain rate should be reduced as possible. In fact, the value used was 5 mm/min.

The sample dimension of the probes were: active length $lo = 100$ mm and width $Wo = 25$ mm

Figure 3 and Fig. 4 show respectively the evolution of the top-sheet during tensile test of the longitudinal and the transverse direction oriented specimens.

Figure 3 shows the variation of the specific force as a function of the strain over the entire tensile test, which is composed of four parts of both of the directions as represented in the Table 1.

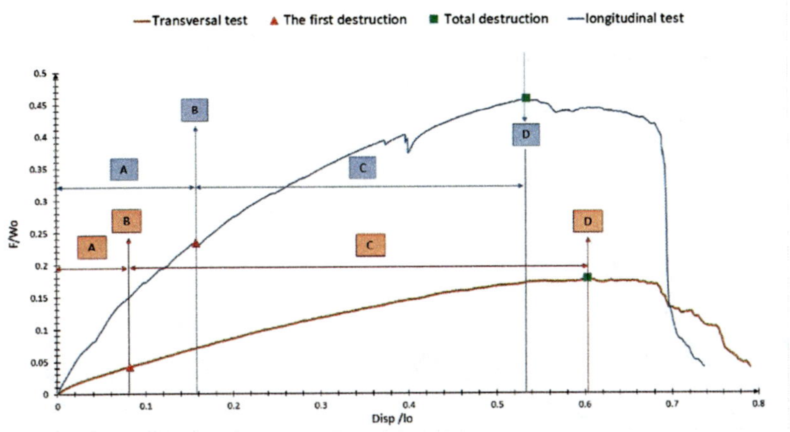

Fig. 3. Evolution of the specific force (N/mm) as a function of the strain during the tensile test of the longitudinal and the transversal direction

Table 1. Different evolution parts of the tensile tests

	A Elastic behavior zone	B First destroyed fibers	C Plastic behavior zone	D Total rupture of the fabric
Longitudinal				
Transversal				

The results presented in the Table 2 show that the value of the energy absorbed by the non-wovens fabrics during the longitudinal tensile tests is higher than the transversal test. The same trend for the load of the first destruction but it is the opposite for the value of the maximum displacement. These comparative results correspond to the curves of Fig. 5. The tensile strength is higher in the longitudinal direction than the transverse one with a maximum breaking load equal to $F_{max,L} = 11.41\ N$ and $F_{max,T} = 4.42\ N$.

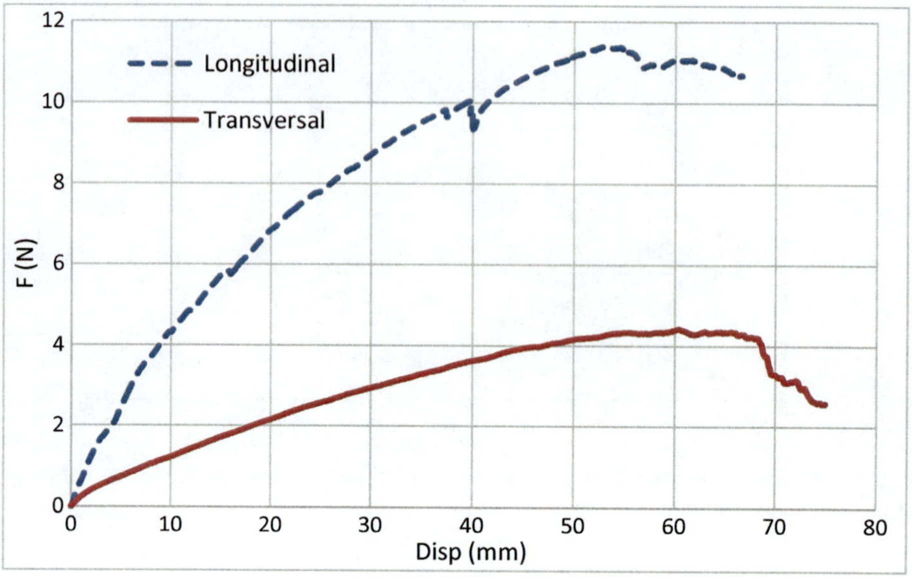

Fig. 4. Comparison of tensile test between longitudinal and transverse direction

Table 2. Tensile test results

	Longitudinal direction	Transverse direction
W_{abs} (J)	0.579	0.228
F_{1st} (N)	5.81	1.06
$disp_{max}$ (m)	0.0534	0.0603

Fig. 5. Compression test of baby diaper

4 Mechanical Characterization of the Baby Diaper

Compression testing is a widely used technique for measuring the stiffness of any materials [5]. Also, the diapers should be characterized concerning the rigidity, additionally the compressive stiffness should be correlated to the tactile perception. For this reason, different compressions tests were done. Figure 6 using a compression speed: $V = 5$ mm/min. until a minimum thickness of $e_{min} = 1$ mm. was reached. The results of these tests are represented in the Fig. 7.

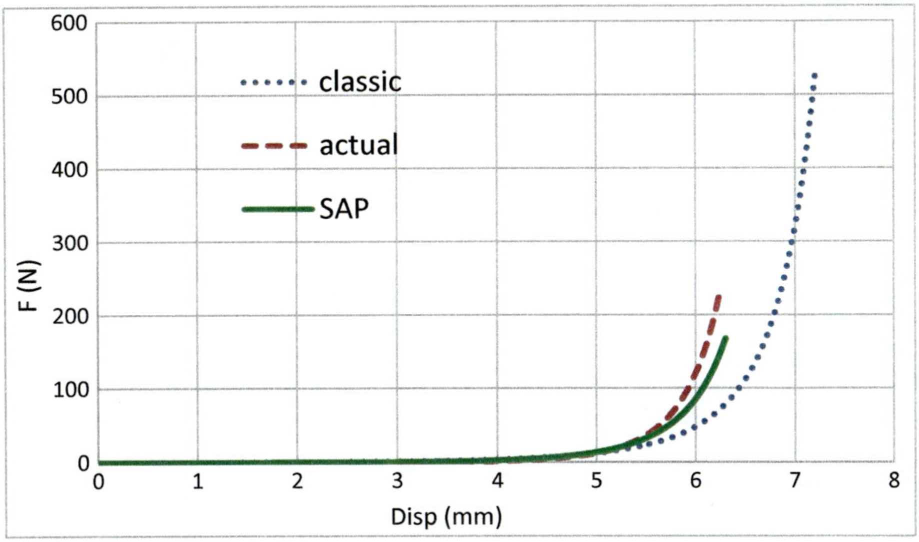

Fig. 6. Compression curves of different types of baby diaper

Knowing that the three types of baby diapers have different thickness, which are represented in the following Table 3, so the compression tests represent the stiffness of the diapers. However, to discuss this parameter of each type of baby diaper, an interval of the load was fixed between 50 N and 150 N as shown in Fig. 7. The linearization of these curves gives an idea about the values of the rigidity of each diaper.

The rigidity of the classic, the SAP paper and the actual baby diaper are respectively as follows 161.07 N/mm, 193.55 N/mm and 228.58 N/mm. As a result, the actual baby diaper has the greatest value.

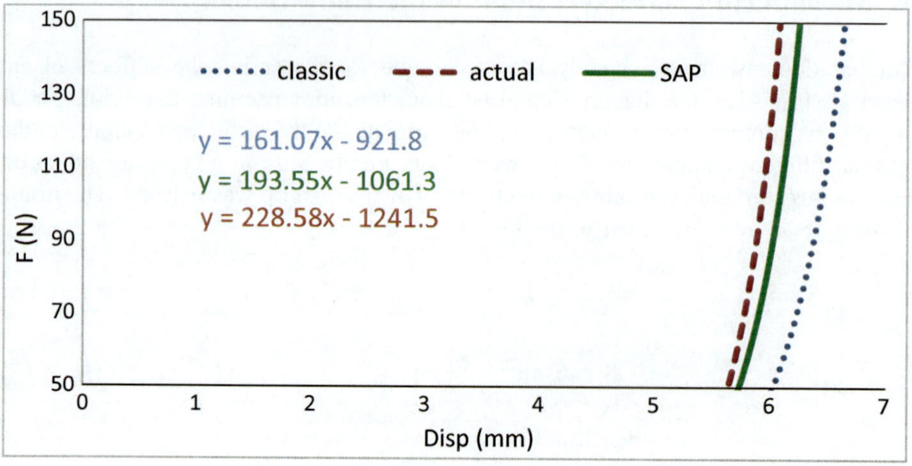

Fig. 7. Linearization of the compression curves

Table 3. Thickness of different types of baby diaper

	CLASSIC	Actual	SAP paper
Thickness (10^{-3} m)	6.15	4.2	4.1

5 Conclusion

The main objective of this paper is to understand the mechanical behavior of the baby diaper and the non-wovens fabrics used in its construction in order to increase the performance of this product.

Based on the results, it can be concluded first of all that the research into the mechanical characterization of the top-sheet demonstrates the resistivity of the top-sheet of the baby diaper on the longitudinal direction. This can be explained by the orientation of the majority of the fibers in the longitudinal direction but the microscopic observations couldn't show it clearly.

With an arrangement of the more to the least rigid, the actual baby diaper is in the first class, then the SAP and finally the classic one.

In our future research, we intend to concentrate on the rubbing contact of the skin/baby diaper.

Acknowledgement. Thanks to the Centre for Mechanical Engineering, Materials and Processes of the University of Coimbra.

Thanks to the Metrology Laboratory at the school of Engineering of the University of Cadiz.

Thanks to the Hygiene Products Company (HYPCOM) for the support and the encouragement.

This work was financed by the National Agency for the Promotion of Scientific Research of Tunisia (ANPR).

References

1. Stamatas, N., Tierney, N.K.: Diaper dermatitis: etiology, manifestations, prevention, and management georgios. Pediatric Dermatol. **31**(1), 1–7 (2014)
2. Necla Yaman, M., Senol, F., Ebru Tayyar, A.: Alternative test methods for assessing mechanical properties of disposable diapers consist. Fibres Text. East. Eur. **15**(2), 80 (2007)
3. Sayeb, S., Ben Hassen, M., Sakli, F.: Study of some nonwoven parameters influence on the absorption kinetics of liquid. Open Text. J. **3**, 1–5 (2010)
4. Gnaba, I., Omrani, F., Ferreira, M., Soulat, D., Wang, P., Vroman, P.: Study of the mechanical behavior of non-woven Lin/PP at reinforcement and composite scale. In: 23rd French Congress of Mechanics, Univ. Lille Nord of France (2017)
5. Zhao, S., Arnold, M., Ma, S., Abel, R.L., Cobb, J.P., Hansen, U., Boughton, O.: Standardizing compression testing for measuring the stiffness of human bone. Bone Joint Res. **7**(8), 524–538 (2018)

Numerical Modelling of Undulatory Elastic Behavior of Metallic Glasses Ribbon

M. A. Yousfi[1,2(✉)], C. Ammari[1], K. Hajlaoui[1,3], and Z. Tourki[1]

[1] Laboratoire de Mécanique de Sousse LMS, ENISo, Université de Sousse, Technopole de Sousse, BP 264 Cité Erriadh, 4023 Sousse, Tunisia
moyousfi@kku.edu.sa, cheima.ammari@gmail.com, zbrtourki@gmail.com, khalil.hajlaoui@eniso.rnu.tn

[2] Department of Mechanical Engineering, College of Engineering King Khalid University, Guraiger, Abha 62529, Kingdom of Saudi Arabia

[3] Al Imam Mohammad Ibn Saud Islamic University (IMSIU), Riyadh, Kingdom of Saudi Arabia

Abstract. In this work a numerical study is developed via Abaqus software that describes the undulatory elastic behavior of arc-shaped metallic glass (MG) ribbon subjected to normal progressive loading. The pre-deformed ribbon is prepared based on the experimental condition which should ensure the initial amplitude *h*over the boundary fixation *L* less than ½. The objective is to estimate numerically the maximum number of undulation that the arc-shaped ribbon could reach before plasticity. The control and estimation of the elasticity becomes very difficult when the miniaturization is requested. This result is very interested when the elastic undulation response of MG ribbon will be used as a flat spring or electromechanical switch for micro-machines and Micro-Electromechanical Systems (MEMS). The obtained numerical results are compared to the experimental available data and we have found a good agreement. This numerical study is done for the first time on the metallic glass ribbon and the improvement of the obtained results is in progress.

Keywords: Metallic glasses ribbon · Undulatory elastic behavior · Modelling

1 Introduction

Metallic Glasses (MGs) or amorphous metals represent a new class of materials which have interrested large technological fields (Peker et al. 1993; Inoue, 2000; Wang et al. 2004 and Wu et al. 2008)". They are also finding increased uses as functional materials in many industrial fields such as medical applications, MEMS (Micro-Electro Mechanical Systems) and NEMS (Nano-Electro-Mechanical Systems), spring and Electronic casing applications (Panagiotopoulos et al. 2016; Yousfi et al. 2017a, b). The high elastic strain limit of metallic glasses combined with a high yield strength, and fracture toughness make metallic glasses ideal for applications where the storage of high density of elastic energy is needed. Newly developed, metallic glasses have shown outstanding properties which allow them to conquer their crystalline counterparts in several industrial domains (such as sports, surgical equipment, radioactive containers,

M. Kharrat et al. (Eds.): ICAMEM 2019, LNME, pp. 320–326, 2020.
https://doi.org/10.1007/978-3-030-52071-7_44

magnets and transformers cores…). However, in the research field, they are still under development and study so they can be better exploited specially for their elasticity range (ten times higher thancrystalline materials elasticity). Our recent work on the exploitation of the elastic wavy response of metallic glass ribbons for the manufacture of flat micro-springs and electromechanical switches intended for micro-machines and MEMS (Micro-electromechanical-systems) (Peker et al. 1993; Nguyel et al. 2010 and Grech et al. 2014). The left problem concerns the control of elastic strain when the metallic glass ribbon is used as very small dimensions for miniaturization. The aim of this work is to propose a numerical model able to simulate the wavy behavior of metallic glasses ribbon in order to depict the elastic strain limit before plasticity.

2 Experimental Procedure and Results

Normal Force was applied on top of the arc shaped Fe-based metallic glass ribbon with composition $Fe_{90.65}B_{3.9}Cr_{2.75}Si_{2.7}$ using a Tinius Olsen H10kS compression machine. The used dimensions ribbon are: d = 19 μm thickness, 25 mm width and boundary distance 30 (L = 30 mm, h = 7.5 mm and W = 25 mm, named $L_{30}h_{7.5}W_{25}$). The pre-deformed arc with the progressive undulation and the correspondent force versus displacement are shown in the following figure:

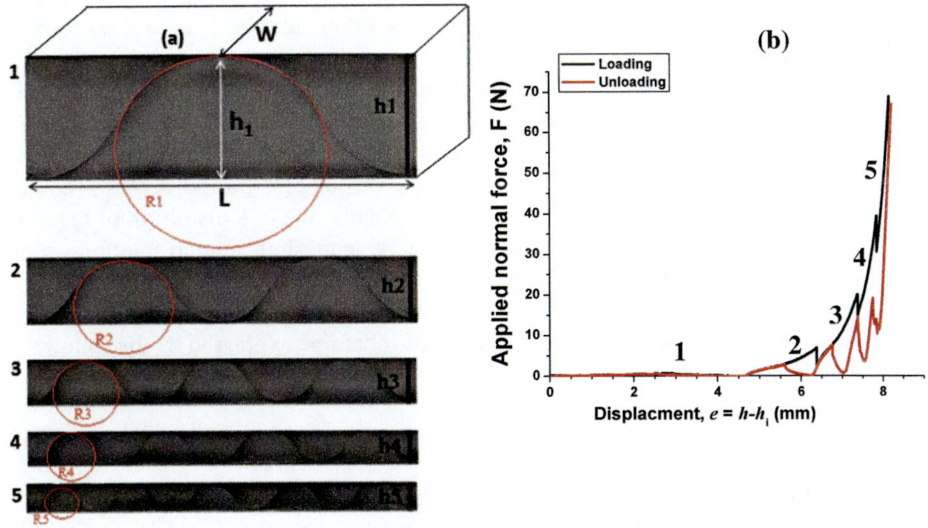

Fig. 1. (a) Arc-shaped metallic glass ribbon with the progressive formed undulations under normal applied loading and (b) Correspondent Applied force F versus the displacement e (e = $h - h_i$: *i indicates number of undulation*).

Figure 1(a) shows the curve of applied force versus displacement during loading and unloading. We note that the curve for unloading is different to loading because when the force is released the ribbon is subjected only to its elastic return. The objective of this work is to simulate the undulatory elastic behavior of pre-deformed metallic glass ribbon in order to estimate the maximum number of undulation reached before plasticity. This result will be useful for miniaturization process.

3 Numerical Modeling of Undulatory Elastic Response

The numerical simulation was performed via Abaqus software using its various options. The initial arc is obtained by applying a longitudinal force P to get a progressive deviation that finally leads to an initial arc with length L and an initial amplitude h. The used ribbon have the geometrical parameters $L_{30}h_{7,5}W_{25}$. By applying a normal load to the initial arc shaped metallic glass ribbon, the transition from the initial arc to next waves is done via a deformation transition defined by a boundary condition corresponding to a critical effort F_c which characterizes the state of equilibrium before the formation of following waves. The following parameters are used:

Young Modulus E = 190 GPa, Poisson ratio ν = 0,36, the boundary distance L = 30 mm, Thickness *d* = 19 μm and the width W = 25 mm.

The following figure shows the obtained numerical results as well as the experiment results for comparison.

Under progressive normal force Y(s) and X(s) represent the variable amplitude *h* and the boundary distance L respectively. According to Fig. 2 until tow formed undulations the numerical results show a good agreement with the experiment. The 3D numerical plots show a stress concentration in the folded zone of the ribbon. This results corroborate our previous work (Yousfi et al. 2017a) that we have proved experimentally that the first failure of the ribbon starts in this zone when it will be subjected to cyclic loading. The calculation code (Abaqus Software) diverges just after the transition of two waves. This divergence is probably due to a distortion of the nuts during the deformation. We're working on the adaptive mesh technique solution which could resolve this problem to get the maximum number of undulation.

In our recent work and based on the evolution of the applied force versus displacement of Fig. 1(b) the evolution of R_i versus h_i could be expressed by the following linear equation (Yousfi et al. 2019):

$$R = n * h - m \tag{1}$$

Where *n* and *m* are parameters that could be identified from the experiment related to $L_{30}h_{7.5}W_{25}$ pre-deformed ribbon and they are found to be equal to 0,814 and 0,145, respectively. The elastic strain ε_{elas} of the used ribbon could be calculated via the following equation:

$$\varepsilon_{elas} = \frac{d}{2 * R} \tag{2}$$

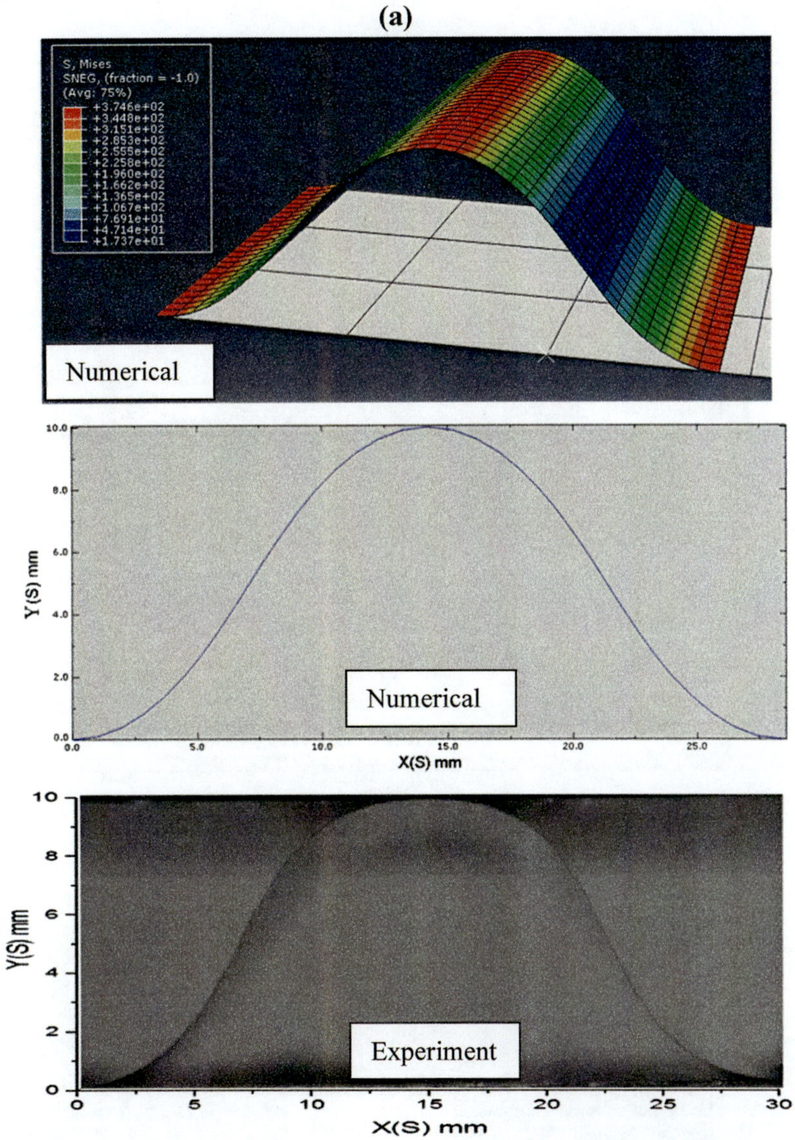

Fig. 2. Numerical results of the undulotary elastic response of metallic glass ribbon compared with the experiment: (a) First arc shaped, (b): transition of deformation and (c) tow formed undulations.

(b)

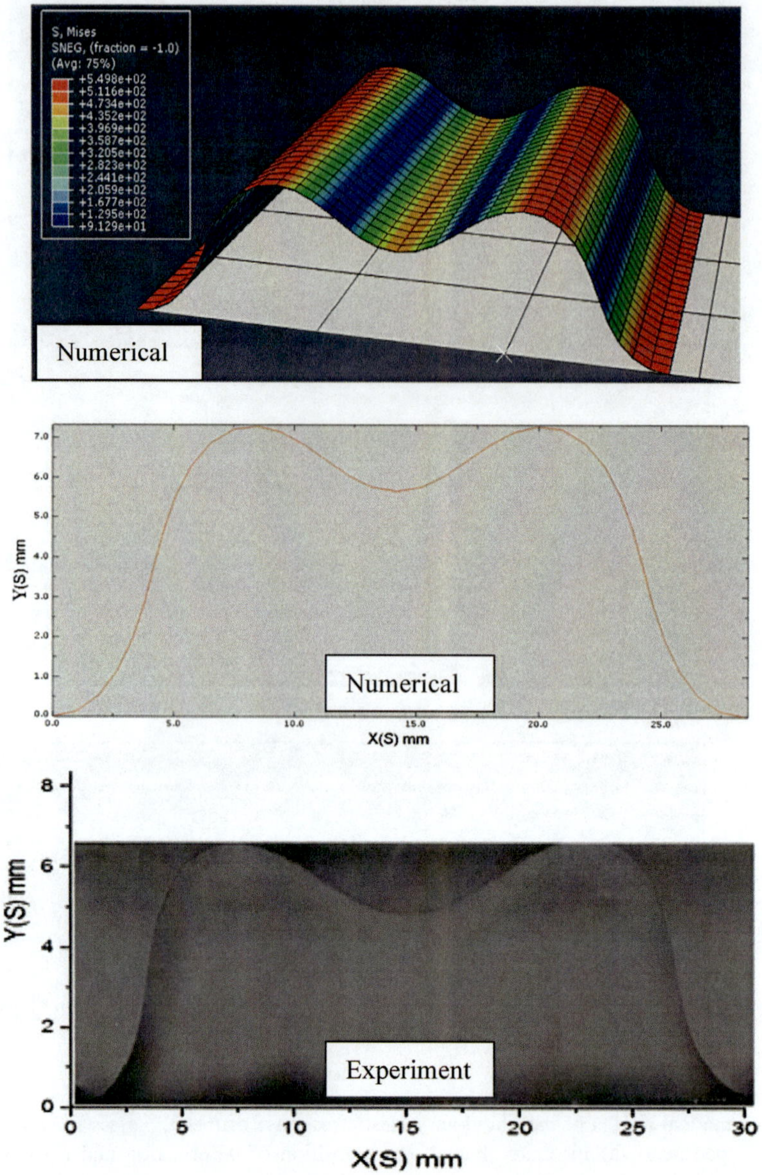

Fig. 2. (*continued*)

(c)

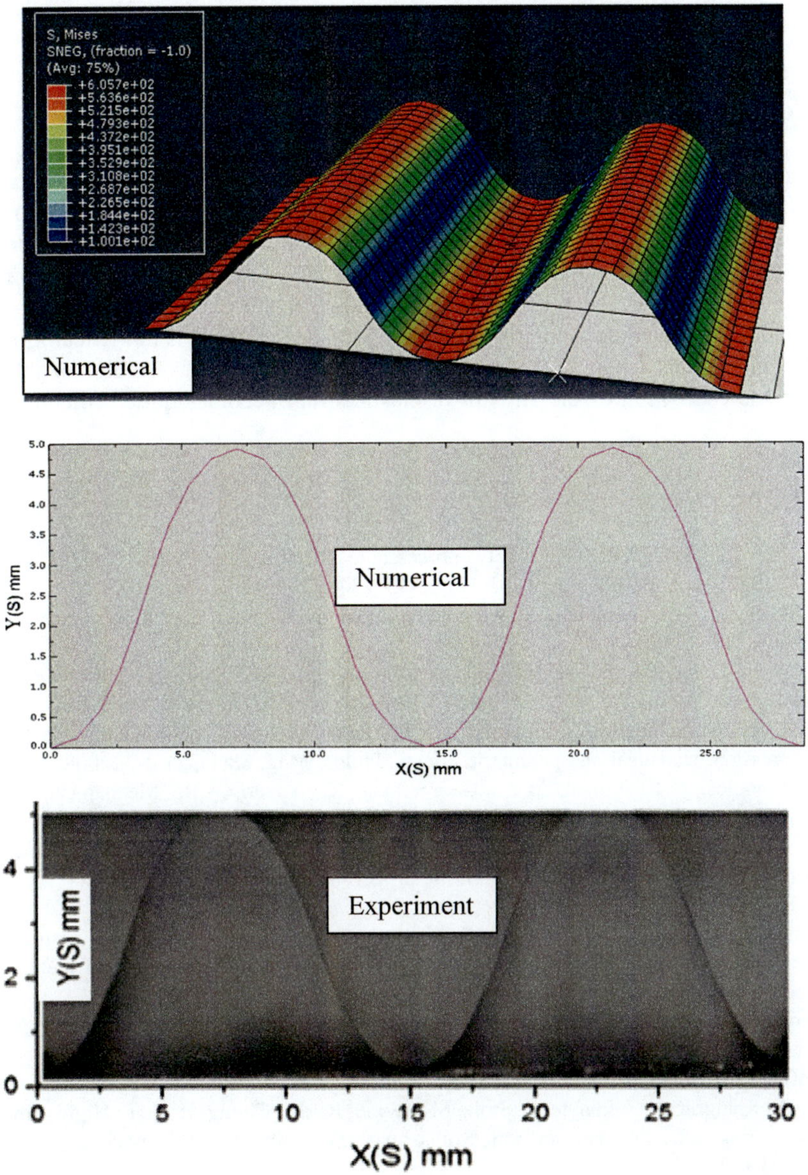

Fig. 2. *(continued)*

For the given boundary distance L and initial amplitude $h_1$the numerical results of Fig. 2 associated with Eq. 1 and Eq. 2 are able to predict the experimental behaviour as well as the elastic deformation following the calculation steps:

MG Ribbon (L, h_1) → amplitude Y(s) or h → curvature radius R → elastic strain ε_{elas}.

4 Conclusion

The undulatory elastic behavior of pre-deformed metallic glass ribbon under progressive normal loading has been investigated via numerical modelling. The aim of the proposed numerical model is to predict the amount of elastic strain associated to maximum number of undulation for a given boundary distance L and initial amplitude h_1. The numerical arc shaping profiles are close to the experiment and the elastic deformation ribbon could be predicted with a good accuracy. This numerical modelling is very interesting when the miniaturization of metallic glass ribbon is requested because the estimation of the amount of elastic strain becomes quite difficult.

References

Peker, A., Johnson, W.L.: A highly processable metallic glass: Zr41.25Ti13.75Cu12.5-Ni10.0Be22.5. Appl. Phys. Lett. **63**(17), 2342–2344 (1993)

Inoue, A.: Stabilization of metallic supercooled liquid and bulk amorphous alloys. Acta Mater. **48** (1), 279–306 (2000)

Wang, W.H., Dong, C., Shek, C.H.: Bulk metallic glasses. Mater. Sci. Eng., R **44**(2–3), 45–89 (2004)

Wu, W.F., Li, Y., Schuh, C.A.: Strength, plasticity and brittleness of bulk metallic glasses under compression: statistical and geometric effects. Philos. Mag. **88**(1), 71–89 (2008)

Panagiotopoulos, N.T., Yousfi, M.A., Georgarakis, K., Yavari, A.R.: Mechanically induced waves in metallic glass foils. Mater. Design **90**, 1110–1114 (2016)

Yousfi, M.A., Panagiotopoulos, N.T., Jorge Junior, A.M., Georgarakis, K., Yavari, A.R.: Novel micro-flat springs using the superior elastic properties of metallic glass foils. Scr. Mater. **131**, 84–88 (2017a)

Yousfi, M.A., Ammari, C., Hajlaoui, K., Panagiotopoulos, N.T., Georgarakis, K., Tourki, Z.: On a new electromechanical switch using the reversible wavy elastic response of metallic glass ribbons. J. Comptes rendus Mecan. **345**, 797–804 (2017b)

Nguyen, D.S., Halvorsen, E., Jensen, G.U., Vogl, A.: Fabrication and characterization of a wideband MEMS energy harvester utilizing nonlinear springs. J. Micromech. Microeng. **20**, 125009 (2010). (11 p.)

Grech, D., Kiang, K.S., Zekonyte, J., Stolz, M., Wood, R.J.K., Chong, H.M.H.: Highly linear and large spring deflection characteristics of a quasi-concertina MEMS device. Microelectron. Eng. **119**, 75–78 (2014)

Yousfi, M.A., Ammari, C., Hajlaoui, K., Loukil, H., Tourki, Z.: Constitutive modelling of wavy elastic response of metallic glass ribbon. Phylos. Mag. Lett. **99**, 414–423 (2019)

Application of the Acoustic Emission Technique for Damage Monitoring of 3D C/C

Jacqueline Saliba[1(✉)], Jalal El Yagoubi[1], and Ludovic Hallo[2]

[1] University of Bordeaux, CNRS, Arts et Métiers Institute of Technology, Bordeaux INP, INRAE, I2M Bordeaux, 33400 Talence, France
{Jacqueline.saliba, jalal.el-yagoubi}@u-bordeaux.fr
[2] CEA-CESTA, 33114 Le Barp, France
Ludovic.HALLO@CEA.FR

Abstract. This paper deals with the detection and monitoring of damage of a three-dimensional carbon/carbon composite (3D C/C) under tensile loading. In fact, due to the manufacturing process, the existing cracking network of 3D C/C at the initial state may be quite extended which may affect the mechanical behavior and in some cases lead to the premature rupture under loading. The objective of this work is to investigate the validity of the acoustic emission as an advanced health monitoring system to provide warning of the distress and characterize the damage state during loading. Load-unload-reload tensile tests have been conducted and monitored in parallel with the Digital Image Correlation (DIC) method and the acoustic emission (AE) technique. The AE technique allows to monitor the evolution of damage inside the volume of the material and the growth in situ. This is very interesting since the other non-destructive techniques usually used allow only a posteriori evaluation of damage. The influences of stress states on tensile properties and the dependency with the AE activity have been analysed. Different stages of damage have been identified based on the acoustic emission activity and AE signals parameters. In addition, the AE technique provides accurate location of damage based on the density of AE events location maps.

Keywords: 3D C/C · Damage · Acoustic emission · Digital Image Correlation

1 Introduction

Three-dimensional carbon/carbon composites (3D C/C composites) are characterized by their ultra-high temperature potential since they can maintain high fracture resistance at high temperatures due to their 3D reinforcement architecture and thus a high structural integrity. In recent years, they received an increasing attention in different fields as aviation, aerospace, naval industry space vehicles and defense ultimate structures as structural materials (Tang et al. 2015; Este et al. 2016). Even though the mechanical behavior of some kinds of 3D C/C composites have been described in the literature, however a better comprehension of the failure process is necessary to simulate the mechanical response under different variations such as loading rate, loading directions, stress, defect sensitivity, temperature, … In addition, few research studied

M. Kharrat et al. (Eds.): ICAMEM 2019, LNME, pp. 327–332, 2020.
https://doi.org/10.1007/978-3-030-52071-7_45

the damage mechanisms assessment of 3D C/C during loading that leads to failure (Almnasour et al. 2015; Breede et al. 2015). Thus, the aim of this study is to propose an experimental investigation that allows the identification of different damage mechanisms under tensile loading and the corresponding AE response/signature. Based on those results, the validity of the AE technique as a structural health monitoring of fracture development under loading will be evaluated (Saliba and Mezhoud 2019). The DIC and the AE techniques are simultaneously applied in the course of the load-unload-reload tensile tests to monitor the fracture process.

First, information about the 3D C/C composite are presented and the experimental program for tensile tests is exposed. Then, the results corresponding to the mechanical response and the influence of the stress states on the AE activity are analyzed.

2 Experimental Procedure

2.1 Uniaxial Tensile Tests

Tensile tests are realized on rectangular specimens which has a length of 64 mm, a width of 16 mm and a thickness of 3.2 mm. Specimens for uniaxial loadings are cut in the X direction considering the same block of the 3D C/C composite in order to minimize the influence related to different batches on the variability of the material properties (Fig. 1a). The surfaces submitted to loading have been polished in order to have a smooth finish. Two aluminium tabs have been fixed at the extremity of the specimens with a resin as strengthening plates. Three similar specimens are at least prepared and tested to evaluate the repeatability. Specimens have been loaded with a displacement control with a speed of 0.5 mm/min using a servo-hydraulic machine with a capacity of 160 kN. Deformations in the longitudinal and transversal direction have been measured with the DIC method. The load and the AE activity data have been registered until failure occurs. The architecture of the material and the experimental setup is shown in Fig. 1b.

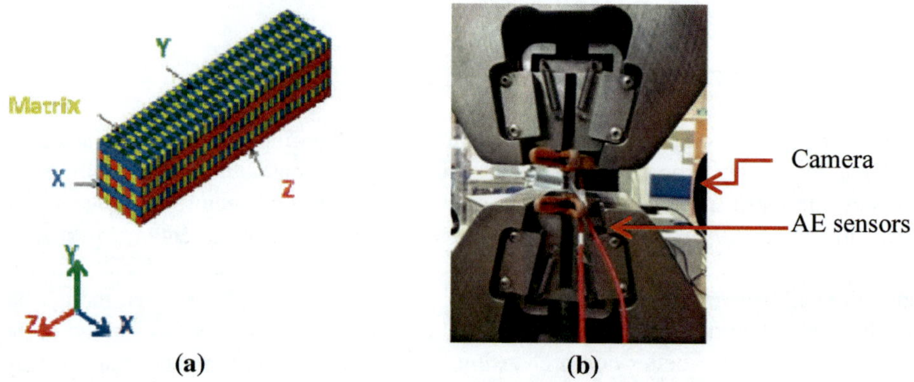

Fig. 1. a) Architecture of the composite b) Specimen mounted in tensile test and monitored with DIC and AE methods

2.2 Digital Image Correlation Technique

The captured images have been registered continuously with a frequency rate of 1 image each five seconds during loading. In order to capture the displacement field of the two opposite faces of the specimen, two cameras have been used. The digital cameras had a macro lens of 100 mm. Both surfaces of the specimens have been covered with a white paint speckle pattern to improve the quality of the images as 256 levels of gray are considered. The cameras presents a resolution of 5184 × 3456 pixels. The resolution obtained was 1 pixel = 86 μm. Two lamps have been also used to increase the luminosity. The images are analyzed afterwards using a commercial software VIC-2D.

2.3 Acoustic Emission Method

The device is constituted of an AEwin system with eight channels, a general-purpose interface bus and a computer for data treatment and saving. The localization of AE events has been performed using 1D algorithm. For this, 2 piezoelectric sensors (nano 30) presenting a frequency range of 125–750 kHz have been used with a resonance frequency of 300 kHz. The sensors have been positioned at a distance of 32 mm. A silicon grease have been used to ensure a good coupling at the surface. The detected signals are amplified with a differential amplifier with a gain of 40 dB. The amplitude of AE signals ranged between 0 and 100 dB. To avoid the background noise evaluated before loading, the threshold of wave's acquisition has been set at 35 dB in order. Before each test, the acquisition is calibrated with the HSU-NIELSEN procedure. In order to locate the source in a 1D analysis, the wave must reach the two sensors. The attenuation of acoustic signals and the effective velocity C have been also evaluated using the distance between the AE transducers ΔL and the time of arrival of the first and second wave Δt before the beginning of the test:

$$C = \frac{\Delta L}{\Delta t} \tag{1}$$

The effective velocity was measured in the XY and Z direction and was equal to 6750 m/s and 8000 m/s respectively. Note here that the velocity is highly related to the initial elastic modulus and the architecture of carbon fibers. This speed has been considered to determine the location of AE events. The signal parameters as the duration, the rise time, the amplitude, the counts, the energy, etc. are also calculated.

3 Results

The formation of a crack is responsible of the emission of energy in the form of an elastic wave. This wave propagates from the micro cracking position in the material and is captured by the AE transducers at the surface. The AE events localization maps are calculated based on the difference between the arrival times of waves at each sensor and the effective velocity. In this study, only the AE events that were detected between the sensors have been considered. The load-unload-reload stress time curve, the corresponding cumulative number of events and the duration of AE hits are plotted during the tensile loading in Fig. 2. In the considered test, the total number of events reached 3200. The damage occurred throughout the different loading levels. In addition, a good correlation has been observed between the AE activity and the loading level. Few AE events have been detected in the beginning with signal of very low energy and duration. The number of the detected AE events increased with increasing the stress and signals presented higher duration and energy. At 80% of the maximum load, the rate of the acoustic emission activity increased indicating an important damage development and then increased drastically due to the brittle failure of the specimens. Note that the final events represents 40% of the total number and the cumulated energy of the detected AE hits. During the unloading and reloading stages, very few AE activity occurred with very low AE energy due to Kaiser effect.

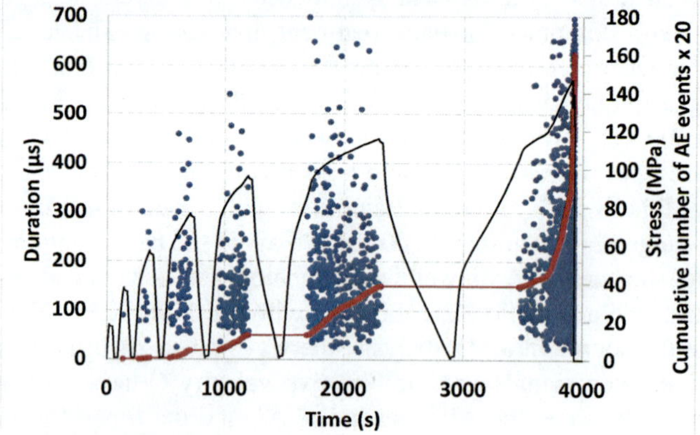

Fig. 2. The stress, the duration and the cumulated number of events during loading/unloading/reloading tensile tests

The density has been also evaluated based on the 1D localization maps of events (Fig. 3a). The number of events is counted in a mesh of 1 cm and every 100 s. The AE events density brings additional information on the zones of higher acoustic activity in function of time. The evolution of the absolute energy of AE events has been also plotted along the x direction and the time for a better evaluation of the fracture rate in parallel (Fig. 3b). In fact, the initiation, propagation and mechanisms of cracking are generally correlated to the parameters of AE signals as the amplitude and absolute acoustic energy (McCrory et al. 2015). The absolute acoustic energy is one of the most important parameters for damage characterization. It is calculated as the integral of the squared voltage over the duration of the signal divided by the resistance of 10 KΩ. The events are classified considering six levels of energy.

The results show that AE events of lower energy (<300 aJ) dominated during tensile loading with a diffuse damage along the x direction. After 60% of the loading level, AE events with higher energy have been detected and localized mainly at the middle due stress concentration near the notch. The different energy levels of AE

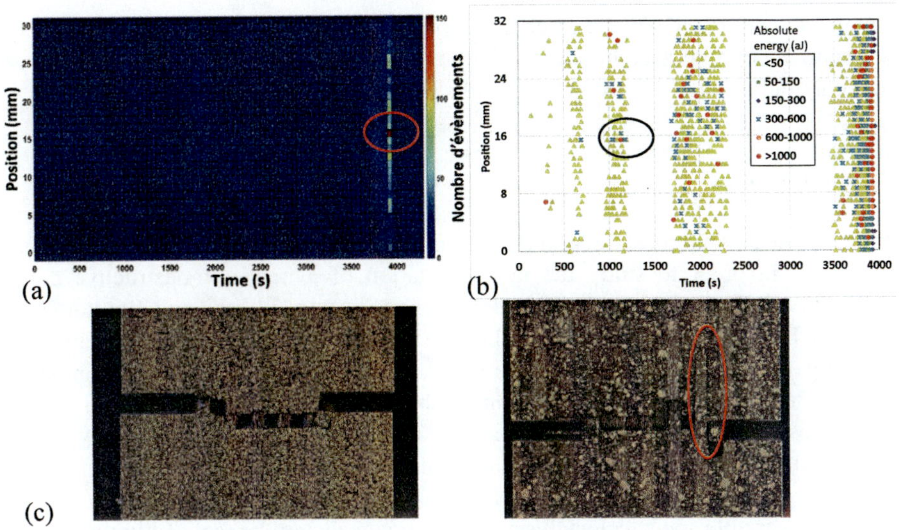

Fig. 3. a) Density map of AE events b) Evolution of the absolute energy of AE events at different loading levels c) Typical failure of specimens under tensile loading.

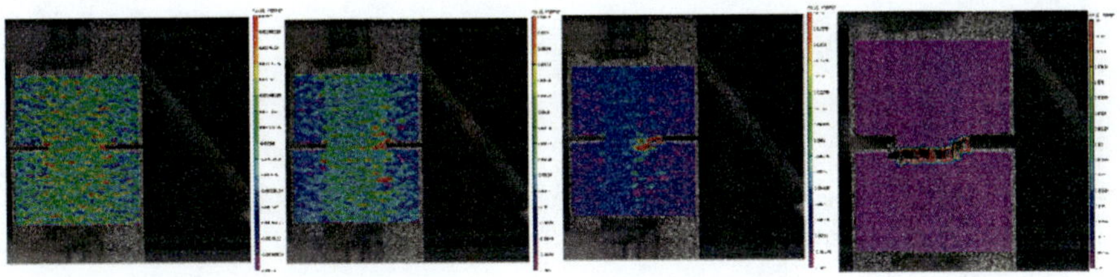

Fig. 4. Vertical deformation obtained with DIC at different loading levels at 40%, 80%, 95% of the maximal strength and at the rupture.

events are related to different rate and source of damage as micro cracks in the areas of high porosity in the matrix that may cause stress concentrations, or the decohesion at the interface between the fibers and the matrix or even the rupture of fibers and friction (Hatta et al. 2005). A Multivariable data clustering analysis can be realized in order to separate the damage mechanisms considering their AE signature (Saliba et al. 2014). At the moment of failure, AE events with energy higher than 1000 aJ have been detected indicating the rupture of the Carbone fibers. Note that with some specimens, the rupture of fibers may not occur at the notch level but rather with a pull out action of the fibers which may also generate AE events due to friction (Fig. 3c). The AE events density shows that the failure occurred at the notch level.

Figure 4 shows the vertical deformation obtained with DIC at 40%, 80%, 95% of the maximal strength and at the rupture. The results show that the deformation is small at low loading levels and is mainly concentrated at the central part of the specimen from the notch limit. Then, the crack initiation and localization begins mainly at 80% of the maximal strength and propagates from the notch level till the rupture indicating a good correlation with AE technique results.

4 Conclusion

This paper deals with the detection and monitoring of damage of a 3D C/C. It has been carried out in order to study the implementation of a new non-destructive control system for the 3D C/C materials. Tensile tests have been conducted and monitored with AE and DIC techniques. The results showed that the descriptors of AE signals varied according to the loading level with higher values as the load increases indicating different damage sources. The location of the crack has also been determined with acceptable accuracy based on the localization maps of the detected acoustic events. All those results show that the acoustic emission technique is a promising technique for monitoring the damage evolution in 3D C/C.

A multivariable statistical analysis of the AE signals will be realized in the future in order to discriminate the damage mechanisms responsible of the appearance of the cracks and the corresponding AE signature.

Acknowledgements. The authors would like to thank the CEA-Cesta who financially supported this work.

References

Este, A., Toson, B., Saliba, J., El Yagoubi, J., Mindeguia, J.-C., Martin, E., Morel, S.: A new approach to simulate interface damage in brittle matrix composites. Procedia Struct. Integrity **2**, 2456–2462 (2016)

Tang, Y., Zhou, Z., Pan, S., Xiong, J., Guo, Y.: Mechanical property and failure mechanism of 3D Carbon-Carbon braided composites bolted joints under unidirectional tensile loading. Mater. Des. **65**, 243–253 (2015)

Almnasour, A., Maillet, E., Ramasamy, S., Morscher, G.N.: Effect of fiber content on single tow SiC minicomposite mechanical and damage properties using acoustic emission. J. Eur. Ceram. Soc. **35**, 3389–3399 (2015)

Breede, F., Koch, D., Maillet, E., Morscher, G.N.: Modal acoustic emission of damage accumulation in C/C-SiC composites with different fiber architectures. Ceram. Int. **41**, 12087–12098 (2015)

Saliba, J., Mezhoud, D.: Monitoring of steel-concrete bond with the acoustic emission technique. Theoret. Appl. Fract. Mech. **100**, 416–425 (2019)

McCrory, J.P., Al-Jumaili, S., Crivelli, D., Pearson, M.R., Eaton, M.J., Featherston, C.A., Guagliano, M., Holford, K.M., Pullin, R.: Damage classification in carbon fiber composites using acoustic emission: a comparison of three techniques. Compos. Part B **68**, 424–430 (2015)

Hatta, H., Goto, K., Aoki, T.: Strengths of C/C composites under tensile, shear, and compressive loading: role of interfacial shear strength. Compos. Sci. Technol. **65**, 2550–2562 (2005)

Saliba, J., Loukili, A., Grondin, F., Regoin, J.-P.: Identification of damage mechanisms in concrete under creep by the acoustic emission technique. Mater. Struct. **47**, 1041–1105 (2014)

Composite Materials in Epoxy Resin Matrix Using Curaua Fibers

Gilberto Garcia Del Pino[1(✉)], Abderrezak Bezazi[2], Haithem Boumediri[2], Antonio Claudio Kieling[1], Jose Luis Valin Rivera[3], Jamile Dehaini[4], and Francisco Rolando Valenzuela Díaz[5]

[1] Department of Mechanical Engineering, State University of Amazonas, Manaus, AM, Brazil
gpino@uea.edu.br, antonio.kieling@yahoo.com

[2] Laboratoire de Mécanique Appliquée des Nouveaux Matériaux (LMANM), Université 8 Mai 1945, Guelma, Algeria
ar_bezazi@yahoo.com, bmdhaithem@yahoo.fr

[3] Pontificia Universidad Católica de Valparaíso, Escuela de Ingeniería Mecánica, Avenida Los Carrera, 01567 Quilpué, Valparaíso, Chile
jose.valin@pucv.cl

[4] Department of Meteorology, Amazonas State University, Manaus, AM, Brazil
jdehaini@uea.edu.br

[5] Department of Materials Engineering and Metallurgy, University of Sao Paulo, Sao Paulo, Brazil
frrvdiaz@usp.br

Abstract. Vegetable fibers have good mechanical properties, low density, no toxicity, are renewable and have low cost among other properties. The use of lignocellulosic fibers as reinforcement in polymeric composite materials in substitution of synthetic fibers is in the direction of sustainability. In the current investigation, an evaluation of biocomposite materials elaborated using curaua fibers as reinforcement in an epoxy resin is carried out. The objective of this work is to highlight the influence of the quantity of fiber and the interaction between the fibers and the resin with different chemical treatments looking for a greater adhesiveness of the fibers in the matrix leading to an increase of the composite tensile properties. The treatments with NaOH consist of varying the concentration and the fiber immersions time in the solutions. Steel molds were used to manufacture the test bodies with dimensions according to the standards for mechanical tensile tests. A Taguchi design of experiment is used to identify the effects of fiber fraction, alkaline treatment concentration and treatment time. The data obtained were statistically evaluated, obtaining good results for Curauá fiber compounds, which may in the future replace glass fiber.

Keywords: Composites · Curauá · Epoxy resin · Vegetable fibres

M. Kharrat et al. (Eds.): ICAMEM 2019, LNME, pp. 333–340, 2021.
https://doi.org/10.1007/978-3-030-52071-7_46

1 Introduction

Ensuring environmental sustainability is required to preserve the ecological balance. The latter is witnessing a sharp deterioration at all levels, especially after the extensive industrial and technological development that led to the spread of pollution that affects the environment and threatens the lives of all living things. In this context, the scientific community is trying to find ways to reduce the use of industrial materials by the available natural resources in order to develop materials and products that respect the environment requirement to maintain its balance. Indeed, in the last decade a big interest of the industrial and researchers is recorded in the literature in biocomposite materials using natural resources as a component or as reinforcement (Ku et al. 2011; Ramesh et al. 2017; Garcia del Pino et al. 2020). Lignocellulosic fibers are one of the most studied natural resources and this is due to its multiple advantages, the most important are: biodegradable, abundant, renewable, exist in fibrous form, low cost, lightness and good mechanical properties. These reasons made them strong candidates to replace the synthetic fibers especially for composites used as secondary structures (Oushabi 2019; Boumediri et al. 2019; Bezazi et al. 2020). The use of vegetable fibers covers a wide domain of applications, from the most traditional to the most complex. They are well known for their importance in the textile industry, but have gained increasing participation in the reinforcement of polymeric materials. The automotive industry, for example, has invested in sisal, coconut, jute and curaua fibers as substitutes for synthetic fibers, because they are lighter, cheaper and safer, since they do not generate sharp edges. On the other hand, the use of vegetable fibers has some disadvantages and challenges to be overcome: low processing and use temperature, low dimensional stability, variability of mechanical properties, cross sections of complex geometries, sensitivity to environmental effects of temperature and humidity, etc. (Silva et al. 2009; Abdul Khalil et al. 2012).

Natural fibers have low compatibility with a polar polymeric matrices and high moisture absorption. The contact surfaces between the phases can act as defects or discontinuities in the material, causing a negative effect on their properties (Marinelli et al. 2008). In order to improve the mechanical properties of the compounds containing these two types of materials, a modification in the fiber surface is necessary through, for example, alkaline treatment (Fiore et al. 2016; Oushabi 2019). This procedure removes part of the lignin, wax and oils that protect the exterior of the fiber, leading to the depolymerization of cellulose. This causes a reduction in fiber hardness and increased surface roughness, increasing its adhesion to the polymer matrix. (Brahmakumar et al. 2005; Li et al. 2007; Mohanty et al. 2001; Porras et al. 2016).

The objective of the work was to determine the influence of fiber quantity and chemical treatment of fiber with Sodium Hydroxide on the mechanical resistance of compounds in epoxy resin matrix. Taguchi experimental design is used for finding the best level for each factor: fiber fraction, alkaline treatment concentration, and treatment time. The data obtained were statistically evaluated using Minitab statistical software and showed good results for Curauá fiber compounds, which may in the future replace glass fiber.

2 Materials and Methods

The fibers used in this work were supplied by the SENAI of Santarém, State of Pará, Brazil, and the resin was supplied by the REDILIS Company of the State of Sao Paulo in Brazil. First, the fibers were benefited and combed to untangle them and remove impurities and were separated into groups, as can be seen in Fig. 1a. L8 Taguchi method is used to evaluation of the effect of the factors and levels which minimize the number of experiments, as showed in Table 1, by the study of the noise impact as presented in Table 2. Recently, many researchers have used this method to analyze and evaluate the mechanical properties of different manufactured biocomposites (Kumar et al. 2014). Subsequently, the fibers were subjected to chemical treatment with sodium hydroxide with different concentrations and different immersion times of the fibers in the solutions. Figure 1 (b and c) shows the preparation of the solutions and the immersion of the fibers in the solutions. The quantity of fiber, the concentrations of the solutions and the immersion times of the fibers in the solutions are shown in Table 1 which corresponds to the Taguchi matrix.

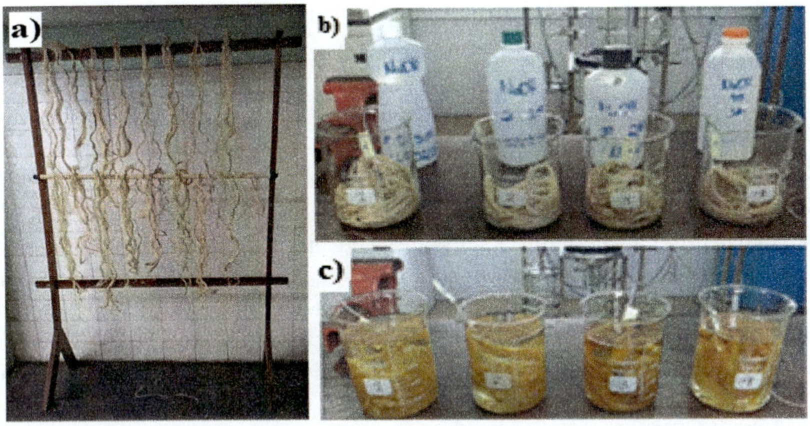

Fig. 1. Fiber treatment: a) Curauá fiber, b) Placement the fibers in the Baker, c) Immersion of the fibers in the solutions

After removing the fibers from the solutions, they were washed many times using distilled water aiming to have a neutral PH and then it was dried at room temperature in a closed room for 24 h and after, in an oven for 8 h at temperature of 60 °C. Before using the fibers in the composites, they were dried again for 20 min at 100 °C in an oven, with the aim of totally eliminating moisture.

The samples for tensile tests were manufactured in a mold with a female cavity in a plate and another male plate that penetrates inside the cavity to be able to apply pressure on the composite material as can be seen in Fig. 2. A cavity plate has 15 mm parallel grooves for fixing the fibers and for the excess resin output. The depth of the grooves in the mold is limited, so that 3 mm depth of the cavity is maintained

Table 1. Factors and levels in the Taguchi matrix

Factor	Type	Levels	Values
Quantity of fiber (% W)	Fixed	4	5, 10, 15, 20
NaOH (%)	Fixed	2	2.5, 5.0
Immersion time (h)	Fixed	2	2, 4

Table 2. Results obtained in tensile tests for different amounts of fibers (Q), different concentration of NaOH (N) and different immersion time (I).

Test bodies	Q (%W)	N (%w)	I (h)	Tensile stress (Mpa)	SNRA1	MEAN1
1	5	2.5	2	34.44 ± 0.44	30.74	34.44
2	5	5.0	4	51.96 ± 0.88	34.31	51.96
3	10	2.5	2	55.55 ± 1.32	34.89	55.55
4	10	5.0	4	61.99 ± 1.54	35.85	61.99
5	15	2.5	4	62.54 ± 2.3	35.92	62.54
6	15	5.0	2	64.55 ± 2.09	36.20	64.55
7	20	2.5	4	95.83 ± 3.23	39.63	95.83
8	20	5.0	2	98.56 ± 3.56	39.87	98.56

throughout the area that will be covered with the composite material (Thickness of the test bodies). In order to laminate the composite material with the fibers, a release agent is first passed through the entire cavity area and the male, then an aluminum foil is placed to guarantee the removal of the composite (see Fig. 2a). The corresponding amount of resin and hardener is introduced into a Baker and stirred by mixing the contents for 15 min and then a layer of resin is placed in the cavity, and after that the fibers are placed in the cavity (see Fig. 2b). At the end of placing the fibers, the rest of the corresponding resin is introduced, it is expected to stand for 5 min and in gel point is placed the lid (Male) and pressure is applied. Then the mold with the test bodies is placed in an oven at the temperature of 80 °C for 8 h to carry out the resin curing process (Fig. 2c) and after they are kept at room temperature for 24 h to eliminate internal stress levels before being subjected to tensile tests.

After the curing process, all samples were subjected to tensile tests to determine the mechanical strength of each composite and study the influence of the fiber quantity and their treatment, on the tensile strength of the compound. Tensile tests were performed using an INSTRON-5984 universal electromechanical testing machine, with a capacity load cell of 150 kN. Tensile tests were performed with a speed of 5 mm/min. Table 2 shows the results of Taguchi's experimental design.

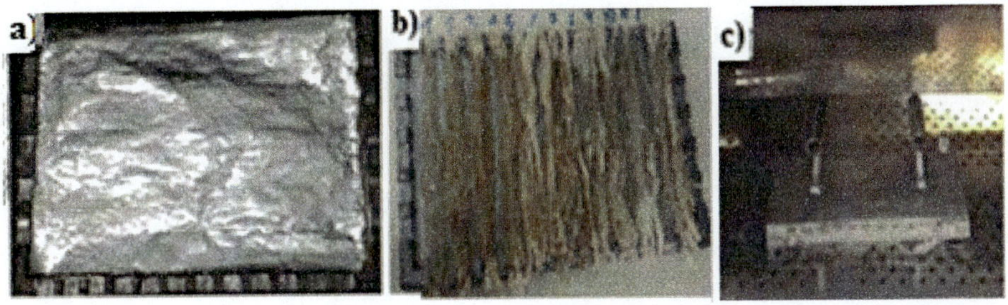

Fig. 2. Manufacturing of test bodies: a) Placement of aluminum foil, b) Placement of Curaua fibers, c) Curing process in a oven.

3 Results and Discussion

In order to analyze the results of the tensile stress response to establish the optimal combination of the parameters, the data was converted to a signal noise (S/N) ratio where the higher is the better (see Table 3).

Table 3. Signal to noise ratios

Level	Signal to noise ratios			Means		
	Amounts of fiber (%W)	NaOH (%W)	Immersion time (h)	Quantity of fiber (%W)	NaOH (%W)	Immersion time (h)
1	32.53	35.30	35.43	43.20	62.09	63.27
2	35.37	36.56	36.43	58.77	69.27	68.08
3	36.06			63.55		
4	39.75			97.19		
Delta	7.22	1.26	1.00	53.99	7.17	4.80
Rank	1	2	3	1	2	3

The Analysis of variance (ANOVA) for the mean tensile stress is presented in Table 4, where P-value highlights the response influences when it is less than or equal to 0.05 and the F-value indicates the factor contributions to the response. In other words, lower is P-value, greater is the influence leading to a large F-value.

From the ANOVA table, it can be shown that the amount of the fibers has the lowest P-value of 0.015 and the highest F-value of 67.09 which is a very influential factor in the tensile strength of the biocomposite and remains the main factor affecting the response. The regressions R-sq, the adjusted and the predicted ones have values of 99.59%, 97.14% and 73.84%, respectively (see Table 5). Regression analysis was performed by developing a mathematical model established on the relationship between the control factors and the output response. Equation 1 shows the developed regression model for tensile stress and the experimental data with the predicted value presented in Fig. 3, which shows a very good agreement.

Table 4. ANOVA analysis for tensile stress

Source	DF	Adj SS	Adj MS	F-Value	P-Value
Quantity of fiber (% W)	3	3101.70	1033.90	67.09	0.015
NaOH (%)	1	102.96	102.96	6.68	0.123
Immersion time (h)	1	46.18	46.18	3.00	0.226
Error	2	30.82	15.41		
Total	7	3281.66			

Table 5. Model summary

S	R-sq	R-sq (adj)	R-sq (pred)
3.66281	99.59%	97.14%	73.84%

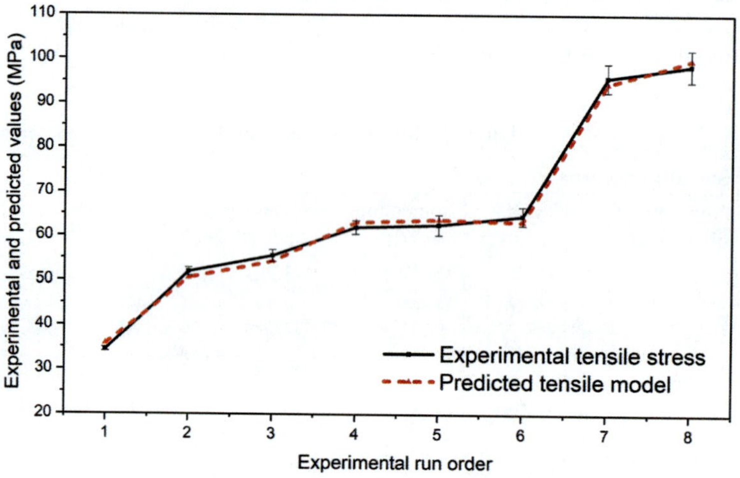

Fig. 3. Experimental data compared to the predicted ones using Eq. 1.

Regression Equation

$$\text{Tensile strength} = 84,8 + 2,17Q - 23,29N - 20,0I + 0,1808QQ + 7,93NI - 0,590QI \quad (1)$$

Tensile tests were performed with epoxy resin in pristine condition (without fibers), to compare with the results obtained, and these results are shown in Table 6.

Table 6. Tensile stress pristine material

Test bodies	Q (%W)	N (%w)	I (h)	Tensile stress (Mpa)
Pristine condition	0	0	0	23.73 ± 0.05

4 Conclusion

Biocomposite materials made of epoxy resin and *Curauá* fibers were investigated and their mechanical properties evaluated using Taguchi method. The Analysis of variance (ANOVA) reveals that the amount of the fibers is the main factor affecting the ultimate tensile stress, followed by the sodium hydroxide concentration and, finally, the immersion time. The best ultimate tensile stress of 98.56 MPa obtained for biocomposite having 20% of *Curauá* fibers treated with 5% NaOH for two hours, which represents an increases of 415.34% compared to the epoxy resin having an average value of 23.73 MPa.

Acknowledgments. The authors would like to thank the collaboration of the University of the State of Amazonas (UEA) and the São Paulo State University (USP).

References

Abdul Khalil, H.P.S., Bhat, A.H., Ireana Yusra, A.F.: Green composites from sustainable cellulose nanofibrils: a review. Carbohydr. Polym. **87**, 963–979 (2012)

Bezazi, A., Amroune, S., Scarpa, F., Dufresne, A., Imad, A.: Investigation of the date palm fiber for green composites reinforcement: quasi-static and fatigue characterization of the fiber. Ind. Crops Prod. **146**, 112135 (2020)

Boumediri, H., Bezazi, A., Del Pino, G.G., Haddad, A., Scarpa, F., Dufresne, A.: Extraction and characterization of vascular bundle and fiber strand from date palm rachis as potential bio-reinforcement in composite. Carbohydr. Polym. **222**, 114997 (2019)

Brahmakumar, M., Pavithran, C., Pillai, R.M.: Coconut fibre reinforced polyethylene composites: effect of natural waxy surface layer of the fibre on fibre/matrix interfacial bonding and strength of composites. Compos. Sci. Technol. **65**, 563–569 (2005)

Garcia del Pino, G., Kieling, A.C., Bezazi, A., Boumediri, H., de Souza, J.F.R., Valenzuela Díaz, Francisco, Valin Rivera, J.L., Dehaini, J., Panzera, T.H.: Hybrid polyester composites reinforced with curauá fibres and nanoclays. Fibers Polym. **21**(2), 399–406 (2020)

Fiore, V., Scalici, T., Nicoletti, F., Vitale, G., Prestipino, M., Valenza, A.: A new eco-friendly chemical treatment of natural fibres: effect of sodium bicarbonate on properties of sisal fibre and its epoxy composites. Compos. Part B Eng. **85**, 150–160 (2016)

Ku, H., Wang, H., Pattarachaiyakoop, N., Trada, M.: A review on the tensile properties of natural fibre reinforced polymer composites. Compos. Part B Eng. **42**, 856–873 (2011)

Li, X., Tabil, L.G., Panigrahi, S.C.: Chemical treatments of natural fiber for use in natural fiber-reinforced composites. A review. J. Polym. Environ. **15**, 25–33 (2007)

Kumar, R., Kumar, K., Bhowmik, S.: Optimization of mechanical properties of epoxy based wood dust reinforced green composite using Taguchi method. Process. Mater. Sci. **5**, 688–696 (2014)

Marinelli, A.L., Monteiro, M.R., Ambrósio, J.D.: Desenvolvimento de Compósitos Poliméricos com Fibras Vegetais Naturais da Biodiversidade: Uma Contribuição para a Sustentabilidade Amazônica. Polímeros: Ciência e Tecnologia **18**(2), 92–99 (2008)

Mohanty, A., Misra, M., Drzal, L.: Surface modifications of natural fibres and performance of the resulting biocomposites: an overview. Compos. Interfaces **8**(5), 313–343 (2001)

Porras, A., Maranona, A., Ashcroft, B.A.: Optimal tensile properties of a Manicaria-based biocomposite by the Taguchi method. Compos. Struct. **140**, 692–701 (2016)

Ramesh, M., Palanikumar, K., Reddy, K.H.: Plant fibre based bio-composites: sustainable and renewable green materials. Renew. Sustain. Energy Rev. **79**, 558–584 (2017)

Silva, R., Haraguchi, S.K., Muniz, E.C., Rubira, A.F.: Aplicações de Fibras Lignocelulósicas na Química de Polímeros e em Compósitos. Quim. Nova **32**(3), 661–671 (2009)

Oushabi, A.: The pull-out behavior of chemically treated lignocellulosic fibers/ polymeric matrix interface (LF/PM): a review. Compos. B **174**, 107059 (2019)

Composite Cross-Ply Laminates Stacking Sequence Effect on Post Flexural Fatigue Residual Strength

Abderrezak Bezazi[1(✉)], Abderrahim El Mahi[2], Boudjema Bezzazi[3], Gilberto Garcia Del Pino[4], and Fabrizio Scarpa[5]

[1] Laboratoire de Mécanique Appliquée des Nouveaux Matériaux (LMANM), Université 8 Mai 1945, Guelma, Algeria
ar_bezazi@yahoo.com, bezazi.abderrezak@univ-guelma

[2] Laboratoire d'Acoustique LAUM UMR CNRS 6613, Avenue Olivier Messiaen, 72085 Le Mans Cedex 9, France
abderrahim.elmahi@univ-lemans.fr

[3] Research Unit Materials, Processes and Environment, University M'Hamed Bougara, 35000 Boumerdes, Algeria
bbezzazi@yahoo.fr

[4] Department of Mechanical Engineering, State University of Amazonas, Manaus, AM, Brazil
ggdelpin@usp.br

[5] Department of Aerospace Engineering, University of Bristol, Queens Building, University Walk, Bristol BS8 1TR, UK
F.Scarpa@bristol.ac.uk

Abstract. This paper concerns an experimental approach to the stacking sequence influence on cross-ply laminates behaviour. Effects of fatigue on the residual strength and stiffness are investigated through performing post fatigue static test after specific numbers of cycles. Residual strengths were measured by testing specimens to a certain number of cycles then static tests to rupture were performed after the specimen had been unloaded without being removed. The results obtained indicate that as well as there is a drop in the specimen's stiffness, there is a drop in the strength of the laminate as the number of cycle increases. However, this is also accompanied by an increase in the displacement mainly at higher number of cycles. The hysteresis loops revealed that a significant decrease in the maximum fatigue load was occurring as the number of cycles increased. It is also noted that the slope of the hysteresis loops decreased with respect to the number of cycles. However, a less significant drop in the minimum fatigue load than the maximum fatigue one was also observed.

Keywords: Cross-ply laminate · Stacking sequence · Flexural · Residual strength

M. Kharrat et al. (Eds.): ICAMEM 2019, LNME, pp. 341–347, 2020.
https://doi.org/10.1007/978-3-030-52071-7_47

1 Introduction

Composite materials are considered to be one of the most important and advanced discoveries due to their numerous advantage characteristics such as lightweight, non-corrosive, good resistance to fatigue, high specific strength and specific stiffness, so they are easily manufactured and can be adapted to performance requirements Bezazi et al. (2003). These multiple advantages have contributed to the development and progress of many technological sectors such as aerospace, marine engineering, defence, energy, oil, and gas, automotive, rail and sporting goods, etc. Sanjay et al. (2015), Agarwal et al. (2017). The use of composites in transportation reduces weight by more than 20%, which results in lower fuel consumption and CO_2 emission. Also, the increased use of composite materials has developed new rapid automated manufacturing techniques, which have reduced the overall manufacturing costs, making the end product affordable to a broader audience Beaumont et al. (2017).

At the beginning of the commercial aircraft manufacturing, composite materials were not used or used in small quantities; Boeing 707 was manufactured with only 2% of all the materials. However, in the latest commercial aircraft as Boeing 787, the composite materials are exceeding 50% of the total mass with an average weight savings of around 20% Zhang et al. (2017).

The fatigue behaviour of composite materials has been studied in the past because it is an important parameter when choosing or manufacturing them for any application. Several investigators have studied the cumulative damage models (CD), also known as shock models. According to Hashin (1985), the CD can be represented in terms of a damage function which must obey certain conditions. In the literature, different residual life theories can be obtained simply by assuming a general functional dependence of the damage function on fatigue variables or by assuming reasonable mathematical forms of residual strength curves.

Caprino et al. (2005) have used acoustic emission (AE) technique for monitoring flexural fatigue damage in glass fibre reinforced plastic specimens, pre-fatigued up to predetermined portions of their fatigue life. They found that the total number of AE events recorded at maximum stresses correlated well empirically with the material residual strength. They then suggested an empirical relationship between the AE events count and the residual strength, which can be a useful way of assessing the composite integrity if the damage is continuously monitored.

Tserpes et al. (2004) used C-scan technique in their experimental investigation of CFRP composite laminates and used progressive damage modelling to assess fatigue damage accumulation and residual strength of glass fibre reinforced plastic laminates under static and fatigue loading.

Currently, many researchers have done much work in the area of composite fatigue behaviour. Nikforooz et al. (2018) have investigated the stiffness degradation and fatigue life of unidirectional and cross-ply composite (glass/epoxy and glass/polyamide. The authors found that the composite $[(0/90)_4]_s$ has greater stress and displacements under static loading than $[0_2/90_2)_s]_s$. However, the later resists better to fatigue. In other words, the static tests are not sufficient to characterize the strength of the composite material. Additionally, the influence of the stacking sequence lay-up configuration on compressive, tensile, flexural and interlaminar shear properties (ILSS) of glass fiber reinforced epoxy (GFRE) having a lay-up of [0°]s, [90°]s, [0°/90°]s and randomly oriented [45°/45°]s has been studied by Seshaiah and Vijaya Kumar Reddy (2018). However, recent studies have focused on studying the use of Glass fiber reinforced plastics (GFRPs) as it is used in various applications because of its cost-effectiveness. Suzuki et al. (2019) have shown a new model of phenomenological fatigue based on the degradation of rigidity in order to predict the lifespan of GFRPs under random ocean waves. The authors found that the predicted values were more precise than those obtained by conventional methods (Palmgren-Miner's rule). Similar research was also performed by Liu et al. (2020).

From this study, two types of cross-layered GFRP laminates are tested in a three-point bending test to assess fatigue strength and residual bending stiffness, which is necessary to understand their fatigue behaviour. The objective of this research is to study the residual properties as a function of the stress levels and the number of fatigue load cycles, which makes it possible to understand the mechanical behaviour caused by the fatigue load. The samples were tested at a number of cycles to measure the residual resistances and then static tests to rupture were performed after the specimen had been unloaded without being removed.

2 Experimental Procedure

2.1 Materials

Two types of cross-ply laminates made up of epoxy resin and glass fibers having the same number of plies oriented at 0° and 90°, differentiated by their layers thicknesses. The laminates have a lay-up of $[0/90)_4]$s and $[0_2/90_4/0_2]$s. The laminates are prepared in vacuum by moulding to obtain 300 × 300 mm^2 plates. The plies are laminated and impregnated at room temperature, then cured under vacuum pressure (30 kPa) for 10 h. The plates were cut out with a diamond saw according to ASTM standard D 790-8a.

2.2 Experimental Setup and Test Procedure

The investigation of the specimens was carried out under 3-point bending using a universal hydraulic monotonic testing machine (INSTRON model 8516 of capacity ±100 kN) whose control and data acquisition was performed by a computer. The applied load was monitored with a 5 kN load cell, the displacement by an LVDT sensor, and the strain using an extensometer. The three supports were of the cylindrical shape of 10 mm diameter. A minimum of five tests was carried out for each type of specimen, the loading rate in the static tests being 2 mm/min. The flexural fatigue was performed under displacement control at a mean displacement of 45% of the ultimate static failure displacement with a sinusoidal waveform of 10 Hz. Residual strengths were measured by testing specimens to a certain number (1, 5 10^3, 10 10^3, 30 10^3, 70 10^3, 100 10^3 and 400 10^3) of cycles then static tests to rupture were performed after the specimen had been unloaded without being removed.

3 Results and Discussion

The flexural stress versus displacement is presented in Fig. 1. This figure shows that the rigidity of both laminates is equivalent. However the flexural stress is more higher for the laminate $[(0/90)_4]$s then that for $[0_2/90_4/0_2]$s. This is due to the different rupture mechanisms occurred during the loading, i.e. the failure of the former is less brutal than the one of the latter.

During fatigue tests, the evolution of the maximum load F versus the number of cycles N has been recorded. To obtain stiffness degradation the maximum load F was normalized to that obtained at the first cycle F_0 (Fig. 2). The analysis of stiffness degradation constitutes one of the methods most widely used to follow the progression of the damage by the fatigue of the composites. Figure 2 shows that the stiffness degradation is more pronounced for the laminate $[(0/90)_4]$s than for $[0_2/90_4/0_2]$s, i.e. the later resists better to fatigue.

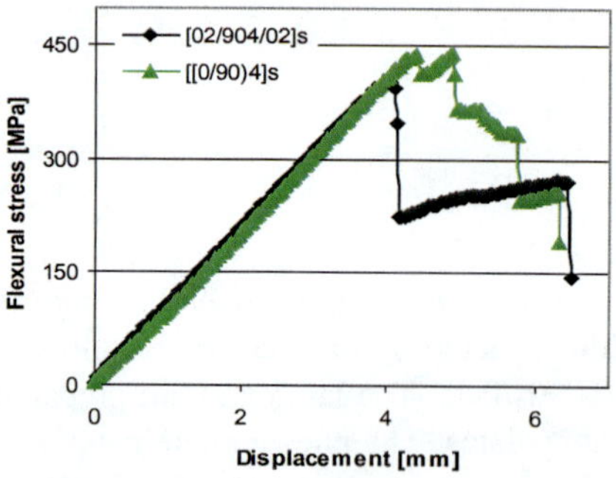

Fig. 1. Static tests.

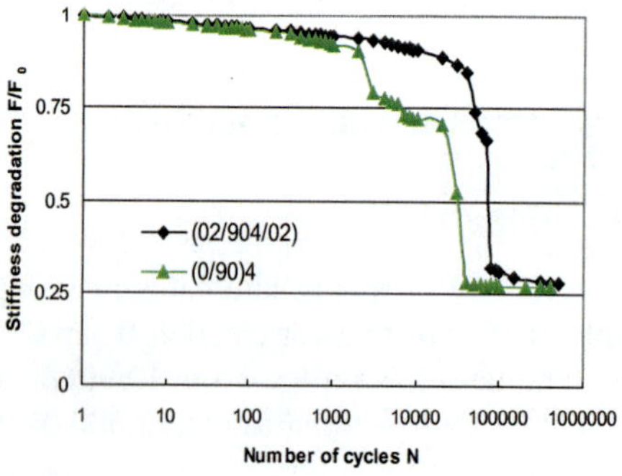

Fig. 2. Stiffness degradation versus number of cycles.

The hysteresis loops for both laminates presented in Fig. 3 reveal that a significant decrease in the maximum fatigue load is occurring as the number of cycles increases. It is also noted, from Fig. 3, that the slope of the hysteresis loops decreases in respect to the number of cycles. However, a less significant drop in the minimum fatigue load than the maximum fatigue one is also observed.

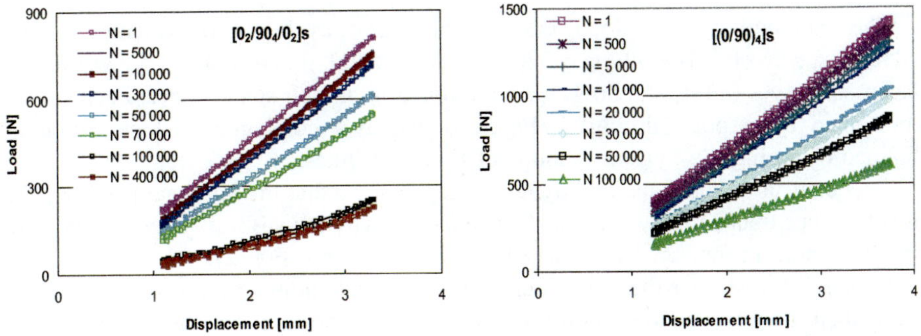

Fig. 3. Hysteresis loop versus number of cycles N for both studied laminates.

Figure 4 shows the load-displacement curves of both laminates tested under static and post fatigue static loading for specimens subjected to different fatigue loading. The curves indicate that as well as there is a drop in the specimen's stiffness, there is a drop in the strength of the laminate as the cycle number *N* increases. However, this is also accompanied by an increase in the displacement mainly at a higher number of cycles. The bending strength of the pre-fatigued specimen decreases as *N* increases relative to the monotonic one. The results are summarised in Table 1.

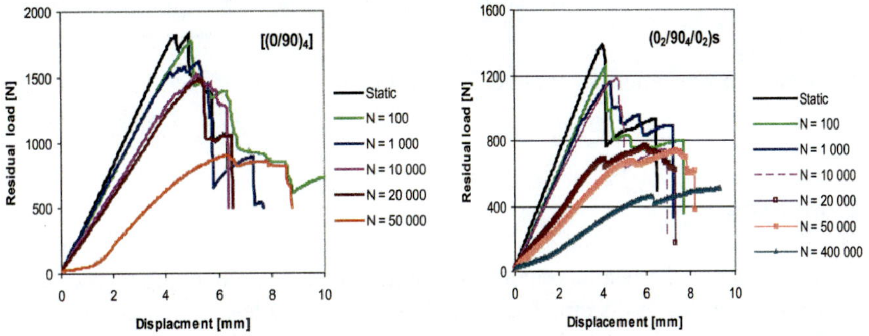

Fig. 4. Residual load versus displacement.

Table 1. Reduction of the strength to a certain number of cycles for both studied laminates.

Laminate	Reduction of the strength (%) at different number of cycles N					
	100	1000	10000	20000	50000	4×10^5
$[(0/90)_4]s$	3.16	11.40	17.05	18.47	50.86	–
$[0_2/90_4/0_2]s$	9.75	15.30	16.02	46.75	47.33	63.35

The static and dynamic slops versus the number of cycles are represented in Fig. 5 and 6, respectively. The fitting of the linear parts of the residual load versus displacement curves of Fig. 4, allows the determination of their static slopes. Each slope is represented by a point in Fig. 5. Figure 6 displays the dynamic slops which are obtained by fitting the hysteresis loops of Fig. 3 by lines.

The analysis of the static and dynamic slops shows that the laminate $[(0/90)_4]s$ has greater values than $[0_2/90_4/0_2]s$ one. This explains why the stiffness degradation of the former is more pronounced than the one of the latter (see Fig. 2).

Figure 6 shows that the behaviour of the dynamic slops is the same until 10000 cycles then there is a sudden drop for $[0_2/90_4/0_2]s$, however it decreases gradually for $[(0/90)_4]s$. This explains the behaviour already mentioned in static tests (see Fig. 1).

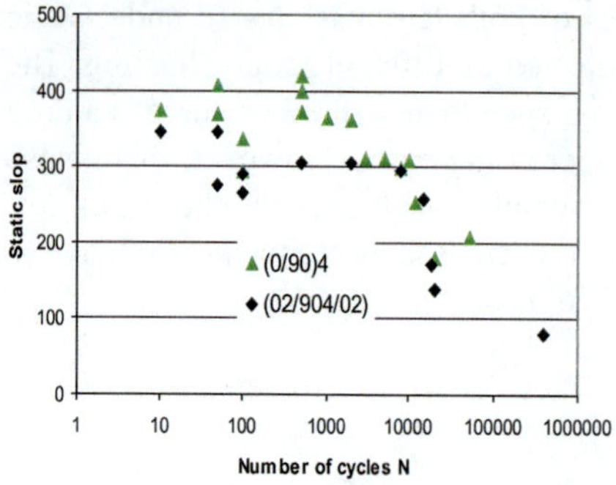

Fig. 5. Static slop versus number of cycles N.

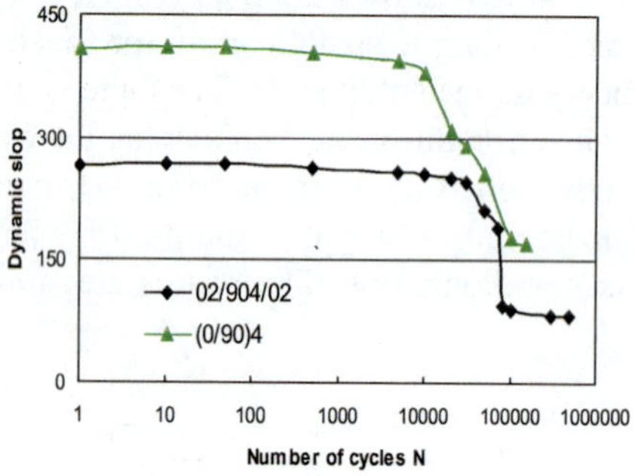

Fig. 6. Dynamic slop versus number of cycles N.

4 Conclusion

From the analysis of the results obtained of monotonic 3-point bending tests carried out on glass/epoxy laminates specimens, pre-fatigued up to predetermined portions of their fatigue life, the following conclusions are drawn:

- The static tests are not sufficient to characterize the strength of the composite material because $[(0/90)_4]s$ has greater stress and displacements than $[0_2/90_4/0_2]s$. However; for the fatigue tests the former has lower stiffness degradation and static and dynamic slops, i.e. that the latter resists better to fatigue.
- The laminate $[0_2/90_4/0_2]s$ has a brutal failure in static and in fatigue compared to the $[(0/90)_4]s$ one.

References

Bezazi, A., El Mahi, A., Berthelot, J.-M., Bezzazi, B.: Flexural fatigue behaviour of cross-ply laminates. An experimental approach. Strength Mater. **35**(2), 66–83 (2003)

Sanjay, M.R., Arpitha, G.R., Yogesha, B.: Study on mechanical properties of natural-glass fibre reinforced polymer hybrid composites: a review. Mater. Today: Proc. **2**(4–5), 2959–2967 (2015)

Agarwal, B.D., Broutman, L.J., Chandrashekhara, K.: Analysis and Performance of Fiber Composites. Wiley, Hoboken (2017)

Beaumont, P.W.R., Soutis, C., Johnson, A.: Fitness considerations for contemporary composite materials: (who's afraid of the composite micro-crack?). Appl. Compos. Mater. **24**(6), 1265–1285 (2017)

Zhang, B., Soltani, S.A., Le, L.N., Asmatulu, R.: Fabrication and assessment of a thin flexible surface coating made of pristine graphene for lightning strike protection. Mater. Sci. Eng., B **216**, 31–40 (2017)

Liu, Z., Yang, Z., Chen, Y., Yu, Y., Wei, Y., Li, M., Huang, C.: Dynamic tensile and failure behavior of bi-directional reinforced GFRP materials. Acta. Mech. Sin. **36**(2), 460–471 (2020)

Hashin, Z.: Cumulative damage theory for composite materials: residual life and residual strength methods. Compos. Sci. Technol. **23**, 1–19 (1985)

Caprino, G., Teti, R., de Iori, I.: Predicting residual strength of pre-fatigued glass fibre-reinforced plastic laminates through acoustic emission monitoring. Compos.: Part B **36**, 365–371 (2005)

Tserpes, K.I., Papanikos, P., Labeas, G., Pantelakis, S.: Fatigue damage accumulation and residual strength assessment of CFRP laminates. Compos. Struct. **63**(2), 219–230 (2004)

Nikforooz, M., Montesano, J., Golzar, M., Shokrieh, M.M.: Fatigue behavior of laminated glass fiber reinforced polyamide. Procedia Eng. **213**, 816–823 (2018)

Seshaiah, T., Vijaya Kumar Reddy, K.: Effect of fiber orientation on the mechanical behavior of e-glass fibre reinforced epoxy composite materials. Int. J. Mech. Prod. Eng. Res. Devel. (IJMPERD), 379–396 (2018)

Suzuki, T., Mahfuz, H., Takanashi, M.: A new stiffness degradation model for fatigue life prediction of GFRPs under random loading. Int. J. Fatigue **119**, 220–228 (2019)

Impact Simulation of PA66-GF Composites Using Finite Solid Shell Element

S. Koubaa[1(✉)], A. Chaker[1,2], J. Mars[1], A. Vivet[2], and F. Dammak[1]

[1] Electro-Mechanical System's Laboratory (LASEM), National Engineering School of Sfax, University of Sfax, B.P 1173, 3038 Sfax, Tunisia
{sana.kouba, fakhreddine.dammak}@enis.tn, aidacheker811@gmail.com, jamelmars@yahoo.fr
[2] Normandie Univ, ENSICAEN, UNICAEN, CEA, CNRS, CIMAP, 14000 Caen, France
alexandre.vivet@unicaen.fr

Abstract. This study is focusing on the finite element simulation of low velocity impact applied to glass fiber-reinforced composites. Elasto-visco-plasticity is adopted taking into account strain rate effect, with isotropic hardening. To avoid numerical locking, hexahedral solid shell finite element SS is developed. The formulation is based on the mixed Assumed Natural Strain (ANS) and the Enhanced Assumed Strain (EAS) methods. A geometrical non linear condition is applied. The finite element is implemented in UEL subroutine via the commercial software ABAQUS/Standard. The numerical element proved its efficiency comparing to conventional element and previous experimental results in (Mars et al. 2018). Numerical simulations with solid element are conducted using Abaqus/standard via user material UMAT subroutine. In terms of computation time, the solid shell element is less time consuming when comparing to conventional solid element. This is due to the fact that a single element through the thickness direction is adopted. Element efficiency is assessed for various fibers contents.

Keywords: Hexahedral solid shell element · EAS/ANS · PA66-GF · Impact · Abaqus · UEL · UMAT

1 Introduction

Known for their advanced mechanical properties (lightweight, high strength, resistance to corrosion), PA66 plastic composites are becoming the dynamic based components in automotive, aerospace market and security (Ku et al. 2011; AlMaadeed et al. 2012). It is well known that, when subjected to dynamic impact solicitations, PA66-GF composite structures are susceptible to high strain causing internal damage evolution until failure (Mouti et al. 2013). This presents a major concern in several works (Chebbi et al. 2016; Mars et al. 2018; Thomason et al. 2009). In most of these researches, it was assessed through experimental, analytical and numerical studies that resistance to impact loading is strongly depending on several aspects. One can distinguish the material properties and geometry effects (mainly fiber contents and thickness) as well

M. Kharrat et al. (Eds.): ICAMEM 2019, LNME, pp. 348–354, 2020.
https://doi.org/10.1007/978-3-030-52071-7_48

as impactor properties (nose shape) and process parameters influence (initial kinematic velocity, energy, contact conditions).

In previous study, (Mars et al. 2018) have conducted experimental and numerical simulations of impacted glass fiber-reinforced polyamide behavior using Abaqus/standard when varying fiber content. The conventional 8-node linear 38350 finite elements was used with 10 elements through the thickness to avoid numerical locking problems. However, this solution proved a time simulation consuming. An attractive alternative to conventional solid and shell elements is the use of free-locking 8-nodes solid-shell element.

Solid shell element is projected by (Klinkel et al. 2006) and (Vu-Quoc and Tan 2003). The formulation is based on the mixed Assumed Natural Strain (ANS) and the Enhanced Assumed Strain (EAS) methods. The first method (Dvorkin and Bathe 1981) is used to solve transverse shear locking. The second one (Simo and Refai 1990) is used to solve Poisson thickness and volumetric locking.

Based on the literature survey, one can deduce that there are limited studies on the low velocity impact response of short glass fiber-reinforced PA66. Moreover, there is a lack of interest on computational efficiency of the developed formulations. The application of an efficient free-locking solid shell element is restricted to metal and FGM materials (Hajlaoui et al. 2012; Chalal and Abed-Meraim. 2018). The majority of researchers used home code to implement the solid shell formulation and no further application on contact problem, mainly impact, can be found.

This paper is focusing on the finite element simulation of low velocity impact applied to glass fiber-reinforced composites. Elasto-visco-plasticity is adopted taking into account strain rate effect, with isotropic hardening. To avoid numerical locking, hexahedral solid shell finite element SS is developed based on mixed EAS/ANS methods. The FSDT is used (Koubaa et al. 2019; Chaker et al. 2020). The SS element is implemented in UEL subroutine via ABAQUS/Standard software. A single element is considered through the solid thickness for efficiency.

2 Formulation of Solid Shell Element

The developed solid shell element in this work is an eight nodes hexahedral finite element with translation degrees of freedom. The formulation of this element (Klinkel et al. 2006; Vu-Quoc and Tan 2005) is obtained by combining the Assumed Natural Strain (ANS) and the Enhanced Assumed Strain (EAS) methods in order to solve transverse shear and volumetric locking problems caused by the conventional finite element. The ANS method was proposed by (Dvorkin and Bathe 1984) in order to solve transvers shear locking problem. This method consists on enhancing the transverse and shear terms (E^c_{33}, E^c_{13}, E^c_{23}) of the compatible strain tensor $\mathbf{E}^c$ as:

$$E^c_{33} = \sum_{P=1}^{4} \frac{1}{4}(1 + \xi_P\xi)(1 + \eta_P\eta)E^P_{33} \tag{1}$$

$$\begin{bmatrix} 2E^c_{13} \\ 2E^c_{23} \end{bmatrix} = \begin{bmatrix} (1-\eta)E^B_{13} + (1+\eta)E^D_{13} \\ (1-\xi)E^A_{23} + (1+\xi)E^C_{23} \end{bmatrix} \tag{2}$$

A_i (i = 1, 2, 3, 4) and A, B, C, D are the transverse interpolation points.

However, the EAS method consists on enhancing the compatible Green-Lagrange strain tensor $\mathbf{E}^c$ with an incompatible one $\tilde{\mathbf{E}}$. This method was developed by (Simo and Rifai 1990) for the purpose of solving volumetric and thickness locking problems. The Green-Lagrange strain tensor **E** is expressed as function of these tensors:

$$\mathbf{E} = \mathbf{E}^c + \tilde{\mathbf{E}} \tag{3}$$

Where the incompatible strain tensor is expressed, as function of the Jacobian matrix **J**, the transformation matrix between the local and global bases at the center of the element $\mathbf{T}^0_S$ and the incompatible strain tensor at the local bases, as:

$$\tilde{\mathbf{E}} = \frac{\det \mathbf{J}_0}{\det \mathbf{J}} \left(\mathbf{T}^0_S\right)^{-T} \bar{\tilde{\mathbf{E}}} \tag{4}$$

Referring to this method, the three field Hu-Whashizu variational principle is expressed as function of the incompatible strain tensor $\tilde{\mathbf{E}}$, the displacement vector **u** and the second Piola-Kirchoff stress tensor $\tilde{\mathbf{S}}$ as:

$$\Pi(\mathbf{u}, \tilde{\mathbf{E}}, \tilde{\mathbf{S}}) = \int_V \left(\psi(\mathbf{E}^c + \tilde{\mathbf{E}})\right) dV - \int_V \left(\tilde{\mathbf{S}} : \tilde{\mathbf{E}}\right) dV - \int_V \mathbf{F}_V.\mathbf{u} dV - \int_{\partial V_f} \mathbf{F}_S.\mathbf{u} dA \tag{5}$$

Where ψ is the strain energy function. $\mathbf{F}_V$ and $\mathbf{F}_s$ present respectively the body and the surface forces. The weak form presents nonlinear terms. To solve this problem, the Newton-Raphson method is applied. After applying the static condensation, the elementary stiffness matrix and residual vector are expressed as the following:

$$\mathbf{R}_e = \mathbf{L}^T \mathbf{H}^{-1} \mathbf{h} + \mathbf{f}^{ext} - \mathbf{f}^{int} \tag{6}$$

$$\mathbf{K}_e = \mathbf{K} - \mathbf{L}^T \mathbf{H}^{-1} \mathbf{L} \tag{7}$$

3 Visco-Elastoplastic Model

In this work, the composite polyamide reinforced with glass fiber is considered with an isotropic visco-elastoplastic behavior (Mars et al. 2018). However, it can be assumed that the low velocity impact on this composite is ranked at the level of the small deformations. The Green strain tensor and the second Piola-Kirchoff stress tensor will be noted thereafter respectively by $\boldsymbol{\varepsilon}$ and $\boldsymbol{\sigma}$. Applying the Von-Mises criterion, the plastic surface is defined as:

$$f = \sqrt{3/2}\,\|S\| - \sigma_p \leq 0 \tag{8}$$

Where S is the deviatoric stress tensor and σ_p is the yield stress. It can be defined when adopting the isotropic hardening and the visco-plastic model of Johnson-Cook:

$$\sigma_p = \left[\sigma_y + Q(1 - e^{-\beta\varepsilon^p}) + k\varepsilon^p\right]\left[1 - C\,Log\left(\frac{\dot{\varepsilon}}{\dot{\varepsilon}_0}\right)\right] \tag{9}$$

k, Q and β represents the hardening parameters. C is the sensitivity parameters of the strain rate and $\dot{\varepsilon}_0$ is the reference strain rate. In this work, the numerical algorithm of Euler implicit is adopted to estimate the plastic part of the tangent modulus. The details of this formulation are depicted in (Koubaa et al. 2017). To form the fundamentals of constitutive modelling of PA66-GF composites, a set of material tests was carried out. Uniaxial tension tests provided material parameters related to elasto-visco-plastic with isotropic hardening response by correlating experimental results with numerical simulations (Mars et al. 2018).

4 Numerical Results

A clamped circular plate of PA66-GF is subjected to initial impact velocity, $v_0 = 2\ ms^{-1}$. The impactor consists on hemispherical nose shape projectile with a mass equal to 1.265 kg (Fig. 1). The diameter of the impactor is 6.35 mm. The composite plate has the radius of 45 mm and thickness equal to 4 mm. The impactor is modeled as analytic rigid body. Owing to symmetry, simulations are performed for the quarter parts of the plate. The plate is meshed with 4500 SS elements (a single element in thickness). A hard contact law is applied (Koubaa et al. 2017). The simulations were carried out via ABAQUS/Standard with UEL subroutine for solid shell element integration. For the case of solid element, Abaqus/standard with UMAT subroutine was developed. The impact force time evolution was determined. A first track of results is validated with (Chaker et al. 2019; Chen et al. 2019) for the case of the isotropic materials using aluminum circular plates.

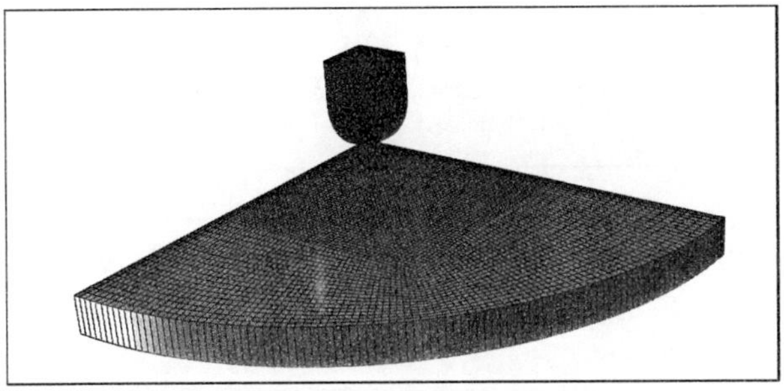

Fig. 1. 3D impact computational problem of circular PA66-GF plate with rigid hemispherical impactor-SS element with single element through thickness

For comparison purposes, and to check the efficiency of the developed solid shell element, simulations are carried out using the conventional solid mesh element C3D8. In this second case, the plate is meshed with 45000 8-node brick elements (10 element in thickness). Temporal response of impact force is depicted in Fig. 2 for polyamide composite with 10% of fibers contents. Results are compared and validated with the experimental findings of Mars et al. (2018). According to Table 1, one can deduce that the developed SS element provides accurate results and allow to save six time computational times (CPU) compared to conventional solid elements.

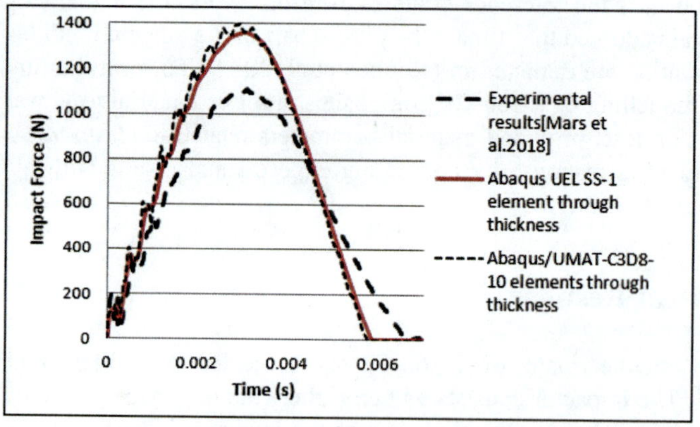

(a)

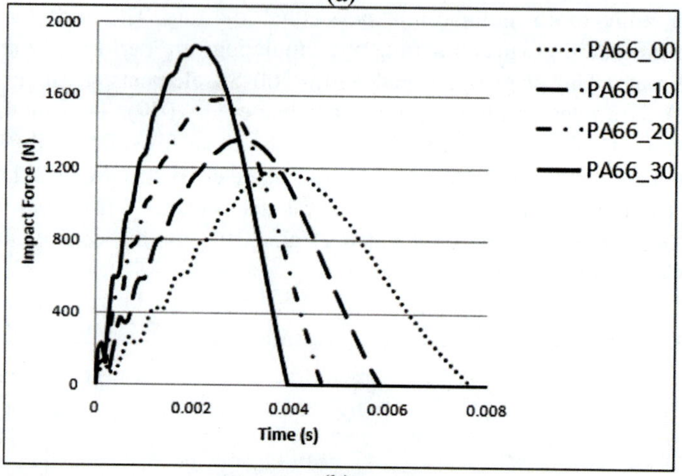

(b)

Fig. 2. (a) Impact force evolution PA66-GF 10%, (b) effect of fiber contents

Table 1. CPU time when applying solid shell element and solid element

Element	Description	CPU	Saving time
C3D8	Hexahedral solid element of ABAQUS- 45000 FE-10 elements through thickness	318.066 min	–
SS	Solid shell element enhanced with 5 incompatible modes-4500 FE-1 element through thickness	56.315 min	**82.29%**

5 Conclusion

In spite of the large interest in composite structure impact of sheet metals, few studies are interesting in PA66-GF dynamic behavior. Moreover, there is a lack of interest on computational efficiency of the developed formulations. In this paper, an elasto-visco-plastic modelling of composite behavior when subjected to low velocity impact is conducted. Abaqus/standard software is used with UEL and UMAT subroutine, for solid shell and conventional solid elements, respectively. The main contribution is the application of 8-nodes solid shell element to avoid numerical locking when simulating impact of glass fiber reinforced composites. Simulation time is less expensive comparing to solid element as only one element is considered through thickness. The developed finite element prove its efficiency for different fiber contents of PA66-GF. Sensitivity analysis will be conducted in further track of work for an optimal design of composite structure with performant numerical tool.

References

AlMaadeed, M.A., Kahraman, R., Khanam, P.N., Madi, N.: Date palm wood flour/glass fibre reinforced hybrid composites of recycled polypropylene: Mechanical and thermal properties. Material **42**, 289–294 (2012)

Chaker, A., Koubaa, S., Mars, J., Vivet, A., Dammak, F.: An Abaqus implementation of a solid-shell element: Application to low velocity impact. Chapter Book Springer. Springer (2019). https://doi.org/10.1007/978-3-030-27146-6_84

Chaker, A., Koubaa, S., Mars, J., Vivet, A., Dammak, F.: An efficient ABAQUS solid shell element implementation for low velocity impact analysis of FGM plates. J. Eng. Comput. (2020). https://doi.org/10.1007/s00366-020-00954-8

Chalal, H., Abed-Meraim, F.: Quadratic solid–shell finite elements for geometrically nonlinear analysis of functionally graded material plates. Materials **11**(6), 1046 (2018)

Chebbi, E., Mars, J., Wali, M., Dammak, F.: Fatigue behavior of short glass fiber reinforced polyamide 66: experimental study and fatigue damage modelling. Period. Polytech. Mech. Eng. **60**, 247–255 (2016)

Dvorkin, E.N., Bathe, K.J.: A continuum mechanics based four-node shell element for general non-linear analysis. Eng. Comput. **1**(1), 77–88 (1984)

Hajlaoui, A., Jarraya, A., Kallel-Kamoun, I., Dammak, F.: Buckling analysis of a laminated composite plate with delaminations using the enhanced assumed strain solid shell element. J. Mech. Sci. Technol. **26**, 3213–3221 (2012)

Klinkel, S., Gruttmann, F., Wagner, W.: A robust non-linear solid shell element based on a mixed variational formulation. Comput. Methods Appl. Mech. Engrg. **195**, 179–201 (2006)

Ku, H., Wang, H., Pattarachaiyakoop, N., Trada, M.: A review on the tensile properties of natural fiber reinforced polymer composites. Compos. Part B Eng. **42**, 856–873 (2011)

Koubaa, S., Mars, J., Wali, M., Dammak, F.: Numerical study of anisotropic behavior of aluminum alloy subjected to dynamic perforation. Int. J. Impact Eng **101**(105), 114 (2017)

Koubaa, S., Mars, J., Dammak, F.: Efficient numerical modelling of functionally graded shell mechanical behavior. J. Appl. Comput. Sci. **15**(1), 84–94 (2019)

Mars, J., Chebbi, E., Wali, M., Dammak, F.: Numerical and experimental investigations of low velocity impact on glass fiber-reinforced polyamide. Compos. Part B: Eng. **146**, 116–123 (2018)

Mouti, Z., Westwood, K., Long, D., Njuguna, J.: An experimental investigation into localised low-velocity impact loading on glass fibre-reinforced polyamide automotive product. Compos. Struct. **104**, 43–53 (2013)

Simo, J.C., Rifai, M.S.: A class of mixed assumed strain methods and the method of incompatible modes. Int. J. Numer. Methods Eng. **29**(8), 1595–1638 (1990)

Tan, X.G., Vu-Quoc, L.: Efficient and accurate multilayer solid-shell element: non-linear materials at finite strain. Int. J. Numer. Methods Eng. **63**(15), 2124–2170 (2005)

Thomason, J.L.: The influence of fibre length, diameter and concentration on the impact performance of long glass-fibre reinforced polyamide 6,6. Compos. Part A Appl. Sci. Manuf. **40**, 114–124 (2009)

Application of Factorial Design to Study the Effect of Recycling of HDPE on Rheological and Mechanical Properties

Karama Elfehri[1(✉)], Ameur Chtourou[2], Sana Koubaa[2], and Basma Samet[3]

[1] Electrochemistry and Environment Laboratory (LEE), National Engineering School of Sfax, University of Sfax, B.P 1173, 3038 Sfax, Tunisia
karama.fehri@enis.tn

[2] Electro-Mechanical System's Laboratory (LASEM), National Engineering School of Sfax, University of Sfax, B.P 1173, 3038 Sfax, Tunisia
ameur.chtourou@yahoo.com, sana.kouba@enis.tn

[3] Advanced Materials Laboratory, National Engineering School of Sfax, University of Sfax, B.P 1173, 3038 Sfax, Tunisia
basma.samet@enis.tn

Abstract. In the plastic industry, scrap is common and can reach important amounts in some applications. For that reason, reprocessing in the format of mechanical recycling is pretty common among plastic industry for its simplicity and efficiency. However, the stated type of reprocessing causes the degradation of the proprieties of plastic. Hence, recycled plastic is mixed with virgin one. The objective of this work is the investigation of the effect of the variation of the fraction of the recycled plastic (X1) in virgin/recycled blends and number of recycled cycles (X2) using factorial design for HDPE. To attend this objective, two sets of specimens were injected according to 2^2 factorial design. Mechanical and rheological proprieties were characterized in terms of melt flow index, Vicat softening temperature, elastic tensile strength at 5% (σ_e) and tensile strength at yield (σ_m). Using nemrodw software, effects plots were plotted. It was depicted that X1 and X2 have no significant effect on σ_e, σ_m and on Vicat softening temperature. Nevertheless, they have a significant effect on MFI.

Keywords: High density polyethylene · Mechanical recycling · Factorial design · Mechanical properties · Rheological properties

1 Introduction

With the overgrowing population, demands on plastic products has been increased dramatically in the last decades.

Plastics, which are made from less than 20% to almost 100% of polymer, are inexpensive, lightweight, and durable materials, which can be formed into a wide variety of products. Therefore, plastic materials find use in a great range of domestic and industrial applications (Francis 2017). As a consequence, the amount of plastics' wastes is rather important even in production phase. As an example, the scrap during

M. Kharrat et al. (Eds.): ICAMEM 2019, LNME, pp. 355–362, 2020.
https://doi.org/10.1007/978-3-030-52071-7_49

production in extrusion blow-molding technology can reach 40% and can surpass that of the injected part by a factor of 100 in the case of micro-injection molding (Jeggy (2004).

For the mentioned cases, re-use of plastics' production waste is important. It reduces both production trash and utilization of raw materials, and this can be done through recycling. According to the Office of International Recycling one ton of plastic saves 5774 kWh of energy, 2604 l of oil and 22 m^3 of landfill. Thus, recycling helps to conserve the nonrenewable fossil oils due to the fact that plastic production consumes about 8% of world's oil production, in order to decrease the consumption of energy and the amounts of solid waste going to landfill (Benoit et al. 2017).

The most used form of recycling is mechanical recycling due to its simplicity and low cost. Plastic is grinded and reprocessed using the appropriate physical process (Goodship 2007; Grigore 2017).

Although it seems perfect, recycling is by no mean perfect as with each recycling cycle the mechanical and rheological proprieties decrease, in order to avoid the reduction in the properties of the final product, the recycled plastic scrap is mixed with virgin polymer. However, those results required a lot of testing and are based on mere observation.

For an industrial application, precision and low-cost manufacturing are priorities. For instance, using a more precise way to investigate the stated degradation is quite important.

In line with what have been said, this work investigates the main correlation effect of recycled portion (X1) in virgin/recycled blends and number of recycling cycles (X2) on the mechanical and rheological properties of HDPE as well as the interactions among the stated variables using 2^k factorial design where k = 2 the number of variables in question. For this matter, two sets of specimens made from four blends were prepared using injection molding process.

2 Materials and Methods

2.1 Material

In this study, we considered the commercial virgin high-density polyethylene (HDPE0) type 17450 N, produced by The Dow Chemical Company. The density of this material is 0.95 g/cm^3 and the melt flow index is 8 g/10 min under 2.16 kg at 190 °C.

2.2 Material Processing

Virgin HDPE was processed by injection molding in a Supermaster SM120 injection molding machine (Chen Hsang, Taiwan) with a screw diameter of 35 mm and an L/D ratio of 25. The temperature of fives heating zones was set at 230, 210, 200 and 190 °C. In the process, some specimens were removed for characterization, the rest were grinded and injected again.

2.3 Methodology

The purpose of this work is to study the effect of two factors namely portion of recycled polymer in virgin/recycled blends (X1) and number of recycling cycles (X2) on melt flow index, Vicat softening temperature, elastic tensile strength at 5% (σ_e) and tensile strength at yield (σ_m), in order to determine the best experimental conditions allowing the optimization of these responses. For predicting the optimal conditions, a second order polynomial function is fitted to correlate relationship between independent variables and response. For two factors this equation is:

$$\widehat{y}_i = b_0 + b_1 X_{i1} + b_2 X_{i2} + b_{12} X_{i1} X_{i2} \quad (1)$$

With

$$X_{ij} = \frac{u_{ij} - u_j^0}{\Delta u_j} \quad (2)$$

Where:
$\widehat{y}_i$ is the calculated response value at the i th experiment
X_{ij} is the value of the coded variable j at the ith experiment
b_j and b_{ij} are the model coefficients
U_{ij} is the value of real variable j at the ith experiment
u_j^0 is the value of the real variable j at the center point
Δu_j is the step of variation of the real variable j
The measured response y_i for the ith experiment is:

$$y_i = \widehat{y}_i + e_i (e_i \text{ is the error}) \quad (3)$$

The experimental matrix of 2^2 factorial design is presented in Table 1.

Table 1. Design matrix corresponding to the 2^2 factorial design

N° Exp	X1	X2
1	−1	−1
2	+1	−1
3	−1	+1
4	+1	+1

X1 is the portion of recycled polymer in virgin/recycled blends varying between 30 and 70%. X2 is the number of recycling cycles varying between 2 and 6.

Each experience was repeated twice. The experimental design is presented in Table 2.

Table 2. Experimental design

N°Exp	X1	X2
1	30	2
2	30	2
3	70	2
4	70	2
5	30	6
6	30	6
7	70	6
8	70	6

The samples are denoted throughout the papier as HDPEX1NX2 as elaborated in Table 3.

Table 3. Virgin/recycled HDPE blends and number of recycling cycles of each blend

Denomination	Recycled (%)	Number of recycling cycles
HDPE30N2	30	2
HDPE70N2	70	2
HDPE30N6	30	6
HDPE70N6	70	6

2.4 Melt Flow Index (MFI)

The MFI measurements were carried out on a on a Melt Indexer MFI-450 (SCITEQ A/S). All experiments were performed according to ASTM D1238. The specimens were analyzed at a temperature of 190 °C and a load of 2,16 kg.

Three samples of each blend were tested and an average value was calculated. The melt flow index was calculated using Eq. (4), were m is the average mass (in grams) of the cut-offs; t is the cut-off time-interval (in seconds), and t_{ref} is 600 s.

$$MFI = \frac{m}{t} \times t_{ref} \tag{4}$$

2.5 Vicat Softening Temperature (T_{vicat})

Vicat softening temperature measurements were made on a standard vicat/HDT station reflex (metrotec S.A) following ISO-306 A50 under a load of 10 N and temperature increasing rate of 50 °C/h. Three samples of each blend were tested and an average value was calculated.

2.6 Tensile Test

The tensile tests were performed by means of HENSGGEAND testing machine with 5KN load cell at a cross head speed of 10 mm/min. The specimens were prepared according to the iso 3167 standard. Average values of the elastic tensile strength at 5% ($\sigma_{e5\%}$) and the tensile strength at yield (σ_m) from three samples were selected as the representative values.

3 Results and Discussion

For the sake of comparison, we present the stress-strain plot (Fig. 1) for virgin HDPE, from which we can deduct the elastic tensile strength at 5% and the tensile strength at yield that are equal to σ_m = 13,83 MPa and $\sigma_{e5\%}$ = 13 MPa that are very close to those of the prepared blends presented in Table 4. Furthermore, MFI of virgin material is equal to 8,4 g/10 min and T_{vicat} is equal to 115 °C.

It is clear from Table 4 that the T_{vicat} stays unchanged for the different blends of virgin/recycled HDPE. However, the MFI is increased with each recycling cycle.

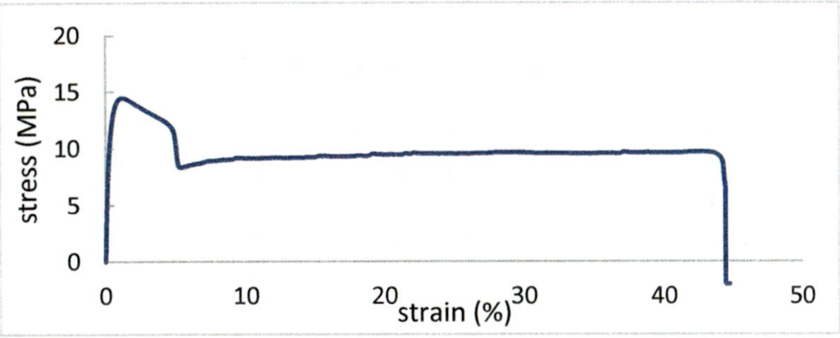

Fig. 1. Stress-strain plot of virgin HDPE

Table 4. σ_m, σ_e T_{Vicat} and MFI for of virgin/recycled HDPE blends during multiple reprocessing for the two prepared sets.

Denomination	σ_m (MPa)	σ_e (MPa)	T_{Vicat} (°C)	MFI (g/10 min)
HDPEN0	13.8 ± 0.2	13 ± 0,1	115 ± 1.2	8.4 ± 0.3
HDPE30N2 (set 1)	13.5 ± 1.3	12.9 ± 1.4	113 ± 1.4	10.6 ± 0.9
HDPE30N2 (set 2)	14 ± 0.3	13 ± 0.4	115 ± 1.4	10.1 ± 1.2
HDPE70N2 (set 1)	13.5 ± 0.3	12.6 ± 0.5	109 ± 0.7	11.1 ± 1
HDPE70N2 (set 2)	14.6 ± 0.1	13.7 ± 0.3	110 ± 0.7	11.3 ± 1.2
HDPE30N6 (set 1)	13.3 ± 0.6	12.4 ± 0.5	114 ± 1.4	13.4 ± 0.2
HDPE30N6 (set 2)	14.6 ± 0.2	13.6 ± 0.1	112 ± 1	13.3 ± 0.2
HDPE70N6 (set 1)	12.9 ± 0.1	12.1 ± 0.1	114 ± 0.7	14.4 ± 0.6
HDPE70N6 (set 2)	14.7 ± 0.1	13.8 ± 0.1	110 ± 0.7	14 ± 0.8

3.1 Effect of Factors and Their Interactions on σ_m, σ_e, T_{vicat} and MFI

The analysis of variance (ANOVA) is an informative tool that states the significance of variables which are in our case X1 and X2 the coded variables of the fraction of recycled polymer and the number of recycling cycles in virgin/recycled blend, on the responses which are σ_m, σ_e, T_{vicat} and MFI noted y_1, y_2, y_3 and y_4 respectively. The mathematical models for each response and the ANOVA have been calculated using Minitab and nemrodw software (Mathieu et al. 2000), respectively.

The models calculated by Minitab are as follows:

$$\hat{y}_1 = 13.897 + 0.048X1 - 0.008X2 - 0.108X1 * X2 \quad (5)$$

$$\hat{y}_2 = 13.016 + 0.049\mathbf{X1} - 0.021\mathbf{X2} - 0.079\mathbf{X1} * \mathbf{X2} \quad (6)$$

$$\hat{y}_3 = 112.125 - 1.375\mathbf{X1} + 0.375\mathbf{X2} + 0.875\mathbf{X1} * \mathbf{X2} \quad (7)$$

$$\hat{y}_4 = 12.2837 + 0.4138\mathbf{X1} + 1.5187\mathbf{X2} + 0.0038\mathbf{X1} * \mathbf{X2} \quad (8)$$

According to ANOVA (Table 5) the only response that is affected by the variation of X1 and X2 is MFI. However, the interaction between X1 and X2 is not significant hence the model for y_4 is:

$$\hat{y}_4 = 12.2837 + 0.4138\mathbf{X1} + 1.5187\mathbf{X2} \quad (9)$$

Those results are represented also using main effect plots in Fig. 2. The analyses of these plots show that X1 and X2 are statistically insignificant and have no effect on σ_m and σ_e. (coefficients are lesser than 0.7 and greater than −0.7) which is in concordance with what Oblak et al. (2015) and Benoit et al. (2017) found for the first ten recycling cycles of HDPE. the authors explained these results by the fact that the degree of crystallinity of the recycled material has not changed compared to virgin HDPE. In addition, those variables have no significant effect on T_{vicat} (coefficients are lesser than 1.6 and greater than −1.6). However, they have a significant influence on the MFI (coefficients are greater than 0.25) (Fig. 2(d)). One can notice as well that X2 has a greater influence on MFI than X1 and that the interaction between X1 and X2 has no effect on the melt flow index. This can be explained by the fact that with each recycling cycle the molecular chains degrade.

Table 5. ANOVA of the responses y_1, y_2, y_3 and y_4

Sources	DF	Adj SS	Adj MS	F-Value	Sig
y_1					
X1	1	0.018	0.0180	0.02	N.S
X2	1	0	0	0	N.S
X1*X2	1	0.092	0.0924	0.12	N.S
Error	4	3.18	0.795		
Total	7	3.29			
y_2					
X1	1	0.019	0.019	0.03	N.S
X2	1	0.003	0.0036	0.01	N.S
X1*X2	1	0.049	0.0496	0.07	N.S
Error	4	2.847	0.7119		
Total	7	2.92			
y_3					
X1	1	15.125	15.125	4.84	N.S
X2	1	1.125	1.125	0.36	N.S
X1*X2	1	6.125	6.125	1.96	N.S
Error	4	12.5	3.125		
Total	7	34.875			
y_4					
X1	1	1.36	1.36	26.22	***
X2	1	18.45	18.45	353.25	*
X1*X2	1	0	0	0	N.S
Error	4	0.209	0.052		
Total	7	20.031			

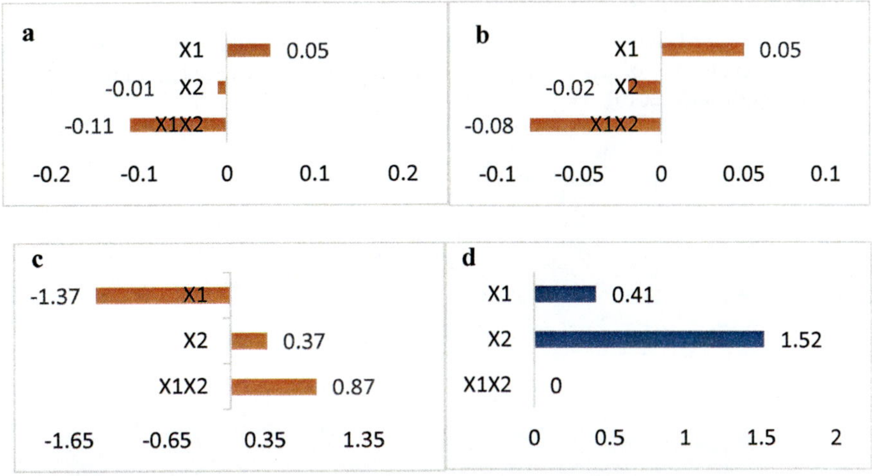

Fig. 2. The effect of X1 and X2 on (a) σ_m, (b) σ_e, (c) T_{vicat} and (d) MFI.

4 Conclusion

In the current study, effect of the variation in the fraction of the recycled plastic in virgin/recycled blends and number of recycled cycles using factorial design was investigated. It was assessed that variables X1 and X2 have no effect on T_{vicat} softening temperature and tensile properties. Nevertheless, the mentioned variables have indeed an influence on MFI.

A further track of work will be the optimization of the virgin/recycled blend for industrial applications. The optimal blend will be mixed with olive core in order to obtain a product which almost made of wastes.

Acknowledgments. We would like to express our gratitude to "polymer industry" Company in SFAX, TUNISIA that provided the funding for this research work.

References

Benoit, N., González-Núñez, R., Rodrigue, D.: High density polyethylene degradation followed by closed-loop recycling. Prog. Rubber Plast. Recycl. Technol. **33**(1), 17–38 (2017)

Francis, R. (ed.): Recycling of Polymers: Methods, Characterization and Applications. Wiley-VCH Verlag, Weinheim (2017)

Goodship, V.: Introduction to Plastics Recycling, 2nd edn. Smithers Rapra, Shawbury (2007)

Grigore, M.: Methods of recycling, properties and applications of recycled thermoplastic polymers (2017)

Jeggy, C.: Micro-Injection Moulding: From Process to Modelling. Louvan la neuve: Presses universitaries de Louvain (2004)

Mathieu, D, Nony, J, Phan-Tan-Luu, R.: Génération de matrices d'expériences en fonction des objectifs et traitement des réponses expérimentales, Nemrod software, LPRAI, Marseille, France (2000)

National Institute of Industrial Research: Complete Book on Biodegradable Plastics and Polymers: Recent Development, Properties, Analysis, Materials & Processes. Delhi, India: Asia Pacific Business (2006)

Oblak, P., Gonzalez-Gutierrez, J., Zupančič, B., Aulova, A., Emri, I.: Processability and mechanical properties of extensively recycled high density polyethylene. Polym. Degrad. Stab. **114**, 133–145 (2015)

Viscoelastic Mechanical Behavior of Fresh Human Bone of Lower Limbs

Saida Benhmida[1(✉)], Dorra Salhi[1], Montassar Zrida[2], Ahmed Hichem Hamzaoui[2], and Hamza Essaddam[1]

[1] Laboratory of Biophysics and Medical Technologies, Research Laboratory of Biophysics and Medical Technologies (LRBTM), Tunis, Tunisia
benhmidasaida@gmail.com, dorrasalhi@hotmail.com, essaddam@gnet.tn

[2] National Center for Research in Materials Science (CNRSM), Laboratory of Valorisation of Useful Materials, Tunis, Tunisia
montassar.zrida@gmail.com, mdmihil0@gmail.com

Abstract. Despite significant developments in orthopedic medicine, prostheses and joint implants have an unlimited life span and various problems citing loosening. Indeed, they are made from a very rigid material which has a mechanical behavior different from that of the human bone. This work is about the study of mechanical behavior of the human bone like dynamic viscosity and elasticity modulus. Accentuating on the viscoelastic nature of the bone, the samples of fresh bone specimens are used directly after their removal. The dynamic viscosity of the cancellous bone was determined by relaxation test. Thirteen human cancellous bone samples of the knee joint are used. The dynamic viscosity is in the order of 8683.3 MPa.s. The dynamic viscosity of human cancellous bone can be used in femoral cement filling during bone grafting.

Keywords: Fresh human bone · Lower limbs · Mechanical behavior · Dynamic viscosity · Elasticity

1 Introduction

Osteoarthritis is a degenerative joint disease. It does not settle abruptly, but it evolves slowly and worsens over several years. It presents a significant public health problem since it affects a large number of people around the world (Felson et al. 1995). Indeed, there is a massive increase in the number of total knee and hip prostheses and joint implants performed worldwide each year (Davis et al. 1988; Kurtz et al. 2007). The materials used in the fabrication of prostheses are very varied, rigid and do not exhibit the actual mechanical behavior of the human bone. Of these materials; metallic alloys such as stainless steels and cobalt-based alloys, polymers whose primary use is polyethylene and ceramics such as alumina Al2O3 and Zirconia ZrO2 (Biot 1941). In reality, these different materials do not have the same mechanical characteristics as the bone. If the mechanical properties of these materials are well known, we cannot say the same about bones. Bone is a viscoelastic material when a stress is applied to a bone sample; its deformation depends on time (Morgan and Keaveny 2001). The viscoelastic

M. Kharrat et al. (Eds.): ICAMEM 2019, LNME, pp. 363–369, 2020.
https://doi.org/10.1007/978-3-030-52071-7_50

nature of the bone has been shown by Biot (Biot 1941) to be the result of fluid flow in bone pores. The viscoelastic behavior of bone is related to the composition of the bone in collagen and water (Lawrence et al. 1998). According to Carter and Hayes (Carter and Hayes 1977) it is the presence of the interstitial fluid at the level of bone porosity which gives the bone the viscoelastic aspect. Recently, authors have emphasized the influence of bone marrow on the mechanical behavior of the bone (Charlebois et al. 2008; Halgrin et al. 2012). Viscoelasticity is due to bone marrow and fluid (Halgrin et al. 2012). When subjected to a low-intensity mechanical force, it adapts through its architecture by exchanging bone marrow and fluid (Guedesa et al. 2006). The bone is considered as an elastic mechanical material to simplify most of his mechanical modelings, it obeys the law of the Wolf and behaves like the other materials to the application of a force (Bousson et al. 2011). In order to improve the orthopedic medicine and the quality of life of persons carrying prostheses and joint implants, it is necessary to determine the actual mechanical characteristics of the human bone in its fresh state, notably its dynamic viscosity and its elasticity modulus.

2 Materials and Methods

2.1 Samples Collection and Preparation

The samples used are from lower limbs; knee joint (femoral condyles and tibial trays). An ethics committee has authorized the use of these samples obtained after the application of joint prostheses or in the case of amputation caused by road accidents. Firstly, this work is about the study of the dynamic viscosity of fresh human cancellous bone of the knee joint by relaxation test. Thirteen cancellous bone samples were used, including seven samples of men aged 28 to 64 years old and 6 samples of women aged 60 years old. Secondly, this work is about the study of the elasticity modulus of fresh human bone by a compression test. Ninety samples are used to determine the elasticity modulus of cancellous bone. The samples are classified according to the anatomical location of the inner or the outer compartment of the knee joint. Forty-five samples from the outer compartment and 45 samples from the inner compartment are tested. The samples collected belong to both sexes. The ages ranged from 19 to 83. Accentuating on the viscoelastic nature of the bone, the bone samples are fresh with the preservation of the bone marrow and the fluid thus containing in the bone pores. The mechanical tests were carried out on the same day of their sampling (maximum 4 h). The samples were transported in compresses soaked with physiological serum, and without additives, in order to preserve as much as possible their physical characteristics of the bone. The cancellous bone was cut into small cubes of side 10 mm for relaxation test and 20 mm for compression test.

2.2 Testing Machine

A tensile/compression testing machine “LlOYd” EZ50 was used for carrying out compression and relaxation tests. Two particular small size cylindrical compressing trays were designed (diameters equal to 50 mm) using aluminum material as shown in (Fig. 1).

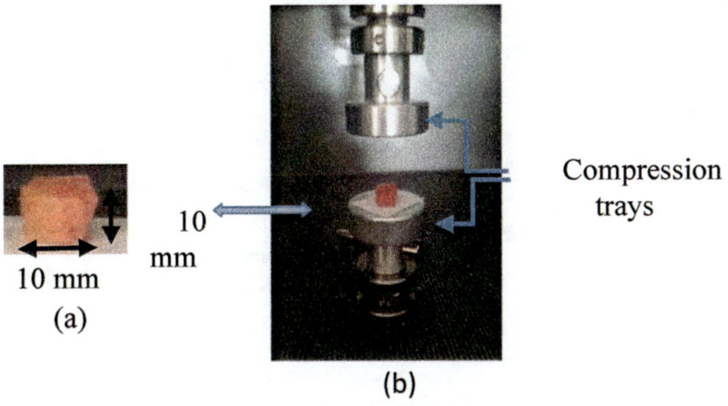

Fig. 1. (a) Cubic bone sample stops 10 mm; (b) LIOYd type EZ50 tensile/compression testing machine

2.3 Relaxation Test

The relaxation test was used to determine the viscous properties of the cancellous bone. It consists of causing a strain to the sample without exceeding the elastic domain of the cancellous bone and follow the evolution of the variation of the stress σ as a function of time. In these tests, a cross-head speed of 1 mm/min was imposed, followed by a relaxation time of 15 min. After the loading phase, the tension was stopped at a cross-head of 1.5 mm corresponding to a strain of 0.15. The dynamic viscosity η is configured by Maxwell's differential equation:

$$d\varepsilon/dt = (1/E) \cdot (d\sigma/dt) + (\sigma/\eta) \quad (1)$$

Where ε is the strain, σ is the stress and E is the elasticity modulus.

3 Results and Discussion

The relaxation curve has a decreasing exponential shape part corresponds to the viscous flow of the bone and which takes a duration of time to stabilize (Fig. 2). In this figure, the normalized stress, denoted normalized, is given by:

$$\sigma\,\text{normalized} = \sigma/\sigma\,\text{max} \quad (2)$$

Where σ max is the maximum stress obtained before the beginning of the relaxation phase.

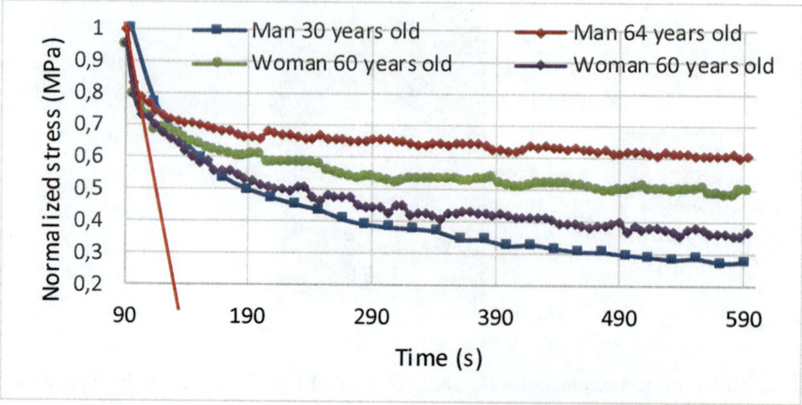

Fig. 2. Experimental curve obtained during a relaxation test.

Figure 2 present four relaxation curve corresponding to four bone cancellous relaxation test in the same conditions to assess the degree of repeatability. There is a slight difference due to the porosity of this material. The relaxation time was determined experimentally. It is the intersection of the slope at the origin with the horizontal axis. We can also determine the value of the viscosity, indeed the relaxation time which is written in the form of

$$\tau = \eta/E \tag{3}$$

The relaxation time is of average value equal to 123.58 s and the viscosity is of 8683.3 MPa.s (Table 1).

Table 1. The dynamic viscosity of cancellous bone

N	η (MPa.s)	τ (s)	σ (MPa)	E (MPa)
13	Mean = 8683.3 Median = 8672.61 Min = 7607.7 Max = 9865.1	Mean = 123.58 Median = 113 Min = 106.9 Max = 166	Mean = 2.34 ± 0.3 Median = 2.2 Min = 1.1 Max = 4.5	Mean = 71.74 Median = 71.14

The elasticity modulus of the human cancellous bone of the knee joint was determined for 90 samples. The elasticity modulus gives a range of values between 27.86 MPa and 163.43 MPa. The elasticity modulus of the knee joint is classified by the anatomical location of the specimens: inner or outer compartment. The obtained average values are 99.48 MPa for the inner compartment and 41.19 MPa for the outer compartment. The highest value is located in the inner knee compartment. The mean stress value of inner compartment is 10.93 MPa, for the outer compartment, it is

22.41 MPa. A significant difference in elasticity modulus was found according to the anatomical seat of the knee joint with a "p" value <0.05 (Table 2). This proves the peculiarity of the bone as an anisotropic material.

Table 2. Mechanical properties of knee cancellous bone

Anatomical seat	N	E(MPa)	Standard deviation	P value	σ max (MPa)
O.C	45	Mean = 41.19 ± 0.91 Max = 57 Min = 27.86	6.09	$P = 1.67 \times 10^{-22}$	Mean = 10.93
I.C	45	Mean = 99.48 ± 3.32 Max = 163.43 Min = 70.06	22.24		Mean = 22.41

Indeed, for (Carter and Hayes 1977) the elasticity modulus of frozen fresh bone is from 1.4 to 79 MPa. (Ciarelli et al. 1986) found values between 1 and 448 MPa. According to Hvid and Hansen (Hvid and Hansen 1985), the elasticity modulus is from 4 to 4430 MPa. The value of the modulus of elasticity determined by Martens et al. (Martens et al. 1983) is in a range of 1000 to 9800 MPa. The causes of these differences are many and varied. Some depend on the nature of the samples and how they are stored. Several authors have used bone samples from animals (Guedesa et al. 2006) and demineralized human cadavers (Ibrahim et al. 2010).

Fresh human bone is much more elastic, which may be related to its fluid and bone marrow-rich nature. It is obvious that fresh human bone is more elastic. This confirms that the strength of the bone is obtained not only its elasticity but also its viscous nature. Viscosity has a very important role in the strength and mechanical behavior of the bone.

4 Conclusion

This work allows us to determine the actual mechanical characteristics of the human bone and the dynamic viscosity and the elasticity modulus.

The dynamic viscosity of the cancellous bone was determined by relaxation test; it is of the order of 8683.3 MPa.s. The dynamic viscosity of human cancellous bone can be used in femoral cement filling during bone grafting.

The study of compression tests of the knee cancellous bone shows many peculiarities. Cancellous bone has anisotropic properties depending on the anatomical location of the inner or outer compartment of the knee joint. The highest modulus of elasticity is located in the inner compartment, which is equal to 99.48 MPa whereas it is 41.19 MPa in the outer compartment. The difference in these results from the published work is due to the nature of the bone samples used which are fresh and not preserved.

The purpose of determining the mechanical characteristics of cancellous bone is to test individual orthopedic implants and their effects on cancellous bone. It is necessary to review the materials used for the fabrication of prostheses, which are generally made of stainless steel or ceramic, which have a modulus of elasticity very high compared to the human bone.

The mechanical characteristics of the human bone thus described in this work can be used in the numerical modeling of the mechanical behavior of bone as well as in the modeling of joint prostheses to determine the resistance and the mechanical behavior of these two materials as well as their biocompatibility.

Acknowledgments. The authors acknowledge the valuable help given by the staff of Orthopedic surgery department of Rabta Hospital Tunisia (Dr. Maher Barsaoui, Mr. Hassen Hamrouni and Mrs. Soumaya Essid) and the technicians of the Laboratory of Valorization of Useful Materials Technologic park of Borj Cedria.

References

Felson, D.T., Zhang, Y., Hannan, M.T.: The incidence and natural history of knee osteoarthritis in the elderly the Framingham osteoarthritis study. Arthritis Rheum. **38**(10), 1500–1505 (1995)

Davis, M.A., Ettinger, W.H., Neuhaus, J.M.: The role of metabolic factors and blood pressure in the association of obesity with osteoarthritis of the knee. J. Rheumatol. **15**(12), 1827–1832 (1988)

Kurtz, S., Ong, K., Lau, E.: Projections of primary and revision hip and kneearthroplasty in the United States from 2005 to 2030. J. Bone Joint Surg. Am. **89**(4), 780–785 (2007)

Biot, M.A.: General theory of three-dimensional consolidation. J. Appl. Phys. **12**, 155–164 (1941)

Morgan, E., Keaveny, T.: Dependence of yield strain of human trabecular bone on anatomic site. J. Biomech. **34**, 569–577 (2001)

Lawrence, R.C., Helmick, C.G., Arnett, F.C.: Estimates of the prevalence of arthritis and selected musculoskeletal disorders in the United States. Arthritis Rheum. **41**, 778–799 (1998)

Carter, D., Hayes, W.: The compressive behavior of bone as a two-phase porous structure. J. Bone Joint Surg. **59**(7), 954–962 (1977)

Charlebois, M., Haunschmid, C., Zysset, P.: Confined and unconfined compression tests of human trabecular bone up to very large strains. J. Biomech. **41**(S1) (2008)

Halgrin, J., Chaari, F., Markiewicz, E.: On the effect of marrow in the mechanical behavior and crush response of trabecular bone. J. Mech. Behav. Biomed. Mater. **5**, 231–237 (2012)

Guedesa, R.M., Simoesb, J.A., Morais, J.L.: Viscoelastic behaviour and failure of bovine cancellous bone under constant strain rate. J. Biomech. **39**, 49–60 (2006)

Bousson, V., Wybier, M., Petrover, D., Parlier, C., Chicheportiche, V., Hamzé, B., Sverzut, J.M., Daguet, E., Wyler, A., Thabet, J., Bossard, P., Laredo, J.D.: Les fractures de contrainte, Revue Iconographique/Ostéoarticulaire. J. Radiologie **92**, 188–207 (2011)

Ciarelli, M.J., Goldstein, S.A., Dickie, D., Ku, J.L., Kapper, M., Stanley, J., Flynn, M.J., Matthews, L.S.: Experimental determination of the orthogonal mechanical properties, density, and distribution of human trabecular bone from the major metaphyseal regions utilizing materials testing and computed tomography. Trans. Orthop. Res. Soc. (1986)

Hvid, I., Hansen, S.L.: Trabecular bone strength patterns at the proximal tibial epiphysis. J. Orthop. Res. **3**, 464–472 (1985)

Martens, M., Van Audekercke, R., Delport, P., De Meester, P., Muelier, J.C.: The mechanical characteristics of cancellous bone at the upper femoral region. J. Biomech. **16**, 971–983 (1983)

Ibrahim, K., Jérôme, C., Samir, A., David, M., Geneviève, G., Olivier, G., et al.: Mechanical properties of glenoid cancellous bone. Clin. Biomech. **25**, 292–298 (2010)

Insights into Researches on the Tribological, Microstructural and Micromechanical Properties of Thermoplastic Based Composites

Basma Ben Difallah[1(✉)], Mabrouka Akrout[1(✉)], Mohamed Kharrat[1], Maher Dammak[1], and Guy Monteil[2]

[1] Laboratory of Electromechanical Systems, National Engineering School of Sfax, University of Sfax, 3038 Sfax, Tunisia
basmaben@yahoo.fr, mabrouka.akrout55@gmail.com, {mohamed.kharrat,maher.dammak}@ipeis.rnu.tn
[2] Laboratoire FEMTO-ST, Ecole Nationale Supérieure de Mécanique et des Microtechniques, 25030 Besançon, France
guy.monteil@ens2m.fr

Abstract. Two disjoint filler materials in a powder form are added to a polycarbonate (PC) matrix. The micro sized additives are Graphite (Gr) and molybdenum disulfide (MoS_2) introduced as layered solid lubricants. Two distinct development methodologies are employed in order to prepare the composites; the first process is industrial by using directly an injection molding machine, while the second process is composite preparation in a laboratory scale. The experimental results of the two process are compared. From friction and wear tests, it is established that MoS_2 particles improved the wear resistance and reduced the coefficient of friction of steel/PC contact with both development methodologies. On the contrary, Gr decreases the tribological properties of PC. Fracture morphologies and x-ray micro-tomography are performed to characterize the microstructural aspects of PC/solid lubricant composites. Micrographs show microcracking, filler agglomeration and the presence of voids with direct injection molding. Besides, the composites present poor scratch resistance properties. The different results are improved with the second development methodology including the powders surface treatment by an amino-silane and mechanical mixing with the PC matrix.

Keywords: Polycarbonate · Solid lubricant powders · Tribological properties · Scratch resistance · Microstructural aspects

1 Introduction

Thermoplastic composites with various types of solid lubricant fillers are approved in tribo-technical applications due to the potentiality of developing required material performances. A large number of previous studies dealt with the ability of layered solid lubricants such as graphite, molybdenum disulfide, and boron nitride on improving the tribological aspects of thermoplastic matrixes (Ben Difallah et al. 2012, 2014, 2020; Puértolas et al. 2019). The studies showed a decreased tendency in friction and/or wear

M. Kharrat et al. (Eds.): ICAMEM 2019, LNME, pp. 370–378, 2020.
https://doi.org/10.1007/978-3-030-52071-7_51

behaviors of the composites which depended on the type and the amount of the additives. In this field, several industrial applications can be improved by the incorporation of such fillers (Deng et al. 2011; Mnif et al. 2013). The self-lubrication capacity of graphene nanoplatelets (GNPs) incorporated to a Poly(etheretherketone) matrix was studied by Puértolas et al. (2019). The authors found that the hardness and the tribological behavior of the composites was improved. Similarly, Su et al. have investigated the role of various solid lubricant fillers in the friction and wear properties of Nomex fabric composites. The authors reported that 10% (volume fraction) graphite or MoS_2 was found ineffective in reducing the friction and wear results of the Nomex materials (Su et al. 2006).

The present work proposes an evaluation of the ability of Gr and MoS_2 in improving the friction and wear behaviors of PC composites elaborated by two different methodologies. It also intends to characterize the microstructural and the micromechanical properties of the composites. With this target, several sliding tests and scratch tests were combined to microscopical techniques in order to enlighten the role of the fillers in the composite's properties.

2 Experimental Set up

2.1 Materials and Composites Preparation

Polycarbonate was purchased from Altech (14 g/10 min fluidity index, 1.2 g/cm^3 density) and dried under vacuum at 110 °C for 4 h before use. The used filler materials were graphite powder (1.9 g/cm^3 density, 25 μm maximum particle size) and 95% purity MoS_2 powder (5.06 g/cm^3 density, 6 μm average size). Different weight fractions were considered in this work and classified into two categories as a function of the processing methodology. The PC based composites were filled separately with MoS_2 or Gr in the range of 0 wt% to 10 wt%.

As a first development methodology (M1), PC and solid lubricant fillers were directly injected molded using Demag Ergotech 125/475 industrial machine. A second composites preparation methodology (M2) was used. In fact, the surface of the solid lubricant powder was chemically treated in a suspension containing 3-Aminopropyltriethoxy silane (AMPTES) as a coupling agent. The solution was stirred mechanically and dried in a vacuum chamber. The treated solid lubricant powder and the PC granules were then mixed in a Brabender®-Plastograph EC W50EHT internal mixer at 120 rpm and 200 °C for 20 min until complete homogenization. The blends were injection molded at a barrel temperature of 250 °C, an injection pressure of 120 MPa and an injection rate of 20 cm^3/s (ARBURG Allrounder 220 S).

2.2 Friction and Wear Characterizations

Tests were carried out using a ball on flat microtribometer in order to measure the friction coefficient evolutions. The home-made device was described in a previous publication (Ben Difallah et al. 2014). The thermoplastic based composites slid against a high chromium steel ball (19 mm diameter). The ball frequency was of 1 Hz, the normal load was of 5 N and the measurements were obtained after 10 000 sliding cycles over a displacement magnitude of ±1 mm. At least, three tests were performed in each type of composite.

Wear quantification was carried out by using the Alicona System as an optical 3D surface topography software program (Infinitefocus Alicona® laboratory system). The wear scars were ultrasonically cleaned to completely remove the debris on wear volume measurements. A scanning electron microscope (SEM by JEOL JSM 6400F) was employed to characterize the composites wear mechanisms. The samples were coated with gold before SEM observations.

2.3 Micromechanical and Microstructural Characterizations

Scratch tests were employed in order to analyze the micromechanical behavior (Revetest Automatic Scratch Tester). A conical diamond indenter with a continually increasing load was used at a speed rate of 2.5 mm/min. The initial loading value was 1 N, the final loading value was 30 N with a rate of 14.5 N/min. The tip radius of the indenter is 138 μm and the included angle is 120°. Different experimental data were measured and stored as a data file. The principal parameters used are the critical load (L_c), the penetration depth (R_p), the residual depth (R_h) and the viscoelastic recovery (f %) calculated as follow:

$$f\,(\%) = [1 - (R_h/R_p)] \times 100 \tag{1}$$

Fractographic analysis were also performed for the purpose of completing the morphological characterization of adhesion between the PC matrix and the solid lubricant particles. A freeze-fracture was prepared by plunging the samples in vacuum air and then exposing them to liquid nitrogen at a freezing room (Oxford CT1500 HF freeze–fracture apparatus). Then, fractures were done by a scalpel and the selected fracture surfaces were gold-coated and examined with the SEM.

The microstructural aspects of the composites were investigated by x-ray μtomography reconstructions based on the x-ray refractions of the specimen and images formation. This technique is based on phase-contrast imaging method which provide the visualization of details such as voids and cracks in weakly absorbing materials. It also determines spatial distribution and particle size of the filler in the thermoplastic based composites. The experimental conditions related to the Phoenix-Xray device used (vtome x-S) are an amorphous detector based on silicon material with 1920 × 1536 pixels, the x-ray tube voltage is about 80 kV with 3 μm resolution and an average scanning time of approximatively one hour.

3 Results and Discussion

3.1 Friction and Wear Behavior

Figure 1 shows the friction and wear volume loss results for the different composites after 10 000 cycles. It is clear from Fig. 1-a that the directly injected and molded PC with graphite inclusions (PC/Gr_M1) increased the frictional coefficient as compared to the unfilled PC. Similarly, Su et al. have investigated the effect of different solid lubricant fillers in the friction and wear properties of Nomex fabric composites. The authors reported that 10% (volume fraction) graphite or MoS_2 was found ineffective in reducing the friction and wear behavior of the Nomex materials (Su et al. 2006). The frictional results with Gr composites seem to be improved by the second methodology. The trend of the friction coefficient with MoS_2 particles is also shown. The decrease in the friction coefficient is clearly observed with the two developments methodologies of PC/MoS_2 composites. With direct injection molding, the friction coefficient of unfilled PC was initially 0.70, it decreased monotonically throughout the test. At 10 wt% of MoS_2 filler, the frictional coefficient approached 0.54.

As shown in Fig. 1-b, the wear measurements of PC/Gr_M1 has a sudden rise in comparison with the neat PC. The reasons for the above results may be as follows. On the one hand, electrostatic repulsions can be introduced between the polymer macromolecules and the graphite structure (Hedayati et al. 2012). On the other hand, the graphite particle size is almost 3 times higher than that of MoS_2, leading to inferior dispersion of graphite particles in the polymer, and thus the weakening of the matrix material. It can also be noticed that the second methodology contributes to the improvement of the wear response in the case of PC/Gr composites. The wear volume loss ranged from 6.15×10^{-3} mm^3 with neat PC to 5.63×10^{-3} mm^3 with PC + 10 wt% Gr. On the other hand, when 10 wt% MoS_2 is incorporated to PC, the wear volume loss decreases by a factor of 2 with both processing methodologies. Actually, its value was remarkably reduced down to 3.54×10^{-3} mm^3 with the second methodology. The MoS_2 particles improve the wear performance of PC when compared to graphite.

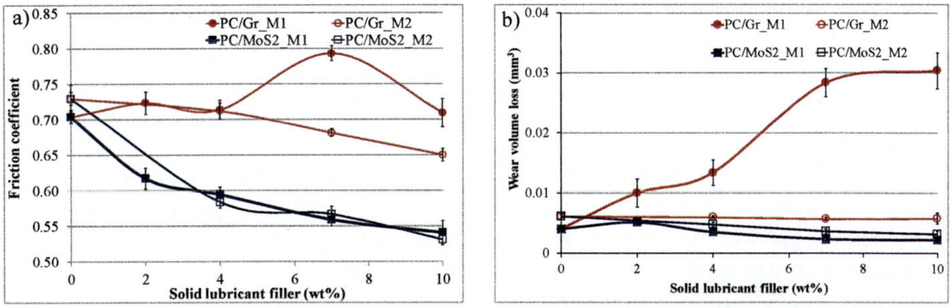

Fig. 1. Evolutions of the (a) mean friction coefficient (b) and the wear volume loss measured after 10 000 cycles for PC based composites.

Figure 2 shows typical profiles for pure PC and the composites with 10 wt% solid lubricant fillers prepared with the first methodology. In Fig. 2-a, the PC surface profile is highly irregular with a forth and back transfer from one surface to another in an adhesive wear mode. In contrast, the PC/MoS_2 surface profile (Fig. 2-b) is markedly different. The pits, pockets and valleys in the surface still remain. The hills are worn down and the surface is relatively smooth compared to the neat PC. The PC/Gr profile describes almost the same PC topography of the surface with an extremely larger pattern (Fig. 2-c).

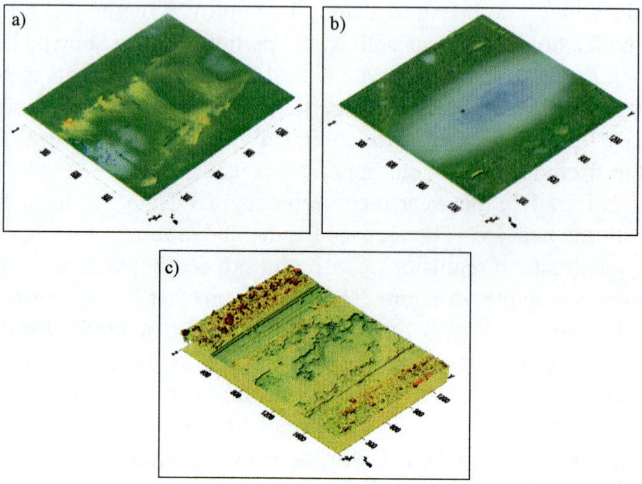

Fig. 2. 3D topography of the wear scars after the tests recorded by using the optical scanning microscopy technique for (a) PC_M1, (b) PC + 10% MoS2_M1 and (c) PC + 10% Gr_M1

The SEM image of pure PC shows a non-uniform wear track (Fig. 3-a). Moreover, the PC debris are dragged out from the matrix and easily compacted by the external force. The abundant particles pulled out from the matrix stick together because of the repeated ploughing mechanism during sliding, therefore, severe plastic deformation and deep cracks are observed (Arribas et al. 2009). It has been demonstrated that pure PC has poor tribological performance characterized by the production of a great number of wear particles in an adhesive-abrasive mechanism (Carriòn et al. 2008). Figure 3-b presents the features of the worn surface in the case of PC with 7 wt% MoS_2. The surface is relatively smooth marked by very fine linear grooves. The agglomerated layer at the edge of the stripe is less thick and adherent than that observed with pure PC. The incorporation of MoS_2 powder to the PC matrix promoted the abrasive component of wear (Ben Difallah et al. 2014; Carriòn et al. 2007; Arribas et al. 2009). The worn surface of PC with 10 wt% graphite is given in Fig. 3-c. It can be noted that a large area of the wear track is flaked away and the material is seriously damaged. At higher magnifications, the clear evidence of particle-polymer debonding and severe plastic deformation can be seen on the surface. The addition of graphite powder to a PC matrix did not impart a good wear behavior. On the contrary, a damaging effect can be seen. It has been suggested that the bonding forces between the matrix and graphite particles directly affect the structural integrity of the composites and have a detrimental effect on the tribological and mechanical response of materials (Su et al. 2005). The weak interfacial bonding deteriorates the tribological properties

(Chen et al. 2007(a), (b)). A rise in the abrasive component of graphite particle-reinforced PC matrix is also believed to result in an increase in the friction coefficient and the wear rate (Carriòn et al. 2007).

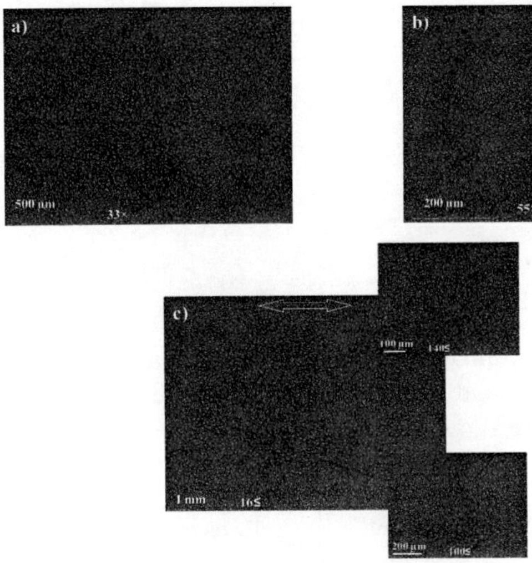

Fig. 3. SEM of the wear scars of (a) pure PC, (b) 7 wt% MoS_2 filled PC and (c) 10 wt% graphite filled PC, the sliding direction is indicated by arrows.

3.2 Scratch Test Results

Table 1 lists the different scratch test results for the PC based composites. It is worthwhile noting that the samples prepared with 10 wt% graphite under the first methodology have a combination of undesirable results. These samples have both low critical load and high residual scratch depth with a low elastic recovery of the material. The critical load of PC/Gr composites is significantly improved with the second processing methodology. The composites with 4 wt%_M2 and 10 wt%_M2 are the best performers because they possess a higher resistance to permanent deformation with excellent elastic recovery of samples. The insertion of the solid lubricant powder to the PC matrix was proven to drastically decrease the critical load for all the samples. Nevertheless, PC/MoS_2 shows significantly higher critical load than the graphite composite samples.

Table 1. Scratch resistance results for PC based composites as a function of the processing methodology and the composition

	1st Methodology					2nd Methodology			
	PC	% Gr		% MoS_2		% Gr		% MoS_2	
		4	10	4	10	4	10	4	10
R_h (μm)	30.0	88.9	92.4	61.2	75.6	65.9	74.8	48.4	55.7
R_p (μm)	126.7	176.4	168.2	143.8	157.8	142.7	153.0	131.4	144.2
f (%)	76.2	49.6	45.0	57.4	52.06	53.8	51.0	63.1	61.3
L_c (N)	26.16	8.54	6.05	17.9	14.8	16.7	15.4	20.0	19.05

3.3 Microstructure/Fracture Behavior

Figure 4 shows typical x-ray images obtained for PC composites at 10 wt% MoS_2 obtained with untreated (M1) and treated filler (M2), respectively. Porosity level and size and filler distribution are shown in the composites.

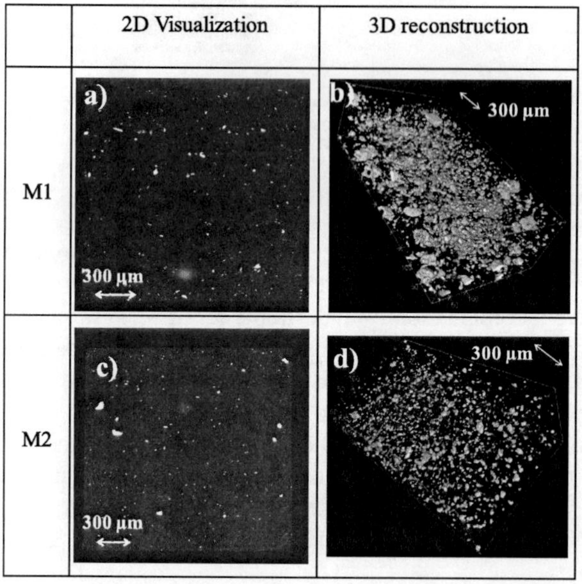

Fig. 4. 2D images and 3D constructions obtained with X-ray μ-tomography technique for PC + 10 wt% MoS_2.

The voids cover the particles of solid lubricant powder and are widely distributed in the matrix. The 3-dimensional views in Figs. 4-b and 4-d manifest a highly dispersed morphology of the filler particles in the composite prepared with M2 relative to the dispersion using the first one. The difference in the aspect between the fractured surfaces of the two fillers incorporated to the PC matrix is shown in Fig. 5. Figures 5-a and 5-b show the SEM images of PC + 10 wt% MoS_2_M1 and PC + 10 wt% Gr_ M1. Both fracture surfaces have a brittle appearance, and the solid lubricant filler is badly bonded to the matrix. On the contrary, no pronounced signs of fracture and debonding are seen with the composite using silane treated filler (Figs. 5-c and 5-d). The adhesion between filler and matrix is clearly improved.

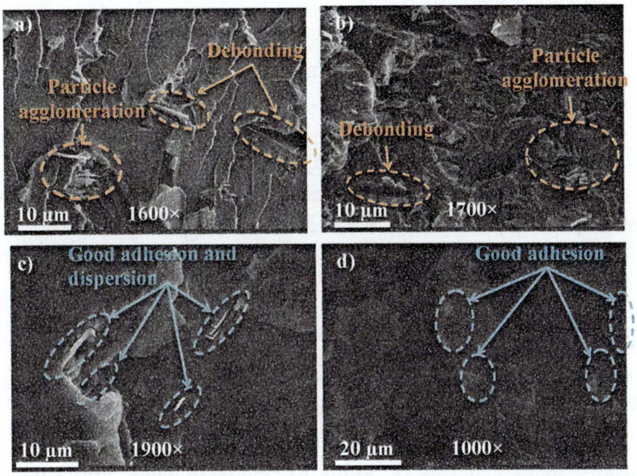

Fig. 5. SEM of fractured surfaces of PC + 10 wt% a) MoS_2_M1 b) Gr_M1 c) MoS_2_M2 and d) Gr_M2.

4 Conclusion

Addition of Gr powder to a PC matrix did not impart a good friction and wear behaviours by direct injection moulding (M1). On the contrary, a damaging effect can be seen. The weak interfacial bonding deteriorates the tribological properties (Hedayati et al. 2012; Chen et al. 2007(a), (b)). The second processing methodology enhanced the tribological parameters, the scratch resistance and the microstructural aspects of the PC/Gr composites. On the other hand, the MoS_2 filler improved the tribological properties of PC with both development methodologies. SEM observations of the fractured surfaces and x-ray µ-tomography micrographs of the composites injected molded directly throw further light on features such as debonding of filler from the matrix, microcracks, voids and filler agglomerations. A poor scratch behavior was also noticed. These microstructural features and the weak adhesion were attenuated by the second methodology.

References

Arribas, A., Bermúdez, M.D., Brostow, W., Carrión-Vilches, F.J., Olea-Mejía, O.: Scratch resistance of a polycarbonate + organoclay nanohybrid. Express Polym. Lett. **3**, 621–629 (2009)

Ben Difallah, B., Cornuault, P.H., Kharrat, M., Dammak, M., Monteil, G.: On the role of solid lubricant fillers in the tribological, micromechanical and morphological properties of PA 66 composites. J. Tribol. **142**(5) (2020). https://doi.org/10.1115/1.4045804

Ben Difallah, B., Kharrat, M., Dammak, M., Monteil, G.: Improvement in the tribological performance of polycarbonate via the incorporation of molybdenum disulfide particles. Tribol. Trans. **57**(5), 806–813 (2014)

Ben Difallah, B., Kharrat, M., Dammak, M., Monteil, G.: Mechanical and tribological response of ABS polymer matrix filled with graphite powder. Mater. Des. **34**, 782–787 (2012)

Chen, H., Jacobs, O., Wu, W., Rudiger, G., Schadel, B.: Effect of dispersion method on tribological properties of carbon nanotube reinforced epoxy resin composites. Polym. Testing **26**, 351–360 (2007a)

Carriòn, F.J., Arribas, A., Bermùdez, M.-D., Guillamon, A.: Physical and tribological properties of a new polycarbonate-organoclay nanocomposite. Eur. Polymer J. **44**(4), 968–977 (2008)

Carriòn, F.J., Sanes, J., Bermùdez, M.-D.: Influence of ZnO nanoparticle filler on the properties and wear resistance of polycarbonate. Wear **262**(11), 1504–1510 (2007)

Chen, H., Jacobs, O., Wu, W., Rudiger, G., Schadel, B.: Effect of dispersion method on tribological properties of carbon nanotube reinforced epoxy resin composites. Polym. Testing **26**(2), 351–360 (2007b)

Deng, J., Song, W., Zhang, H., Yan, P., Liu, A.: Friction and wear behaviors of the carbide tools embedded with solid lubricants in sliding wear tests and in dry cutting processes. Wear **270** (9–10), 666–674 (2011)

Hedayati, M., Salehi, M., Bagheri, R., Panjepour, M., Naeimi, F.: Tribological and mechanical properties of amorphous and semi-crystalline PEEK/SiO2 nanocomposite coatings deposited on the plain carbon steel by electrostatic powder spray technique. Prog. Org. Coat. **74**, 50–58 (2012)

Mnif, R., Ben Jemaa, M.C., Hadj Kacem, N., Elleuch, R.: Impact of viscoelasticity on the tribological behavior of PTFE composites for valve seals application. Tribol. Trans. **56**(5), 879–886 (2013)

Puértolas, J.A., Castro, M., Morris, J.A., Rios, R., Ansón-Casaos, A.: Tribological and mechanical properties of graphene nanoplatelets/PEEK composites. Carbon **141**, 107–122 (2019)

Su, F.-H., Zhang, Z.-Z., Guo, F., Wang, K., Liu, W.-M.: Effects of solid lubricants on friction and wear properties of Nomex fabric composites. Mater. Sci. Eng., A **424**(1–2), 333–339 (2006)

Su, F.-H., Zhang, Z.-Z., Wang, K., Jiang, W., Liu, W.-M.: Tribological and mechanical properties of the composites made of carbon fabrics modified with various methods. Compos. Part A Appl. Sci. Manuf. **36**(12), 1601–1607 (2005)

Mechanical Characterization of PP/Alfa Bio-composite Obtained by Thermocompression

Ismahen Zaafouri(✉), Montassar Zrida, Khaled Labidi, Ahmed Hichem Hamzaoui, and Moufida Borni

Useful Materials Valorization Laboratory,
National Centre of Research in Materials Science,
Technologic Park of Borj Cedria, B.P. 73, 8027 Soliman, Tunisia
esmahenzaafouri@gmail.com, montassar.zrida@gmail.com,
khaled.labidil@gmail.com, mdmihil0@gmail.com,
borni.moufida58@gmail.com

Abstract. This study contributes to the research of new materials from natural resources. It aims to exploit the natural fibres extracted from the Alfa plant in combination with a thermoplastic polymer matrix of polypropylene type in order to develop bio-composite materials. The biocomposite are composite materials made from natural or biofibre and non-biodegradable polymers such as polypropylene (PP), polyethylene (PE) and epoxies, or with biopolymers such as polylactic acid (PLA). The biocomposite made in this study was classified as eco-friendly and green since natural fibre and a renewable resource-based polymer were used. Various approaches have been used, including retting in salt water and chemical alkaline treatment. The extraction fibres was improved according to a two-step investigation. Firstly, the processing temperature, duration and the sodium hydroxide (NaOH) concentration, together with the usefulness of the post-treatment, were studied based on visual examinations. Morphological analyses carried out on the fibres obtained show an increase in the cellulose content and a decrease in the size of the treated fibres. Polypropylene/Alfa biocomposites have been formulated from zero to 30% in mass of fibres using the thermocompression molding process followed by injection molding. Tensile tests show the increase in Young's modulus and mechanical strength as a function of cellulose content and fibre dimensions. Chemical treatment with the sodium hydroxide (NaOH) solution with a concentration of 3.25 M at 100 °C during 2.5 h with 6-day retting together with applying the post-treatment has the best mechanical properties.

Keywords: Chemical treatment · Polypropylene · Alfa · Retting · Bio-composite material

1 Introduction

Several studies examined the use of natural fibres (flax, hemp, kenaf, cotton, jute, sisal, coconut, abaca, wood,...) as an alternative to the mineral fibres traditionally used as reinforcements in thermoplastic polymer matrix.

M. Kharrat et al. (Eds.): ICAMEM 2019, LNME, pp. 379–384, 2020.
https://doi.org/10.1007/978-3-030-52071-7_52

The natural fibres are used in various application sectors such as transport, packaging, construction, etc. However, their use as reinforcements in composite materials often pose problems of compatibility with the matrix. Indeed, the difference in polarity between natural hydrophilic fibres and conventional hydrophobic matrix results in poor charge transfer ability at the interface, and therefore in poor mechanical resistance. For that reason, it is necessary to modify the matrix or fibres by physical or chemical methods. The most commonly used technique is the alkaline treatment [1]. This treatment removes non-cellulosic compounds from the fibres, such as lignin, hemicellulose, pectin and impurities. Thus, it gives a roughness surface and leads to a better impregnation of the matrix [1].

2 Experimental Study

2.1 Materials

The natural fibres used in this study are fibres extracted from Alfa (Stipa tenacissima L) to reinforce thermoplastic polymer.

The matrix chosen is a copolymer polypropylene designated by its commercial quality "PPC7712" supplied by Total Petrochemicals in the form of granules. It is recyclable with good mechanical properties and excellent impact strength and is easy to inject [2].

2.2 Alkaline Extraction

Before starting the extraction of the fibres from the plant, a preliminary work is necessary in order to better prepare the stems for the various pre-treatments such as retting in salt water in an open environment. It consists in immersing the raw fibres in a tank of salt water (35 g/L) exposed to the open air, for six days.

For the chemical treatment, the fibres were treated with NaOH solution of 3.25 M concentration at a temperature of 100 °C for 2.5 h [3].

2.3 Elaboration of Bio-composite Material

The process used to make the composites is thermocompression molding which is the most suitable technique to achieve a high fibre content with good impregnation [4, 5]. It is carried out by compressing the material between two hot plates under pressure. Then, the tensile specimens for the mechanical characterization of the PP/Alfa bio-composite is prepared by the injection moulding process using a RAY RAN injection micro press.

2.4 Characterization

The chemical compositions of the fibres were determined after a succession of extractions to isolate the main components contained in the fibres, namely cellulose, lignin and hemicellulose [6]. The percentages of lignin, cellulose and hemicellulose were calculated.

In order to control the dimensions of Alfa fibres after each treatment, an optical microscopy study was carried out. This technique is used to determine the length and diameter of the extracted fibres. To determine the effect of fibres extraction parameters on the mechanical properties of the bio-composite material developed, tensile tests were carried out using a universal tensile testing machine (LLOYD EZ50).

3 Results and Discussions

3.1 Chemical Analysis

The chemical composition of raw and processed Alfa fibres is summarized in Table 1. The percentages of cellulose, hemicellulose and lignin are indicated.

Table 1. Chemical composition of Alfa fibres

Fiber	Lignin (%)	Hemicellulose (%)	Cellulose (%)
Raw stems	24.83	21.67	46.84
Extracted by NaOH	5.24	3.28	79.76
Extracted by NaOH with retting 6 days	4.40	3.07	83.61

From the analysis of the chemical composition of the fibres, we can draw the following conclusions:

- Comparing to the raw Alfa fibres, after chemical treatment with sodium hydroxide, the hemicellulose and lignin contents decreases while the cellulose contents increases.
- The retting also increases the cellulose content of the fibres.

3.2 Fibres Size Analysis

Table 2 shows the lengths (L), diameters (d) and aspect ratios (L/d) of the Alfa fibres extracted by NaOH without and with retting for six days.

Table 2. Diameters, lengths and aspect ratios for extracted fibres.

Extraction	Diameter (μm)	Length (μm)	Aspect ratio (L/d)
NaOH without retting	10.84	1104.85	101.92
NaOH with retting 6 days	10.31	1176	114.06

Based on the fibres size analysis, an increase in length and a decrease in diameter were observed when the fibres undergo a 6-day retting period. This result can be attributed to the improvement of fibres quality through retting.

3.3 Mechanical Properties

Tensile tests are carried out on a universal tensile testing machine of the brand LLOYD EZ50. The strain will be measured using an extensometer. The load is measured using a load cell of 1 kN. At the end of the tests, the Young's modulus, mechanical strength and elongation at break will be determined (Fig. 1).

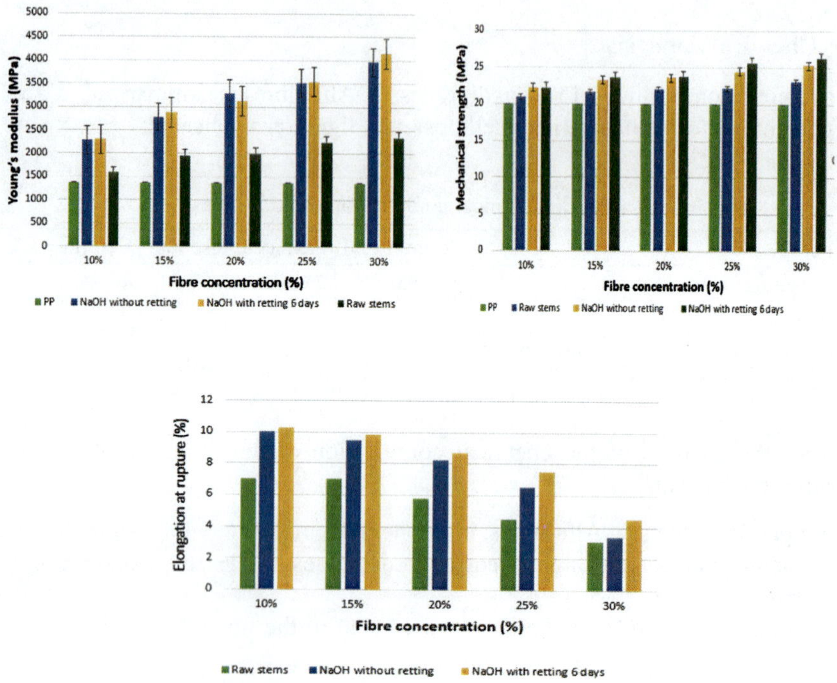

Fig. 1. Mechanical properties of PP/Alfa bio-composites

These results show that fibre extraction has an influence on the mechanical behavior of bio-composites. The incorporation of fibres was beneficial mainly for Young's modulus where treated fibres were more efficient than raw fibres. In this context, fibres extracted by NaOH with six-day retting showed the best results. The reason behind this is the improved adhesion between fibres and matrix as shown by micrographic analysis. In addition, the increase in the mass concentration of the fibres has led to an improvement in Young's modulus and mechanical strength. Break strain was significantly reduced from about 400% for PPC7712 to minor than 10% for the various bio-composites.

3.4 Morphological Analysis

In order to qualitatively the adhesion between Alfa fibres and matrix, we have proceeded to a micrographic analyses by SEM as shown in Fig. 2.

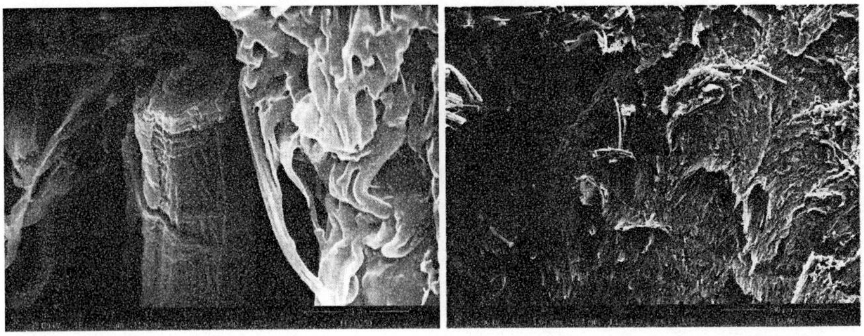

Fig. 2. SEM micrographs of the failure zone of fibre bio-composites extracted by NaOH with retting for 6 days.

We observe a homogeneous dispersion of fibers. They are uniformly distributed in all directions. In addition, some medium-length fibres appear fractured at their cross-section confirming that the adhesion in contact with the components is sufficient to transfer the stresses from fibres to the matrix. These pictures show that the PP/Alfa fibre adhesion extracted by NaOH with six days retting is good, indeed the alkaline extraction of the fibres has improved the quality of the interface. This improvement would be at the origin of the mechanical strength increment and Young's modulus.

4 Conclusion

In this study, the incorporation of Alfa fibres into the polypropylene matrix significantly changed its mechanical properties. So, fibres extraction method has an influence on the mechanical behaviour of bio-composites. Indeed, the passage through a retting stage in salt water makes it possible to obtain a better quality of the fibres and subsequently the most important mechanical properties. In this context, the biocomposite obtained by thermocompression elaboration method have shown good results. The reason behind this is the improved adhesion between fibres and matrix as shown by micrographic analysis.

References

1. Faulstich de Paiva, J.M., Frollini, E.: Unmodified and modified surface sisal fibers as reinforcement of phenolic and lignophenolic matrices composites: thermal analyses of fibers and composites. J. Mater. Eng. **291**(4), 405–417 (2006)
2. Zrida, M., Laurent, H., Rio, G.: Numerical study of mechanical behaviour of a polypropylene reinforced with Alfa fibres. J. Compos. Mater. **50**(21), 2883–2893 (2016)
3. Labidi, K., Zrida, M., Korhonen, O., Borghei, M., Hamzaoui, A.H.: Alfa fiber: alkaline extraction together with structural and morphological characterization. J. Cell. Chem. Technol. **52**(9–10), 701–709 (2018)

4. Satyanarayana, K.G., Arizaga, G., Wypych, F.: Biodegradable composites based on lignocellulosic fibers—an overview. J. Prog. Polym. Sci. **34**(9), 982–1021 (2009)
5. Shah, D.U.: Developing plant fibre composites for structural applications by optimising composite parameters: a critical review. J. Mater. Sci. **48**(18), 6083–6107 (2013)
6. Jayaramudu, J., Guduri, B.R., Rajulu, A.V.: Characterization of natural fabric sterculia urens. Int. J. Polym. Anal. Charact. **14**(2), 5–125 (2009)

Modeling of Iron Based Shape Memory Alloys Behavior Within Finite Strain Formulation

Achref Sallami[1(✉)], Walid Khalil[2], Tarak Bouraoui[1], and Tarak Ben Zineb[3]

[1] Université de Monastir, Ecole Nationale d'Ingénieurs de Monastir, LGM, LR99ES32, 5000 Monastir, Tunisie
{achref.sallami,tarak.bouraoui}@enim.rnu.tn

[2] Université de Gabès, Institut Supérieur des Systèmes Industriels de Gabès, 6032 Gabès, Tunisie
walid4m@yahoo.fr

[3] Université de Lorraine, CNRS, Arts et Métiers ParisTech, LEM3, 54000 Nancy, France
tarak.ben-zineb@univ-lorraine.fr

Abstract. This paper presents a finite-strain thermomechanical constitutive model for Iron based Shape Memory Alloys (Fe-SMAs). The model considers the multiphysical coupling between the phase transformation and the plasticity which distinguishes Fe-SMAs from classical SMAs (Ni-Ti and Copper based ones). Based on the multiplicative decomposition of deformation gradient into elastic, transformation and plastic components, the model is developed within a thermodynamically consistent framework. Model constitutive equations associated with the internal variables (volume fraction of martensite χ and the accumulative plastic strain γ) are derived from the reduced form the constructed Lagrangian, which is the sum of the Helmholtz free energy function and the inner potential, and the Clausius-Duhem inequality. The model is used to simulate tensile experiments on Fe-SMAs specimens at different thermomechanical loadings. The good agreement of the model predictions demonstrates the capabilities of the proposed model to well the specific behavior of Fe-SMAs.

Keywords: Finite strain constitutive model · Multiplicative decomposition · Volume fraction of martensite · Accumulative plastic strain

1 Introduction

Shape Memory Alloys (SMAs) presents a specific behavior represented by the Shape Memory effect (SME), in which the original configuration can be recovered by heating, and the Pseudo Elasticity (PE), in which the original configuration can be recovered after unloading [1]. Fe-SMAs are distinguished from

M. Kharrat et al. (Eds.): ICAMEM 2019, LNME, pp. 385–391, 2020.
https://doi.org/10.1007/978-3-030-52071-7_53

classic SMAs by a coupling form between phase transformation and plasticity [2]. Developed models for Fe-SMAs are constructed with respect to the small strain hypothesis [3–5]. But, Fe-SMAs can reach high levels of deformation which contract that a small strain model is not a suitable formulation in order to model these materials. This was the main idea to construct a finite strain model for Fe-SMAs.

The development of finite strain solids behavior modeling devotes attention in recent years. Using this formulation allows to model small and large strain behavior laws. Unlike the small strain formulations which are limited in front of solid large strain, displacements and rotations, the finite strain ones can manage those conditions via the gradient of deformation. Basic hypothesis of finite strain materials can be found in the works of [6,7].

This work is an extension of the small strain behavior model [4]. In order to satisfy the specific behavior of Fe-SMAs represented by the coupling between phase transformation and plasticity. This model was inspired from works developed in finite strain formulation to take into account the plasticity (Elastoplastic materials) and the phase transformation (Classical Shape Memory Alloys).

The elastoplastic behavior models are based on the multiplication of the deformation gradient into elastic and plastic ones, $\mathbf{F} = \mathbf{F}_{el}\mathbf{F}_{pl}$ [8–11]. The same technique has been adopted by researcher in order to predict classical SMAs behavior in large strain [12–15]. Those models are based on the multiplicative form of the deformation gradient into elastic and phase transformation ones as, $\mathbf{F} = \mathbf{F}_{el}\mathbf{F}_{tr}$.

In order to satisfy the coupling form between phase transformation and plasticity on the case of Fe-SMAs, and based on the small strain behavior model, the finite strain models developed for classical SMAs and elastoplastic materials, a tri-partite decomposition of the deformation gradient has been carried out as, $\mathbf{F} = \mathbf{F}_{el}\mathbf{F}_{tr}\mathbf{F}_{pl}$. The constitutive equations have been constructed from a thermomechanical consistent formulation. Those equations have been implemented on a Matlab software in order to verify the model capability to predict large strain levels. Results presents a well-being compliance with experimental data.

2 Thermomechanically Coupled Constitutive Model

In this section, a tri-partite decomposition of the total deformation gradient into, elastic, phase transformation and plasticity as follows:

$$\mathbf{F} = \mathbf{F}_{el}\mathbf{F}_{tr}\mathbf{F}_{pl} \tag{1}$$

Based on this multiplicative form a thermodynamically consistent framework and a Helmholtz free energy potential are introduced. Then, a Lagrangian $\Im$ has been defined as the sum of the Helmholtz free energy Ψ and the inner Ψ_{IC} potentials. Then, Thermomechanically coupled constitutive model equations are derived to describe Fe-SMA behavior. The Helmholtz free energy has been defined as the sum of hyper elastic and inelastic parts as follows:

$$\Psi = \Psi_{el} + \Psi_{ine} \tag{2}$$

with

$$\Psi_{el} = \frac{\mu}{2}\left(I_{\mathbf{C}_{el}} - 3\right) - \mu \ln J + \frac{\lambda}{2}(\ln J)^2 \tag{3}$$

$$\Psi_{ine} = c_v\left[(T - T_0) - T\ln\left(\frac{T}{T_0}\right)\right] + B\,(T - T_0)\,\chi + G_g\left(\mathbf{E}_{ine}\right) + \left[G_v\left(\chi\right) + (1-\chi)\,G_s\left(\gamma\right) + G_{sv}\left(\gamma,\chi\right)\right] \tag{4}$$

The inner potential has been defined in order to stabilize the physical conditions on the volume fraction of martensite χ . This latter should be less then one and grater then zero. To obey this condition, two Lagrangian multiplier, ℓ_1 and ℓ_2 , which obey to the Kuhn-Tucker conditions, have been defined. The inner potential Ψ_{IC} has the following form:

$$\Psi_{IC} = -\ell_1\chi - \ell_2(1-\chi) \tag{5}$$

The Lagrangian is defined as follows:

$$\Im = \frac{\mu}{2}\left(I_{\mathbf{C}_{el}} - 3\right) - \mu\ln J + \frac{\lambda}{2}(\ln J)^2 + c_v\left[(T - T_0) - T\ln\left(\frac{T}{T_0}\right)\right] + B\,(T - T_0)\,\chi + G_g\left(\mathbf{E}_{ine}\right) + \left[G_v\left(\chi\right) + (1-\chi)\,G_s\left(\gamma\right) + G_{sv}\left(\gamma,\chi\right)\right] - \ell_1\chi - \ell_2(1-\chi) \tag{6}$$

where μ and λ are equivalent Lamé constants, $I_{\mathbf{C}}el$ being the first invariant of $\mathbf{C}_{el}$, J is the deformation Jacobean, T is the current temperature, c_v , B and T_0 are the material parameters and G_v, G_s, G_{sv} are, respectively, the interaction function between martensite variants, the interaction function between plastic slip systems and the interaction functions between martensite variants and plastic slip systems .

3 Clausius Duhem Inequality

Using Clausius–Duhem inequality form:

$$-\left(\dot{\Im} + \xi\dot{T}\right) + \mathbf{S} : \dot{\mathbf{E}} - \mathbf{q}\frac{\mathbf{grad}\,T}{T} \geq 0 \tag{7}$$

where $\mathbf{S}$ being the second Piola-Kirchhoff stress tensor, $\mathbf{E}$ being the Green-Lagrange strain tensor, ξ presents the entropy and $\mathbf{q}$ is the heat flux.

After mathematical procedure and taking into account that the temperature is homogenous in the VER, the final form is defined as follows:

$$\Lambda_{dr}^{\chi}\dot{\chi} + \Lambda_{dr}^{\gamma}\dot{\gamma} \geq 0 \tag{8}$$

with Λ_{dr}^{χ} and Λ_{dr}^{γ} are respectively the driving forces for phase transformation and plasticity, defined as follows:

$$\Lambda_{dr}^{\chi} = \bar{\varepsilon}_{\max}^{tr}\left(\mathbf{M} + H_g(\mathbf{E}_{ine} : \mathbf{E}_{ine})^{n_g}\mathbf{F}_{ine}\mathbf{E}_{ine}\mathbf{F}_{ine}^{T}\right) : \mathbf{N}^{tr} - B\,(T - T_0) - H_v\chi^{n_v} + \frac{H_s}{(n_s+1)}\gamma^{n_s+1} - H_{sv}\gamma(\chi\gamma)^{n_{sv}} + \ell_1 - \ell_2 \tag{9}$$

$$\Lambda_{dr}^{\gamma} = (1-\chi)\left(\mathbf{M} + H_g(\mathbf{E}_{ine} : \mathbf{E}_{ine})^{n_g}\mathbf{F}_{ine}\mathbf{E}_{ine}\mathbf{F}_{ine}^T\right) : \mathbf{N}^{pl} - (1-\chi)\, H_s\gamma^{n_s} - H_{sv}\chi(\chi\gamma)^{n_{sv}} \quad (10)$$

with $\bar{\varepsilon}_{\max}^{tr}$ being the saturation phase transformation equivalent tensor, $\mathbf{M}$ is the so called Mandel stress, $\mathbf{N}^{tr}$ and $\mathbf{N}^{pl}$ are the direction vectors, H_g, H_v, H_s and H_{sv} are the interaction parameters and n_g, n_v, n_s and n_{sv} are the Coefficients to integrate the nonlinearities of interactions.

The critical forces for different Fe-SMA responses (Forward and reverse transformation, and plasticity) are defined, respectively, as follows:

$$\Lambda_y^{tr,m} = \sigma_y^m \bar{\varepsilon}_{tr}^{\max} + \zeta_m (T_0 - M_s) - H_\gamma \gamma^2 \quad (11)$$

$$\Lambda_y^{tr,a} = \sigma_y^a \bar{\varepsilon}_{tr}^{\max} + \zeta_a (T_0 - A_s) - H_\chi \chi \quad (12)$$

$$\Lambda_y^{pl} = \sigma_y^\gamma - K_T T \quad (13)$$

with ζ_m and ζ_a are control parameters, M_s being the martensite start temperature, A_s being the austenite start temperature, σ_y^m, σ_y^a, and σ_y^γ are, respectively the critical stress of, direct and reverse phase transformation and plasticity and K_T represents the evolution slope of the critical stress of plasticity with the temperature.

In addition, we emphasize that the evolution equations satisfy the loading functions for the direct phase transformation Υ_{tr}^m , reverse phase transformation Υ_{tr}^a and plasticity Υ_{pl}^a

$$\Upsilon_{tr}^m = \Lambda_{dr}^\chi - \Lambda_y^{tr,m} \quad (14)$$

$$\Upsilon_{tr}^a = -\Lambda_{dr}^\chi - \Lambda_y^{tr,a} \quad (15)$$

$$\Upsilon_{pl}^a = \Lambda_{dr}^\gamma - \Lambda_y^{pl} \quad (16)$$

The loading functions satisfy the Kuhn–Tucker conditions:

$$\begin{cases} \Upsilon_{tr}^m \leq 0, & \dot{\chi} \geq 0, & \dot{\chi}\Upsilon_{tr}^m = 0 \\ \Upsilon_{tr}^a \leq 0, & -\dot{\chi} \geq 0, & -\dot{\chi}\Upsilon_{tr}^a = 0 \\ \Upsilon_{pl}^a \leq 0, & \dot{\gamma} \geq 0, & \dot{\gamma}\Upsilon_{pl}^a = 0 \end{cases} \quad (17)$$

The solving of this equations may us to know the specific behavior of the Fe-SMA at the discrete time.

4 Uniaxial Tests Model Validation

Small strain simulations have been compared to experimental results [2] and small strain model [4]. Table 1 presents the model parameters. Those parameters were equilibrated to the Fe-$Mn_3$1.6-Si_6.45-C_0.018 (wt%). Figure (1a) shows a comparison between current model, experimental results and small strain model response. The model shows a good agreement. Looking to the evolution variables (Fig. (1b)), only the volume fraction of martensite evaluates. At 20 °C

Table 1. Model parameters for Fe-Mn$_{31.6}$-Si$_{6.45}$-C$_{0.018}$(wt%)

Parameter	Value	Parameter	Value
σ_y^m (MPa)	137.7	A_s (°C)	96
σ_y^a (MPa)	125.2	T_0 (°C)	79.4
σ_y^γ (MPa)	210	H_v (MPa)	27.2
ν	0.3	H_s (MPa)	632.5
μ (MPa)	65385	H_{sv} (MPa)	1000
λ (MPa)	98077	H_g (MPa)	627
$\bar{\varepsilon}^{tr}_{\max}$	0.048	n_v	0.44
ζ_m (MPa/°C)	0.086	n_s	0.55
ζ_a (MPa/°C)	0.036	n_{sv}	0.45
B (MPa/°C)	0.104	n_g	3.07
K_T (MPa/°C)	0.45	H_γ (MPa)	500
M_s (°C)	−25	H_χ (MPa)	−3

the material behavior is purely phase transformation Fig. (2a) shows a comparison between different numerical simulations and experimental results at 50 °C. The experimental curve is consistent with the current model comparing to the small strain one which presents a linear hardening. Figure (2b) shows the evolution of χ and γ as a function of the total strain at 50 °C. At the start, the phase transformation is activated. At 0.45% of strain, the plasticity starts and a coupling form activates between the phase transformation and the plastic slip. The evolution slope of martensite volume fraction exchange when the plastic behavior activates, validating that the activation of the plasticity stops drastically the phase transformation evolution. At this temperature, the inelastic strain comprises the phase transformation and the plastic slip. furthermore, the

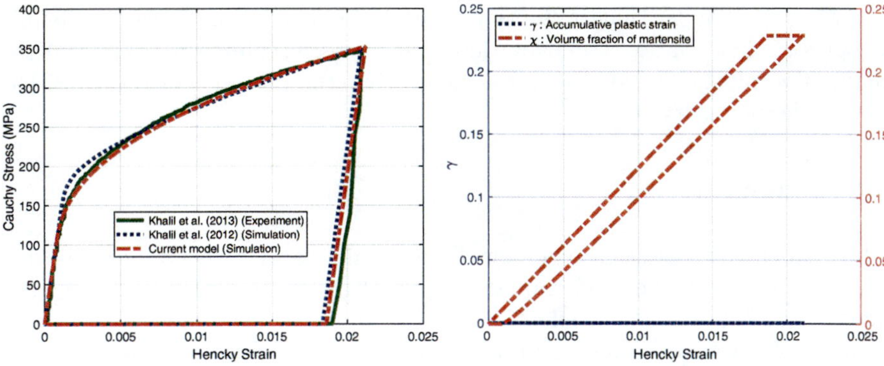

Fig. 1. (a) Tensile test at 20 °C; (b) Evolution of internal variables

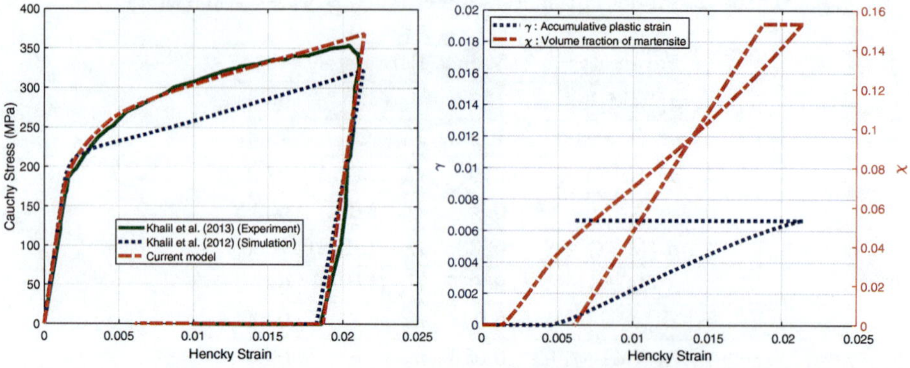

Fig. 2. (a) Tensile test at 50 °C; (b) Evolution of internal variables

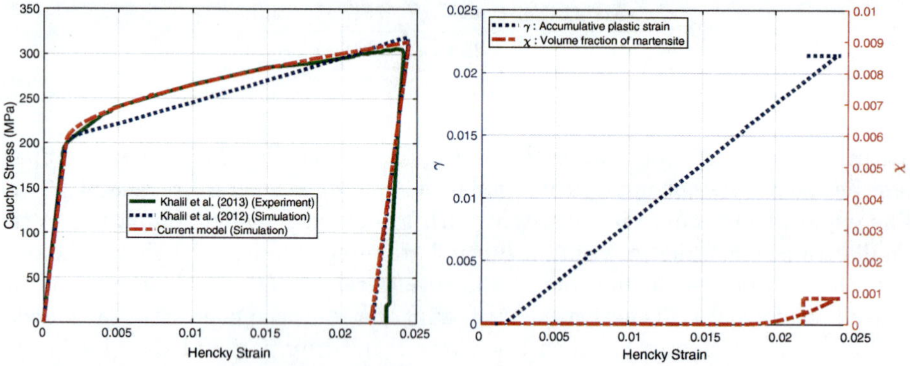

Fig. 3. (a) Tensile test at 130 °C; (b) Evolution of internal variables

comparison of numerical and experimental results in Fig. (3a) appearances that the current model and experimental data at 130 °C are steady comparing with the small strain model that presents the same problem of the linear hardening when plasticity activates. The evolution of martensite volume fraction and the accumulative plastic strain (Fig. (3b)) shows that the behavior of the Fe-SMAs is purely the plasticity.

5 Conclusion

In this paper, we investigate a new finite-strain constitutive model for Fe-SMAs. A consistent thermodynamic framework and a Lagrangian have been defined as the sum of Helmholtz free energy function including super elastic and inelastic interaction and an inner potential, from which constitutive were derived. Two important characteristics of Fe-SMA behavior are considered in formulating the model. Those characteristics introduce a well-defined form for model variables. Model constitutive equations have been solved implicitly by the

Newton–Raphson method. The obtained results show a good agreement with experimental data. More then that, this new model enhances the small strain model by taking into account nonlinear hardening.

References

1. Cissé, C., Zaki, W., Ben Zineb, T.: Int. J. Plast. **76**, 244–84 (2016)
2. Khalil, W., Saint-Sulpice, L., Arbab Chirani, S., Bouby, C., Mikolajczak, A., Ben Zineb, T.: Mech. Mater. **63**(1), 1–11 (2013)
3. Jemal, F., Bouraoui, T., Ben Zineb, T., Patoor, E., Bradaï, C.: Mech. Mater. **41**, 849–856 (2009)
4. Khalil, W., Bouby, C., Mikolajczak, A., Ben Zineb, T.: J. Intell. Mater. Syst. Struct. **23**, 1143–1160 (2012)
5. Cissé, C., Gu, X., Zaki, W., Ben Zineb, T.: Mech. Mater. **107**, 1–21 (2016)
6. Taylor, G.I.: J. Inst. Metals **62**, 307–24 (1938)
7. Malvern, E.L.: Introduction to the Mechanics of a Continuous Medium. Prentice-Hall, Englewood Cliffs (1969)
8. Lee, E.H.: J. Appl. Mech. **36**, 1–6 (1969)
9. Sidoroff, F., Dogui, A.: Int. J. Solids Struct. **38**, 69–78 (2001)
10. Vladimirov, I.N., Pietryga, M.P., Reese, S.: Int. J. Plast **26**, 659–678 (2010)
11. Grillo, T.J., Vladimirov, I.N., Valente, R.A.F., Reese, S.: Arch. Appl. Mech. **86**, 147–163 (2016)
12. Christ, D., Reese, S.: Int. J. Solids Struct. **46**, 3694–3709 (2009)
13. Evangelista, V., Marfia, S., Sacco, E.: Int. J. Numer. Methods Eng. **81**, 61–85 (2009)
14. Baghani, M., Arghavani, J., Naghdabadi, R.: Mech. Mater. **73**, 1–10 (2014)
15. Wang, J., Moumni, Z., Zhang, W., Zaki, W.: Int. J. Eng. Sci. **117**, 51–77 (2017)

Surface Properties and Adhesion Behavior of PVD Thin Film Using Multi-cyclic Nanoindentation and Scratch Test

Kaouther Khlifi[1,2(✉)], Hafedh Dhiflaoui[1], Najoua Barhoumi[1,2], and Ahmed Ben cheikh Larbi[1]

[1] Laboratoire de Mécanique, Matériaux et Procédés LR99ES05, Ecole Nationale Supérieure d'Ingénieurs de Tunis, Université de Tunis, 5 Avenue Taha Hussein, Montfleury, 1008 Tunis, Tunisia
Kaouther.khlifi@ipeiem.utm.tn

[2] Institut Préparatoire aux Etudes d'Ingénieurs d'El Manar, Université de Tunis - El Manar, 2092 Tunis, Tunisia

Abstract. PVD thin film based on chromium nitride was deposited on XC100 steel substrate with reactive radio frequency magnetron sputtering. Thin film was characterized in terms of both surface properties and adhesion behavior. The characterization of surface properties was based on the measurement of mechanical properties coupled with a preliminary microstructural study using SEM and AFM. Fatigue resistance was considered as an important surface properties and it was investigated using multi-cyclic nanoindentation test. Results indicated that chromium-based coatings have a smooth, dense and compact microstructure with a grain size of about 109 nm and a roughness of about 11 nm. CrSiN showed a high adhesion quality, critical loads were LC1 = 16 N, LC2 = 20 N and LC3 = 25 N. PVD coatings exhibited a hardness of about 30.52 GPa and Young's modulus of about 338.32 GPa. Moreover, multi-cycle nanoindentation technique, with variation of loading mode (progressive and quasi-static) and loading rate (from 200 mN/mn, 400 mN/mn and to 600 mN/mn) was used to analyze superficial fatigue behavior and surface resistance. Whatever, the cyclic loading mode, an elasto-plastic behavior was observed and an appearance of hysteresis loops in the unloading-reloading curves witch was explained by a fatigue effect. In addition, a hardening phenomenon was found when the loading rate increase.

Keywords: PVD coating · Surface properties · Adhesion · Multi-cycle nanoindentation

1 Introduction

Chromium nitride PVD coating has been used as a hard protective coating, which has a combination of a good adhesion [1] high wear resistance [2] and good corrosion-resistance [3]. Many studies were performed on the use of classical indentation test to study the surface properties [4, 5]. But, film thickness was limited to a few micrometers, the use of micro indentation technique is not simple because of the contribution

M. Kharrat et al. (Eds.): ICAMEM 2019, LNME, pp. 392–401, 2020.
https://doi.org/10.1007/978-3-030-52071-7_54

of substrate properties [6]. In addition, the surface properties were not limited to hardness and Young's modulus but rather it requires a study of fatigue resistance. Due to very small thickness of PVD coating, fatigue test should be performed on a small scale. Recently, few studies have been conducted on fatigue test using multi-cyclic nanoindentation [7]. But the study of the influence of cyclic nanoindentation parameters, such as loading mode, loading rate and cycle number, was not well discussed. The investigation of the film adhesion is always necessary, a coating having high mechanical properties but does not adhere to its substrate has no importance. The principal objective of this work is to study the film adhesion and the surface properties by microstructural and morphological analysis and by measurement of mechanical properties and superficial fatigue resistance.

2 Experimental Procedure

2.1 Film Deposition

CrSiN PVD films were deposited on XC100 steel by magnetron sputtering. The separately applied targets are Cr (99.99%) and Si (99.99%). The films were deposited in Ar and N_2 mixture introduced into the deposition chamber. Before the deposition, substrates were ultrasonic-cleaned using acetone and ethanol. The base pressure in the deposition chamber was 2×10^{-3} Pa, while the total pressure of N_2 and Ar during the CrSiN deposition was adjusted to 1.0 Pa. A layer of pure Cr (for 10 min in Ar atmosphere) was deposited as the transition layer to improve the adhesion. Thickness film was 2,5 μm. The deposition process is schematized by Fig. 1.

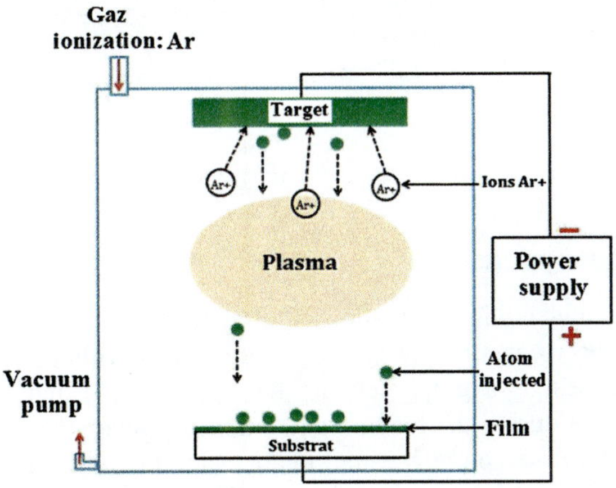

Fig. 1. Schematic illustration of the PVD Vacuum process

2.2 Film Characterization

CrSiN surface morphology was explored using AFM in contact mode. Microstructural properties were investigated using scanning electron microscopy (SEM). Mechanicals properties were measured with monotonic nanoindentation test developed by CSM instruments Switzerland (Anton Paar) using a diamond Berkovich tip. The evaluation of the interface behavior of the PVD film was performed using scratch test. Cyclic nanoindentation test were realized in order to investigate the superficial fatigue resistance. This repetitive test was performed at different loading mode (progressive and quasi-static loading) and loading rate.

3 Results and Discussion

3.1 Morphological and Microstructural Properties

The microstructure of the CrSiN thin film was examined by (SEM) and the surface morphology was analyzed with AFM. Results are given in Fig. 2. It was found a smooth, dense and compact microstructure with a small grain size witch was about 109 nm and a roughness of about 11 nm.

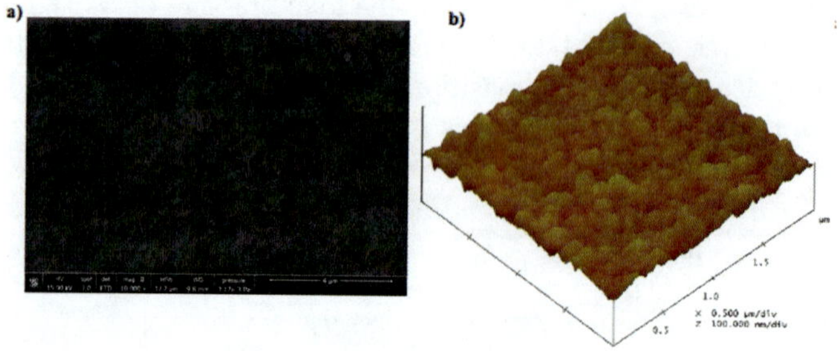

Fig. 2. SEM microstructure a) and AFM image b) of CrSiN thin film

3.2 Mechanical Properties

Measuring by nanoindentation test allows the application of some nano-newton in order to evaluate the mechanical properties of thin films without substrate influence. The output of this test is a graph relating the applied load and the corresponding indenter displacement during the loading and unloading phases. The analysis of load–displacement graphs, based on Oliver and Pharr method [8], lead to the determination of the coating mechanical properties: Hardness (H) and Young's modulus (E) of the coating. Figure 3 show a schematic representation of nanoindentation process, typical loading unloading curve as well as the Oliver and Pharr model. The loading unloading curve shows a first part, it is the loading phase in which a plastic deformation has occurred until a maximum load (P_{max}) for a maximum penetration depth (h_{max}). Then, when the indenter was completely removed, the residual depth h_f was obtained. S was the contact stiffness. (dP/dh), β and ε are parameters related to indenter geometry. CrSiN coating showed an elasto-plastic behavior. Penetration depth increases as a function of normal load. It was found that the hardness and Young's modulus were 30.52 GPa and 338.32 GPa respectively.

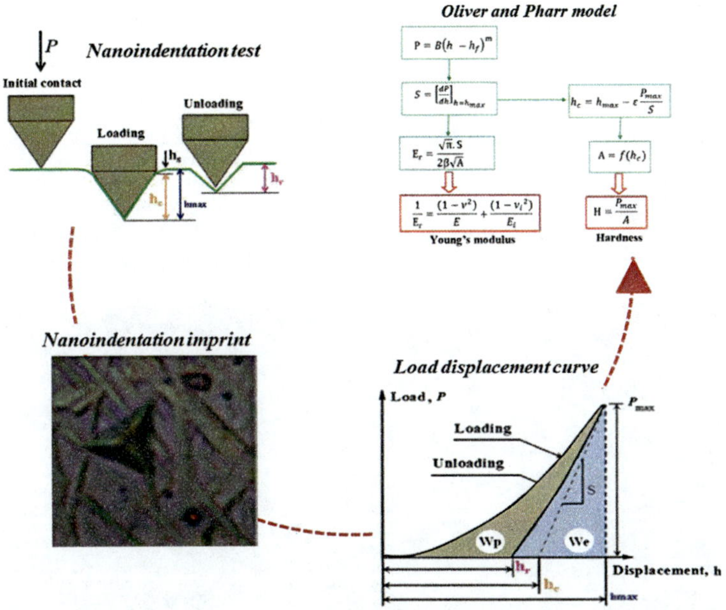

Fig. 3. Schematic representation of nanoindentation process, typical loading- Displacement curve and Oliver and Parr model

3.3 Adhesion Behavior

Adhesion behavior was evaluated using scratch technique. During this test, normal applied load increases gradually and film was more and more solicited. The output of the scratch test was a graph relating the normal applied load, the friction load and the corresponding scratch depth as indicated in Fig. 4. Scratch track was then analyzed and exploited in order to identify the film damage (Fig. 5).

Figure 4 and 5 show that progression of scratching was accompanied by a successive degradation of the film according to two stages of damages: the first one is a cohesive failure witch was indicated by occurrence of a micro cracks at the edge of the scratch tracks (Fig. 5a) and 5b)). This damage occurs at a first critical load called L_{C1}. The second stage was defined, at a second critical load L_{C2}, by a local chipping of the coating from its substrate. This stage of failure was followed by a total damage of the film/substrate interface as indicated in 2 a). It was found that critical load were $L_{C1} = 16$ N, $L_{C2} = 20$ N and $L_{C3} = 20$ N. CrSiN coating has a good adhesion quality comparing to other PVD hard coating studied in our previous works [9].

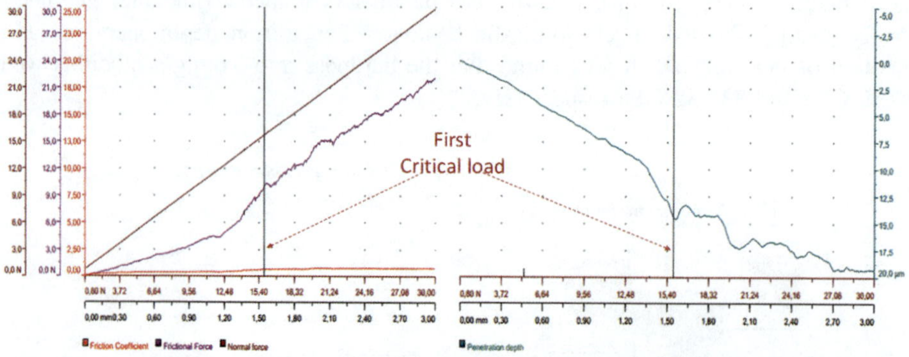

Fig. 4. Friction coefficient, friction load and penetration depth versus normal load

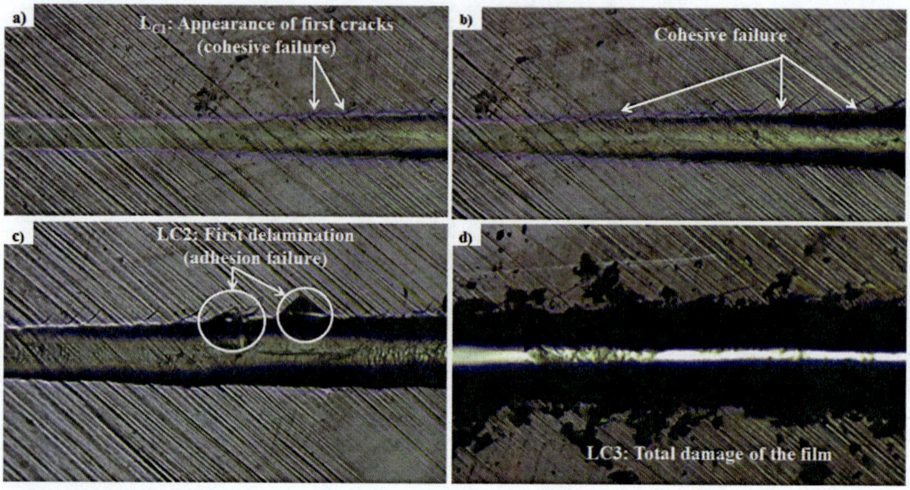

Fig. 5. Optical micrograph of scratch track: Cohesive failure a) and b), Adhesive failure c) and total damage d)

3.4 Superficial Fatigue Behavior

Surface properties of thin film were not limited to the hardness and young's modulus but rather to resistance to cyclic solicitation and fatigue effect. However, this fatigue test should be performed with a few nano-load because of low film thickness. Consequently, cyclic nanoindentation tests present an adequate technique that has allowed the cyclic nanoindentation with a small scale. In this work, multi-cyclic nanoindentation test were performed on CrSiN coating with different loading modes (Progressive loading and Quasi-static loading) and different loading rate. Progressive loading consists of applying multiple loading-unloading cycles with increasing of the maximum load at each cycle. Quasi-static loading consists of repeating the charge-discharge cycle at the same conditions. The number of cycles (load-unload) was 50 cycles and the maximum load was 100 mN. Test was achieved with an increment of 20 mN for Progressive mode, to reach 100 mN at the last load. The loading speed range was 200, 400 and 600 mN/min. The loading mode and the loading rate effect are the subject of the rest of this work.

3.4.1 Effect of Mutli-cyclic Loading Mode

3.4.1.1 Progressive Nanoindentation

Figure 6 show the load-displacement curves under conventional and multi-cycle progressive nanoindentations for 5 cycles Fig. 6a) and 50 cycles Fig. 6b).

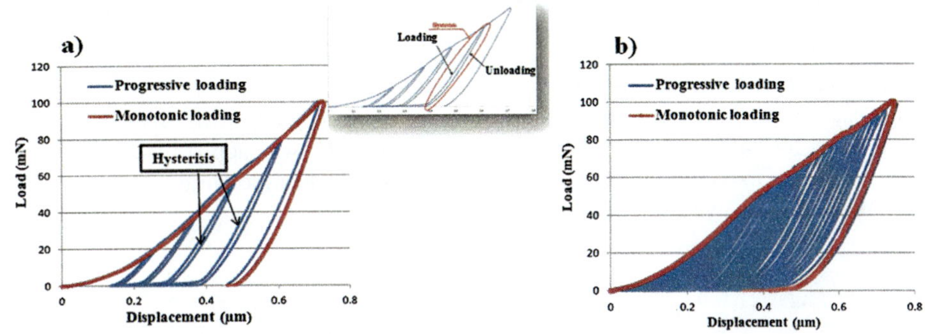

Fig. 6. Load-displacement curves under conventional and progressive nanoindentations for 5 cycles 6 a) and 50 cycles 6 b)

The superposition of the load-displacement graph of the two types of loading (monotonic and progressive) showed an elasto-plastic behavior with the same penetration depth (0,7 µm). This finding is true whether for 5 cycles or 50 cycles. Under progressive loading, hysteresis loops were produced which signifies that resistance to deformation becomes greater during reloading path comparing to the unloading path. This hysteresis have been found and confirmed by other works [10–12].

3.4.1.2 Quasi-statique Nanoindentation

Figure 7 shows a superposition of the two curves obtained under progressive and quasi-static loading after 5 cycles.

Hysteresis loops were found for both loading conditions but the unloading profile of the last progressive cycle was superimposed with the unloading profile of the first quasi-static cyclic. In addition, the total penetration depth is greater for quasi-static (0,8 µm) loading compared to that achieved for the progressive loading (0,5 µm). The indenter depth increases with the repetitions of the loadings-unloading and coating resistance to penetration decreases. Results confirmed fatigue effect that has been found under cyclic loading, in other works [9].

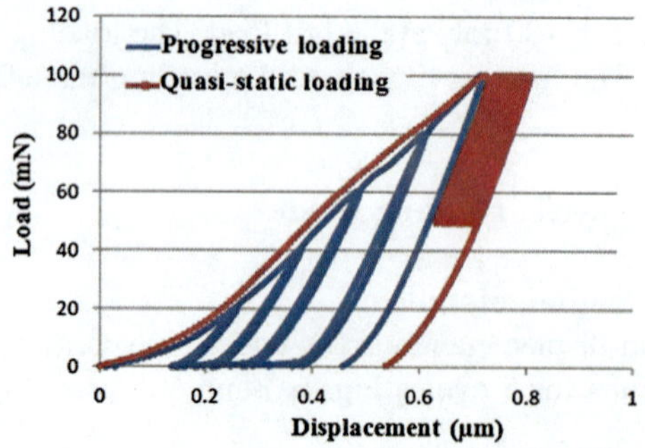

Fig. 7. Load-displacement curves of CrSiN under progressive and quasi-static nanoindentation

3.4.2 Effect of Loading Speed

Progressive and quasi-static nanoindentation were carried out with varying loading speed from 200, 400 to 600 min/min. Progressive loading was applied with a step of 20 mN to reach maximum load of 100 mN. For each speed and loading mode, a profile (load-displacement) was recorded as shown in Fig. 8.

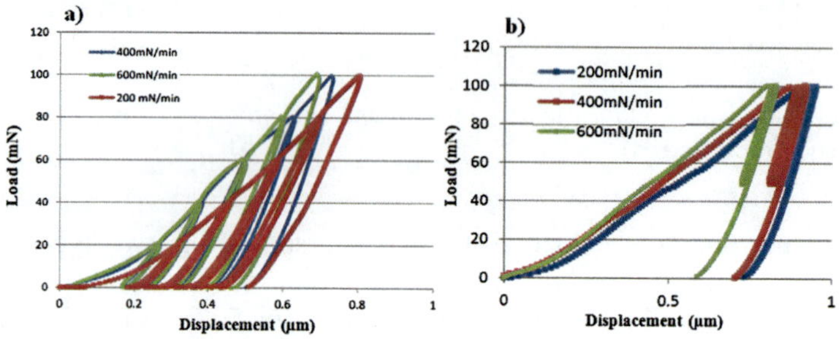

Fig. 8. Load–displacement curves of progressive a) and quasi-static loading b) at different loading rate

When the loading speed increases, the indentation depth increases regardless of the loading mode. The maximum penetration depth reaches 0.80 μm for a loading speed of 200 mN/min and 0.71 μm for 400 mN/min whereas it does not exceed 0.68 μm for 600 mN/min. In addition, we note that whatever the loading speed, the loading-unloading curve has hysteresis loops and the hysteresis width increases with increasing of the loading speed. This finding presents an indication of hardening which accompanies the cyclic nanoindentation. This result has confirmed the hardening phenomenon that occurs when the loading speed increases.

3.4.3 Degradation of Mechanical Properties

To quantify the variation of the hardness at each loading cycle, an approach was used to determine specific hardness values. The hardness was proportional to the stress equal to P/A, where P is the applied load and A is the indentation area calculated for Berkovich indenter, $A = 24.56\ h^2$ [13]. Figure 9 illustrates the variation of hardness as a function of displacement of indenter for progressive cyclic loading at 5 cycles Fig. 9a) and 50 cycles Fig. 9b).

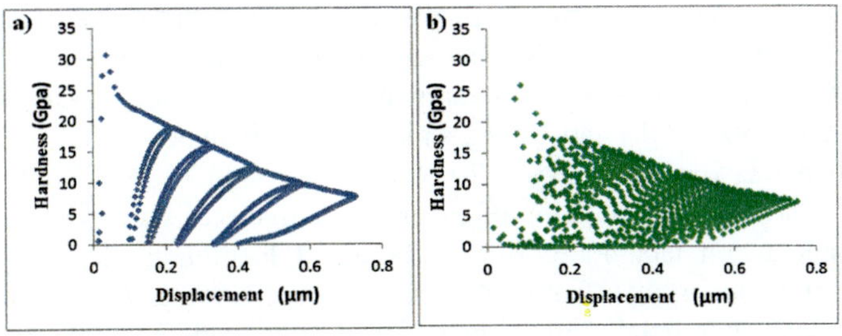

Fig. 9. Variation of hardness as a function of displacement for progressive loading at 5 cycles 9a) and 50 cycles 9b)

The hardness decreases with the increase of penetration depth. Hardness-displacement profile reproduces the hysteresis loops that have been noted in the curve load-displacement (Fig. 6). This result indicates a loss of hardness from one cycle to other and confirmed the fatigue effect. For Young's modulus, it was represented as a function of cycle number as shown in Fig. 10. A decrease of the Young's modulus with the increase of the cycle number was noted.

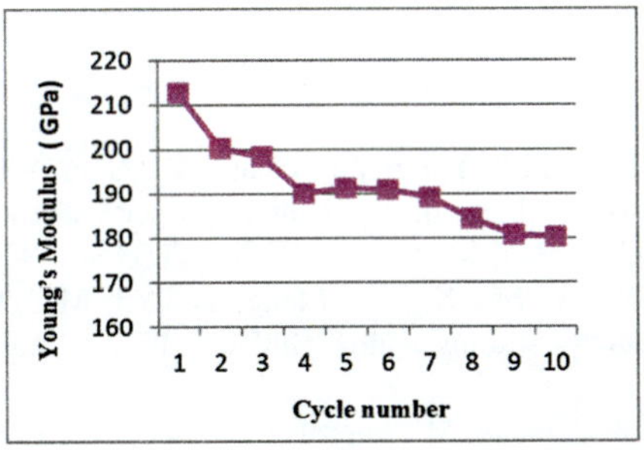

Fig. 10. Variation of Young's modulus as a function of cycle number

It has been found a degradation of mechanical properties with cyclic loading. This was explained, on the one hand by a substrate effect: when the indentation number increase, the penetration depth increase and the substrate influences more and more the mechanical properties. On the other hand, this degradation of mechanical properties was explained by a fatigue effect indicated by a loss of rigidity.

4 Conclusion

CrSiN PVD thin films were deposited on XC100 steel substrate using magnetron sputtering system. Coating was characterized in terms of both adhesion behavior and surface properties. Analysis of the microstructural and morphological properties showed a smooth, dense and compact microstructure with a grain size of about 109 nm and a roughness of about 11 nm. Mechanical properties and adhesion of PVD coating was evaluated. Chromium-based coatings revealed high mechanical properties, a hardness of about 30.52 GPa and Young's modulus of about 338.32 GPa. Critical loads were important, L_{C1} = 16 N, L_{C2} = 20 N and L_{C3} = 25 N. The study of surface properties extends to the evaluation of superficial fatigue resistance. Therefore, multi-cycle nanoindentation technique, with variation of loading mode (progressive and quasi-static) and loading rate (from 200 mN/mn, 400 mN/mn and to 600 mN/mn) was used to analyze superficial fatigue behavior. Cyclic nanoindentation leads to an increase of penetration depth and a decrease of hardness and Young's modulus from one cycle to another. Therefore, the study of the loading mode effect has shown that there is a significant effect on surface behavior for quasi-static loading. Regarding the effect of loading rate, we noticed a phenomenon of hardening related to the increase in loading speed. In addition, as the cycle number increase, a degradation of mechanical properties was founded and explained by fatigue effect. Other important result in this work that we confirmed the efficacy and the performance of multi-cyclic nanoindentation to investigate the superficial fatigue resistance of thin films.

Acknowledgments. The work described in this paper has been funded by the LMMP (Laboratoire Mécanique Matériaux et Procédés) at ENSIT, University of Tunis.

References

1. Teng, Y., Guo, Y.Y., Zhang, M., Yang, Y.J., Huang, Z., Zhou, Y.W., Wu, F.Y., Liang, Y.S.: Effect of Cr/CrNx transition layer on mechanical properties of CrN coatings deposited on plasma nitrided austenitic stainless steel. Surf. Coat. Technol. **367**, 100–107 (2019)
2. Zhou, F., Chen, K., Wang, M., Xu, X., Meng, H., Yu, M., Dai, Z.: Friction and wear properties of CrN coatings sliding against Si3N4 balls in water and air. Wear **265**(7–8), 1029–1037 (2008)
3. Li, H., Zhang, C., Liu, C., Huang, M.: Improvement in corrosion resistance of CrN coatings. Surf. Coat. Technol. **365**, 158–163 (2019)
4. Vecchione, N., Wasmer, K., Balint, D.S., Nikbin, K.: Characterization of EB-PVD yttrium-stabilised zirconia by nanoindentation. Surf. Coat. Technol. **203**, 1743–1747 (2009)
5. Kanders, U., Kanders, K., Maniks, J., Mitin, V., Kovalenko, V., Nazarovs, P., Erts, D.: Nanoindentation response analysis of Cu-rich carbon–copper composite films deposited by PVD technique. Surf. Coat. Technol. **280**, 308–316 (2015)
6. Khlifi, K., Dhiflaoui, H., Zoghlami, L., Ben Cheikh Larbi, A.: Study of mechanical behavior, deformation, and fracture of nanomultilayer coatings during microindentation and scratch test. J. Coat. Technol. Res. (2015)

7. Chicot, D., Tilkin, K., Jankowski, K., Wymysłowski, A.: Reliability analysis of solder joints due to creep and fatigue in microelectronic packaging using microindentation technique. Microelectron. Reliab. **53**, 761–766 (2013)
8. Oliver, W.C., Pharr, G.M.: An improved technique for determining hardness and elastic modulus using load and displacement sensing indentation experiments. J. Mater. Res. **7**(6), 1564–1583 (1992)
9. Khlifi, K., Ben cheikh Larbi, A.: Mechanical properties and adhesion of TiN monolayer and TiN/TiAlN nanolayer coatings. J. Adhes. Sci. Technol. **28**, 85–96 (2013)
10. Tekaya, A., Benameur, T., Labdi, S.: Cyclic Nanoindentation Studies on TiN and Ti-TiN Multilayer Coatings Deposited on Zr-Based Metallic Glasses (2013)
11. Kaouthar, K., Dhifelaoui, H., Zoghlami, L., Ben Cheikh Larbi, A.: Multi-cycle nanoindentation studies on TiN/TiAlN nano-multilayer coating: effects of mechanical properties and coating structure. Appl. Mech. Mater. **798**, 367–371 (2015)
12. Tekaya, A., Ghulman, H.A., Benameur, T., Labdi, S.: Cyclic nanoindentation and finite element analysis of Ti/TiN and CrN nanocoatings on Zr-based metallic glasses mechanical performance. J. Mater. Eng. Perform. **23**, 4259–4270 (2014)
13. Khan, M.K., Hainsworth, S.V., Fitzpatrick, M.E., Edwards, L.: Application of the work of indentation approach for the characterization of aluminium 2024–T351 and Al cladding by nanoindentation. J. Mater. Sci. **44**, 1006–1015 (2009)

Fluid Mechanics

Investigating the Free-Surface Flow Behavior Due to Sluice-Gate Maneuvers

Ali Triki(✉)

High Institute of Applied Sciences and Technology of Gabès, University of Gabès, Gabès, Tunisia
ali.triki@issatgb.u-gabes.tn

Abstract. This paper explored the free-surface wave behavior in prismatic open-channel. The one-dimensional Boussinesq model was used for transient flow description, along with the (2/4)-dissipative scheme implemented for numerical computations. Investigation addressed the flow behavior in a rectangular open-channel supplied by a constant depth reservoir and equipped with a sluice gate at its outlet inlet. The determination of the fundamental pulsation of the hydraulic system followed from the water-hammer concept, used in pressurized-pipe flows. Two kinds of boundary conditions were investigated; involving the abrupt closure and/or the sine-excitation of the sluice-gates. Results showed that the sine excitation case, using the fundamental pulsation of the hydraulic system, induced a significant depth rise; which may lead to the flood of the channel. Furthermore, it was found that the free-surface wave pattern was driven by the sine excitation using submultiple value of the fundamental pulsation; however, this pattern was controlled by the abrupt closure maneuver for the case using submultiple values of the fundamental pulsation.

Keywords: Boussinesq model · Free-surface · (2/4)-dissipative scheme · Open-channel · Propagation · Pulsation · Rectangular reflection · Wave

1 Introduction

Canals constitute major systems for delivering water for irrigation and power production. These systems are frequently subjected to flow disturbances due to either accidental or deliberate operations; such as rapid changes in sluice-gate settings (Fig. 1). These disturbances are relevant to many architectural and engineering technological applications, including the efficient design of sluice-gates and dams.

Concerning free-surface flow forecasts, the shallow water assumption constitutes a beneficial simplification of the full Navier-Stokes equations set. In this context, the resulting one-dimensional (1-D) St. Venant model provides acceptable results for subcritical flow regimes (Garcia-Navarroa and Savirona 1992); however, this model is no longer accurate in describing supercritical flow regimes, as they are strongly sensitive to the variations of vertical velocity during the transient event (Treske 1994). In this line, Boussinesq proposed a second alternative to the St.-Venant model, in order to account for vertical velocity effects in rapidly varied flow regimes. Precisely, Boussinesq proposed an add-on term to the 1D St.-Venant momentum equation,

M. Kharrat et al. (Eds.): ICAMEM 2019, LNME, pp. 405–411, 2020.
https://doi.org/10.1007/978-3-030-52071-7_55

considering a linear variation of the vertical velocity (Gharangik and Chaudry 1991; Chaudhry et al. 2008). It should be delineated herein that the Boussinesq equations (BWEs.) have third-order terms. Consequently, it is necessary to employ at least a third-order accurate numerical model to solve the BWEs.

Fig. 1. Illustration of a sluice-gate.

Previous study of the author (Triki and Hajtaieb 2010; Triki 2013, 2014a, b, 2016, 2017) focused on the free-surface wave behavior due to sluice-gate maneuvers. Basically, this study follows from the water-hammer concept in pressurized-pipe flow. The author used a McCormack-based numerical solver to compute the hydraulic parameters. However, this numerical solver is second-order accurate in space and time (Garcia-Navarroa and Savirona 1992); and therefore the Boussinesq term is dropped from the numerical solution. Alternatively, we planned in this study to re-examine the free-surface flow behavior due to sluice-gate maneuvers using a more appropriate numerical solver. Specifically, the (2/4)-dissipative scheme is selected in this study to discretize the BWEs.; which is a second-order accurate in time and fourth-order in space explicit finite-difference scheme.

2 Materials and Methods

Referring to Gharangik and Chaudhry (1991) and Triki (2017), the 1D-BWEs, may be expressed as follows:

$$\frac{\partial \mathrm{U}}{\partial t} + \frac{\partial \mathrm{G(U)}}{\partial x} = \mathrm{H(U}, x) \tag{1}$$

In which: $\mathrm{U} = \{q \quad A\}^{\mathrm{T}}$; $\mathrm{G(U)} = \{q \quad (q^2/A) + gA^2/(2T) + \mathrm{B}\}^{\mathrm{T}}$; $H(\mathrm{U}, x) = \{q_l \quad Ag(s_0 - s_f)\}^{\mathrm{T}}$; q denotes the flow-rate discharge; $d = q/A$ designates the flow-depth; $A = T \times d$ stands for the cross-section area of the open-channel; T corresponds to the free-surface width; q_l is the lateral discharge; s_0 designates the slope of the open-channel bed; $s_f = n^2u^2 \Big/ R_h^{4/3}$ is the flow resistance; u is flow velocity; R_h denotes the hydraulic-radius; n is the Manning's roughness; $\mathrm{B} = \left\{ \left(A^3/3T^2\right) \left[\partial^2(q/A)/\partial x \partial t + (q/A)\left[\partial^2(q/A)/\partial x^2\right] - \left(\partial(q/A)/\partial x\right)^2\right]\right\}$ is the Boussinesq term; g refers to gravity acceleration; and t and x denote the time and distance in the direction of the open-channel bed, respectively.

As per Gharangik and Chaudhry (1991) and Triki (2017), the (2/4)-dissipative scheme applied to Equations set (1), may be summarized in three-steps:

(a)-Predictor-step:

$$\overline{\mathrm{U}}_i = \overline{\mathrm{U}}_i^k + \frac{\sigma}{6}\left[\mathrm{G}_{i+2}^k - 8\mathrm{G}_{i+1}^k + 7\mathrm{G}_i^k\right] + \frac{\Delta t}{2}\left[\mathrm{H}_{i+1}^k + \mathrm{H}_i^k\right] \tag{2}$$

(b)-Corrector-step:

$$\overline{\overline{\mathrm{U}}}_i^{k+1} = \frac{1}{2}\left[\overline{\overline{\mathrm{U}}}_i^k + \overline{\mathrm{U}}_i\right] + \frac{\sigma}{12}\left[-7\overline{\mathrm{G}}_i^k + 8\overline{\mathrm{G}}_{i-1}^k - \overline{\mathrm{G}}_{i-2}^k\right] + \frac{\Delta t}{2}\left[\overline{\mathrm{H}}_{i+1}^k + \overline{\mathrm{H}}_i^k\right] \tag{3}$$

(c) Updating step of the flux vector:

$$\mathrm{U}_i^{k+1} = \overline{\overline{\mathrm{U}}}_i^{k+1} + \kappa_{i+(1/2)}\left(\overline{\overline{\mathrm{U}}}_{i+1}^{k+1} - \overline{\overline{\mathrm{U}}}_i^{k+1}\right) - \kappa_{i-(1/2)}\left(\overline{\overline{\mathrm{U}}}_i^{k+1} - \overline{\overline{\mathrm{U}}}_{i-1}^{k+1}\right) \tag{4}$$

in which, $\sigma = \Delta t/\Delta x$ is grid-mesh ratio; i $(1 \le i \le N_s)$ refers to the grid points in the x-direction; and k denote the time step; $N_s = L/\Delta x$ refers to the number of reaches of the open-channel; L stands for the open-channel length; the single and the double over bars denote the hydraulic parameter values, computed in the prediction and the correction steps, respectively; and $\kappa_i = \left|\left|\mathrm{U}_{i+1}^k - 2\mathrm{U}_i^k + \mathrm{U}_{i-1}^k\right| / \left(\left|\mathrm{U}_{i+1}^k\right| - 2\left|\mathrm{U}_i^k\right| + \left|\mathrm{U}_{i-1}^k\right|\right)\right|$ and $\kappa_{i+(1/2)} = \kappa/\sigma \mathrm{Max}(\kappa_{i+1}, \kappa_i)$ are updating coefficients (Jameson et al. 1981).

The stability of the numerical algorithm is satisfied basing on the Courant-Friedrichs-Lewy condition (i.e.: $\Delta t \le \mathrm{Cr} \times \Delta x / \left(|u| + \sqrt{gA/T}\right)_{\max}$; in which, $\mathrm{Cr} \le 2/3$ is the Courant number (Jameson et al. 1981)).

3 Application, Results and Discussion

Test case (Fig. 2) corresponds to a rectangular open-channel ($L = 305$ m; $T = 3.6576$ m; $s_0 = 0.001$; $n = 0.014$ (Wylie et al. 1993)) supplied by a reservoir at its inlet and discharging under a sluice-gate, placed at the downstream extremity. The initial steady-state flow relates to a uniform flow regime defined by: $d_0 = 2.4$ m and $q_0 = 63.407$ m$^3/s$.

The fundamental pulsation of free-surface waves may be inspected following the concept of the water-hammer phenomenon in pressurized-pipe flows. Specifically, the channel is experienced to a sudden closure of the downstream sluice-gate; while a constant free-surface level is maintained at the upstream reservoir (i.e.: $d|_{x=0}^{t \succ 0} = d_0$ and $q|_{x=L}^{t \succ 0} = 0$).

Figure 3 illustrates the depth signals predicted for the forgoing transient scenario at different cross sections of the channel. Referring to the water-hammer theory, the fundamental period of free-surface wave corresponds to the period of the downstream depth oscillation (i.e.: $T_f \approx 252.373$ s). In other words, the corresponding fundamental pulsation of free-surface wave is equal to: $\omega_f = 2\pi/T_f = 0.0248\text{rd/s}$.

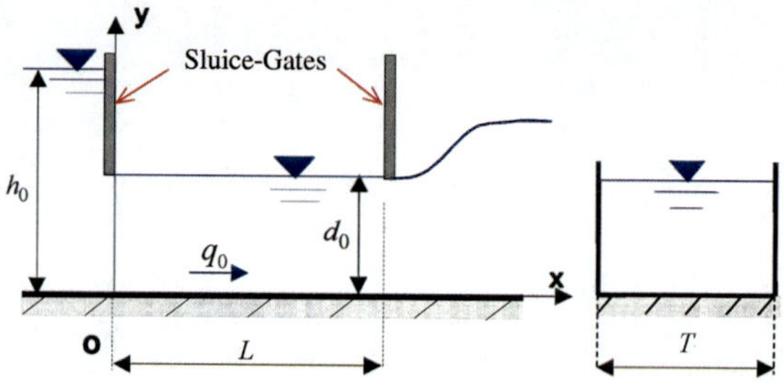

Fig. 2. Hydraulic system setup.

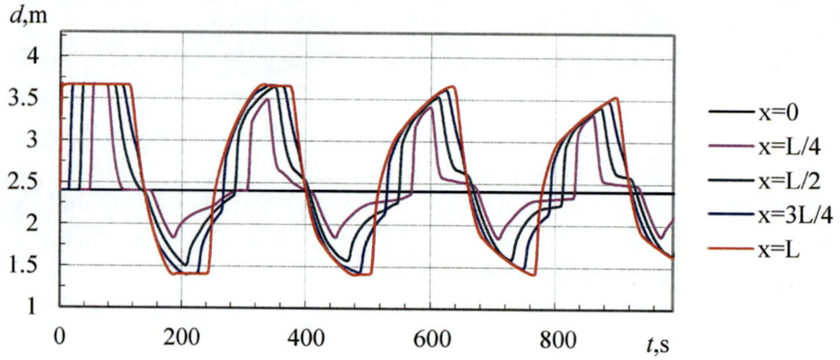

Fig. 3. Depth signals, versus time for a downstream gate closure.

In the following, the transient flow depth resulting from the sin-excitation of the upstream sluice-gate and the abrupt closure of the downstream sluice-gate (i.e.: $d|_{x=0}^{t\succ 0} = d_0 + \delta d\ \sin(\omega \times t)$ and $q|_{x=L}^{t\succ 0} = 0$; wherein, $\delta d = 0.3$ m) is investigated.

Figure 4 compares the depth signals along the channel for an abrupt closure of the downstream sluice-gate and a sinusoidal excitation of the upstream depth with the pulsation ($\omega = \omega_f$).

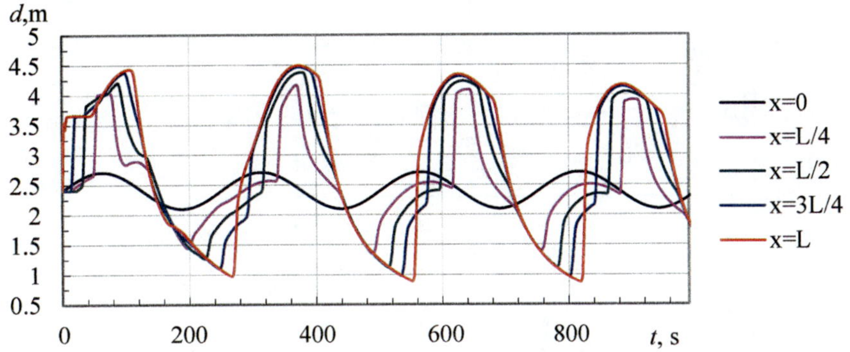

Fig. 4. Depth signals, versus time, involved by the superposition scenario.

Figure 4 suggests that the effect of the sine excitation induces important depth rise along the open-channel. In particular, for this case (i.e.: $\omega = \omega_0$) the maximal depth value, observed at the downstream extremity of the open-channel, is equal to: $d_1 =$ 4.5 m (at $t_1 = 100.94$ s); while it was equal to 3.9 m (at $t = 98.86$ s) for the case without excitation $\omega = 0$ (Fig. 3).

Figure 5 illustrates the depth signals resulting from the sine excitation of the upstream depth and the closure of the downstream sluice-gate. These curves indicate that the wave issued from the closure maneuver is predominating, for the case using multiple values of the fundamental pulsation (Fig. 5-a); however the sine excitation is predominating for the case using submultiples values of the fundamental pulsation (Fig. 5-b).

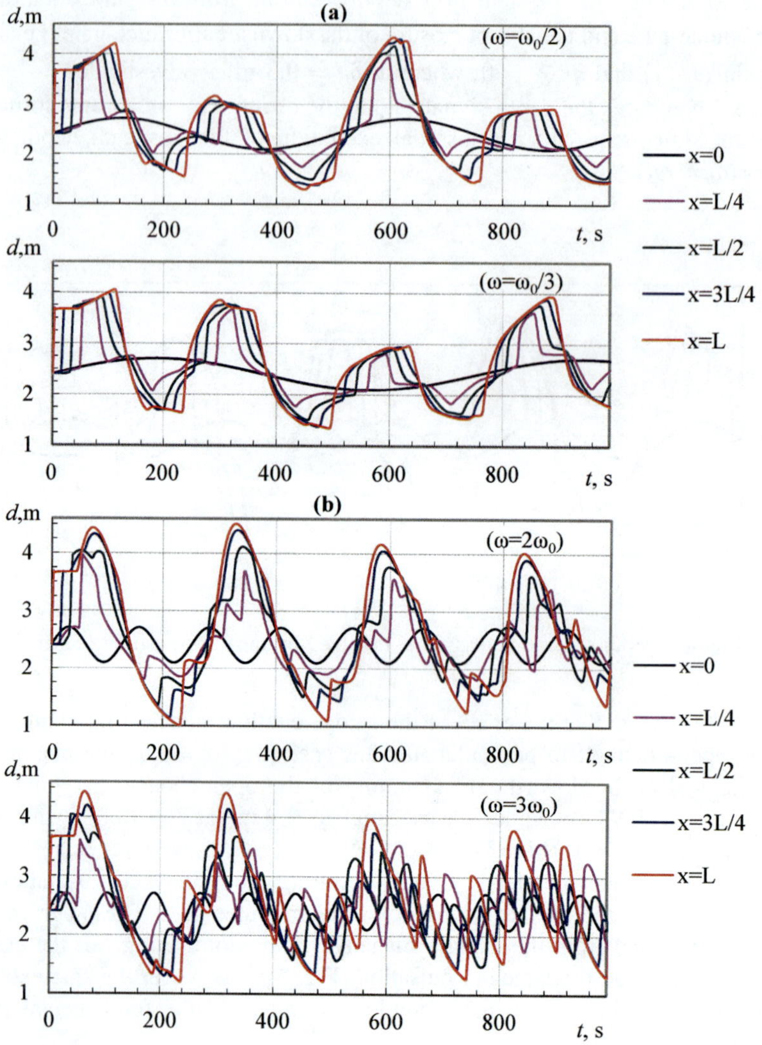

Fig. 5. Depth signals, for the sine excitations of upstream depth using multiple or submultiples values of the fundamental pulsation.

4 Conclusion

In summary, this study explored the propagation and reflection mechanisms of free-surface waves induced by sluice-gates maneuvers into a rectangular open-channel. Results highlighted that the sine excitation of the flow depth, using the fundamental pulsation of the hydraulic system, induced significant depth rise; which may lead to the flooding of the channel. It was also found that the free-surface wave pattern was driven by the sine excitation, for the case using submultiple values of the fundamental pulsation; however, this pattern was driven by the abrupt closure maneuver for the case using submultiple value of the fundamental pulsation.

References

Chaudhry, M.H.: Open-Channel Flow, 2nd edn. Springer, Heidelberg (2008)

Garcia-Navarroa, P., Savirona, J.M.: McCormack's method for the numerical simulation of one-dimensional discontinuous unsteady open channel flow. J. Hydraulic Res. **30**(1), 95–105 (1992). https://doi.org/10.1080/00221689209498949

Gharangik, M.A., Chaudhry, M.H.: Numerical simulation of hydraulic jump. J. Hydraul. Eng. **117**(9), 1195–1211 (1991). https://doi.org/10.1061/(ASCE)0733-9429(1991)117:9(1195)

Jameson, A., Schmidt, W., Turkel, E.: Numerical solution of the Euler equations by finite volume methods using Runge Kutta time stepping schemes. In: 14th Fluid and Plasma Dynamics Conference, AIAA, Palo Alto, CA (1981)

Treske, A.: Undular (favre-waves) in open channels-experimental studies. J. Hydraulic Res. **32** (3), 355–370 (1994). https://doi.org/10.1080/00221689409498738

Triki, A., Hajtaieb, E.: Simulation numérique des écoulements transitoires à surface libre provoqués par la superposition de manœuvres de vannes. Houille Blanche **3**, 71–80 (2010). https://doi.org/10.1051/lhb/2010037

Triki, A.: A finite element solution of the unidimensional shallow-water equation. J. Appl. Mech. ASME **80**(2), 021001 (2013). https://doi.org/10.1115/1.4007424

Triki, A.: Multiple-grid finite element solution of the shallow water equations: water-hammer phenomenon. Comput. Fluids **90**, 65–71 (2014a). https://doi.org/10.1016/j.compfluid.2013.11.007

Triki, A.: Resonance of free-surface waves provoked by floodgate maneuvers. J. Hydrol. Eng. **19**, 1124–1130 (2014b). https://doi.org/10.1061/(ASCE)HE.1943-5584.0000895

Triki, A.: Erratum for "resonance of free-surface waves provoked by floodgate maneuvers" by Ali Triki. J. Hydrol. Eng. **21**(7), 08216003 (2016). https://doi.org/10.1061/(ASCE)HE.1943-5584.0001408. https://doi.org/10.1061/(ASCE)HE.1943-5584.0001408

Triki, A.: Further investigation on the resonance of free-Surface waves provoked by floodgate maneuvers: negative surge waves. Ocean Eng. **133**, 133–141 (2017). https://doi.org/10.1016/j.oceaneng.2017.02.003

Wylie, E.B., Streeter, V.L., Suo, L.: Fluid Transients in System. Prentice Hall, Englewood Cliffs (1993)

Effect of Anchor Conditions on Structural Responses During Fluid Transients in Pipelines

Abdelaziz Ghodhbani(✉), Ezzeddine Haj Taieb, and Mohsen Akrout

Laboratory of Applied Fluid Mechanics, Process Engineering and Environment, Department of Mechanical Engineering, National Engineering School of Sfax, University of Sfax, Sfax, Tunisia
abdelaziz.ghodhbani1@stud.enis.tn,
{ezed.hadj,mohsen.akrout}@enis.rnu.tn

Abstract. This paper deals with the evolution of structural responses accompanying water hammer in quasi-rigid straight pipelines with emphasis on the effect of the support rigidity. The coupled discrete vapour cavity model (DVCM) is used to simulate axial stress and axial displacement of the pipe in presence of column separation. Transient is caused by a fast-downstream valve closure. The method of characteristics (MOC) with wave-speed adjustment scheme is used in calculation. The numerical results obtained for rigid and viscoelastic supports are performed and compared. The cavity collapse involves greater stress peak at the valve with rigid support, but the stress spikes are reduced in case of viscoelastic support. The axial displacement is more important for this last case because of junction coupling effect. The results obtained are assumed to be important in hydraulic engineering.

Keywords: Water hammer · Column separation · Axial vibration · Fluid-structure interaction (FSI) · Poisson coupling

1 Introduction

Pipeline systems especially full-filled liquid pipes are usually subjected to water hammer events that may lead to severe industrial accidents. Consequently, both liquid and structure behaviour should be accurately predicted in order to install safety devices therein. The classical theory of water hammer deals only with the pressure and the velocity of the fluid in the pipeline. In order to accurately predict structural responses during water hammer, engineers need to use appropriate software packages that consider fluid-structure interaction (FSI), such as the four-equation model. Tijsseling (1993) investigated the four-equation model in axial, radial and torsional motion. Ghodhbani and Haj Taïeb (2017) proposed and validated the four-equation friction model which considers unsteady friction coupling. Their model was applied later in prediction of column separation in pipelines by providing the coupled DVCM (Ghodhbani et al. 2019). This latter is more suitable than the classical DVCM although junction coupling mechanism is ignored. Structure responses, such as axial stress, axial velocity and axial displacement of the pipe can be simulated using the coupled DVCM.

M. Kharrat et al. (Eds.): ICAMEM 2019, LNME, pp. 412–419, 2021.
https://doi.org/10.1007/978-3-030-52071-7_56

However, the authors noted that this simulation may be improved by considering support rigidity and axial pipe stiffness.

This paper investigates structural responses of quasi-rigid straight pipelines with presence of column separation. The emphasis is made on the effect of the rigidity of the supports and the axial vibration of the pipe on these structural responses at different localizations along the pipeline.

2 Materials and Methods

The coupled DVCM was proposed by Ghodhbani et al. (2019) in order to give accurate prediction of column separation in quasi-rigid straight pipeline where transient is caused by a fast-downstream valve closure. This model is obtained by incorporating the classical DVCM into the four-equation water hammer model. The emphasis was made on the effect of FSI on pressure at specific localizations in the pipeline. Dynamic coupling was investigated by considering Poisson and friction coupling; junction coupling was ignored by assuming the valve to be fixed. Unsteady friction was considered and tested against steady friction. Moreover, the variable wave-speed (VWS) method was proposed and used to improve the numerical results. The full-MOC with wave-speed adjustment scheme was used in calculation. Four unknowns are calculated at each time step. While cavitation does not occur, the unknowns are the piezometric head H (or the pressure p) and the velocity V for the liquid and the axial stress σ_z and the axial velocity $\dot{u}_z$ for the pipe. When the pressure drops below the vapour pressure of the liquid p_v, then the algorithm switches to the calculation of the upstream discharge Q_u, the downstream discharge Q_d, the axial velocity of the pipe $\dot{u}_z$ and the axial stress σ_z.

In this work, the coupled DVCM is used to predict stress, velocity and displacement for axial direction of a quasi-rigid straight pipe. Transient is caused by a fast-downstream valve closure fixed at the end of the pipe. Two types of anchor conditions are considered: the rigid support and the viscoelastic support. Thus, two types of boundary conditions (BC) at the valve can be described. If the valve is rigidly fixed, then there is no junction coupling and the BC at the valve are modelled by

$$(Q_d)_{z=L} = 0 \text{ and } (\dot{u}_z)_{z=L} = 0 \tag{1}$$

If junction coupling is considered, the axial velocity at the valve is no longer equal to zero and the equation of motion should be used (Tijsseling 1993; Henclick 2018). Obviously, the vibrating downstream valve of mass m is subjected to pressure force, elastic force of the pipe and equivalent stiffness and damping forces generated by the pipe and the support. Tijsseling (1993) gave the equation of motion as

$$m\ddot{u}_z + c\dot{u}_z + ku_z = Ap - A_p\sigma_z \tag{2}$$

with c is the equivalent viscous damping coefficient, k is the stiffness and A and A_p are, respectively, cross sections of the fluid and the pipe. Noting that c and k are in fact equivalent properties of the system composed by the pipeline and the supports. The determination of the damping coefficient of the pipeline can be obtained from the simple oscillator theory as $c = 2\xi A_p\sqrt{E\rho_p}$ (Budny et al. 1991) with ξ is the damping ratio and E and ρ_p are, respectively, the Young's modulus of elasticity and the density of the pipe-wall. The spring-stiffness of a pipeline of a length L is simply obtained by $k = EA_p/L$. Accurate integration of Eq. 2 can be obtained thanks to the Newmark method (Henclick 2018). By taking β = 1/4 in this method, the differentiation with a backward finite-difference scheme allows

$$(\ddot{u}_z)_{t+\Delta t} = \frac{(\dot{u}_z)_{t+\Delta t} - (\dot{u}_z)_t}{\Delta t} \text{ and } (u_z)_{t+\Delta t} = (u_z)_t + \frac{\Delta t}{2}\left[(\dot{u}_z)_{t+\Delta t} + (\dot{u}_z)_t\right] \quad (3)$$

Thus, after incorporating Eq. (4) in Eq. (3), the second BC can be written with respect to the unknowns of the four-equation model

$$-A(p)_{t+\Delta t} + \left(\frac{m}{\Delta t} + c + \frac{k\Delta t}{2}\right)(\dot{u}_z)_{t+\Delta t} + A_p(\sigma_z)_{t+\Delta t} = \left(\frac{m}{\Delta t} - \frac{k\Delta t}{2}\right)(\dot{u}_z)_t - k(u_z)_t \quad (4)$$

3 Results and Discussion

The physical properties of the experiment of Bergant and Simpson (1995) are used. A copper straight 37.23 m long sloping pipeline of 22.1 mm internal diameter and 1.6 mm wall thickness connects two pressurized tanks fulfilled by a demineralized water. The initial velocity of water is equal to 0.3 m/s so that column separation occurs at the valve following water hammer. The above experiment does not indicate whether the pipe supports are rigid or viscoelastic, and unfortunately there are no experimental records for structural responses.

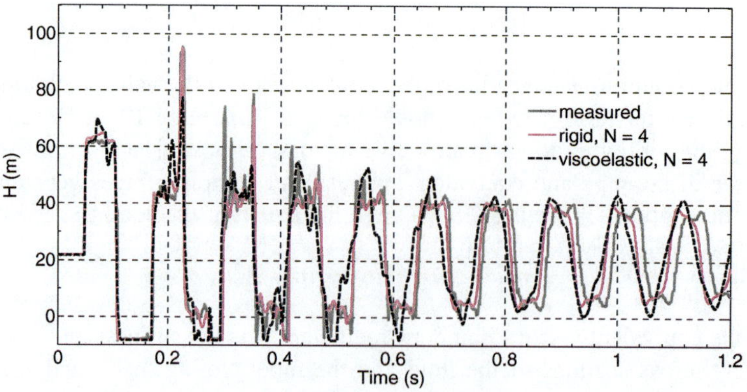

Fig. 1. Piezometric head at the valve

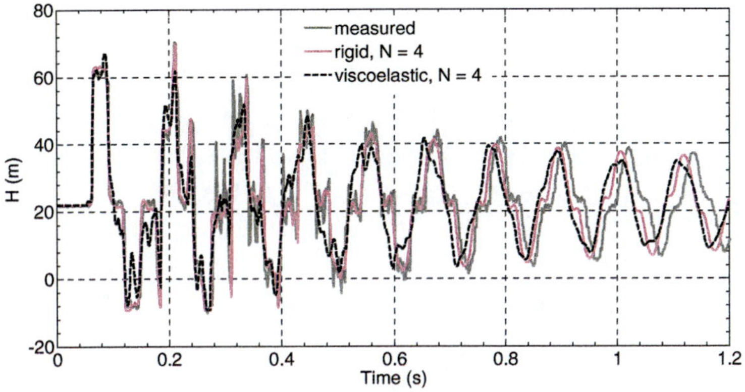

Fig. 2. Piezometric head at the midpoint.

Figures 1 and 2 show the piezometric heads, respectively, at the valve and at the middle of the pipe for a small number of reaches, i.e. $N = 4$ and compare the viscoelastic support to the rigid support. It is observed that the rigid support solution agrees with the experimental result better than the viscoelastic one.

Figure 3 shows the piezometric head at the valve in case of the quasi-rigid support. This latter corresponds to the freely moving valve boundary condition where the equivalent stiffness k is amplified compared to the viscoelastic support, but the equivalent viscous damping c is conserved. It is observed that the quasi-rigid solution

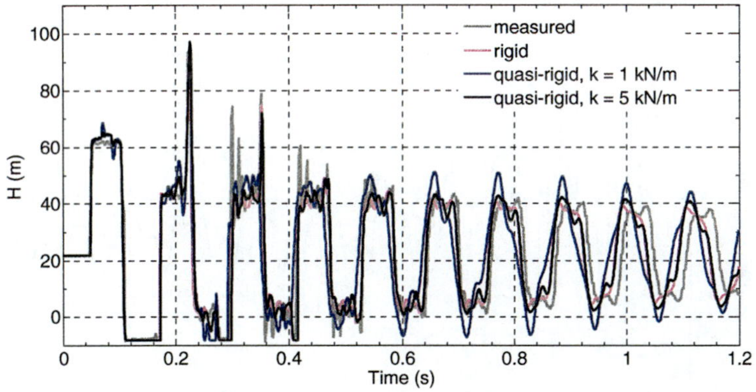

Fig. 3. Piezometric head at the valve and comparison between quasi-rigid and rigid supports.

converges to the rigid solution (fixed valve) when the equivalent stiffness increases. Hence, when the stiffness is so high, the viscous damping effect is no longer observed.

The structural responses at the valve are displayed in Fig. 4 whereas Fig. 5 shows those at the midpoint. The effects of both rigid and viscoelastic supports are simulated and compared. The former corresponds only to Poisson coupling while the latter allows

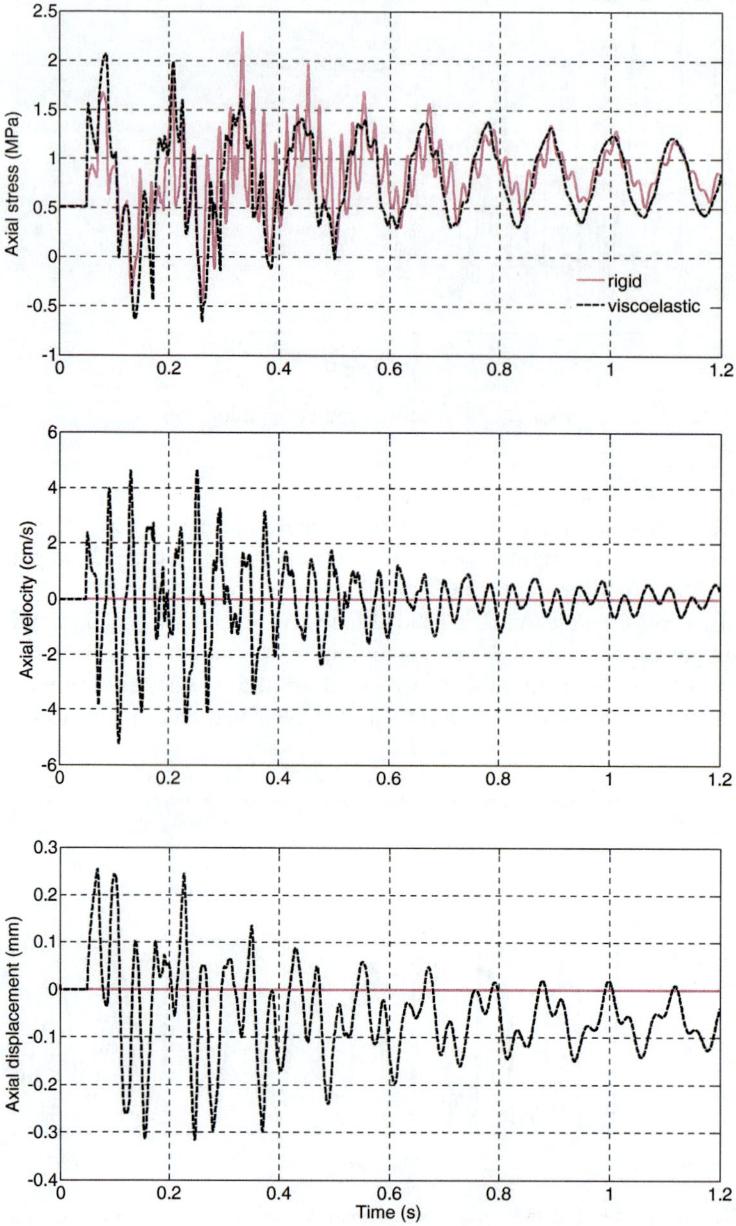

Fig. 4. Structural responses at the valve.

both Poisson and junction coupling. Ghodhbani et al. (2019) provided only axial stress at the valve, which is assumed to be fixed, and they considered that their simulation is of minor spikes since neither support stiffness nor spring-stiffness of the pipe are considered. Heinsbroak and Tijsseling (1994) established that FSI is as important as the

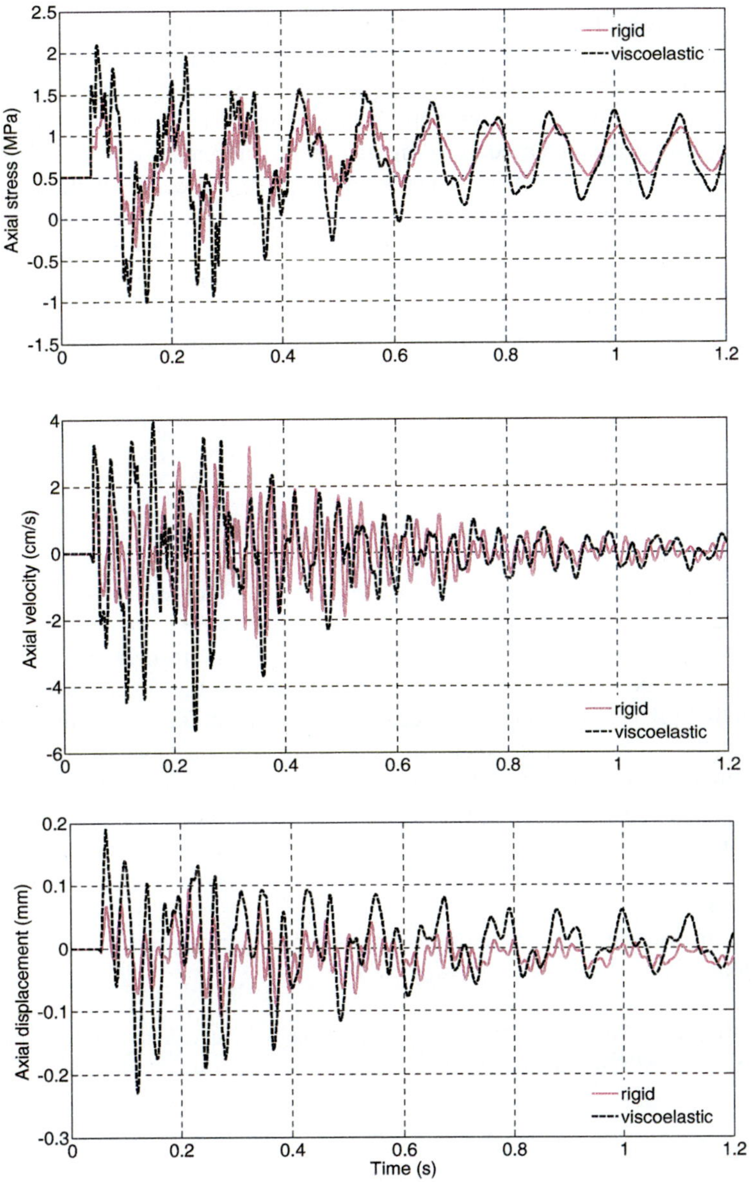

Fig. 5. Structural responses at the midpoint.

ratio of the support stiffness to the spring-stiffness of the pipe is low. If the support stiffness is approximated by 1 kN/m, the above condition is satisfied since the calculated spring-stiffness of the pipe is equal to 361.67 kN/m. Moreover, by taking a damping ratio ξ equal to 5%, the equivalent stiffness of the system formed by the pipe and a unique anchor at the valve can be approximated (Budny et al. 1991). A damping ratio of 2% is assigned to the pipe while 3% corresponds to the anchor which is modelled as linear spring. Due to the short duration pressure pulse following the cavity collapse the stress peak involved by the rigid support at the valve exceeds that of the viscoelastic support (Fig. 1). However, the comparison is inversed at the midpoint (Fig. 2) because of the increased axial. In other word, although the viscoelastic support can lead to more load reduction at the freely moving extremity of the pipe, severe deformation can occur at the midpoint. The accuracy of the predicted structural responses depends on the accuracy of stiffness and damping coefficient introduced in calculation. The correct is the equivalent stiffness and damping coefficient, the accurate is the prediction of the structural responses. Henclick and Maurin (2019) gave and validated an analytical method for the calculation of stiffness matrix of pipeline support.

4 Conclusion

In the paper, the coupled DVCM is applied to simulate structural responses accompanying water hammer with vaporous cavitation in quasi-rigid straight pipelines. The physical properties of the experiment of Bergant and Simpson (1995) is used. The stiffness and damping forces used in boundary conditions are equivalent properties for the pipe and the support together. The calculation was performed for rigid and viscoelastic pipe supports. Poisson coupling corresponds to the former while the latter allows both Poisson and junction coupling. The simulation shows that the cavity collapse at the valve involves axial stress peak greater than that of the viscoelastic support. However, it is observed that junction coupling reduces stress spikes at interior points like the middle of the pipe. Moreover, the viscoelastic support condition involves more displacement in the whole pipe. However, high-frequency responses involved by the rigid support condition are visibly reduced thanks to the viscoelastic consideration.

Further investigation of structural responses can be performed in case of one-elbow pipe system with and without cavitating flow.

Acknowledgments. The authors thank all the members of the Laboratory of Applied Fluid Mechanics, Processes Engineering and Environment of the National Engineering School of Sfax.

References

Bergant, A., Simpson, A.R.: Water Hammer and Column Separation Measurements in an Experimental Apparatus. Adelaide (1995)

Budny, D.D.: The influence of structural damping on internal pressure during a transient pipe flow. ASME J. Fluids Eng. **113**(9), 424–429 (1991)

Ghodhbani, A., Haj Taïeb, E.: A four-equation friction model for water hammer calculation in quasi-rigid pipelines. Int. J. Press. Vessels Piping **151**(3), 54–62 (2017)

Ghodhbani, A., Akrout, M., Haj Taïeb, E.: Coupled approach and calculation of the discrete vapour cavity model. J. Fluids Struct. **91**, 1–21 (2019)

Heinsbroek, A.G.T.J., Tijsseling, A.S.: The influence of support rigidity on water hammer pressures and pipe stresses. In: Proceedings of 2nd International Conference on Water Pipeline Systems, pp. 17–30 (1994)

Henclick, S.: Numerical modeling of water hammer with fluid-structure interaction in a pipeline with viscoelastic supports. J. Fluids Struct. **76**, 469–487 (2018)

Henclick, S., Maurin, A.: Determination of the stiffness matrix of flat springs for modelling of the boundary condition of a pipeline support. Mech. Syst. Sign. Process. **123**, 102–116 (2019)

Tijsseling, A.S.: Fluid-structure interaction in case of water hammer with cavitation, Ph.D Thesis, Delft (1993)

Investigation of Blade Exit Angle Effects on the Performance of a Centrifugal Pump

Faouzi Omri[(✉)], Sami Elaoud, Noura Bettaieb, Issa Chalghoum, and Ezzeddine Hadj Taieb

Laboratory of Applied Fluid Mechanics, Process Engineering and Environment, National School of Engineering of Sfax, 3038 Sfax, Tunisia
omrifawzi@yahoo.fr

Abstract. To achieve the goal of developing energy-efficient mechanical systems, centrifugal pump designers often encounter problems of selecting and optimizing pump impeller geometry. This paper presents a theoretical approach to investigate the blade exit angle effects on the performance of a low specific speed centrifugal pump with standard closed impeller. The governing equations of the flow through pump impeller have been numerically solved by the one-directional method of characteristics of specified time intervals. The mathematical model is validated with the pump experimental performance curves. The numerical results have clearly shown the influence of impeller blade exit angle on the head, input power and efficiency of the considered pump. Indeed it has been found that both head and shaft power increased with the increase of the blade exit angle. Best efficiency is obtained for impellers with 15° and 30° exit angle.

Keywords: Centrifugal pump · Blade exit angle · Method of characteristics · Slip factor · Shock losses

1 Introduction

Centrifugal pumps are used in wide range of applications including the production of electricity, nuclear industry, naval, mining, etc. With the continuous increase of energy demand, even a small increase in the pump efficiency is very important. Various studies have been devoted to investigate the parameters affecting the pump performance. It was found that among the pump components, the impeller geometry has the major impact on the pump efficiency (Gülich 2014).

For instance, Varley (1961) experimentally studied the effect of impeller blade exit angle on the performance of a double suction centrifugal pump with low specific speed and different blade exit angle. The results showed that both head and flow rate increased with the increase of the blade outlet angle [4]. Shojaeefard et al. (2012) conducted a numerical study on the effects of a certain geometric characteristics of a centrifugal pump impeller. The flow analysis indicated that with the modification of the original geometry of the pump, at an outlet angle 30° and a passage width of 21 mm, the pump head and efficiency increased compared to the other cases. This improvement is the result of losses reduction that arises from the generation of eddies

M. Kharrat et al. (Eds.): ICAMEM 2019, LNME, pp. 420–427, 2020.
https://doi.org/10.1007/978-3-030-52071-7_57

in the passage and outlet of the impeller [3]. Sayed Ahmed Imran and Abdus Samad (2015) presented a 3D flow simulation using Reynolds-averaged Navier–Stokes (RANS) equations to study a crude oil pumping behavior by considering different blade angles and impeller surface roughness. They reported that the blade angles have significant influence on the head input power and efficiency of the impeller [2]. Therefore both numerical and experimental studies have clearly shown that the impeller blade exit angle have notable effects in a pump hydraulic performance. Indeed, by selecting the convenient impeller shape, energy costs may be reduced.

The purpose of the present work is to numerically investigate the influence of the blade exit angle on the performance of a low specific speed centrifugal pump with standard closed impeller. The case study investigated by (Issa et al. 2016) [1] was considered for the numerical analysis.

2 Pump Model

The hydraulic test facility (TE 83), depicted in Fig. 1, was used to study the impeller exit angle effect on the pump performance. The pump rotational speed is 2900 rpm. The total maximal head is about 16.8 m. The maximal flow rate is equal to 11.34 m^3/h. Figure 2 illustrates the drawing definition of the pump impeller.

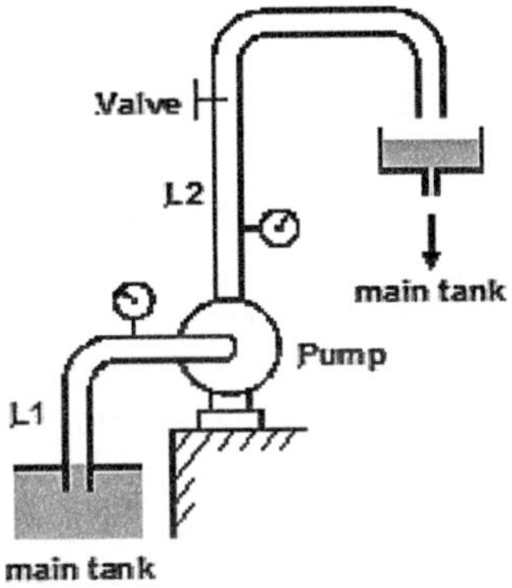

Fig. 1. Hydraulic installation

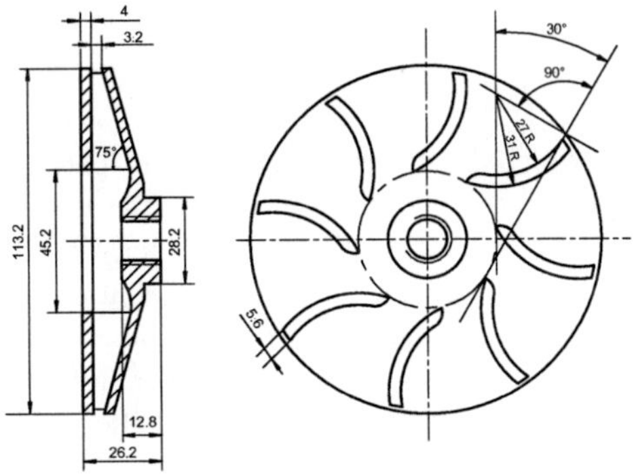

Fig. 2. The pump impeller drawing

2.1 Fluid Governing Equations

The incompressible flow through the rotating impeller was assumed as a flow through a rotating duct with a diameter equivalent to the mean hydraulic diameter of the impeller passage. The basic equations of continuity and momentum, in a reference-rotating frame related to the impeller passage can be expressed as follows:

$$\frac{\partial \rho}{\partial t} + \rho \frac{\partial w_j}{\partial x_j} = 0 \tag{1}$$

$$\rho\left(\frac{\partial v_i}{\partial t} + v_j\frac{\partial v_i}{\partial x_j}\right) = -\frac{\partial p}{\partial x_i} + \mu\frac{\partial^2 v_i}{\partial x_j \partial x_j} + \rho fi \quad (2)$$

where ρ is the fluid density, $\vec{w}$ is the relative velocity, $\vec{v}$ represents the absolute velocity of the flow and μ is the dynamic viscosity. The previous set of equations can be expressed in a frame of reference that rotates with the impeller. Hence, the following expressions are obtained:

$$\frac{\partial H}{\partial t} + \frac{C^2}{gA}\frac{\partial Q}{\partial r} = 0 \quad (3)$$

$$\frac{1}{A}\frac{\partial Q}{\partial t} + g\frac{\partial H}{\partial r} + \lambda\frac{Q|Q|}{2D_h A^2} - r\Omega^2 = 0 \quad (4)$$

where A is the impeller passage section, C is the pressure wave speed, Q is the fluid discharge, H is the head, λ is the Darcy Weisbach friction coefficient and D_h is the hydraulic diameter of the section A. It should be noted that the term $r\Omega^2$ is a result from the centrifugal forces of the rotating impeller that generate the dynamic pressure developed by the pump. Meantime, to calculate the theoretical head of the pump, the equation of Euler head is used:

$$H_i = \frac{U_2 V_{u2} - U_1 V_{u1}}{g} = \frac{U_2}{g}(U_2 - Q/Atg\beta_2) \quad (5)$$

in which U_2 is the outlet impeller peripheral velocity (m/s), V_{u2} is the tangential component of absolute speed, Q is the flow rate and A is the impeller outlet area. It should be noted that the flow velocity at the exit of the impeller does not follow correctly the profile of the rotor blade. It presents an angular deflection, which depends on the blade number, impeller exit angle, and the flow rate. A better explanation is presented in Fig. 3, which shows the difference between the real (in blue) and theoretical (in red) triangles of the impeller output velocity. The slip factor is used to specify the flow effect at the exit of a centrifugal pump impeller. Several tests and

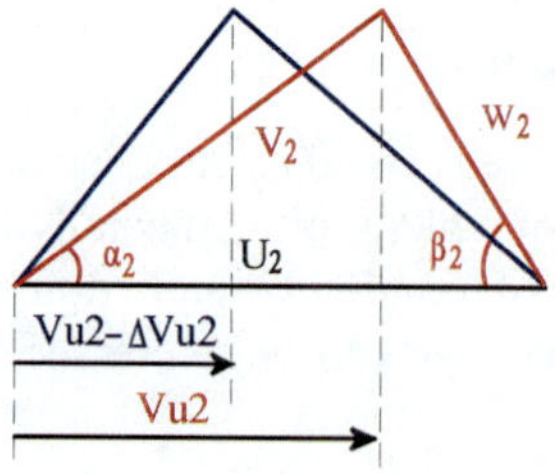

Fig. 3. The impeller sliding effect in the outlet triangle velocity

correlations were analyzed to obtain formulas for the slip factor. In this study, the following correlation is used (Brennen 1994):

$$\sigma = \frac{U_2 - \Delta V_{u2}}{U_2} = 1 - \frac{\Delta V_{u2}}{U2} \tag{6}$$

The slip factor of a centrifugal pump impeller is frequently estimated by means of the Weisner's or Stodola's formulas.

$$\sigma = 1 - \frac{\sqrt{\sin \beta 2}}{Z^{0.7}} \tag{7}$$

where β_2 is the impeller blade exit angle and Z is the number of impeller blades. As a result, the pump head is estimated by the following relation (Japikseet al. 1997), which include the slip factor:

$$H = \sigma \frac{U_2}{g}(U_2 - Q/Atg\beta_2) \tag{8}$$

2.2 Impeller Losses

To investigate the performance of the pump impeller, an accurate loss correlation analysis should be used. In what follows, only two types of losses are considered:

Impeller Inlet Shock Losses. The assumption that the absolute velocity at the inlet is radial is not always verified. The tangential component of the absolute speed at the inlet is equal to zero only at the rated flow rate Q_0 of the pump. Shock losses are given by the following equation:

$$\Delta h_{sh} = k_s(Q - Q_0)^2 \tag{9}$$

Impeller Friction Losses. Impeller blades are assumed as rotating ducts with the mean hydraulic diameter of the blade cross section, thus impeller friction losses can be obtained using the Darcy Weisbach expression:

$$\Delta h_i = \lambda \frac{Q|Q|}{2D_h A^2} = 0 \tag{10}$$

The shaft power can be obtained by the following equation:

$$P_m = P_0 + \rho.gH_i Q \tag{11}$$

where P_0 is the input shaft power at the pump shut off head, which is determined experimentally. To compute the overall pump efficiency, the following expression is used:

$$\eta = \frac{\rho.gHQ}{P_0 + \rho.gH_iQ} \tag{12}$$

3 Numerical Model

In this section, the characteristic equations are derived to transform the partial differential Eq. (3) and (4) to a system of ordinary differential equations grouped and identified as C+ and C– equations:

$$C^{\pm}\begin{cases} \pm\frac{gA}{C}dH + dQ + JAdt - r\Omega^2 Adt = 0 \\ dr = \pm Cdt \end{cases} \tag{13}$$

where $J = \lambda Q|Q|/2D_hA^2$ is the head loss by impeller length unit.

These ordinary differential equations are numerically solved in order to determine the flow and the head at any interior point of the impeller. To get the complete solution, boundary conditions must be defined at different junctions. The transient flow calculation in the pipes is performed using the same numerical model without considering the term related to the pump angular velocity. The effect of impeller exit angle is numerically analyzed by introducing Eq. (5) and Eq. (7) in the numerical solution of the characteristic equations.

4 Results and Discussions

The test facility investigated herein is the same used by (Issa et al. 2016). In the cited study, it was supposed that the pump motor speed evolves in a linear manner during the startup procedure. In the present work, we will introduce the real profile of the speed and the engine torque given by the numerical resolution referring to the model of motor and pump.

4.1 Numerical Model Validation

The numerically obtained results were compared with the experimental data to validate the accuracy of the previously presented numerical model. For the given rotational speed N = 2900 rpm, the head-discharge curve is obtained experimentally by changing the valve opening position. Figure 4 (a) depicts a comparison between the experimental and numerical pump steady state head curves. The shift between the two curves is a result of the impeller shape approximation and the neglecting of the volute hydraulic losses.

It is clear form Fig. 4 (b) that the numerical model gives approximately the same pump power results especially when the pump operates in partial flow rate conditions. Figure 5 (a) also reveals a good agreement between the experimental and numerical pump input power at different flow rate values, where the input power is calculated with the Eq. 5 and Eq. 11. The difference is mainly due to the impeller shape

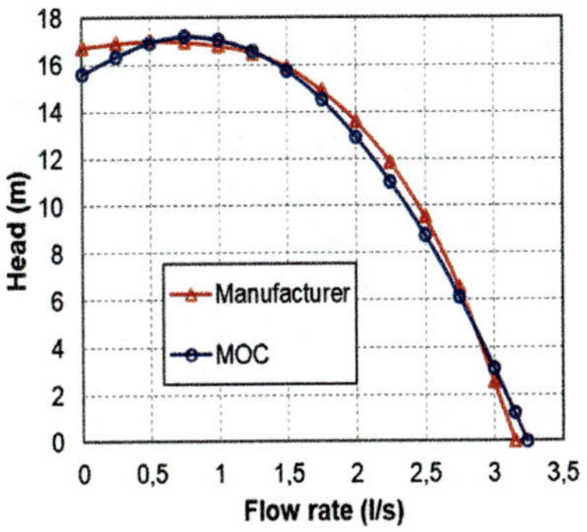

(a) Head curve **(b)** Power curve

Fig. 4. Comparison between experiment and theoretical Head and Power curves for N = 2900 rpm

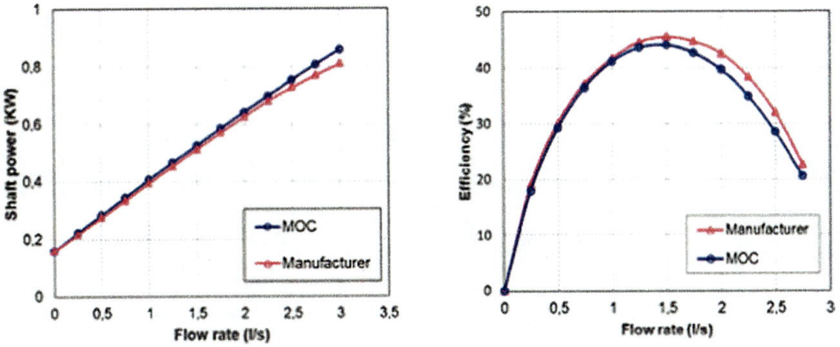

(a) Shaft power curve (b) Efficiency curve

Fig. 5. Comparison between experiment and theoretical shaft power and total efficiency curves for N = 2900 rpm

approximation and the neglecting of the volute hydraulic losses and the approximation in the pump shaft power.

4.2 Effect of Blade Exit Angle on the Pump Performance Curves

This subsection is devoted to study the effect of the blade exit angle value on the pump performance. The following results are obtained by varying the value of β_2 in the head equation. A close look to Fig. 6 (a) reveals different values of the pumping head under the same pump flow rate. A further observation indicates that the maximal pumping head is obtained when the blade exit angle is equal 15°, 30° and 120°.

The increase in the pumping head is also accompanied with an increase in the pump power. Figure 6 (b) clearly indicates that higher pump power values are required for

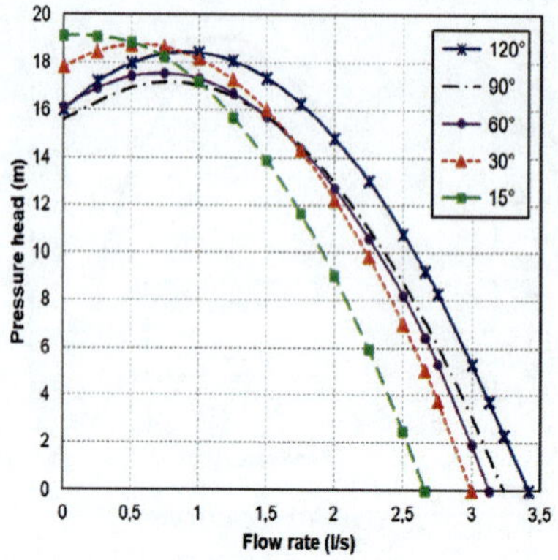

(a) Head curves

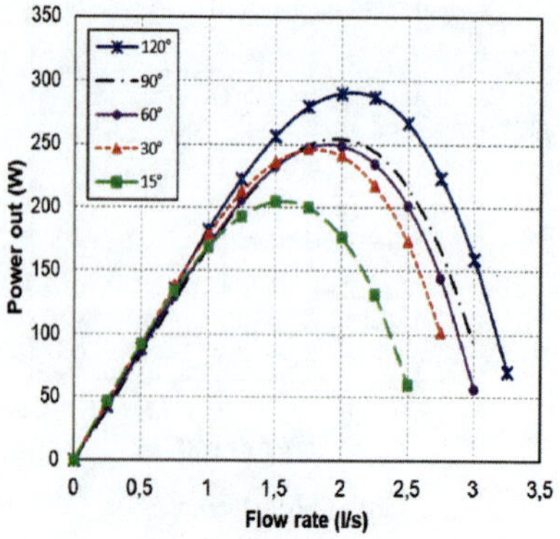

(b) Power curves

Fig. 6. Effect of blade exit angle β_2 on the head and power curves

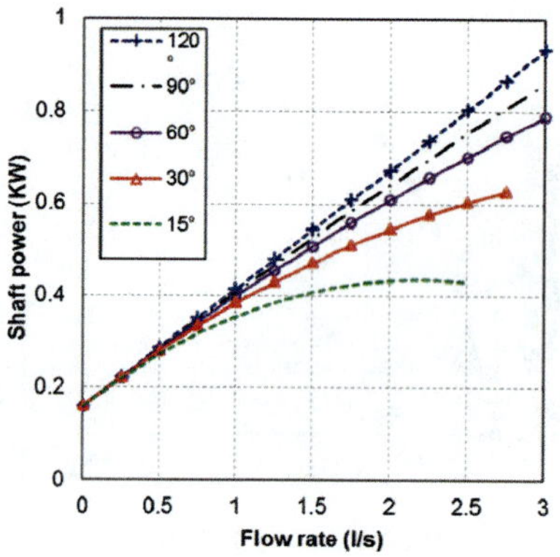

(a) Shaft power curves

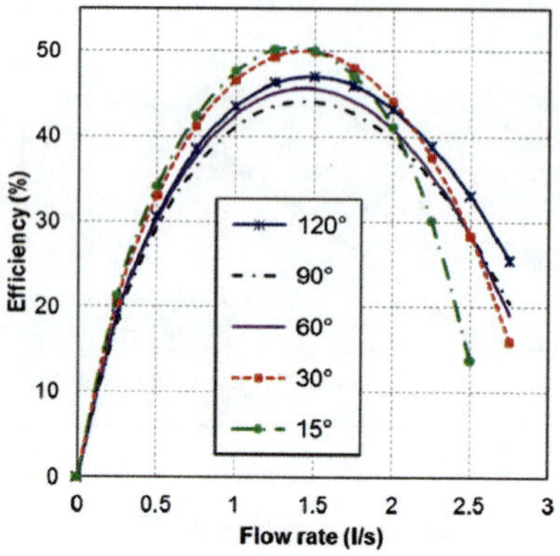

(b) Efficiency curves

Fig. 7. Effect of blade exit angle β_2 on the torque and efficiency curves

cases with blade exit angle greater than 60°, especially in the overload flow rate conditions. Furthermore, the shaft power increases as demonstrated in Fig. 7 (a). The impact of increasing the blade exit angle on the pump performance is illustrated in Fig. 7 (b). The highest pump efficiency is obtained for pump impeller with exit blade angles 15° and 30°. However, it should be noted that, for the case of 15°, the best efficiency point corresponds to a flow rate lower than the nominal flow rate.

5 Conclusion

In the present study, the effect of impeller blade exit angle on the centrifugal pump performance was investigated. A numerical model was developed based on the momentum and continuity equations. Following the validation of the numerical model based on the pump experimental results, the pump performance was investigated under different values of the blade exit angle. It was demonstrated that pump best efficiency is obtained with 30° exit angle and the energy consumption has been widely reduced. Even though the blade exit angle 15° achieved the highest pump efficiency, the flow rate for the highest efficiency was lower than the nominal pump flow rate.

References

1. Chalghoum, I., Elaoud, S., Akrout, M., HadjTaieb, E.: Transient behavior of a centrifugal pump during starting period. Appl. Acoust. **109**(2016), 82–89 (2016). https://doi.org/10.1016/j.apacoust.2016.02.007
2. Imran, S.A., Samad, A.: Pumping crude oil by centrifugal impeller having different blade angles and surface roughness. J. Petrol Explor. Prod. Technol. **2016**(6), 117–127 (2015). https://doi.org/10.1007/s13202-015-0173-y
3. Shojaeefard, et al.: Numerical study of the effects of some geometric characteristics of a centrifugal pump impeller that pumps a viscous fluid. Comput. Fluids **60**(2012), 61–70 (2012). https://doi.org/10.1016/j.compfluid.2012.02.028
4. Varley, F.A.: Effects of impeller design and surface roughness on the performance of centrifugal pumps. Proc. Inst. Mech. Eng. **175**(21), 955–969 (1961)

New Proposed Design of Rotor Shaft Equipped with Blades Inside a SSHE

Rabeb Triki(✉), Hassene Djemel, and Mounir Baccar

National School of Engineers of Sfax, Sfax, Tunisia
rabeb.triki.enig@hotmail.fr, hassenedjemel@yahoo.fr,
Mounir.baccar@enis.rnu.tn

Abstract. Thermal treating of highly viscous product is associated with many problems such as fouling, an inhomogeneous treatment and burn-on of product. The efficiency of a SSHE lies behind its simple design; the rotating blades continuously scrape the film of product deposed at the heat transfer surface, thus providing a large heat transfer coefficient. The bulk mixing phenomena is purposely induced in SSHE by adopting a new arrangement of blades noted alternate blade in order to ameliorate the thermal performance within the SSHE. A three dimensional numerical simulation of the coupled hydrodynamic and thermal behavior within a conventional SSHE and an alternate blade SSHE was conducted for comparison purpose. Continuity, momentum and energy equation were solved using the finite volume ANSYSFLUENT 19.0 tool. Results of the simulation show that the new design offer a heat transfer coefficient equal to 647.55 W/m^{2}K, around 63% higher than the heat transfer coefficient within the conventional SSHE design.

Keywords: SSHE · Hydrodynamic and thermal behavior · Blades · ANSYS-FLUENT

1 Introduction

The heat transfer surface is the key component of the exchange between the thermal and the product fluid. During operation, the treated (heat or cool) product is pumped through a flange (flow inlet), and mixed by the blades (sheared zone) until it reaches the outlet flange (flow outlet). The feature of the SSHE with its simple design was the subject of many scientific research papers. The first interest was to understand the complex thermal phenomenon-taking place inside the exchanger. In (1958), Kool proposed a theoretical model noted the penetration theory model, which is based on the transient conduction into a semi-infinite solid of liquid deposed at the heat transfer surface between two scraping action. The model assumes that this film of liquid completely mixes with the bulk of fluid after a scraping action. This famous model associated with a mathematical correlation developed by Latinen (1959) was followed by several thermal heat transfer correlations for a specific design of SSHE treating different Newtonian and non-Newtonian fluids. Luca et al. (2012) proposed a new arrangement of scraping blades that offer higher heat transfer coefficients than conventional SSHE (almost twice). The alternate SSHE (A-SSHE) equipped with six

M. Kharrat et al. (Eds.): ICAMEM 2019, LNME, pp. 428–434, 2020.
https://doi.org/10.1007/978-3-030-52071-7_58

blades compared to continuous SSHE (C-SSHE) equipped with four blades offer a higher heat exchange due to local turbulence inside the exchanger. In this study, we propose a rotor shaft equipped with five couples of scraping blades in order to take advantage of local turbulence associated with the transition from one couple to the next couple. The new design of A-SSHE preserve the same length of scraping blades with a C-SSHE. The attractive feature about this design is to offer a higher heat transfer coefficient without recourse to increase the heat transfer surface. This configuration will offer a solution to treat sensitive product to high shear rate associated with height speed of scraping blade.

2 Computational Domain

In this study, we performed two numerical simulations in order to compare the thermal performance and the hydrodynamic behavior-taking place inside a (C-SSHE) and a SSHE equipped with alternate blades (A-SSHE). Figures 1 and 2 illustrate the rotor shaft equipped with blades within different SSHEs studied in this paper. The rotor shaft of an A-SSHE is equipped with five couples of scraping blades. Each couple contains two scrapers set at the opposite side (180°) and rotated by 45° with respect to the previous couple. The three-dimensional models were designed with SOLIDWORKS tool. Table 1 resume the geometric characteristic of SSHEs.

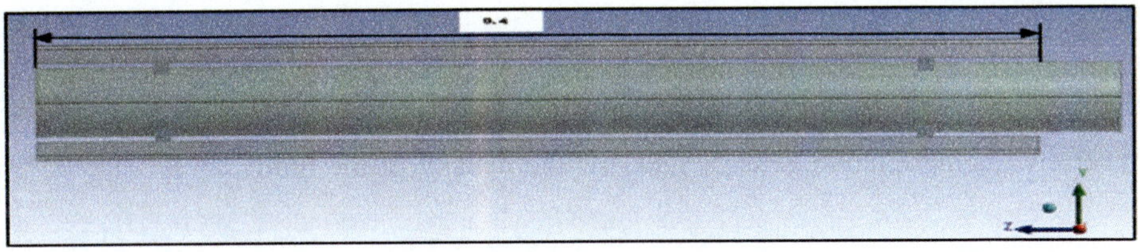

Fig. 1. Longitudinal and transversal cross section of rotor shaft within a C-SSHE.

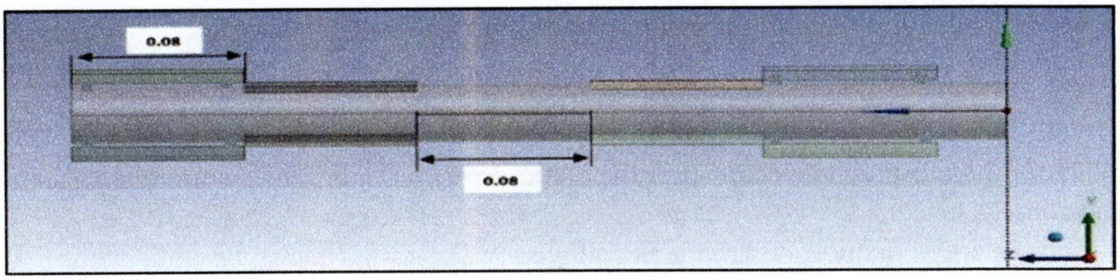

Fig. 2. Longitudinal and transversal cross section of rotor shaft within an A-SSHE.

Table 1. Geometric characteristics of the numerical models of SSHEs.

Parameters	C-SSHE	A-SSHE
Stator diameter	0.065 m	0.065 m
Rotor diameter	0.04 m	0.04 m
Total surface heat exchange	0.088 m^2	0.088 m^2
Blade-stator clearance	130 μm	130 μm
Length of scraper	0.08 m	0.4 m

The ICEM CFD tool is used to generate a complex meshing with refinement near the blade tip in order to capture high gradient of velocity and temperature that take place in the narrow gap between the heat transfer surface and the blade tip.

3 Numerical Model

A numerical investigation of the thermal and hydrodynamic behaviour of SSHE is performed under steady state, non-isothermal laminar flow condition. The fluid tested in this study was a Newtonian, incompressible glycerin fluid. The rotating reference frame formulation is used to induce the rotation of the rotor shaft and blades. The continuity equation in a steady state condition is written as:

$$\nabla.\left(\rho\vec{V}_r\right) = 0 \tag{1}$$

Where $\vec{V}_r$ is the relative velocity and ρ is the density of the fluid.

$$\vec{V}_r = \vec{V} - (\vec{w} \times \vec{r}) \tag{2}$$

With $\vec{V}$ is the absolute velocity variable, $\vec{r}$ is the position vector and $\vec{w}$ is the angular velocity vector of the rotating frame.

The momentum equation is written as follows:

$$\vec{\nabla}.\left(\rho\vec{V}_r\vec{V}_r\right) + \rho\left(2\,\vec{\omega} \times \vec{V}_r + \vec{\omega} \times \vec{\omega} \times \vec{r}\right) = -\vec{\nabla}P + \nabla\vec{\tau}_r + \rho\,\vec{g} \tag{3}$$

Where $\vec{g}$ is the gravity vector, p is the static pressure and $\nabla\vec{\tau}_r$ is the relative extra shear stress tensor.

The energy equation is written as follows:

$$\nabla.\left(\vec{V}_r.\vec{\tau}r\right) = \nabla.\left(-\lambda_f.\nabla T + \rho C_P \vec{V}_r T\right) \tag{4}$$

where T refer to the temperature, C_P and λ_f are respectively the heat capacity and thermal conductivity of the fluid.

The finite volume based solver ANSYSFLUENT 19.0 was used to solve the preceding governing equation. Fluxes on all faces delimiting the control volume are approximated using the second order interpolation scheme. For pressure-velocity coupling a simple prediction-correction iterative algorithm was adopted.

4 Boundary Conditions

The boundary conditions considered in numerical simulation are:

- At the inlet: the fluid is introduced to the exchanger with a velocity equal to 0.0067 ms^{-1} and a temperature equal to 288 K.
- At the outlet: zero velocity gradient is applied at the outlet.
- At the scraper and rotor: A no slip condition with a constant rotating speed equal to 3 rev s^{-1} are applied at the scraper and rotor walls.
- At the stator: we applied a constant temperature equal to 278 K with no slip condition at the stator surface.

5 Results and Discussions

The Results of numerical simulations are presented in this section.

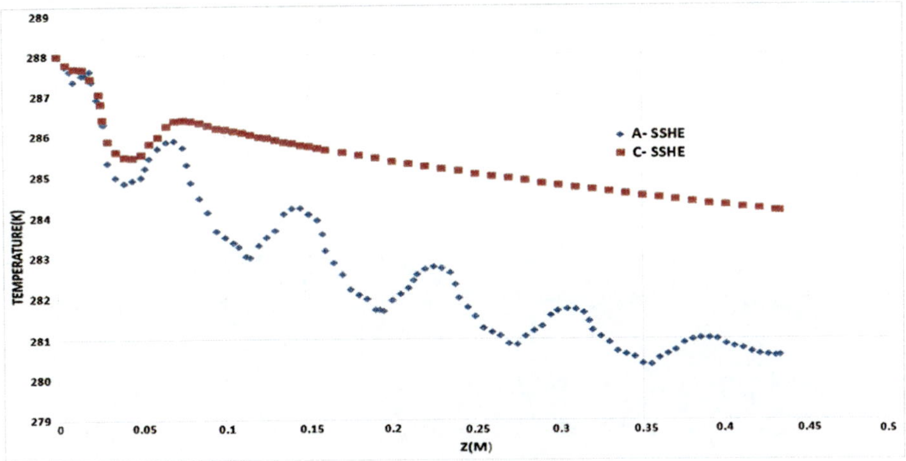

Fig. 3. Axial temperature profile within an A-SSHE and a C-SSHE.

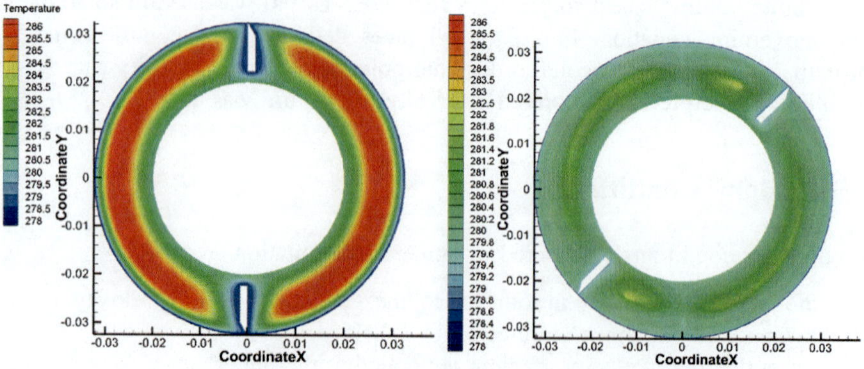

Fig. 4. Temperature contour in the (r-θ) plane for z = 0.35, left: C-SSHE, right: A-SSHE.

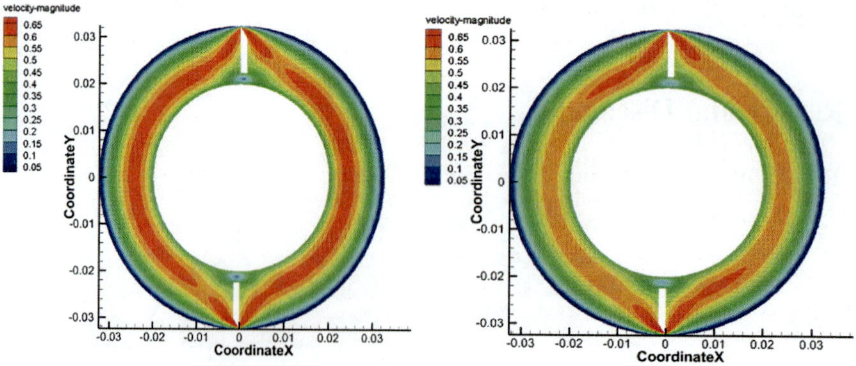

Fig. 5. Velocity contour in the (r-θ) plane for z = 0.4, left: C-SSHE, right: A-SSHE.

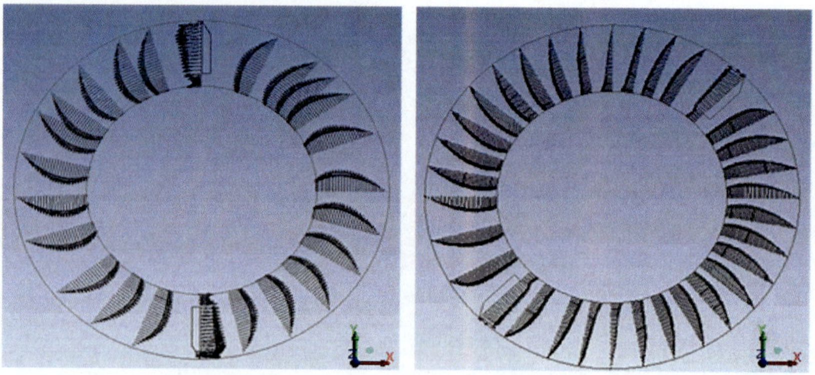

Fig. 6. Velocity contour in the (r-θ) plane for z = 0.35, left: C-SSHE, right: A-SSHE.

Figure 3 illustrates the temperature profile along z-axis. The A-SSHE adopted in this study present many local bulk mixing (local fluctuation) manifested by an over-shoots and undershoots of temperature just before the transition zone from one couple of blades to the next couple. The bulk mixing consists in the displacement of cool fluid towards the bulk, which lead to enhance the driving force for heat transfer respectively to improve the heat transfer coefficient. The fluid cools down more rapidly inside the A-SSHE; this can be explained by higher bulk mixing at each beginning of couple of scrapers. Figure 4 illustrates the temperature contour in the (r-θ) plane for z = 0.35, for different simulated HE. The results of simulations indicate very low temperature value around the scraper (zone with height velocity values) due to scraping action that keeps the heat transfer exposed to fresh product. Figure 4b illustrates two zones with low values of temperature set at 45° of each scraper, due to high bulk mixing that takes place at the passage of fluid from the previous couple located at this zone to next one. The cooling process inside the exchanger is controlled by scraping action. Table 2 resumes the values of heat transfer coefficient for the two adopted geometries of SSHE (Fig. 5).

Table 2. Numerical result of C-SSHE and A-SSHE.

Parameters	T_{oulet} (°C)	Q (W)	h (W/m^{2}K)
C-SSHE	284.131	−163.42	233.97
A-SSHE	280.582	−313.32	647.55

The A-SSHE offer a heat transfer coefficient around 63% higher than the one given by the C-SSHE. The high thermal efficiency of the A-SSHE is explained by the perturbation of the flow just before the beginning of each couple of blades that leads to ameliorate the local heat transfer coefficient thus enhancing the global heat transfer coefficient. This new design could be a promising technology, especially for sensitive product where high scraper rotation associated with high heat transfer coefficient may deteriorate the organoleptic properties of the product.

The fluid flow within a C-SSHE is controlled by scraper rotation. Maximum velocity values are located around the scraper; this can be explained by the important radial mixing due to the continuous scraping action of the product contained in the annular space. Near the rotor, we note the existence of an unmixed layer of product due to very poor radial mixing. Velocity vector in the (r-θ) plane are presented in Fig. 6. Far from the blades zone, we notice a superposition of Couette-Poiseuille flow behaviors with a maximum velocity located in the core of the annular space. The presence of blades in the annular space greatly affect the tangential velocity.

6 Conclusion

SSHE represent a promising technology in-line of continuous thermal treatment of highly viscous product, with a simple operative process. In this research paper, a numerical simulation of the thermal and fluid dynamic behavior of a new design of scraped surface heat exchangers equipped with alternate blades (A-SSHE) was studied. For comparison purpose, we run the same numerical simulations for a conventional SSHE at the same operating conditions as the A-SSHE. The new design offers higher heat transfer coefficients for the simulated cooling process of glycerin fluid with a fixed rotational and axial Reynolds number, showing values of h equal to 647.55 W/m^2K compared to 233.97 W/m^2K within a C-SSHE. As a perspective, we propose to study a new design of SSHE with more couple of blades in order to take advantage of local bulk mixing that takes place at the beginning of each couple. We propose also to study this design for different axial and rotational Reynolds number.

References

Kool, J.: Heat transfer in scraped surface heat vessels and pipes handling viscous materials. Trans. Inst. Chem. Eng. **36**(4), 253–258 (1958)

Latinen, G.A.: Discussion of correlation of scraped film heat transfer in the votator. Chem. Eng. Sci. **9**(4), 263–266 (1959)

Luca, D., Claudia, C., Francesco, D.N., Roberto, N.: A new arrangement of blades in scraped surface heat exchangers for food pastes. J. Food Eng. **108**, 143–149 (2012)

Different Applied Boundary Condition Within SSHE Treating Non-Newtonian Shear Thinning Fluid

Rabeb Triki(✉), Hassene Djemel, and Mounir Baccar

National School of Engineers of Sfax, Sfax, Tunisia
rabeb.triki.enig@hotmail.fr, hassenedjemel@yahoo.fr, Mounir.baccar@enis.rnu.tn

Abstract. Scraped surface heat exchanger (SSHE) is widely spread in different chemical, food pharmaceutical and petrochemical industry. The attractive feature behind its design is their capability to handle highly viscous, sensitive, sticky product and slurry system. A numerical study of the performance of different cooling process (different applied boundary condition) within SSHE equipped with blades was conducted using ANSYSFLUENT 19.0. Three boundary conditions are examined in this study; a constant applied wall temperature, a constant applied heat flux and a variable wall temperature. The cooling process of 2% Carboxylmethyl cellulose (cmc) is investigated under steady state, non-isothermal, laminar flow condition. Numerical simulation shows that the thermal behavior within SSHE is highly dependent on the boundary condition applied at the stator wall. The hydrodynamics behaviour within SSHE is controlled by scraper rotation. The outlet temperature of fluid reaches the desired cooling temperature for a constant applied heat flux, although it was higher for the other applied boundary condition. It is more effective to cool the fluid gradually by applying a variable fluid temperature.

Keywords: SSHE · Shear-thinning fluid · Thermal behavior · ANSYSFLUENT

1 Introduction

SSHE is mainly used in industry for many applications such as gelatinization, sterilization, heating or cooling, concentration, emulsification, crystallization and other high-fouling applications. The complex heat transfer phenomenon that takes place within the SSHE was described by a simple theoretical model named the penetration theory model proposed by Kool (1958). Based on transient conduction into a semi-infinite solid applied to the liquid adjacent to the heat transfer surface, the model assumes that after a scraping action the stagnant film of fluid immediately and completely mixes with the

M. Kharrat et al. (Eds.): ICAMEM 2019, LNME, pp. 435–441, 2020.
https://doi.org/10.1007/978-3-030-52071-7_59

bulk. The convection of fluid is neglected. Latinen (1959) formulated the penetration theory model into a mathematical model written as follows:

$$\mathrm{Nu} = \frac{2}{\sqrt{\pi}} \mathrm{Pr}^{0.5} \mathrm{Re}^{0.5} \mathrm{n}_{\mathrm{b}}^{0.5} \tag{1}$$

Where Pr is the Prandtl number, Re is the Reynolds number and n_b is the number of blades. Due to it's vast applications in industry, the SSHE received a lot of scientific and industrial attention. Different heat transfer correlations are proposed in literature. The generalized form of the heat transfer correlation is written as follow:

$$\mathrm{Nu} = \mathrm{aRe}_{\mathrm{ax}}^{\mathrm{b}} \mathrm{Re}_{\mathrm{rot}}^{\mathrm{c}} \mathrm{n}_{\mathrm{b}}^{\mathrm{d}} \mathrm{Pr}^{0.33} \tag{2}$$

Where Re_{ax} and Re_{rot} refer respectively to axial and rotational Reynolds number. The variables a, b, c and d are the constants of the equation.

In this study, we investigated the effect of different applied boundary condition on the cooling process inside a SSHE. In the first part, we impose a constant heat flux at the stator wall, in the second part, we tested the case of a constant applied wall temperature and finally we investigate a variable applied wall temperature.

Table 1. Geometric characteristics of the numerical models of SSHE.

Parameters	
Stator diameter	0.065 m
Rotor diameter	0.04 m
Total surface heat exchange	0.088 m^2
Blade-stator clearance	130 μm
Number of blades	2

2 Computational Domain

The three dimensional numerical model of SSHE is designed with SOLIDWORKS 3D CAD tool. Table 1 resume the geometric characteristic of SSHE.

A complex meshing is generated with the ICEM CFD tool. A refinement near the blade tip was achieved in order to capture high gradient of velocity and temperature in the narrow space between the stator and the blade tip.

3 Governing Equations

ANSYS-Fluent 19.0 code based on the finite volume method is used to solve continuity, momentum and energy equation within SSHE. Multiple reference frames formulation (MRF) was employed to render the problem, which is unsteady due to shaft rotation when viewed from a stationary (inertial) frame steady with respect to the moving frame, thus saving a considerable computational effort.

The continuity equation in terms of relative velocity $\vec{V}_r$ is written as:

$$\nabla.\left(\rho\vec{V}_r\right) = 0 \tag{3}$$

where ρ is the density of the product.

The momentum equation is written as follows:

$$\vec{\nabla}.\left(\rho\,\overrightarrow{V_r}\,\overrightarrow{V_r}\right) + \rho\left(2\vec{\omega}\times\overrightarrow{V_r} + \vec{\omega}\times\vec{\omega}\times\vec{r}\right) = -\vec{\nabla}P + \nabla\vec{\tau}_r + \rho\vec{g} \tag{4}$$

where $\vec{r}$ is the position vector, $\vec{\omega}$ is the angular velocity vector of the rotating frame, $\vec{g}$ is the gravity vector, p is the static pressure and $\nabla\vec{\tau}_r$ is the relative extra shear stress tensor.

The energy equation expressed in terms relative velocity is written as follows:

$$\nabla.\left(\vec{V}_r.\vec{\tau}r\right) = -\nabla.\left(-\lambda_f.\nabla T + \rho C_P\vec{V}_r T\right) \tag{5}$$

where T is the temperature, λ_f and C_P refer respectively to the thermal conductivity and heat capacity of the fluid.

4 Boundary Conditions

The boundary conditions considered in this study are specified as follows:

- At the inlet: velocity inlet is imposed and equal to 0.0067 ms^{-1}. The temperature of the fluid introduced in the SSHE was taken equal to 288 K.
- At the outlet: zero velocity gradient was applied at the outlet.
- At the rotor and scraper: a constant rotating angular velocity equal to 4 rev s^{-1} is applied with no-slip condition at the rotor and the blades wall.
- At the stator: The three-boundary conditions investigated in this study are resumed in Table 2.

Table 2. Different boundary conditions applied at the stator wall.

	Wall stator boundary condition
Case 1	A constant Wall Temperature is applied T = 278 K
Case 2	A constant heat flux is applied
	A Variable Wall Temperature is imposed at the wall
Case 3	For 0 < z < 0.144: T = 283 K
	For 0.144 < z < 0.288: T = 278 K
	For the rest of the exchanger: T = 273 K

5 Fluids and Materials Properties

In this study, we investigated the cooling application without phase change of 2% CMC solution that presents a non-Newtonian shear thinning behavior. Table 3 resumes the fluid properties investigated in this study.

Table 3. Fluid properties.

	Viscosity	Thermal conductivity ($w\ m^{-1}\ k^{-1}$)	Heat capacity ($J\ kg^{-1}\ k^{-1}$)	Density ($kg\ m^{-3}$)
2% CMC fluid	$\eta(T,\dot{\gamma}) = K0\dot{\gamma}^{n(T)-1}$ $K0(T) = 0.07768.\exp\left(\frac{1888.1}{T(K)}\right)$ $n(T) = 45.15.\exp\left(\frac{-1528.9}{T(K)}\right)$	0.6	4000	1050

6 Results and Discussions

The hydrodynamics behavior within SSHE is controlled by the frequency of scraping action. High values of velocities are determined around the blade more precisely behind it. Figure 2 shows an intensive flow around the blades due to continuous scrapers rotation. The hydrodynamic behavior within SSHE does not change for different applied boundary condition.

Figures (1, 3, 4) show a non-uniform temperature distribution in the (r-θ) plane. Low values of temperature are located around the blades due to the non-uniform convective heat flux at the stator surface. In fact, the continuous scraping action keeps the heat transfer surface exposed to fresh product, thus high local driving force of heat transfer and high local heat transfer coefficient are determined near the blade tip. In comparison, between the distribution of temperature at two different position (z = 0.2 and z = 0.35) we note that the highly viscous CMC fluid cools down rapidly due to scraping action. This result highlights the feature design behind SSHE compared to tubular exchanger. High values of velocities are located around the blades more precisely behind it. Figure 5 shows the temperature variation along the z-axis inside the SSHE for different applied boundary condition. The cooling process is more effective when applying a constant specific heat flux at the stator wall due to high driving force of heat transfer thus high heat transfer coefficient. The fluid cools down less effectively in the first and the second section of variable temperature boundary condition compared to constant applied boundary condition. In contrary in the last section where higher driving force of heat transfer is determined the fluid cools down more rapidly than constant applied temperature boundary condition and reaches lower outlet temperature. We can conclude that cooling the fluid gradually by imposing a variable wall temperature is more effective and offer a higher heat transfer coefficient.

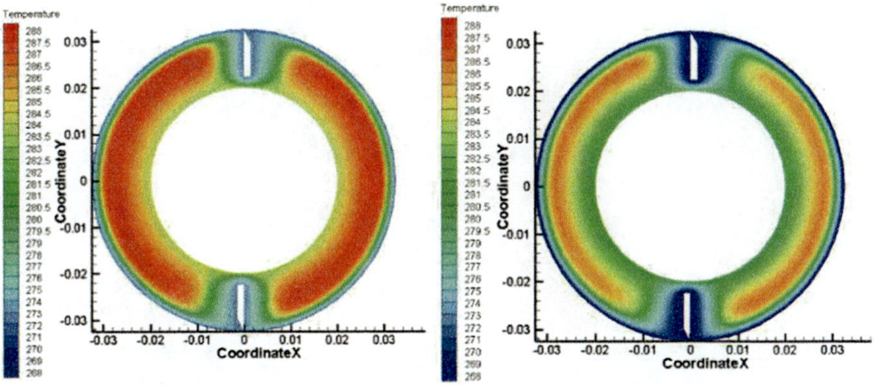

Fig. 1. Temperature contour in the (r-θ) plane for a constant applied heat flux boundary condition; left: (z = 0.2) right (z = 0.35).

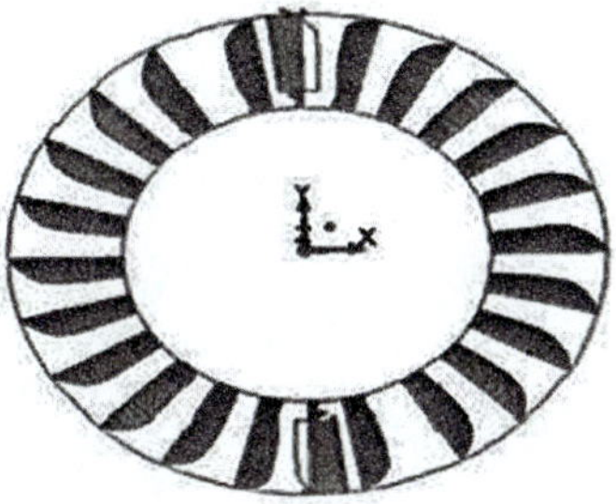

Fig. 2. Velocity vector in the r-θ plane (z = 0.35) for a constant heat flux boundary condition.

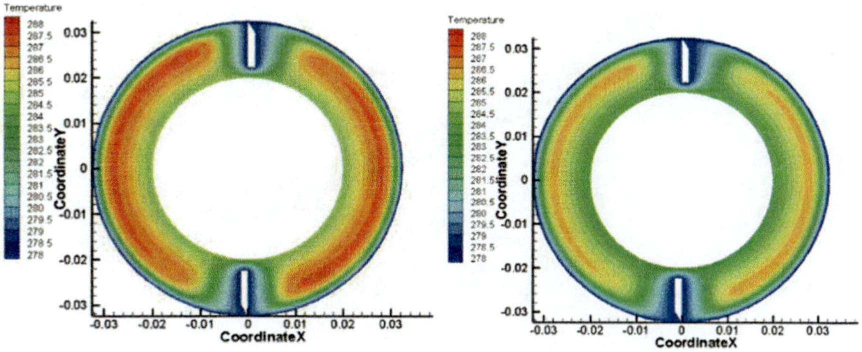

Fig. 3. Temperature contour in the (r-θ) plane for a constant applied temperature boundary condition left: (z = 0.2) right (z = 0.35).

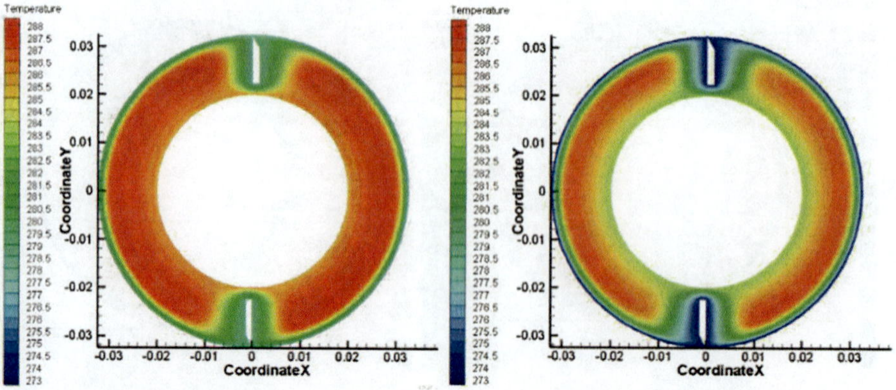

Fig. 4. Temperature contour in the (r-θ) plane for a variable wall temperature boundary condition; left: (z = 0.2) right (z = 0.35).

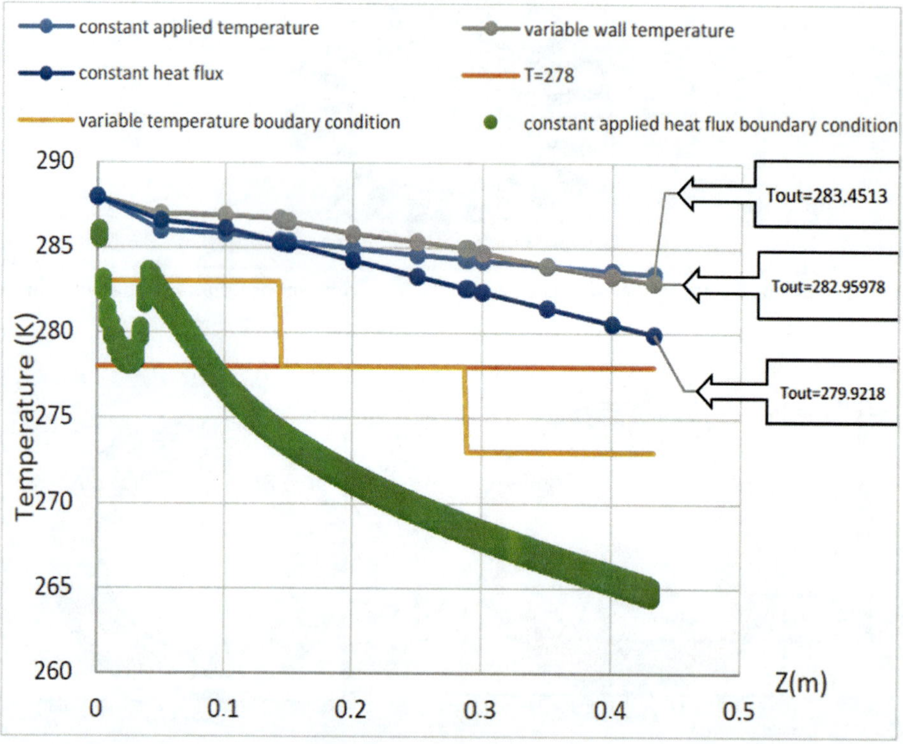

Fig. 5. Fluid temperature distribution along the z-axis for different applied boundary condition.

7 Conclusion

The coupled thermal and hydrodynamic behaviour within SSHE was studied numerically. The non-Newtonian shear thinning 2% CMC fluid was investigated under non-isothermal laminar flow regime for different applied boundary conditions. The hydrodynamic behaviour of fluid does not change with the applied thermal boundary condition. Results of numerical simulation show more effective cooling of product for a constant applied heat flux boundary condition. Cooling the fluid gradually by applying a variable temperature along z-axis is more effective than applying a constant temperature. Further studies should investigate the effect of different scraping design under different operating parameters (axial and rotational Reynolds number) on the hydrodynamic and thermal behavior of the SSHE. An experimental approach will add valuable validation to our numerical results.

References

Kool, J.: Heat transfer in scraped surface heat vessels and pipes handling viscous materials. Trans. Inst. Chem. Eng. **36**(4), 253–258 (1958)

Latinen, G.A.: Discussion of correlation of scraped film heat transfer in the votator. Chem. Eng. Sci. **9**(4), 263–266 (1959)

Experimental Investigation of Air Conditioning in a Bi-climatic Room

Naima Boumediene[1,2](✉), Florence Collet[1], Sylvie Pretot[1], Lazhar Ayed[2], and Sami Elaoud[2]

[1] Laboratory of Civil Engineering and Mechanical Engineering (LGCGM), University of Rennes, Rennes, France
naima.boumediene@etudiant.univ-rennes1.fr, {florence.collet,sylvie.pretot}@univ-rennes1.fr

[2] Laboratory of Applied Fluid Mechanics and Modeling, Engineering School of Sfax, Sfax, Tunisia
lazhar.ayed21@gmail.com, elaoudsa@yahoo.fr

Abstract. This paper is a preliminary study to the investigation of hygrothermal behavior of wall implemented as separating wall of bi-climatic rooms. One room simulates indoor environment while the second simulates exterior climatic solicitations (temperature and relative humidity). Four typical Tunisian winter climate are selected for this study. The aims are to check the homogeneity of ambient conditions in the rooms, the reaching of the set point, the delay to reach it after setting then if the ambient conditions remain constant once they are reached, and finally the response time of the whole system. The results underline that some climate are easy to reach while other may be difficult to obtain due to competitions between climatic systems. This may be solved by acting on the device management. The typical response times are short enough to plan daily cyclic variations in the following steps of the study and ambient conditions are homogeneous in the room.

Keywords: Bi-climatic room · Ambient conditions · Device management

1 Introduction

The study of hygrothermal behavior at wall scale is an intermediate step between the characterization of materials properties and the investigation of hygrothermal behavior of building envelop, which will impact energy needs and indoor comfort.

The first part of this paper is a bibliographic study of experimental investigations performed at wall scale.

The second part introduces the experimental device and metrology that will be used in order to study hygrothermal behavior of wall made of bio-based building materials (Mazhoud et al. 2016, Collet et al. 2019).

The last part investigates the management of the experimental device regarding the response to change in ambient set point. The aims is to find the best way to reach the climates to be simulated (typical French and Tunisian climates for all seasons).

M. Kharrat et al. (Eds.): ICAMEM 2019, LNME, pp. 442–448, 2020.
https://doi.org/10.1007/978-3-030-52071-7_60

2 Bibliographic Study of Wall Scale Tests

Various studies of hygrothermal behavior at wall scale can be found in bibliography. They deal with bio-sourced or conventional materials. They are performed following several ways: (i) with controlled or uncontrolled environments, (ii) on one test-wall, on one experimental building or on a real building …

Among the studies performed at wall scale in laboratories, (Rahim et al. 2017) control ambient conditions on one side of the wall while the other side is exposed to the laboratory ambient conditions, as shown in Fig. 1. In the studies of (Mazhoud et al. 2016; Collet and Prétot 2014), the ambient conditions are controlled on both sides with climatic rooms. In such studies, the hygrothermal behavior of the wall is highlighted thanks to the monitoring of temperature and relative humidity at several depths in the walls regarding variations of ambient conditions. Several kinds of scenarii are considered: isothermal conditions with step in relative humidity, simultaneous square temperature and relative humidity variations and sinusoidal temperature and RH variations. The monitoring highlights that thermal and hygric phenomena are highly coupled. It also allows to analyse inertia effect. These studies can be completed with infra-red thermography to assess the homogeneity of temperature surface of the wall.

Fig. 1. Experimental system of (Rahim et al. 2017) (left) and (Mazhoud et al. 2016) (right)

Among studies performed on experimental building, (Shea et al. 2012) investigate a single zone hemp-lime building with a floor area of 27 m^2. The co-heating test consists in maintaining ambient temperature to 24 °C and measuring energy consumption of the building. The results of the co-heating test are in good agreement with a simple steady-state heat loss prediction model when the steady state is reached. But, for transient conditions, the most representative of building life, the U value alone is not a suitable parameter for evaluating the thermal performance of the building envelop due to the influence of thermal inertia and dynamic hygrothermal behavior.

The study of (Claude 2018) is an example of on-site studies where the effect of retrofitting on hygrothermal performances is investigated. They measure temperature profiles and heat flux and thus can show that the transmission coefficient is reduced, the wall transmission inertia is modified and shifting and damping are increased.

3 Experimental Device and Metrology

The experimental set-up consists in two climatic rooms separated with a test-wall. Each room is 2.35 m deep, 2.78 m wide and 2.4 m high (Fig. 2). The test-wall is 2.4 m wide and 2.1 m high, it can be used as a whole (Collet and Prétot 2014) or split in several parts to compare several kinds of wall simultaneously (Fig. 3). The floor is made of concrete; the external walls and the ceiling are insulated with humidity and air proof polyurethane panels.

Fig. 2. The bi climatic rooms

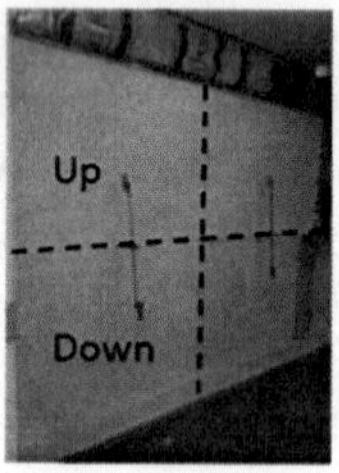

Fig. 3. Test wall with ambient sensors (up and down)

These rooms simulate the indoor and the outdoor climate, with respective temperature and humidity ranges of 18 to 27 °C/−5 to 35 °C and of 30 to 60% RH/30 to 90% RH. Each parameter (T and RH) is regulated by the DR4020 universal controllers, which define the set point and a bandwidth that controls the switching on and off of room devices. Each controller has two outputs: heating or cooling for temperature, humidification or dehumidification for humidity. Heating is provided by convectors, cooling and dehumidification by refrigeration unit, while humidification is provided by ultrasonic humidifiers.

The initial metrology associated to the device consists in temperature and relative humidity sensors. This metrology is completed with heatflux sensors and anemometers.

The hygrothermal behavior of the wall is characterized by temperature and RH variations in the wall regarding ambient variations. The monitoring of temperature and relative humidity is performed with SHT35 Sensirion sensors (Fig. 4), used as ambience sensors and in the wall. According the sensor specifications, the accuracy of this sensor is ±0.1 °C for a range from 20 °C to 60 °C of temperature and ±1.5% of relative humidity up to 80% RH at 25 °C. Previously to the implementation, the RH accuracy of sensors is tested with salt solution method (Fig. 4).

In order to reach new characteristics and to improve study analysis, new sensors supplement this metrology.

Heatflux sensors are added in order to measure the heat flux, and thus to be able to determine the U-value coefficient, using them with surface temperature sensor. A comparative study of commercial sensors was performed to select the most convenient one on

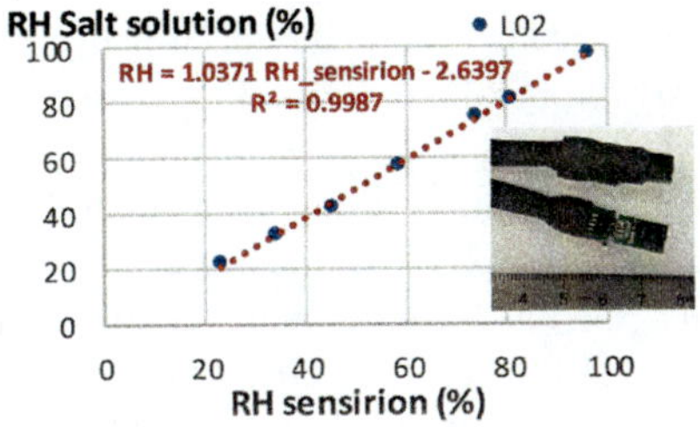

Fig. 4. Temperature and relative humidity sensors (SHT 35 Senirion) and calibration curve

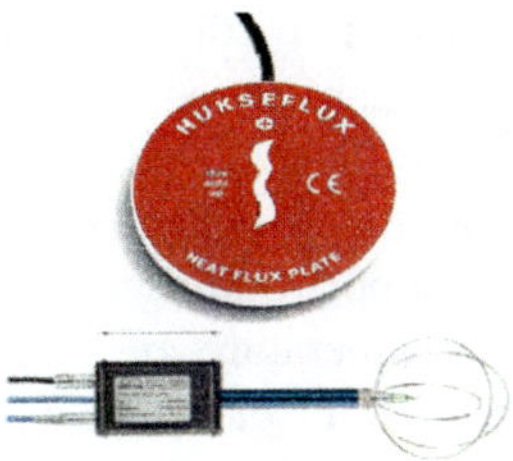

Fig. 5. Heat flux sensors HFP01 (up) and anemometer (down)

measurement range and size, to reduce impacts on hygric transfert, points of view. The HFP01 heatflux sensor from Hukseflux is a thermopile sensor. Its measurement range is ±2000 W/m^2 for an operating temperature between −30 and 70 °C. It is 80 mm in diameter and 5 mm thick (Fig. 5).

Air velocity is measured in order to assess the air mixing in each room and to check that the values remain representative of values met in buildings. Several kinds of anemometer exist; the choice depends on the investigated value. The interesting value here is the absolute value of the air velocity vector, not the direction, so a hot-ball anemometer is chosen (Fig. 5). Its range has to be suitable for building applications.

4 Studied Climate and Bi-climatic Rooms Management

4.1 Studied Climates

The aim is to study the hygrothermal behavior of walls in several climates. In this study, Tunisian winter climates are first considered. Four cities are selected to be representative of several kinds of climate: Mediterranean climate (Tunis), semi-arid climate (Kairouan), desert climate (Tozeur) and island climate (Djerba) (Fig. 6) [https://www.timeanddate.com/].

Fig. 6. Winter climate in Tunisian cities

Previously to the investigation of the wall hygrothermal behavior, it is necessary to ensure that ambient conditions can be easily controlled. From Fig. 6, temperature ranges of the four cities overlap. There is a bit more discrepancy for relative humidity ranges. The management of the outdoor climatic room is tested with several scenarii while the indoor room is maintained at (23 °C, 50% RH). The test scenarii consist in changing temperature or relative humidity or temperature and relative humidity simultaneously in the range of cities climates (Table 1). The objectives are to check (*i*) if the ambient conditions are homogenous in the room, (*ii*) if the set point is reached, (*iii*) the delay to reach the set point after setting, (*iv*) if the ambient conditions remain constant once they are reached. Besides, the aim of this investigation is also to identify the best management of the device to reach the setting point, depending on initial and final point. Actually, competition may appear between devices, for example between dehumidification (with refrigeration unit) and heating or between humidification and cooling (freeze on refrigeration unit).

Table 1. Sequence of outdoor ambient set points

Tests	Test 1	Test 2	Test 3	Test 4	Test 5	Test 6
T	18 °C	13 °C	8 °C	12 °C	18 °C	11 °C
RH	64%	64%	56%	56%	56%	68%

4.2 Results in Climatic Rooms Management

Outdoor ambiences recorded during the test sequence are given Fig. 7 at the two locations shown on Fig. 3.

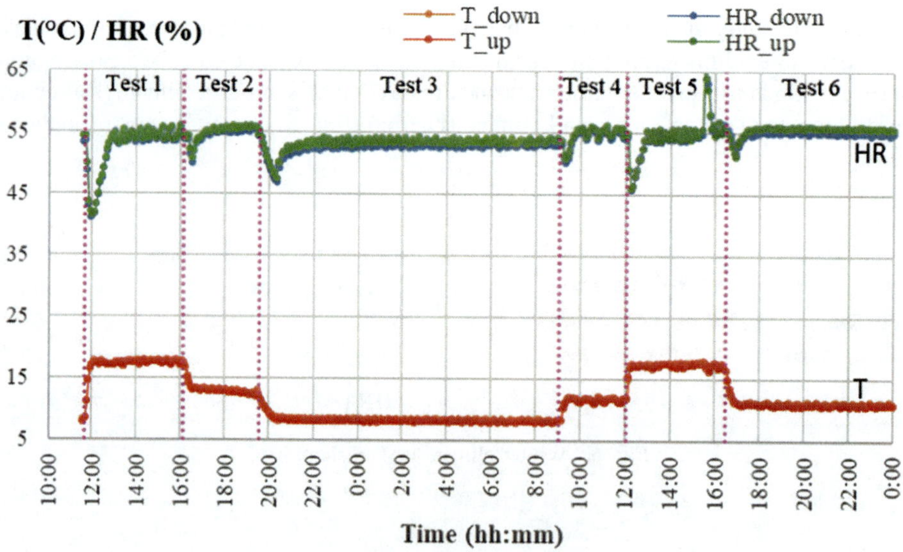

Fig. 7. Recorded outdoor conditions

For all the test, the up and down sensors give the same temperature and relative humidity, with average discrepancies of 0.12 °C in temperature and 0.44% in RH, this is in the range of sensors sensibility. Ambient conditions are thus homogenous in the outdoor climate room.

During the sequence test (Fig. 7), outdoor ambiences converge to stabilized ambient conditions. The target values are reached in temperature in all cases while the relative humidity is reached or nearly reached for the lowest values (56% RH) but not for the highest ones (64% RH).

The temperature variations are fast as well while heating as while cooling. Actually, for temperature variation between 4 and 9 °C the time needed to reach the stabilization value range between 45 min and one hour. For relative humidity, the time needed to reach the stabilization value is higher, about two hours. This is due to several phenomena.

For tests 4 and 5, the change in set point leads to a higher temperature and a same relative humidity, so higher water content. In a first time, the relative humidity decreases and then reach the target value. Actually, the heating induces an increase in temperature which is faster than the increase in water content thanks to humidification. Once the set point is reached, RH and temperature variations are observed due to the regulation of RH which induces refrigeration group action for dehumidification, which induces temperature decrease. Then, heating and humidification are needed for compensation. Such variations could be reduced by acting on the refrigeration group power.

For test 1, the test begins with ambient conditions around (8 °C; 55% RH). The change in set point induces an increase in temperature and in relative humidity. In such scenario, heating and humidification are required as for tests 4 and 5. Similar phenomena are observed, but the set point is not reached in RH. This may be due to temperature regulation which induces dehumidification that is not enough balanced by humidification. This competition could be reduced by acting on temperature band width or on refrigeration group power.

For tests 2, 3 and 6, the change in set point induces a decrease in temperature and in water content, without changing RH or increasing it from the actual point. Firstly, the relative humidity decreases due to the effect of refrigeration unit which acts to reduce temperature. Then humidification acts and allows to reach RH when the value is medium (around 54%), but is not sufficient for higher RH value (64–68% RH). Like previously, the refrigeration group has a wider effect than the humidification system. Maybe, this problem could be reduced by deactivating the refrigeration group at the beginning of the test.

5 Conclusion

This paper investigates the management of climatic systems to control ambient conditions in a bi-climatic room. The investigated ambiances correspond to winter Tunisian climates (Mediterranean, semi-arid, desert climate and island climate).

It is shown that ambient conditions are homogeneous in the room. The typical response times are short enough to simulated daily cyclic variations. Some conditions are easily reached with the current device but other may be difficult to obtain due to the competition between equipments. This may be solved by acting on band width, refrigeration group power, heating power and with priority on temperature or RH regulation.

Before testing hygrothermal behavior of walls, these tests will be completed with investigation of other typical climates (all seasons Tunisian and French climates).

References

Claude, S.: Etude expérimentale et numérique de solutions basées sur les éco-matériaux pour la rénovation thermique du patrimoine bâti urbain. INSA Toulouse (2018)

Collet, F., Prétot, S., Colson, V., Gamble, C.R., Reuge, N., Lanos, C.: Hygric properties of materials used for ISOBIO wall solution. J. Civ. Eng. **37**(2), 349–355 (2019)

Collet, F., Prétot, S.: Experimental highlight of hygrothermal phenomena in hemp concrete wall. Build. Environ. **82**, 459–466 (2014)

Mazhoud, B.: Elaboration et caractérisation mécanique, hygrique et thermique de composites bio-sourcés. Université Rennes1 (2016)

Medjelekh, D.: Caractérisation multi-échelle du comportement thermo hybride des enveloppes hygroscopiques. Université de Limoges (2016)

Rahim, M., Douzane, O., Le Tran, A.D., Promis, G., Langlet, T.: Experimental investigation of hygrothermal behavior of two bio-based building envelopes. Energy Build. **139**, 608–615 (2017)

Shea, A., Loence, M., Walker, P.: Hygrothermal performance of an experimental hemp–lime building. Constr. Build. Mater. **36**, 270–275 (2012)

Numerical Study of Droplets Coalescence in an Oil-Water Separator

Zahreddine Hafsi[1](✉), Sami Elaoud[1], Manoranjan Mishra[2,3], and Ines Wada[4]

[1] Laboratory of Applied Fluids Mechanics, Process and Environment Engineering, National Engineering School of Sfax, University of Sfax, Sfax, Tunisia
zahreddine.hafsi@enis.tn
[2] Department of Mathematics, Indian Institute of Technology Ropar, Rupnagar 140001, Punjab, India
[3] Department of Chemical Engineering, Indian Institute of Technology Ropar, Rupnagar 140001, Punjab, India
[4] Faculté des Sciences, Département de Chimie, UR13ES63-Chimie Appliquée et Environnement, Boulevard de l'Environnement, Université de Monastir, 5000 Monastir, Tunisia

Abstract. Through this paper a numerical modeling of oil-water flow through parallel plates integrated into a rectangular oil-water separator is conducted. Oil droplets of tiny sizes are dispersed in water and rectangular plates are used as coalescing chambers. Results have shown that upgrading a conventional API skimmer by introducing parallel coalescing chambers enhances coalescence of oil droplets by increasing their sizes. Since the buoyancy force that enables oil to float on water surface is proportional to the volume of the submerged oil droplet, droplets of greater sizes are then susceptible to rise to the water surface. Droplets floating on the water surface can be then easily skimmed. COMSOL Multiphysics modeling of two phase flow between two plates of the coalescing chamber has enabled to visualize the coalescence phenomenon and to determine coalesced droplets diameters. Further, the capability of enhancing oil water separation through coalescing chambers was discussed for the studied case.

Keywords: Oil water separator · Droplets coalescence · Phase field method · Numerical modeling · Coalescing chamber

1 Introduction

Separation of oil droplets from a volume of water is a key element in numerous industrial applications. This process can be required to remove oil from water in oil and gas productions operations. Indeed, in petroleum industries oil-water separators are of huge importance in order to enhance the productivity, to reduce transport costs and to respect environment policies (Yu et al. 2017). Indeed, for mature oil field, enhanced oil recovery from reservoir wells are mostly performed through injection of water under high pressure (Arnold et al. 2004). Then, the recovered crude oil presents water droplets that need to be removed through oil-water separation before shipping crude oil for

M. Kharrat et al. (Eds.): ICAMEM 2019, LNME, pp. 449–454, 2020.
https://doi.org/10.1007/978-3-030-52071-7_61

refining (Lee et al. 2018). On the other hand, the need to remove oil droplets from water can be also encountered in petroleum industry where a huge volume of wastewater is produced after oil recovery and that water has to be cleaned up from oil droplets for potential reuse. API (American Petroleum Institute) skimmers are widely used as oil-water separators (Kundu and Mishra 2018).

Oil in petroleum production water can be present under different forms (Wang et al. 2010). Free oil as discrete oil globules with typical droplet size around 150 μm is able to rise due to flotation forces and forms an oil layer on the top of the water. Dispersed oil is characterized by droplet sizes ranging from 120 to 150 μm. Emulsified oil as particles with a diameter less than 20 μm forms a stable suspension in water. Also, dissolved oil may be present and it is the most difficult oil fraction to be removed from water. Hence, measures need to be taken to ensure the coalescence of oil droplets in a way to increase their size. Consequently, oil droplets will be able to float on the water surface thanks to buoyant forces. In fact, Stokes law stipulates that the rising velocity of an oil droplet is proportional to the square of its diameter (Chiesa et al. 2005). To enhance the droplets coalescence, coalescing chambers are designed to be part of oil-water separators. A coalescing chamber is a succession of several parallel plates and the coalescence of oil droplets is achieved through the oil-water flow between each two successive plates.

In this numerical study, COMSOL Multiphysics is used in order to model the oil-water flow between two plates to visualize the coalescence phenomenon and to track different parameters that are likely to influence the efficiency of the separation process. Obtained results prove that the oil-water separator was improved for the studied case by more than 50% due to the use of coalescing chambers.

2 Modeling Parameters

Immiscible, incompressible isothermal liquid-liquid flow is governed by Navier Stokes equations viz. momentum and continuity equations. In addition, to track the evolution of a moving interface between two immiscible fluids, an interface tracking method is required. For instance, phase field method can be used. The method reposes on Cahn Hilliard equation for time evolution of a diffuse interface profile. Surface tension forces ensure coupling between flow field solution (Navier Stokes equations) and interface evolution (phase field model) (Akter and Deb 2017).

A simplified 3D geometry of the oil water separator is represented in Fig. 1, the waste water produced from petroleum industry is flowing along the separator and floating oil droplets on the free surface are then skimmed. The ultimate aim is to eliminate all oil droplets that are greater than *60* μm in diameter to obtain water of an acceptable degree of clarity and that using the separator of Fig. 2 that has as dimensions a length $Ls_1 = 4$ m, a width $Ws_1 = 0.66$ m and a depth $Ds_1 = 0.26$ m.

In order to skim oil droplets, the flow conditions have to allow dispersed droplets rising to the free surface of water. Rising velocity, also known as the vertical velocity, of an oil droplet is expressed through Stokes' law by (Park et al. 2017)

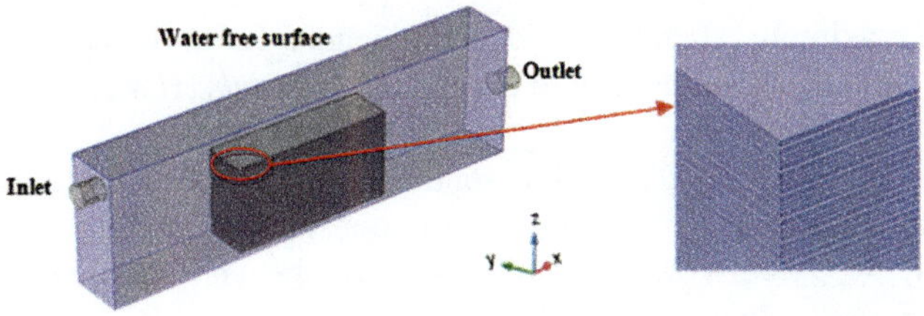

Fig. 1. Geometry of an oil water separator with horizontal coalescing chamber

Fig. 2. 2D geometry of the modeled domain

$$v_v = \frac{2}{9}\frac{\rho 1 - \rho 2}{\mu} g R^2 \tag{1}$$

where prefixes 1 and 2 stand respectively for water and oil. ρ is the density, μ is the dynamic viscosity, g is the gravitational acceleration and R is the radius of the droplet. Flow parameters in the working conditions are given in Table 1.

Table 1. Parameters of the flow in the oil water separator

Parameter	Value	Unit
Inlet designed flow rate Q_d	0.002	m^3/s
Water density ρ_1	1072	kg/m^3
Oil density ρ_2	822	kg/m^3
Water viscosity μ_1	1	cP
Oil viscosity μ_2	6	cP
Oiliness degree of waste water d_{oil}	10%	–
Viscosity of waste water μ	1.35	cP
Droplet diameter D	60	μm
Gravitational acceleration g	9.82	m/s^2
Length to width ratio of the separator r_{lw}	6*	–
Depth to width ratio of the separator r_{dw}	0.4**	–

*r_{lw} must be minimum equal to 5 according to API; **r_{dw} must be between 0.3 and 0.5 according to API*

Using Eq. (1), the minimum vertical velocity to rise a droplet of diameter $D = 60$ µm is $V_v = 0.000363$ m/s. According to API standards, and for a design of oil water separator with maximum efficiency, it is recommended for horizontal velocity to be 15 times of V_v and in all cases it has not to exceed *0.01524* m/s (Hsin et al. 2011) then it is assumed $V_h = 0.0055$ m/s.

For a designed flow rate Q_d, the horizontal and the cross sectional areas A_h and A_c of the separator that is likely to get risen the considered droplets are $A_h = Q_d/V_v = 5.5$ m^2 and $A_c = Q_d/V_h = 0.36$ m^2. Considering the chosen dimensions ratios of Table 1, the required dimensions of the separator are $Ls = 5.74$ m, $Ws = 0.95$ m and $Ds = 0.38$ m. It is then obtained that in order to guarantee the rising of all droplets of diameter greater than *60* μm, a separator with the previous dimensions is required. In some practical cases it is not possible to integrate such separator in the industrial process and smaller one is needed. Nevertheless, the smaller separator of Fig. 2 remains inappropriate and incapable of ensuring the elimination of oil droplets in the considered working conditions. The use of the presented small cost effective separator is only possible if the droplets to be removed will have a greater diameter.

Referring to API standards (Hsin et al. 2011), and by considering the same inlet design flow rate Q_d, a vertical velocity $V_{v1} = Q_d/A_{h1} = 0.075$ cm/s is obtained. The obtained rising velocity is able to rise droplets of diameter greater than $D_1 = 86.25$ μm.

Then, it comes the idea of introducing coalescing chambers to increase droplets sizes and hence improving the capability of oil droplets removing. Coalescing chambers are designed as interceptors of the oil-water flow. Multiple parallel plates (as shown in the zoomed selection of Fig. 1) enhance the coalescence of oil droplets once flowing in the inter-plate area. Larger diameter droplets lead to higher rising velocity and are likely to float on the free surface to be easily skimmed.

3 Numerical Modeling and Results

In order to numerically solve the governing equations of oil water flow in a coalescing chamber and to track the evolution of the interface between the two liquid phases, *Laminar Two-Phase Flow, Phase Field* module of COMSOL Multiphysics is used. The numerical model is built considering a 2D geometry of a tiny rectangular domain of dimensions $L = 2.5$ mm and $H = 0.4$ mm representing a part of an horizontal section of a coalescing chamber where droplets coalescence may occur.

3.1 Initial and Boundary Conditions

As oil droplets are randomly dispersed in the water phase, random initial disposition of circles in the rectangular domain is created through a java code via the *Application Builder* feature of COMSOL Multiphysics. The built java code permits to set the number of droplets to be generated as well as the range of droplet diameters. Also, the code is written to avoid drawing droplets out of the rectangular domain. Although overlapping of two or more droplets may be encountered in reality, this case has been avoided for better illustration of the coalescence. Respecting the Oiliness degree of waste water in the considered domain as reported in Table 1, it is obtained initially 35 non overlapping droplets of diameters $D = 60$ μm randomly dispersed in the rectangular domain as illustrated in Fig. 2.

Additionally, for sake of simplification of the numerical model it is considered that in the inlet side only water is entering. Pure water phase is pushing out the oil water mixture along the length of the rectangular domain. Initial and boundary conditions for the flow variables $u(x,t)$ and $p(x,t)$ are $u(0,t) = u_0$ with $\phi = -1$; $p(L,t) = p_0$; $u(x,0) = 0$ and $p(x,0) = p_0$, where $p_0 = 1\ bar$ and $u_0 = 0.01125$ m/s ($15\ times\ V_{v1}$).

3.2 Results and Discussion

Snapshots of evolution of the volume fraction of both immiscible fluids is reported in Fig. 3 for several time steps. Figure 3 illustrates the coalescence of droplets between initial time and the instant at which the first droplet exits the domain.

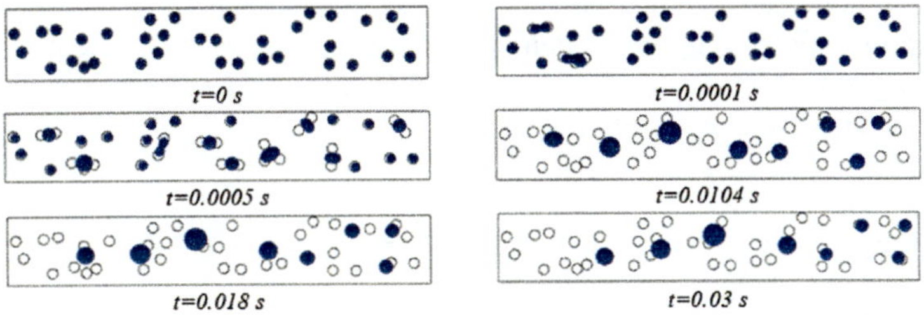

Fig. 3. Time evolution of volume fraction of oil phase (white: water, bleu: oil)

While flowing under the flow conditions, larger droplets are formed up of two or more droplets. The process of droplets coalescence starts by a rapprochement of droplets then a collision for coalescing and finally the formed droplet takes the circular shape. In order to quantify the usefulness of the coalescing chamber, diameters of obtained droplets are measured. It was found that, out of the 35 droplets initially dispersed in the domain, after 0.03 s 23 droplets have coalesced into 8 big ones. Details of obtained results are summarized in Table 2.

Table 2. Summary of obtained results

Parameter	Value	Unit
Total number of droplets	35	–
Number of coalesced droplets	23	–
Minimum coalesced droplets diameter	73.4	μm
Maximum coalesced droplets diameter	130.8	μm
Coalesced droplets of diameter greater than *86.25* µm	04	–

According to previous results it was found that the considered coalescing chamber has improved the oil water separation by about 50% (the volume of oil droplets of diameter greater than $D_1 = 86.25$ µm compared to the total volume fraction of oil initially in the domain). In fact, as detailed previously and under the working conditions, coalesced droplets of diameter greater than D_1 are able to rise to the surface to be skimmed later.

4 Conclusion

Through the presented numerical model, the effect of integrating coalescing chambers in an existing oil water separator is discussed. It is proved numerically that coalescing chambers are likely to enhance the separation process by increasing droplets diameters of the dispersed phase. As reported in this paper, horizontal coalescing chambers have improved the separation process by more than 50% for the studied case. Additionally, the improvement of the separation efficiency through integrating coalescing chambers with a slope angle needs to be more discussed and numerically modeled. Furthermore, for a more realistic model, 3D modeling of coalescing phenomenon is required to be able to easily consider oil with dispersed water droplets entering in inlet side instead of the simplification presented in this work.

References

Yu, L., Han, M., He, F.: A review of treating oily wastewater. Arab. J. Chem. **10**, S1913–S1922 (2017)

Arnold, R., Burnett, D.B., Elphick, J., Feeley III, T.J., Galbrun, M., Hightower, M., Jiang, Z., Khan, M., Lavery, M., Luffey, F., Verbeek, P.: Managing water-from waste to resource. Oilfield Rev. **16**(2), 26–41 (2004)

Lee, H.S., Zheng, C., Yang, B.: Modeling and experimental study on the design of separators for water/oil separations using electric fields. J. Appl. Mech. Eng. **07**(04) (2018). https://doi.org/10.4172/2168-9873.1000313

Kundu, P., Mishra, I.M.: Treatment and reclamation of hydrocarbon-bearing oily wastewater as a hazardous pollutant by different processes and technologies: a state-of-the-art review. Rev. Chem. Eng. **35**(1), 73–108 (2018). https://doi.org/10.1515/revce-2017-0025

Wang, L.K., Shammas, N.K., Selke, W.A., Aulenbach, D.B.: Flotation Technology. Humana Press, Totowa (2010). https://doi.org/10.1007/978-1-60327-133-2

Chiesa, M., Melheim, J., Pedersen, A., Ingebrigtsen, S., Berg, G.: Forces acting on water droplets falling in oil under the influence of an electric field: numerical predictions versus experimental observations. Eur. J. Mech. B/Fluids. **24**(6), 717–732 (2005). https://doi.org/10.1016/j.euromechflu.2005.03.005

Akter, F., Deb, U.K.: Computational modeling and dynamics of the oil and water flow using the Galerkin approximation. J. Comput. Math. **7**, 58–69 (2017). https://doi.org/10.4236/ajcm.2017.71005

Park, S.H., Park, C., Lee, J., Lee, B.: A simple parameterization for the rising velocity of bubbles in a liquid pool. Nucl. Eng. Technol. **49**(4), 692–699 (2017). https://doi.org/10.1016/j.net.2016.12.006

Hsin, H.H., Sebastien, B., Yue, S.: Oil-water separation system for industrial wastewater. Bree 495 Engineering Design III, Final Report, McGill University (2011)

An Alternative Approach for Minimum Independent Loops Identification in Water Distribution Networks

Oussama Choura(✉), Zahreddine Hafsi, and Sami Elaoud

Laboratory of Applied Fluids Mechanics, Process and Environment Engineering, National Engineering School of Sfax, University of Sfax, Sfax, Tunisia
oussema.choura@gmail.com

Abstract. In the management and design of water distribution network, the knowledge of the flows in pipes and pressures in nodes is indispensable. Such knowledge requires a prior analysis of the network topology to be used for equilibrium procedure. The incidence matrix linking pipes to loops is one of the main topological representations of networks necessary for the process. Even though different algorithms have been presented by several researchers, the question of the minimum independent loops, which effects computing complexity, still raises. Different approaches in the selection of these loops have been established. However, few have arrived to determine the minimum independent loops matrix. The aim of this paper is to present an alternative approach for determining the sparsest pipes/loops matrix possible using Matlab® and graph theory. The reliability of the proposed algorithm was tested and verified through a real test case. The performances of the algorithm were also compared to other recent methods available in literature in terms of the sparsity of the independent loops' matrix.

Keywords: Independent loops · Network topology · Water distribution systems · Network resolution

1 Introduction

For years, the interest in developing reliable and computationally efficient algorithms for network resolution has been growing significantly within the technical-scientific community, specifically when it hinges on precious resources such as water (Todini and Rossman 2013). The analysis of Water Distribution Networks (WDNs) is of major importance since it allows to determine the flow characteristics i.e. pressure and flow rates and hence predicting the network response to any sudden perturbation. Nevertheless, and mainly for complex networks topologies involving closed loops, such hydraulic analysis is not straightforward and it requires prior identification of independent loops to be used for equilibrium procedure (Todini 2000).

A steady-state analysis of a looped WDN is based on the resolution of Kirchhoff's equations viz. nodes and loops equations (Krope et al. 2006). Hence, a set of independent loops for which loops equations are to be written has to be determined. It's encountered in literature plenty of research works devoted to algorithms for

M. Kharrat et al. (Eds.): ICAMEM 2019, LNME, pp. 455–462, 2020.
https://doi.org/10.1007/978-3-030-52071-7_62

automatically generate a set of independent loops for a looped network (Alvisi et al. 2011; Creaco and Franchini 2014; Alvarruiz et al. 2015). Added to that, in recent years, the optimization of independent loops was a target function in several research works. The ultimate aim was to define the set of minimum independent loops viz. the independent loops that are built with the minimum number of pipes.

In this paper, and by considering a looped WDN as an oriented graph, an alternative algorithm for the determination of the minimum independent loops is presented. The reliability of the proposed algorithm was verified through test cases going from a simple topology to a real case of a WDN. Also, a comparison with previous related works was conducted.

2 Looped Network Analysis: Basic Formulae

Nodes law traducing mass conservation is written for each node i

$$\left(\sum q_{in} - \sum q_{out}\right)_i = 0 \tag{1}$$

where q denotes the volumetric flow rate passing through the node i.

For a looped network, conservation of momentum is to be considered. Loops law, stipulating that the sum of the pressure drops within a closed-loop is null, needs to be applied. Frictional pressure drop along a pipe is expressed through the Darcy Weisbach formula (Walski et al. 2003) as follows:

$$\Delta H = \frac{8\lambda L}{\pi^2 g D^5} q^2 \tag{2}$$

where λ is the friction coefficient, g is the gravitational acceleration, D and L are, respectively, the diameter and the length of the pipe.

A set of independent loops is crucial to determine to be able to apply Kirchhoff laws to each loop and node. In a looped network and by denoting n the total number of nodes, n_0 the number of source nodes and p the number of pipes, the number of independent loops is then written (Creaco and Franchini 2015):

$$l = p - n + n_0 \tag{3}$$

The prior knowledge of the number of independent loops that features a looped network is primordial to build an algorithm of loops selection and to optimize the set of independent loops to be determined.

Independent loops are represented under matrix form by a matrix denoted M_{31}. M_{31} is an incidence matrix loops-pipes of size [lxp]. $M_{31}(k,i)$ is dependent on the agreement between the assumed flow direction in the pipe i belonging to the loop k and the rotation direction in the same loop.

$$M_{31}(k,i) = \begin{cases} 1 & \textit{if the flow direction in the pipe i is in agreement with the rotation direction in the loop k} \\ -1 & \textit{if the flow direction in the pipe i is not in agreement with the rotation direction in the loop k} \\ 0 & \textit{if there is no incidence between the pipe i and the loop k} \end{cases} \quad (4)$$

3 Independent Loops Selection Algorithm

3.1 Overview on Related Algorithms

One way to maximize the computational efficiency of a steady-state hydraulic equilibrium of a complex water distribution network is to refer to the sparsest independent loops' matrix possible (Creaco and Franchini 2014), i.e., made up of independent loops with minimum weight (lowest number of pipes).

Hafsi et al. (2017) proposed a selection algorithm which refers to the Minimum Spanning Tree for each iteration of determination of an independent loop. The authors stated that this choice may be time-consuming for complex topologies.

Creaco and Franchini (2015) followed a different approach when referring to the original layout in every iteration which, unlike Hafsi et al. (2017), showed to be the optimum choice of independent loops with minimum pipes. Creaco's algorithm compares the ith loop with the previously determined loops to check whether it is repeated or not.

3.2 Steps of the Proposed Algorithm for Minimum Loops Selection

The proposed algorithm, built around these two approaches, updates the reference in each iteration. Its main steps are resumed in the following:

(i) Transform the initial topology of the network to a single source network, if it has multiple source nodes, i.e., connect the source nodes to one reference node. Otherwise, maintain the initial topology.

(ii) The Minimum Spanning Tree, herein noted MST, is determined based on the new topology of the network (single source topology) using either Kruskal (1956) or Prim (1957) algorithms.

(iii) Construct the removed pipes vector, herein noted *V*, containing the indexes of the pipes removed from the full topology to obtain the MST and a searching matrix, herein noted *SM*, containing the elements of the MST.

(iv) Look for the shortest path, in topological viewpoint, linking the two ends of the pipes in the vector *V* by means of the predefined algorithms: Dijkstra's algorithm (1959), Bellman-Ford-Moore algorithm (Bellman 1958) and construct a decision matrix, herein noted *DM*.

(v) Sort *DM* based on the shortest path corresponding to each pipe element and select the first element corresponding to the minimum shortest path of all element, herein noted *P*, in the vector *V*.

(vi) Construct the independent loops matrix row formed by the pipe *P* and its corresponding shortest path.

(vii) Update the vector *V* by deleting the element P and the matrix *SM* by adding the pipe *P*. In other words, the new searching matrix, reference for the next calculations, contains the MST and the already chosen elements *P* of the previous steps.

(viii) Restart from the step (iv) until all the removed pipes are added to the MST, i.e., all independent loops are obtained.

4 Case Studies

The proposed algorithm was tested on a network of simple topology and the real case of the water distribution network of Modena in Italy (Bragalli et al. 2008; Creaco and Franchini 2015).

4.1 Simple Topology Test

The simple topology network is defined as 12 pipes and 9 nodes forming 4 loops (Fig. 1). The test network is single-sourced.

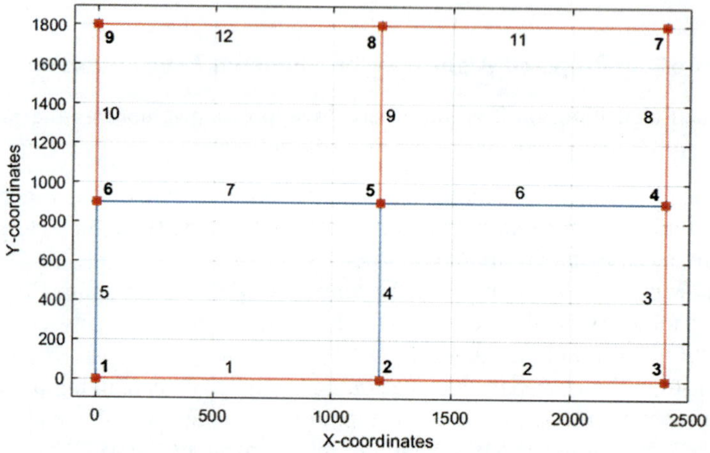

Fig. 1. Simple network topology: the red lines indicate the MST, the normal and bold numbers indicate, respectively, the pipes and nodes indexes.

The algorithm used Kruskal method to determine the MST and Dijkstra algorithm for shortest path delimitation.

The test network contains four independent loops. Then, the determination of the independent loops' matrix took 4 iterations as summarized in Table 1.

Table 1. Algorithm iterations for a simple topology.

Iteration	Search matrix	Removed pipes vector	Decision matrix (sorted)	Selected element
1	$SM = MST$	$V = \begin{bmatrix} 4 \\ 5 \\ 6 \\ 7 \end{bmatrix}$	$DM = \begin{bmatrix} 6 & 4 \\ 7 & 4 \\ 4 & 6 \\ 5 & 8 \end{bmatrix}$	$P = [6]$
	$M_{31} = \begin{bmatrix} 0 & 0 & 0 & 0 & 0 & -1 & 0 & 1 & -1 & 0 & 1 & 0 \end{bmatrix}$			
2	$SM = SM + P$	$V = \begin{bmatrix} 4 \\ 5 \\ 7 \end{bmatrix}$	$DM = \begin{bmatrix} 4 & 4 \\ 7 & 4 \\ 5 & 8 \end{bmatrix}$	$P = [4]$
	$M_{31} = \begin{bmatrix} 0 & 0 & 0 & 0 & 0 & -1 & 0 & 1 & -1 & 0 & 1 & 0 \\ 0 & -1 & 1 & -1 & 0 & 1 & 0 & 0 & 0 & 0 & 0 & 0 \end{bmatrix}$			
3	$SM = SM + P$	$V = \begin{bmatrix} 5 \\ 7 \end{bmatrix}$	$DM = \begin{bmatrix} 7 & 4 \\ 5 & 6 \end{bmatrix}$	$P = [7]$
	$M_{31} = \begin{bmatrix} 0 & 0 & 0 & 0 & 0 & -1 & 0 & 1 & -1 & 0 & 1 & 0 \\ 0 & -1 & 1 & -1 & 0 & 1 & 0 & 0 & 0 & 0 & 0 & 0 \\ 0 & 0 & 0 & 0 & 0 & 0 & -1 & 0 & 1 & -1 & 0 & 1 \end{bmatrix}$			
4	$SM = SM + P$	$V = [5]$	$DM = [5 \quad 4]$	$P = [5]$
	$M_{31} = \begin{bmatrix} 0 & 0 & 0 & 0 & 0 & -1 & 0 & 1 & -1 & 0 & 1 & 0 \\ 0 & -1 & 1 & -1 & 0 & 1 & 0 & 0 & 0 & 0 & 0 & 0 \\ 0 & 0 & 0 & 0 & 0 & 0 & -1 & 0 & 1 & -1 & 0 & 1 \\ -1 & 0 & 0 & 1 & -1 & 0 & 1 & 0 & 0 & 0 & 0 & 0 \end{bmatrix}$			

Following the above-described algorithm, a set of independent loops defined by as M_{31} is produced. All obtained loops have the minimum topological lengths and matrix M_{31} features a minimum number of nonzero elements, i.e., of good sparsity.

4.2 Modena Network

Modena network features 317 pipes and 272 nodes where 4 of them are source nodes (Bragalli et al. 2008; Creaco and Franchini 2015).

As a first step, the proposed algorithm changes the original topology of the network to a single-sourced network (Hafsi et al. 2017). The resulting topology is shown in Fig. 2 (pipes and nodes indexes are not mentioned for clarity reasons).

The independent loops (49 loops) selection process results in an optimal independent loops matrix in terms of nonzero elements. The algorithm gives rise to a 49×317 matrix of good sparsity with 571 nonzero elements (Fig. 3). It's to notice that in the obtained loops matrix M_{31} there are $(n_0 - 1)$ extra loops compared to the loops matrix of similar network topology with a one source node. These extra loops describe the transformation of the multi-source to a single source network that reduces the sparsity of matrix M_{31}. Nevertheless, in the presented algorithm, these extra loops are minimized as they are built following the shortest path from the two sources nodes to be related. One more detailed way to write the expression of the number of

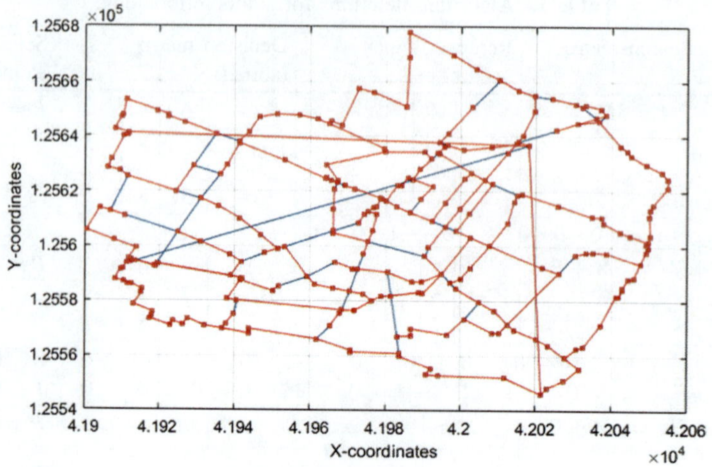

Fig. 2. Modena network topology: the red lines indicate the MST.

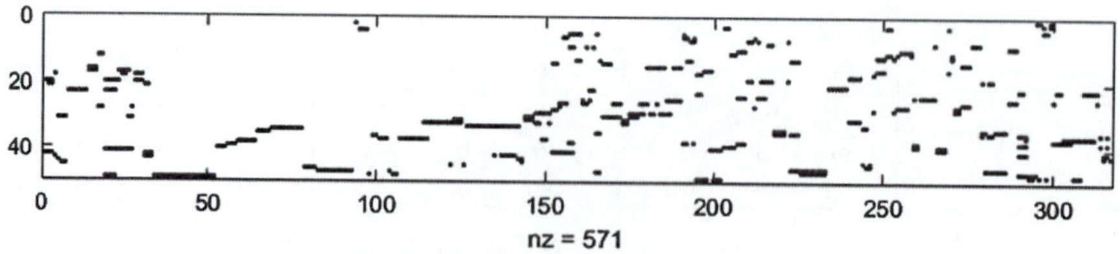

Fig. 3. The sparsity of the independent loops' matrix of Modena network

independent loops (Eq. 3) for a general network topology regardless of the number of source nodes is

$$\underbrace{\underbrace{l = p - n + 1}_{single\ source\ network} + (n_0 - 1)}_{multi-source\ network} \tag{5}$$

It is worthy to note that, as far as the computation time is considered, using Matlab® 2014b, the independent loops matrix is obtained in about 0.49 s.

Table 2 summarizes sparsities of loops matrix obtained using Creaco and Franchini (2015), Hafsi et al. (2017, 2018) algorithms and the above-presented one for Modena network. The difference is significant in matrices sparsities comparing our algorithm with the one proposed by Hafsi et al. (2017) but not to Creaco and Franchini (2015) who stated that his algorithm is designed for determining the optimum choice of independent loops. However, it's hardly noticeable when it comes to CPU times.

Table 2. Comparison of the loops selection algorithm.

Algorithm	The sparsity of loops matric (number of nonzero elements)
Hafsi et al. (2017)	622
Hafsi et al. (2018)	880
Creaco and Franchini (2015)	571
Proposed algorithm	571

5 Conclusion

The hydraulic analysis is easy to accomplish for simple networks. It is, in fact, easy for manually select independent loops. Nevertheless, complexity arises when it comes to analyzing complex looped networks. Prior to any hydraulic analysis, the process of independent loops selection is required and it plays a crucial role in the complexity of calculations, i.e., computational time and memory. Thus, a graph study is performed to identify the optimum set of independent loops as a preliminary step of hydraulic steady-state equilibrium.

In this paper, an alternative algorithm for an optimum independent loops selection is presented. The proposed algorithm generates a sparser possible independent loops matrix, and therefore, a faster and simpler hydraulic analysis.

References

Alvarruiz, F., Martínez-Alzamora, F., Vidal, A.M.: Improving the efficiency of the loop method for the simulation of water distribution systems. J. Water Resourc. Plann. Manag. **141**(10), 04015019 (2015)

Alvisi, S., Creaco, E., Franchini, M.: Segment identification in water distribution systems. Urban Water J. **8**(4), 203–217 (2011). https://doi.org/10.1080/1573062X.2011.595803

Bellman, R.: On a routing problem. Q. Appl. Math. **16**(1), 87–90 (1958). https://doi.org/10.1090/qam/102435

Bragalli, C., D'Ambrosio, C., Lee, J., Lodi, A., Toth, P.: Water network design by MINLP. Rep. No. RC24495. IBM Research, Yorktown Heights (2008)

Creaco, E., Franchini, M.: Comparison of Newton-Raphson global and loop algorithms for water distribution network resolution. J. Hydraul. Eng. **140**(3), 313–321 (2014). https://doi.org/10.1061/(ASCE)hy.1943-7900.0000825

Creaco, E., Franchini, M.: The identification of loops in water distribution networks. Procedia Eng. **119**, 506–515 (2015). https://doi.org/10.1016/j.proeng.2015.08.878

Dijkstra, E.W.: A note on two problems in connexion with graphs. Numer. Math. **1**(1), 269–271 (1959). https://doi.org/10.1007/bf01386390

Hafsi, Z., Elaoud, S., Mishra, M., Akrout, M.: Automated framework for water looped network equilibrium. Water Resour. Manag. **32**(2), 641–657 (2017). https://doi.org/10.1007/s11269-017-1831-2

Hafsi, Z., Elaoud, S., Mishra, M., Akrout, M.: Independent loops selection in a hydraulic looped network. In: Design and Modeling of Mechanical Systems—III Lecture Notes in Mechanical Engineering, pp. 235–246 (2018). https://doi.org/10.1007/978-3-319-66697-6_24
Krope, J., Dobersek, D., Goricanec, D.: Flow pressure analysis of pipe networks with linear theory method. In: Proceedings of the 2006 WSEAS/IASME International Conference on Fluid Mechanics, Miami, Florida, USA, pp. 59–62 (2006)
Kruskal, J.B.: On the shortest spanning subtree of a graph and the travelling salesman problem. Proc. Am. Math. Soc. **7**(1), 48 (1956). https://doi.org/10.1090/s0002-9939-1956-0078686-7
Prim, R.C.: Shortest connection networks and some generalizations. Bell Syst. Tech. J. **36**(6), 1389–1401 (1957). https://doi.org/10.1002/j.1538-7305.1957.tb01515.x
Todini, E.: Looped water distribution networks design using a resilience index based heuristic approach. Urban Water **2**(2), 115–122 (2000). https://doi.org/10.1016/s1462-0758(00)00049-2
Todini, E., Rossman, L.A.: Unified framework for deriving simultaneous equation algorithms for water distribution networks. J. Hydraul. Eng. **139**(5), 511–526 (2013). https://doi.org/10.1061/(ASCE)hy.1943-7900.0000703
Walski, T.M., Chase, D.V., Savic, D.A., Grayman, W., Beckwith, S., Koelle, E.: Advanced Water Distribution Modeling and Management. Civil and Environmental Engineering and Engineering Mechanics Faculty Publications, paper 18 (2003)

Author Index

M. Kharrat et al. (Eds.): ICAMEM 2019, LNME, pp. 463–465, 2021.
https://doi.org/10.1007/978-3-030-52071-7